Economic and Social Commission for Asia and the Pacific (ESCAP)

AGRO-PESTICIDES: PROPERTIES AND FUNCTIONS

IN INTEGRATED CROP PROTECTION

A manual for staff of plant protection services and pesticide distributing companies, for instructors of retailers and professional applicators, and for students of crop production and protection sciences

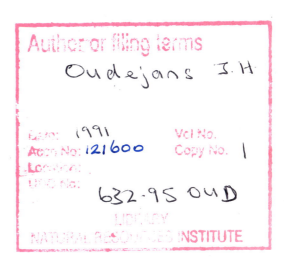

UNITED NATIONS

Bangkok
1991

This publication was written by Jan H. Oudejans
at the request of the ESCAP secretariat

This manual has been published with funds generously provided by the Government of the Netherlands as extrabudgetary assistance to ESCAP.

For additional copies of *Agro-pesticides: Properties and functions in integrated crop protection* and other information, please contact:

FADINAP/ARSAP
Agriculture and Rural Development Division
ESCAP, United Nations Building
Rajdamnern Avenue
Bangkok 10200, Thailand
Fax: 281-2403
Tel: 280-3618

Agro-pesticides: Properties and functions in integrated crop protection
Completely revised edition: 1991
International Standard Book Number (ISBN) 974-88754-8-2

Related publication:

Agro-pesticides: Their management and application
First edition: 1982
Second edition: 1983
Third edition: 1985

ST/ESCAP/975

FOREWORD

ESCAP's first pesticide training manual, written by the same author and published in 1982 under the title *Agro-pesticides: Their Management and Application*, found a huge audience not only in the Asian-Pacific region but across much of the developing world. The manual grew out of experience gained in a regional series of training courses on the safe handling and proper use of pesticides, conducted by the Agricultural Requisites Scheme for Asia and the Pacific (ARSAP) of the ESCAP Agriculture Division. This technical assistance programme and the resulting training manual, both generously funded by the Government of the Netherlands, responded to a widely felt need in developing countries.

When the Green Revolution brought acceptance of high- yielding varieties of wheat and rice, the use of agro-chemicals had begun increasing rapidly. Although fertilizers posed no immediate threat to farmers, pesticides quite definitely did, and the general lack of information and safety training swiftly loomed as a problem of considerable urgency. This was the main reason for the wide popularity that the first training manual enjoyed not only in Asia and the Pacific but in Africa and Latin America as well.

Now, a decade after the launching of ARSAP's pesticide safety training programme, comes this completely revised new edition of the training manual. Its different title reflects changing emphasis and lessons learned, for example, that the efficient application of pesticides should not be the only concern. Though pesticides enabled tremendous gains in agricultural production, the original flawed concept of chemical plant protection gave rise to a number of setbacks and disappointments. Resistance through genetic changes in pests, mass outbreaks due to the destruction of natural pest enemies and consequently increased frequency and dosage of application, leading to damage to man and the environment, have combined to bring about a re-evaluation of plant protection concepts and a more cautious view of pesticides.

Integrated pest management emerged as one of the promising new ideas duly taken into account in this publication. Also reflected are the intensified efforts of international organizations, particularly the Food and Agriculture Organization (FAO) of the United Nations, to protect developing countries from exported products its farmers are ill prepared to handle. The prominence given here to the FAO International Code of Conduct on the Distribution and Use of Pesticides demonstrates our enthusiastic support for this effort.

With this fully revised and updated training manual on responsible and need-based pesticide use, we are pleased to note that ESCAP is making a useful contribution to the protection of mankind and the environment. At the same time, we hope that this publication will provide tangible help in maintaining agricultural production growth at levels at least commensurate with the continuing expansion of the Asian-Pacific population. I wish to express my sincere thanks to the Government of the Netherlands for the financial support that made this publication possible.

S.A.M.S. Kibria
Executive Secretary

PREFACE

Agriculture in Asia has gone through great changes in the second half of the twentieth century. The threat of widespread hunger in Asia has been partially countered by the so-called "green revolution" in food grains. Increases in the production of rice and wheat resulting from plant breeding and modern technology developed by national and international research programmes have enabled some of the developing countries of Asia to match their unabated population growth. But elsewhere in Asia and also in other regions of the world, such as Sub-Saharan Africa and South America, malnutrition and acute shortages of food are still a reality of daily life for hundreds of millions of people. Moreover, world food supplies are generally very low and barely sufficient to meet calamities such as large-scale crop failures and droughts.

There are three major sets of reasons for this sad situation. The first is that agricultural productivity is still very low in many places owing to a lack of development of the people and their environment, to the marginal fertility of soils and poor water supply, to the landlessness of millions of farmers and to the non-availability of improved cultivars of many crops. To foster farm development, farmers must, among others, have access to more knowledge about all aspects of agricultural inputs and farm management.

The second is that the massive poverty of scores of people in many developing countries results in a stagnation of agricultural development. Their insufficient purchasing power even leads to a lack of demand for agricultural products on the world market and, ironically, thereby to a conscious restriction of food production in the normally highly productive industrial countries.

The third major reason is the great loss inflicted by countless pests and diseases on the crops in the field and after they have been harvested. Such crop destruction is a serious problem especially in the warm and humid tropics where animal pests and plant diseases proliferate and weeds are very competitive owing to their luxuriant growth.

Modernization and intensification of agricultural production is generally accompanied by a rapid increase in the use of chemical fertilizers and pesticides, because high yielding cultivars need high doses of fertilizer and because they are more vulnerable to damage from pests and diseases. As chemical pesticides are a relatively cheap and reliable means for controlling most injurious organisms, intensive agriculture as a whole has become largely dependent on chemical methods of crop protection. Excessive use of pesticides, however, has caused an alarming level of resistance in many types of pest and disease organisms as well as a serious contamination of mankind and the environment.

Over the last decade, ever more politicians, scientists and members of the general public have come to realize that in order to make agricultural production sustainable and to protect national resources, intensive farm systems should become less dependent on chemicals.

In this context, great progress has been made in the development of methods of integrated pest management, particularly for rice, cotton and estate crops. Much attention is being drawn to the need for the rational and safe management of pesticides at the national and international level. It is recognized, however, that a more rational use of pesticides and large-scale application of the integrated pest control concept can be achieved only through the training of farmers and all other people professionally engaged in pesticide distribution and application.

To this end, this publication provides practical and comprehensive information to teachers and students of crop production and crop protection sciences as well as to scientists and staff of agricultural services and pesticide distribution companies. The field of plant protection is very complicated because of the numerous interactions between the cultivated crops, the many damaging or beneficial organisms and the variable factors of the environment. The use of chemicals, having an enormously disruptive power on the fragile balance of nature, requires general insight into their properties and effects. Therefore, this source book for further training describes many issues in an introductory way.

I hope that it will stimulate a more rational and safe use of pesticides and that it will boost the implementation of integrated pest management technology in Asia and elsewhere in the world.

This Netherlands contribution to the development of agriculture is a result of progress made in my country, the high agricultural productivity of which is based on intensive collaboration between researchers, extension agents and farmers.

P. Bukman
Minister of Agriculture, Nature Management
and Fisheries of the Netherlands

ACKNOWLEDGEMENTS

In writing this book I have consulted many colleagues in the Netherlands and from abroad. It is impossible for me to acknowledge all of them individually here, but my sincere appreciation is noted for all who contributed to the improvement of the manual and who provided illustrations. I want to express my gratitude to the following individuals in particular for their advice or help in extensively reviewing the manuscript.

Dr. Ir. J.C. Zadoks	Professor of Phytopathology, Agricultural University of Wageningen, the Netherlands
Dr. P.A. Oomen	Phytopharmaceutical Entomologist, Plant Protection Service, Ministry of Agriculture, Nature Management and Fisheries, the Netherlands
Dr. P.S. Tyler	Storage Entomologist, Storage Operation and Management, Overseas Development, Natural Resources Institute, United Kingdom of Great Britain and Northern Ireland
Dr. Keith Moody	Weed Scientist, International Rice Research Institute, Los Baños, the Philippines
Ir. H. Naber	Lecturer of Weed Science, Agricultural University of Wageningen, the Netherlands
Dr. P.E.K. Kenmore	Entomologist, Regional Manager of FAO-Intercountry IPC in Rice Programme, Manila, the Philippines
Ir. F. Meerman	Entomologist, Agricultural University of Wageningen, the Netherlands
Dr. Ir. A. Tempel	Phytopathologist (ret.), Wageningen, the Netherlands

Further, I express my profound gratitude to the Editor, Mr. John P. Loftus, who did so much to make this manual readable and to give it an attractive lay-out, and to Mrs. Dussadee Nunthavichitra, who with painstaking attention to detail took care of all the secretarial and computer typesetting work involved.

The material in this manual comes from a wide variety of sources. Although it is impossible to acknowledge all these sources fully, I want to mention particularly Mrs. Heather Rogers, Editor of Shell Agriculture. Where possible, effort has been made to duly provide references for the material used.

Jan H. Oudejans
Senior Advisor
on Agricultural Development Cooperation
Ministry of Agriculture, Nature Management
and Fisheries of the Netherlands

Note on the Author:

The author, Dr. Jan H. Oudejans, is a Senior Advisor on Agricultural Development Cooperation with the Ministry of Agriculture, Nature Management and Fisheries of the Netherlands. After graduating as M.Sc. from the Agricultural University of Wageningen, the Netherlands, he worked with the chemical industry and in estate agriculture in a number of developing countries for 13 years. After that, he was in charge of the ARSAP Agro-pesticide Training Programme of ESCAP, which provided assistance to 10 developing countries in the Asian-Pacific region.

AGRO-PESTICIDES: PROPERTIES AND FUNCTIONS IN INTEGRATED CROP PROTECTION

CONTENTS

Page

Foreword iii

Preface v

Acknowledgements vii

Chapter 1: Principles of integrated pest control 1

Chapter 2: Classification of pesticides 29

Chapter 3: Pesticide manufacture and formulation 53

Chapter 4: Pesticide toxicity and residues 67

Chapter 5: Symptoms and treatment of pesticide poisoning 87

Chapter 6. Pesticide regulation and registration and the product label . 97

Chapter 7. Plant protection policies and regulatory infrastructure . . 107

Chapter 8: Safe handling, storage and marketing of pesticides 137

Chapter 9: Transport, spills and disposal 153

Chapter 10: Spray equipment and maintenance 163

Chapter 11: Spraying techniques 185

Chapter 12: Control of animal pests 201

Chapter 13: Plant diseases: Causes and integrated control 241

Chapter 14: Integrated control of weeds 263

Chapter 15: Control of storage pests 291

Glossary 313

PRINCIPLES OF INTEGRATED PEST CONTROL

Contents   Page

Agricultural production and losses from pests 2

Factors causing damage and loss in agriculture 4

Control of damaging organisms 5

- Natural control 5

- Applied control 6

Integrated Pest Management 7

- How is Integrated Pest Management defined? 7

- What does Integrated Pest Management mean? 8

- Economic threshold and injury level 8

- Integrated Pest Management in (sub-) tropical countries 9

Examples of Integrated Pest Management projects 10

Components of Integrated Pest Management 15

- Preventive control procedures 16

- Cultural control 16

- Host-plant resistance 17

- Biological control: means and methods 19

- Biotechnology 22

- Chemical control 24

Political views on Integrated Pest Management 27

References 28

Principles of Integrated Pest Control

AGRICULTURAL PRODUCTION AND LOSSES FROM PESTS

Plants provide either directly or indirectly the main source of nutrition for mankind, animals and uncountable masses of lower organisms. An ample supply of food, fibres and other vegetable matter is a prerequisite to health as well as to social and economic development of any human society. Thus, the incessant competition from all kinds of organisms, which cause injury, disease or destruction to man himself, his crops and livestock as well as to his goods and structures, is unwanted. Man is waging a continuous struggle against these injurious and troublesome organisms, called pests, in order to protect his food, shelter and living. The knowledge, methods and means, which have been developed to control pests in agriculture, are the subject of this manual.

Integrated pest management depends largely on preventive actions which enhance natural enemies and on measures which avoid disruption of nature's potential for self-regulation of pest populations. The farmer should, however, pay close attention to the development of his crop and of beneficial and damaging organisms therein in order to be able to make correct decisions on control. (Photo courtesy of the FAO IPC in Rice Programme)

Owing to the spectacular development of agriculture in general during the twentieth century, world food production has grown considerably to the extent that per capita food production has increased in spite of the explosive growth of the world population. However, growth figures differ greatly for continents and individual countries. Since the 1940s, agricultural production in the United States of America has increased three times through intensification measures, despite the fact that the area under cultivation was reduced by one fifth; in Africa and Central America, both total and per capita production declined over the same period. Developing countries provide only about 30 per cent of the world requirement of food, although they are home to more than half of the global population, and agriculture accounts for more than 70 per cent of their national income.

An estimated 34 per cent of the population in developing countries suffer from malnutrition; in 1985, the ratios for malnourished and healthy people were 1:3 in Africa, 1:5 in East Asia and 1:7 in Latin America. The heaviest toll is among children; every week about a quarter of a million children die from illness and malnutrition. In view of the rapid growth of the world population, there can be little doubt that their numbers will rise further. Whereas the production of major food crops in the Asian and Pacific region appears to have stagnated since 1985, its population is expected to have doubled from the 1980 level by the year 2000.

The necessary increase in food production should come from an expansion of the area under cultivation and from agricultural intensification. Among others, research and development should be directed at exploiting the potential of rainfed farming and crop diversification. However, intensification and the significant changes in production technology that it brings about appear to be accompanied by a worsening of the general pest situation. Current estimates put global losses due to animal pests, diseases and weeds at $US 300 billion annually. This is equal to around 30 per cent of potential global food, fibre and feed production. Losses each year in grain crops alone are estimated to exceed the gross and net food grain deficits of the developing world (NRI, 1991). Pests not only cause quantitative losses, but also affect the quality of agricultural produce which may become noticeable

Table 1.1: Statistical data concerning population, available agricultural land and production in Asia in 1987

Country/ area	population (millions)	(%)[a]	(million ha)[b]	(ha/caput)[c]	[d]
Bangladesh	106.7	70.5	9.2	0.12	109.3
China	1,086.3	69.6	97.0	0.13	140.1
Fiji	0.7	41.0	0.24	0.82	93.9
India	802.7	63.8	169.0	0.33	133.4
Indonesia	172.3	47.0	21.2	0.26	138.1
Japan	122.1	7.3	4.7	0.53	102.4
Malaysia	6.2	33.0	4.4	0.33	165.5
Myanmar (Burma)	39.1	48.6	10.1	0.53	146.2
Nepal	17.8	92.1	2.3	1.99	124.8
Pakistan	111.2	54.4	20.8	0.14	140.0
Papua New Guinea	3.7	70.0	0.4	0.35	119.2
Philippines	58.0	48.0	7.9	0.15	108.5
Republic of Korea	42.1	25.4	2.1	0.28	112.3
Sri Lanka	16.6	52.2	1.9	0.22	100.2
Thailand	53.3	62.6	20.1	0.60	120.0
Vanuatu	0.2	47.4	0.15	2.01	101.5
Viet Nam	62.8	62.7	6.5	0.16	136.0
World	**5,026.3**	**46.3**	**1,473.7**	**0.63**	**116.0**

Source: Selected indicators of food and agricultural development in the Asia-Pacific region, 1978-88. FAO/RAPA, Bangkok, 1989.

Notes: [a] = Percentage of agricultural population out of total population

[b] = Arable and permanently cropped land in million hectares (exclusive of pastures and forest land. Double cropped areas are counted only once).

[c] = Ratio of agricultural land to agricultural population (in hectares per capita).

[d] = Gross production of food products; index 1979-1981 = 100

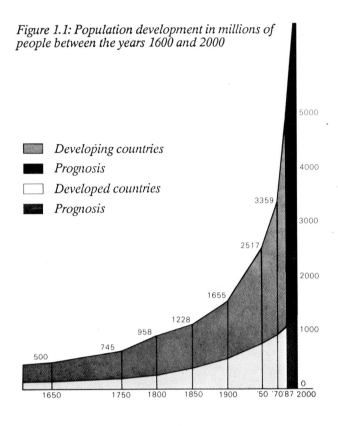

Figure 1.1: Population development in millions of people between the years 1600 and 2000

Legend:
- Developing countries
- Prognosis
- Developed countries
- Prognosis

Chart values: 500, 745, 958, 1228, 1655, 2517, 3359, 5000

Years: 1650, 1750, 1800, 1850, 1900, '50 '70'87 2000

through a detoriation of taste, flavour, fibre strength, processing characteristics, seed germination etc.

The massive damage from pests and diseases indicates that any real progress in agricultural production can be made only if measures aiming at an expansion and intensification of agricultural production are simultaneously supported by an adequate system of integrated crop protection and pesticide management. Evidently, the implementation of such a system requires reinforcing the existing infrastructure for national plant protection and extension services and the formulation of an effective regulatory framework.

The flora and fauna of the humid (sub-)tropical countries generally comprise a much larger number of species of higher and lower organisms than those of the temperate and arid zones. Unfortunately, the number of injurious organisms and viruses threatening mankind, animals and agriculture is also higher and consequently losses tend to be greater overall. For instance, the number of major and minor insect pests of rice in Asia amounts to about 40 as compared with 12 in temperate climates, whereas the number of rice diseases amounts to more than 500 in the tropics compared with 54 in the temperate zone. Because of a higher average temperature and relative humidity as well as the availability of food and shelter throughout a greater part of the year, there are more generations per year and pest populations develop more rapidly and abundantly.

Remarkably, the production centres with the highest yields of important food crops appear to be situated in temperate areas and outside the region of origin (see table 1.2). The lower yields shown are to a certain degree the result of a higher pressure from pests, diseases and weeds. The high virulence of pathogens in the tropics can be illustrated by the example of hybrid varieties of maize, which were developed in the United States, when these varieties are grown in Africa. The hybrid varieties are resistant to a number of diseases if grown in North America, but in Africa they are heavily infected by pathogens that also occur in North America. The degree of resistance appears to be insufficient under conditions which favour an optimum growth of the pathogens. Moreover, these hybrid varieties are sensitive to African maize streak virus and downy mildew, and thus they are unsuitable for African conditions.

FACTORS CAUSING DAMAGE AND LOSS IN AGRICULTURE

One can distinguish two types of factors, namely abiotic and biotic factors, as the causal agents of damage and loss in agriculture, forestry and related environmental activities.

Damaging abiotic factors are conditions of the environment which adversely affect the health and productivity of plants. They include unfavourable climatic conditions such as drought, too high or too low temperature, poor texture or toxic properties of the soil, deficiency or surplus of soil moisture, nutrients and elementary metals, and pollution of the environment.

Generally, measures aimed at the prevention or solution of problems concerning water and nutrient supply of the crops, soil structure and pollution are regarded as belonging to the discipline of crop cultivation technique rather than crop protection. However, as some cultivation practices have a profound influence on the development of weeds, animal pests and pathogens, good agricultural practice and crop protection should be combined in an overall strategy of preventing and controlling damage to plants in the field.

Harmful biotic factors can be distinguished in animal and microbial organisms. Generally, animal organisms which infest plants or stored agricultural products cause damage by feeding that results in a loss of quantity and quality. The infesting animal organisms may also be dispersed through the soil or other substrates.

Table 1.2: Yields of some food crops grown within and outside their region of origin (in ton/ha)

Crop	Origin	Within origin	Outside origin
Cowpea	Africa	0.20	0.70
Barley	Western Asia	1.33	2.05
Maize	Central America	1.22	3.02
Rice	South Asia	1.85	3.31
Soya bean	China	0.87	1.67
Sorghum	North-east Africa	0.70	1.33

Animal pests include insects, mites, nematodes, snails, rodents, birds and, incidentally, larger animals such as rabbits and wild boars.

Animal species, which are chronic pests under prevalent conditions at various growth stages of the crop and cause considerable damage, are called key pests or major pests. Others, which occur irregularly and locally and cause sporadic outbreaks, are regarded as occasional pests. These pests are also called endemic, i.e. the species are permanently available, but cause little damage usually, as their populations are continuously limited by the action of natural enemies and other environmental control factors. Pests which come to the fore as a result of a particular action, such as the spraying of a broad spectrum residual insecticide that also kills their natural enemies are termed secondary, induced, or potential pests.

Examples of key pests are grasshoppers, aphids, cotton bollworms, rice stemborers, Mediterranean fruitflies; occasional pests are desert locusts, armyworms, scale insects; induced pests are cotton whiteflies, brown planthoppers, diamondback moths.

Microbial organisms infect their host by entering

Much of the fertilizer solution that is sprayed onto young cabbage plants by these two farmers will be taken up by the sedge weeds (Cyperus esculentus) which cover these vegetable beds. Weed competition is most detrimental during the early stage of crop growth. (Photo by the author)

and establishing a temporary or permanent relationship. They disrupt physiological processes of normal plant growth, for instance, by blocking the vascular tissues, by deregulating normal cell growth, or by killing cells. A plant which is infected thus contains a parasitic organism or a virus; it is said to be diseased only when symptoms of the infection become evident. The causal vegetative organisms are called pathogens. They include fungi, bacteria, mycoplasmas and viruses. The terms endemic, key pest or secondary pest etc., which indicate their location, frequency and harmfulness, are not only applicable to animal pests but also to diseases.

Weeds are harmful vegetative organisms. They are higher plants that are unwanted by man, because they happen to occur in places where and at times when they have a depressing effect on the yield and quality of planted crops or are otherwise regarded as a nuisance. They not only compete with crop plants for light, nutrients and soil moisture, but as alternative hosts they may also harbour pests and pathogens, hinder cultivation, harvest and processing activities, block irrigation canals, contaminate a harvest of seeds or taint the flavour of dried produce.

CONTROL OF DAMAGING ORGANISMS

The control of an animal pest, disease or weed can generally be achieved by assisting the regulating forces which occur in nature with various old and new control practices applied by man. The control of pests is only one component of a more encompassing activity, the protection of crops and harvest. Crop protection aims at the prevention and limitation of damage and loss from injurious organisms in agriculture, horticulture and forestry. Crop protection science is the study of the factors which threaten planted crops and of the ways and means to provide protection in the pre-harvest and post-harvest stages.

Natural control

Many pest organisms have enormous powers of reproduction and could rapidly overwhelm the world if they were not kept in check by natural control mechanisms that have developed without human intervention.

Natural control is the collective action of environmental factors, physical and biotic, that maintain numbers of pest populations within certain upper and lower limits over a period of time. Important components of natural control are climatologic factors, natural enemies, availability of food and space and evolutionary pressure. Natural enemies include higher predatory animals such as birds and bats, predatory or parasitic species of insects, mites and nematodes, infectious species of fungi, bacteria and protozoa, and viruses.

Plant-feeding (phytophagous) insects and mites often live and feed on the crop plants throughout their entire life cycle. Their population development and

  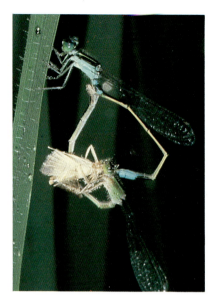

Usually, insect pests are themselves threatened by natural enemies at all growth stages. Examples include eggs of rice yellow stemborer attacked by the egg parasite Tetrastichus schoenobii *Fer. (photo left); a green semilooper larva (*Naranga aenescens*) parasitized by* Cotesia ruficrus *Haliday, the cocoons of which can be seen under its host (photo centre); and a leaffolder moth being caught by the predatory damselfly,* Agriocnemis femina femina, *Brauer (photo right). (All photos courtesy of IRRI)*

density is generally limited by a very complex control mechanism, of which natural enemies constitute a major component. Natural enemies include many species of predatory or parasitizing animal organisms, as well as pathogenic microbial organisms and viruses. Natural enemies have in common the fact that they utilise the pest organisms as their source of food or otherwise cause them harm.

Under favourable physical conditions, such as in a perennial crop environment in the humid tropics, both the pests and their natural enemies usually reproduce throughout the year. Their generations can overlap freely and all stages of growth - eggs, larvae, pupae and adults or spores, mycelium and fruiting bodies etc. - are simultaneously available. Although the population densities of both pests and natural enemies vary all the time, their proportional representation is generally sufficiently balanced to make natural enemy limitation of damaging pests adequate. However, if the natural enemy control fails temporarily, the populations of occasional and secondary pests can build up rapidly, particularly of those species which have a high fecundity rate. Where outbreaks of occasional or secondary pests do occur, the vast majority of insects or mites are of about the same growth stage (instar). Because at the beginning of an outbreak the natural enemies are far outnumbered, it may take some time before natural enemies build up to an effective control density, even though the disruptive condition may have been brief, such as would be the case with a minor change in local conditions. Insect diseases often spread relatively rapidly in an outbreak population bringing down the density of their host population. But even when natural control appears to have been restored, there may be a resurgence of the pest some generation later. The key pests are those in which the limitation of their population growth by natural enemies is generally inadequate.

Applied control

Applied control includes a whole range of practices, developed or modified by man, that becomes necessary when natural control factors fail to control pests.

Age-old farming and cultivation practices can be employed in a manner that makes the environment less well suited for the reproduction and growth of certain pests. These practices include soil tillage, weeding and removal of infested plant parts, adapting the time of planting and harvesting of a crop, spacing, water management, crop rotation with or without fallow and use of resistant varieties. Traditional farming communities, for instance, use several indigenous varieties of food crops with enhanced resistance against pests and diseases, or local varieties which, owing to a shorter or longer duration of growth, can still give acceptable yields if rains arrive irregularly or late. Many of these valuable land races are already lost and hard-won experience is often neglected, when agricultural innovations are introduced and inexpensive synthetic chemical pestides are welcomed as an easier solution to pest and disease problems. However, a whole range of problems associated with pest resurgence, secondary pest outbreaks and resistance, with diminishing yields and with unacceptable poisoning of man and the environment, has surfaced in places where farming communities turned to intensified cropping with the aid of high-yielding, non-resistant varieties and massive use of chemical imputs.

Insecticides tend to disturb the natural equilibrium over a relatively long time. Phytophagous insects usually remain in or near the crop and are the first target to be killed by spraying. Natural enemies range more widely in search of hosts and may become accidental victims through consumption of poisoned insects or contact with contaminated surfaces. Broad spectrum insecticides or those with long residual activity as well as repeated applications are most likely to affect natural enemies and, thus, cause a serious disruption of the control balance over a longer period of time. Whilst the insect population has time to recover rapidly, for instance because the eggs were not killed and fecundity is high, there are not enough natural enemies to stop the pest from reaching outbreak density again. This phenomenon of repeated, pesticide-induced outbreaks is called pest resurgence. Generally, the natural control balance will be restored within a few generations. If applications are repeated regularly (calendar-scheduled spraying), the natural enemies may not recover at all and secondary pests such as cotton whitefly and brown planthopper may get the chance of establishing themselves and of becoming regular key pests.

To date, regular outbreaks of pests and huge yield losses are compelling the international agricultural community to rethink control strategies and to drastically reduce the use of chemical pesticides which destroy natural enemies as well as pests. Breaking loose from the pesticide treadmill entails a revaluation of common control practices applied in traditional agriculture, the development of less damaging or non-hazardous control methods and a search for applicable natural control factors. All this calls for an integrated approach in which, for the first time, scientists of the public and private sectors co-operate on a global scale to find ways of managing pest problems in the longer term rather than applying chemicals as an immediate, but short lasting solution.

Many types of spider, such as the long-jawed spider Tetragnatha virescens *Okuma, are very effective although indiscriminate predators. They are usually fairly numerous in cropping ecosystems but also very vulnerable to most insecticides. (Photo courtesy of IRRI)*

INTEGRATED PEST MANAGEMENT

How is Integrated Pest Management defined?

Integrated Pest Management (IPM) has been defined by the Panel of Experts on Integrated Pest Control, Rome, as: **"A pest management system that, in the context of the associated environment and the population dynamics of the pest species, utilizes all suitable techniques and methods in as compatible a manner as possible and maintains the pest population at levels below those causing economic injury".**

In this definition, the term "pest species" stands for all organisms that through direct action (such as feeding) or indirect action (such as competition) cause damage or loss to crops and harvest (see section above under biotic factors).

The terms "Integrated Pest Managament" (IPM) and "Integrated Pest Control" (IPC) are regarded as having an identical meaning. A concept analogous to IPM is "Integrated Weed Managament" (IWM), which implies that IPM would be used more for describing control measures for animal pests such as insects or for diseases. The economic significance of weeds is not as well recognized as the economic significance of insects and diseases, although, in very many farming systems, weed control may consume up to 50 per cent of a farmer's labour. This may explain the desire to name Integrated Weed Management separately.

Both concepts, IPM as well as IWM, are components of "Integrated Crop Management" (ICM), the latter indicating the sequence of activities needed to grow a healthy crop and to produce a rewarding yield. Cultivation activities, such as soil preparation, irrigation and fertilization, all aim at reaching the maximum yield obtainable within the given potential of the sown cultivar and the environmental conditions. The protection or control measures, such as spraying pesticides, removing infested fruits and weeding, form a distinct set within the total of the crop management activities to the effect that they do not actually increase the yield, but rather prevent loss of yield and quality before and after harvest. Distinguishing between these two concepts may be helpful in assessing the chances of Integrated Pest Management in areas of marginal agricultural productivity.

For instance, in the Sahalien countries of Africa, pilot projects were set up at the farm level in the 1970s to demonstrate improved methods of millet and groundnut production, including better fertilizer use and pest control. Yields could be considerably increased, especially with fertilizer use. However, the combination of a low market value of the crop and unreliable rainfall that is typical of the region made the use of inputs and good crop management often uneconomical and farmers soon became less interested. The conclusion was drawn that in the Sahel there is need for more on-farm research on Integrated Crop Management than on Integrated Pest Management.

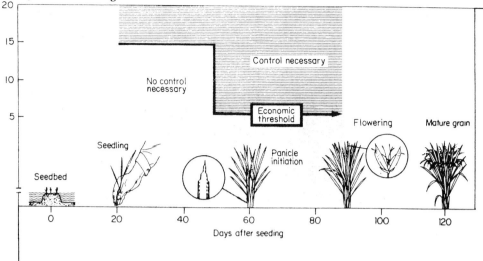

Damaged leaves (%) *Figure 1.2*

Economic threshold levels for rice leaffolders at different growth stages of rice plants. Farmers should increase routine sampling frequency to twice a week when moths are found in the field or in the pheromone traps. Insecticide should be applied when the economic threshold is reached. The economic threshold is lower from the time of panicle emergence until flowering when the flag leaves are present. (Illustration courtesy of IRRI)

What does Integrated Pest Management mean?

Although the term Integrated Pest Management should have one explanation, it now seems to have many shades of meaning, often coloured by the objectives of the user. It is simultaneously used to justify the total chemical control programmes of commercial interests, as well as the pesticide-free practices of organic gardeners (Gutierrez, 1987).

Ideally, Integrated Pest Management is an economically justified and sustainable system of crop protection, consisting of a combination of cultural, biological, genetical and chemical control methods, that aims at maximum productivity with the least possible adverse consequences for the environment.

Thus, one should not view Integrated Pest Management merely as a modified, sophisticated spray programme, because in many control operations some limited use of pesticides remains necessary. Such an impression would be erroneous as it does not recognize the fact that natural mortality of the pests is the most important and cheapest element in avoiding outbreaks. IPC-programmes aim rather at creating optimum conditions for natural mortality factors among which natural enemies play an important role. Careful selection and restrained use of pesticides is essential to avoid harmful interactions.

Economic threshold and injury level

A basic factor of integrated pest management is the recognition of the fact that injurious organisms neither can be nor have to be totally eradicated. They can be tolerated to a certain economic threshold level that has to be determined for each pest separately. This involves a judgement of the extent to which a particular pest population can be allowed to increase before a pesticide must be applied to prevent further crop loss. Various definitions have been proposed to describe the concept of economic threshold. Simply put, it implies that a pest control procedure is merited when the present costs of the procedure are equal to future benefits. The International

Rice Research Institute (IRRI) uses the following definitions:

1. Economic injury level (EIL): the pest population is large enough to cause crop losses costing more than the control.

2. Economic threshold (ET): the pest population at which control measures should be taken to prevent pest numbers from reaching the economic injury level.

Thus, above the economic threshold, economic injury occurs, while below it no control measures are necessary. Note that threshold levels of any pest are expressed either in terms of percentage of damaged plant parts (see fig. 1.2) or of population density, for instance:

- one rice stemborer (white tiller) per rice hill, or
- one *Pyricularia* rice blast fleck per rice leaf.

Economic threshold actually has a variable value that may differ with, among other factors, the variety and age of the crop, its location, previous damage received, simultaneous infestations of other pests and agronomic practices such as plant spacing, fertilizer levels and irrigation etc. Factors such as the market value of the crop, the cost of pesticide application, the expected income of growing another crop instead, and the weight of all negative health, social and environmental effects would also have to be considered. In practice, however, only a few of all the above variables are taken into account when an ET-value is established.

Some IRRI scientists consider unreliable the use of ET-values that are derived from controlled single insect pest studies in the field, because the ET is influenced by the above-mentioned variable factors and the time lapse since the last scouting. Therefore, they recommend instead the use of action thresholds (ATC), which are derived empirically from trial and error verification on farmers' fields. Such ATC-values have been determined for four key insect pests of rice and are said to correspond better with the practical thresholds which farmers have developed themselves. Farmers habitually monitor their fields for the occurrence of pests when they are working or while walking around their fields. They sight caterpillers, moths, sucking insects, egg-masses and signs of damage changing over time and they decide about treatment on the basis of experience and information (Bandong and Litsinger, 1987).

The concept of economic threshold should not be confined to assessing the need for chemical control only, as is often suggested. It is just as valid for evaluating any form of pest control, including biological and cultural control, breeding of resistant varieties and other methods which may often provide the more permanent forms of pest control. In each case, the cost of the activity needs to be weighed against the cost of the expected crop loss.

Integrated Pest Management can provide the cheapest way of controlling pests in the medium to long term. This might not be the case for the short-term costs, i.e. the direct expenses for the chosen control method and the costs of the losses that have to be accepted. It is a summation of the costs on separate control components, which in themselves may not be the cheapest immediate solution, but add up to the lowest cost strategy overall. For, instance, it has been found that very frequent and complete harvesting of cacoa pods is an effective and profitable way of controlling cocoa pod borer (*Acrocercops cramerella*), although costs per ton harvested will rise steeply. In this case, the high market value of cacoa beans amply covers the extra harvesting costs (Mumford, 1986).

There are further complications involved in applying the economic threshold concept. In some instances, a low level of infestation may have a beneficial effect, because it stimulates plant growth or enables a lesser amount of fruit to grow to greater size, thus avoiding the need for chemical fruit thinners. Cereal plants can compensate for dead tillers or missing plants by increased tillering and so eventually produce even higher yields. Cotton normally sheds from 65 to 75 per cent of all of the fruits initiated, irrespective of whether the cotton bolls are being infested by insects. If some more bolls are abscised due to insects, the dry matter may simply be invested in the surviving fruits, often resulting in some enhancement of yield. However, once damage exceeds the crop's potential for compensating losses, a further increase in the size of a pest population results in a progressive reduction of yield.

Working with economic thresholds requires a system of scouting for the actual presence of pests in the crops and of forecasting probable population developments. Forecasting should also include the sampling of natural enemies and predicting their impact on pest development. It is clear that farmers need to be trained in recognizing pests and natural enemies and in the proper counting of pest numbers on the plants. Generally, economic thresholds are determined for major food and cash crops by national crop protection institutions, sometimes in collaboration with international research institutes, and they should be taught to farmers through adequate extension.

Integrated pest management in (sub-)tropical countries

Implementation of integrated pest management throughout the tropics appears to be favoured by the following factors:

* Climate and physical environment in tropical and subtropical areas generally allow populations of both pest organisms and natural enemies to proliferate throughout the year. Regulation of numbers by natural biotic mechanisms thus can function optimally and in relative abundance.

* Agro-ecosystems, which have not yet been subjected to rigorous economic development, generally have been only little disturbed by pesticide applications. They offer a relatively sound base for the development of an integrated pest mangagement system, while farmers do not have to be re-educated.

* If the agricultural production is destined for home consumption, then farmers usually pay little attention to light infestation and some blemishes on the harvested produce; they would hardly spray for cosmetic reasons. However, transformation to market-oriented production introduces the criterion of external quality, which leads to excessive use of insecticides and fungicides on fruits, vegetables and flowers.

Close examination of the crop for the presence of pests and natural enemies and the recording of their development during the growing season provide the data on which the economic threshold principle can be applied and the need for control be determined. (Photo courtesy of FAO IPC in Rice Programme)

* Where new areas are opened up for agricultural production, as may be the case in sparsely inhabited parts of Africa, Asia and South America, there would be an opportunity to introduce Integrated Pest Management from the beginning and to ban unwanted chemical control practices.

The majority of farmers in developing countries own or lease farms of less than one hectare in size and many of them have not got the resources for extensive chemical control. Thus, unless the department of agriculture has initiated a large-scale food production intensification drive with subsidized supply of agronomic inputs, poor farmers use little or no pesticides on food crops. Still, there exists often a strong need for adequate crop protection and more food security. These small-scale farmers are, therefore, a very suitable target for extension of sound and sustainable IPC-programmes. As suitable IPC-technologies are gradually being made available, the attitude of farmers and crop protection authorities is becoming a vital factor in their adoption. An apropriate training message and a motivated, commercially independent extension service is required, but often the same persons providing the technical information to the farmers are also merchandizing pesticides.

EXAMPLES OF INTEGRATED PEST MANAGEMENT PROJECTS

The reasons for developing and propagating IPM are best explained by citing some well-documented historical cases. These examples also show various components of an IPM strategy and the way they may interact.

Example 1: Cotton growing in Canete Valley, Peru (South America).

For many generations, Canete Valley had been a successful cotton growing area with progressive farmers who were sufficiently wealthy to invest in advanced technologies. In 1939, the tobacco bud worm (*Heliothis virescens*) appeared in the valley. The spraying of arsenical insecticides and nicotine sulphate resulted in a rapid build-up of cotton aphid (*Aphis gossypii*) and a worsening *Heliothis* infestation. By 1949, cotton yields (lint) dropped from about 500 kg/ha to 365 kg/ha. As it was realized that the natural enemies had disappeared owing to the insecticide applications, a new programme for pest control was proposed which comprised the following measures: sowing of early maturing varieties, prohibition of ratoon cotton, selection and proper use of insecticides, crop rotation, and fixing by law the planting dates and sanitation measures. Between 1949 and 1956, these recommendations were only partly implemented as many farmers resorted to spraying with newly available organochlorine insecticides. Within three years the number of major pests occurring in the valley soared from 7 to 13 and resistance against BHC, DDT and toxaphene had developed in the cotton aphid, *Heliothis* and leafworm (*Anomis texana*) populations. Cotton

production, after reaching a peak of 728 kg/ha in 1954, droppped to 332 kg/ha, even though 16 sprayings were carried out on average. As the same situation developed in other valleys, a total of 40,000 ha were affected in Peru.

After 1956, the aforementioned proposals were strictly implemented and the use of organochlorine insecticides replaced with the arsenic and nicotine insecticides. Moreover, biological control was introduced with the release of the egg-parasite *Trichogramma*, predatory lady bird beetles and carabid beetles (*Calosoma abreviatum*); also, maize and wheat were planted as hosts for building up natural enemy populations. Within three seasons, cotton production reached 800 kg/ha, whereas the average number of sprayings dropped to 2.3.

Example 2: Cotton growing in the Nile Valley, Egypt.

Egypt is the only African country with a long-time experience of large-scale IPC-programmes in maize, cotton, sugar-cane and rice. As elsewhere, the reasons for undertaking integrated control were the development of insecticide resistance, appearance of secondary pests, ever increasing costs of chemical control and environmental contamination in a very densely populated area. Cotton is economically the most important crop, thus pesticides were applied frequently until pest infestations became severe in the early 1970s. The major pests were pink bollworm (*Pectinophora gossypiella*), cotton leafworm (*Spodoptera littoralis*), and spiny bollworm (*Earias insulana*), whilst the cotton bollworm (*Heliothis armigera*) has come up as a very damaging secondary pest since 1972.

Over the years, the Egyptian authorities have developed a range of very effective IPC-recommendations, consisting of cultural, biological, chemical and regulatory measures, examples of which are:

* The cotton must be sown around 15 March. As pink bollworm larvae hatch mostly during the first half of May, the cotton plants are still too small to suffer much damage.

* Egg-masses of *Spodoptera littoralis* are removed by hand during the first part of the growing season. This very effective control practice has been propagated since 1905.

* Insecticides are not sprayed until early June in order to allow optimum propagation of natural enemies.

* All infested or dry bolls are picked and burned at the end of the growing season to reduce pink bollworm infestation in the following year. The seeds receive a heat treatment to kill the diapausing larvae.

These practices, in combination with some others, have been so successful that the average number of sprays in 1975 and later years was less than one. However, the practices of picking eggs off the plants by hand and collecting the dry and infested bolls are very labour intensive. Therefore, alternative procedures need to be developed if wage levels rise in the future.

Example 3: Cotton growing in the Gezira Irrigation Scheme, Sudan (Africa).

In the Sudan, cotton growing in the extensive Blue and White Nile irrigation schemes makes a very sizeable contribution to the national income. More than 160,000 hectares of cotton are planted annually in the Gezira-Manegil Scheme, which constitutes more than half of the total area planted to cotton each year. In spite of the hard work of good entomologists over decades, insect pest control has gone through a similar vicious circle of ever more insecticide treatments and a steadily deteriorating pest situation. The history of pest control in the Gezira can teach a lesson to anyone involved in plant protection.

Insecticides were initially used in the Sudan for the control of cotton jassids (*Empoasca lybica*). Between 1945 and 1960, the standard application was one insecticide spray, commonly using a DDT-formulation. This practice resulted in a considerable increase in cotton yield, but even then it was observed that spraying DDT could induce outbreaks of the cotton whitefly, *Bemisia tabaci* (Joyce and Roberts, 1959). The use of parathion against whiteflies brought about an increased occurrence of American bollworm (*Heliothis armigera*), the latter becoming a major early-season pest of cotton since 1964. Meanwhile, the number of sprays per season had risen to five on average, and *B. tabaci* established itself as a permanent major pest by each year remaining active until late in the growing season. Stickiness of cotton lint - the result of contamination with "honey-dew" excreted by the huge number of whiteflies and aphids - became a great problem for the ginning mills and the price for graded cotton from the Sudan fell sharply.

Acknowledging that, in spite of the ever-increasing use of insecticides, they were losing their ground against the jassid, whitefly and bollworm-complex, the management of the irrigation schemes introduced the concept of package deals with agrochemical companies in 1972. These were contractual agreements giving the responsibility for all (aerial) pest control operations to the private companies, which in turn guaranteed a certain minimum yield and quality. This system gained rapid acceptance and by 1979 almost half of the Gezira cotton-spraying and large parts of the other Nile schemes were under contract. However, whitefly infestations got worse until, by 1981, the problem had spun out of control and the staggering costs for pest control forced the management to abolish the package deals altogether. For the next move, the use of DDT and DDT-containing mixtures was prohibited in the entire Sudan as from the 1981/1982 season.

In the following years, a definite change towards the use of synthetic pyrethroids and aldicarb (Temik) granules, among others, could be noticed in the insecticide usage pattern. However, the number of applications and the cost of control kept on rising. During the 1985/86 season, the average number of sprays peaked at 8.5, whilst the costs of pest control reached an all-time high of 33 per cent of the total production costs (ref. table 1.3). These figures seem to represent a turning point for cotton pest control in the Sudan, because the number of sprays came down to an average of 5.4 in the following three seasons and costs are declining.

A programme for the development and application of integrated pest control in cotton and rotational food crops is being implemented under the auspices of FAO and financed by the Government of the Netherlands. In the first project phase from December 1979 until September 1983, field studies were performed on host-plant resistance to whitefly and on suitable natural enemies. A resistant cotton variety, Sudac-K, was selected and two aphelinid wasps were identified as effective parasites of *B. tabaci*.

During the second phase of the project from September 1985 until April 1989, the programme of work focused on study and demonstration trials in the large cotton production schemes with the main objective of comparing routine schedules of spraying with reduced and no-treatment schedules. Furthermore, an egg-parasite of the American bollworm, namely *Trichogramma pretiosum*, was introduced into another area, the Rahad Scheme. This parasitic wasp was collected in Texas and is being mass-reared in the Netherlands for release in the Sudan. These large-scale trials in the commercial production areas could be carried out owing to an agreement under which the Netherlands Government guaranteed to compensate farmers for eventual yield losses and in fact, did settle the claims. Remarkably, the initial experiment, involving 320 hectares of cotton left unsprayed in the 1986/1987 season, was made possible through the decision of the Gezira tenant union to underwrite the assurance of yield loss compensation for its members involved in the trial. Since the "FAO/Sudanese IPM in cotton"-research had shown that a reduction in the number of insecticide sprays was really possible, the World Bank made the promotion of research on IPM in cotton a pre-condition to lending $US 75,000 for payment of the costs of pest control in the 1987/88 season as part of its Sudan Gezira rehabilitation programme.

The results obtained during the second phase indicate that no spraying, or less than an average of 2.3 applications, promotes the natural enemy control of whitefly, whereas 4-6 applications cause outbreaks of whitefly. However, the mean population densities of cotton jassids, American bollworm and aphids as well as the percentage of damage from bollworms tend to be lower in fields which have received a higher number (4-6) of applications. Consequently, yields of seed cotton and of lint are also higher in these fields.

11

A preliminary appraisal of the second phase indicates that reducing the number of insecticide applications produced a generally better control of cotton pests in Sudan, but that the economic injury levels (EILs) were set much too low. Therefore, the current third phase of the programme will pursue the following research objectives:

- Increasing EILs and ETs to a more realistic level in order to avoid unnecessary spraying;

- Determining the importance of naturally occuring natural enemies and introducing foreign ones if needed;

- Developing crop rotation schedules which can help to reduce injurious pest organisms and pathogens in and above the soil;

- Intensifying training and extension on IPM in cotton and rotational crops.

To date, emphasis had been directed towards maximizing cotton yields. However, although usually more yield results from 4-6 sprays per season, a higher actual profit appears to be obtained from spraying four times only owing to the saving of the high application costs. Recognizing this, the Sudanese Agricultural Research Corporation has advised the Group-entomologist in charge of determining ETs and spray schedules to let the actual profit principle prevail over higher yield. The National Pests and Diseases Committee of the Sudan has already taken the important political decision to raise the ETs for two commercial production areas of 2,000 ha each in the 1989/90 season. Also, in extension messages, the idea of profit instead of yield maximization needs to be worked out.

Table 1.4: Economic threshold (ET) values for four major insect pests of cotton in the Sudan, 1989

Pest species	Current ET (1989)	Increased ET (1990)
Heliothis armigera	5-10 eggs or larvae per 100 plants	10-20 eggs or larvae per 100 plants
Empoasca lybica	50 nymphs per 100 leaves	100 nymphs per 100 leaves
Aphis gossypii	20 per cent infested plants	40 per cent infested plants
Bemisia tabaci	200 adults per 100 leaves	600 adults per 100 leaves

Source: Stam, Abdel Rahman and Munir, 1989.

Example 4: Perennial estate crops in South-east Asia.

Tropical tree crops are of great economic importance to this subregion because of their contribution to the countries' balance of trade position from export revenues. Consequently, the estate sector is a focus of political attention and preferential financial investment. Generally, Governments are providing much technical and financial assistance to small-holders growing perennial crops and established specialized research stations and nucleus estates with core-processing facilities. The most valuable crops are oil palm, rubber and cocoa, but also coffee, tea, coconut, cloves, fruit and nuts have great national and local importance. In Malaysia, for example, the area under cocoa has expanded from 30,252 ha in 1975 to 242,000 ha in 1984, whereas an estimated 1,458,100 ha were planted with oil palm at the end of 1986.

Table 1.3: Statistics on the cost of crop protection and yields of cotton in the Gezira Irrigation Scheme since 1945, Sudan, 1989

Period or year	No. of sprays[a]	Cost of protection[b] (% of total cost)	Long staple Area[c]	Long staple Yield[d]	Medium staple Area[e]	Medium staple Yield[f]
1945-1954	0.5	-	-	-	-	-
1955-1964	1.6	-	-	-	-	-
1965-1974	5.0	14.5	-	-	-	-
1975-1984	6.9	25.0	83	3.5	17	5.3
1985-1986	8.5	33.3	94	3.4	6	5.5
1986-1987	5.2	27.5	79	4.7	21	5.7
1987-1988	5.7	24.7	62	3.9	38	5.7
1988-1989	5.3	19.3	55	4.1	45	6.5

Notes:
[a] = mean number of insecticide sprays
[b] = costs of crop protection measures as a percentage of total cotton production cost
[c] = area planted with long staple "Barakat" variety as percentage of total area planted to cotton
[d] = yield "Barakat" cotton in *kantar* per *feddan* (1 *kantar/feddan* = 336.5 kg of seed cotton/ha)
[e] = area planted with medium staple "Shambat" variety as percentage of total area planted to cotton
[f] = yield "Shambat" variety in kantar per feddan

Source: Sudan Gezira Board, August 1989.

A traditional preventive control measure against soil-borne diseases is the use of resistant rootstock in nurseries. A Sumatran woman inserts grafts of the high yielding Cinchona ledgeriana *cultivar (a tree producing quinine) onto understems of a* Cinchona *cultivar with resistance against cancer caused by* Phytophthora cinamoni *and protects the wound with wax. When the graft has taken on, the original top of the understem will be cut. (Photo by the author)*

Plantations of perennial crops have for centuries been nursed with great care and many years of research have produced the high-yielding stands of today. Much experience had been collected in integrated cultural and biological control of occasionally very troublesome, but throughout manageable insect pests. Diseases are generally more difficult to deal with than animal pests and they sometimes make it impossible to grow certain crops, e.g. in 1870, coffee leafrust disease (*Hemileia vastatrix*) destroyed the stands of *Coffea arabica* in Sri Lanka (then called Ceylon) and is currently posing a threat to arabica coffee trees in Central America. The traditional cultural and biological control methods started to fail when the area planted to estate crops was vastly expanded and growers changed over to extensive use of broad spectrum pesticides in order to control increasing pest problems and rising labour costs. The fact that broad spectrum, long-residual insecticides were as readily propagated in these valuable crops as in field crops shows that scientists generally did not anticipate their disastrous effects.

Massive outbreaks of pests and diseases, a changing weed spectrum and soil erosion owing to herbicide application forced agronomists to search for ways to restore the natural balance between pests and natural enemies. Because of the diversity of problems and the number of crops involved, Integrated Pest Management in perennial estate crops has come about as a result of a gradual evolution in which new technology has been blended with established practices in a step-by-step process rather then through the introduction of a wholly completed system.

Aspects of integrated pest management in coconut, oil palm and cocoa plantations are discussed in the following sections as examples of IPM in perennial tree crops.

Coconut growing contributes considerably to the income of numerous small-scale farmers in India, Indonesia, the Philippines as well in many other places throughout South-east Asia and the Pacific island countries. Some insect pests, diseases and rat species are causing damage to the coconut palm trees. Of these, the rhinoceros beetle (*Oryctes rhinoceros*) is the most serious pest and it is threatening rehabilitation programmes for replanting over-aged stands with new hybrid coconut varieties. The adult rhinoceros beetles fly to the crown of the palm and bore into the fibrous leaf sheaths encircling the terminal bud. This apex of the palm is its most vulnerable part, and it is often attacked by pests and diseases. The larvae develop in moist, decomposing vegetative materials; rotting stumps and felled trunks are the usual breeding sites. The extent of damage recorded ranges from total loss of young palms to yield depressions up to 100 per cent. A reduction in leaf area of 10, 20, or 30 per cent causes a decrease in nut production of 19, 35, or 51 per cent, respectively.

Chemical control is difficult because of the height of older palms and the cost of using adequate power-sprayers and dusters. Sanitation, consisting of removal of dead fronds, trunks and other debris, should be carried out and the material can be used as fuel. In the late 1960s, two very effective methods of biological control were developed. A baculovirus of *Oryctes* species was discovered in Malaysia in 1963 and artificially introduced in 1967 into rhinoceros beetle populations in Western Samoa, where it spread throughout major palm areas (Huger, 1969). Further introductions of the baculovirus into other South Pacific countries also resulted in an effective control of rhinoceros beetle populations. Regular surveys since 1964 show that a high degree of virus infection persists in Western Samoa. Such surveys on the incidence of virus infection in adults should be

Desert locust Schistocerca gregaria *infected by the green fungus* Metharizium anisopliae. *The potential of parasitic fungi for controlling migratory locusts is being investigated at the International Institute of Biological Control, United Kingdom. (Photo courtesy of CAB International)*

carried out twice a year. However, in the Philippines, the endemic baculovirus of *Oryctes* appears to persist in only 7-12 per cent of adult beetles in Quezon province. Presumably, the virus transmission is slowed down owing to a stagnation in population development during the annual dry season of several months' duration.

The second bio-agent of the rhinoceros beetle is the fungus *Metharizium anisopliae*, a green muscardine fungus that was recognized as a possible control agent as early as 1913. Recently, techniques have been discovered to cultivate the fungus on organic material such as bran and kernels of various kinds of grain. The collected mycelium can be introduced into the habitats of *Oryctes* larvae, from where it is carried along by larvae and adults. Thus, the fungus is spread throughout the population and, under conditions of high relative humidity and regular rainfall, it can remain infective to *Oryctes* larvae for more than one year.

The same muscardine fungus is also very effective against the hispine beetle (*Brontispa longissima*) that was accidentally introduced into the coconut stands of Western Samoa in 1979 and there soon gained a reputation as the country's worst enemy. The *M. anisopliae* fungus is reported to eliminate up to 65 per cent of third and fourth instar larvae of this hispine beetle. The hispine beetle is further parasitized by a newly introduced parasitic wasp, *Tetrastichus brontispae,* that is mass-reared and released in infested coconut plantations throughout Western Samoa. The rate of parasitation of hispine larvae averages 30 per cent.

The sowing of leguminous cover crops such as *Pueraria* spp. and *Centrosoma* spp. also provides a certain degree of protection against rhinoceros beetles in young coconut plantings, as these legumes provide such a thick vegetative cover that egg-laying in the rotting stumps and trunks is greatly reduced.

The integrated control of rhinoceros and hispine beetles and other pests including rats has been evaluated

economically in the coconut plantations of Western Samoa. An estimated 10 per cent loss in nut production is prevented, amounting to $US 2.2 million annually, while the routine cost of the programme - evaluating the infestation level and the percentage of parasitation - amounts to only $US 2,000 per year. The benefit to farmers is on average $US 50 per ha of coconut palms per year without any input at the farm level (Klingauf and Voegele, 1986).

Oil palm plantations in Malaysia were plagued during the period of their rapid expansion in the 1960s by defoliating caterpillars, in particular three species of bagworm (*Psychidae*) and five species of nettle caterpillars (*Limacodidae*). Other pests are cockchafers (*Scarabaeidae*), the rhinoceros beetle (*Oryctes rhinoceros*) and a grasshopper species (*Acrididae*). Local economic loss caused by the defoliators was estimated at 40-50 per cent of the crop yield or about $US 500 per ha. All the aforementioned pests are regarded as occasional pests that are usually kept well under control by a great number of parasites and predators, which have been recorded during outbreaks, but which are as yet unsufficiently studied. The only chronic pests in Malaysian oil palm plantations are rats, namely the Malayan wood rat (*Rattus tiomanicus*), the rice field rat (*R. argentiventer*) and the house rat (*R. diardii*). If the damage from rats reaches a value of about $US 75 per ha per year, control operations should begin (S.S. Liau, 1988).

The integrated control of oil palm pests still depends much on the use of selective pesticides and special application techniques, such as trunk injection of monocrotophos. Using pesticides has become a matter of special concern since, in 1981, an oil-palm pollinating weevil, *Elaeidobius kamerunicus*, was introduced from Africa. This weevil has effectively replaced the need for hand-pollinating, thus saving the industry millions of dollars on labour costs. The weevil has become so well established that pollination is found not to be affected either by occasional aerial applications of methamidaphos, or by trunk injections or fogging with deltamethrin.

Against the defoliating caterpillars, biological agents, such as *Bacillus thuringiensis* and naturally occurring viruses, are recommended. Hand picking of bagworms is an economically effective method of control in young plantings, particularly of the larger species *Mahasena corbetti*. For prevention of rhinoceros beetles, sanitation by removing palm debris as its main breeding site is a common practice. Previously, the palm trunks were felled and left to rot; nowadays, the trunks are mechanically shredded and burnt at a cost of $US 220 per ha, but if burning is not complete, the beetle grubs may still thrive on the charred remnants.

Rat control practices on the Malaysian estates include the periodic removal of rat nests, dead fronds and waste material from the palm crowns, tin sheet baffles and wire guards around the bases of palms, destruction of burrows, and trapping. However, these methods have only a limited effect. The main line of defence is the use of anti-coagulant rodenticides in wax-bound baits.

The introduction of the barn owl (*Tyto alba*) as a predator of rats has a chance of meeting with success, since the owl appears to have spread to all Malaysian States. It is, however, feared that the use of newer, more potent anti-coagulants may prove fatal to the owl as a result of secondary poisoning.

Cocoa plantations in Malaysia suffer from about eight key and 20 occasional insect pests of which only the mirids and pod borer are discussed below.

The major species of mirids attacking cocoa plantings are *Helopeltis theobromae* in Peninsular Malaysia and *H. clavifer* in Sabah. Nymphs and adult bugs feed on pods and young shoots by sucking the plant sap. The toxic saliva injected through the feeding puncture causes a dark spot, that may become infected with fungus. Extensive damage may be inflicted causing up to 85 per cent pod loss, whilst damage to shoots may be intense. Control of *Helopeltis* spp. in Malaysia is based on results obtained in West Africa with the control of cocoa capsids (*Sahlbergella singularis*), an insect of the same Miridae family.

However, the use of insecticides is still the primary technique for control of *Helopeltis* spp. on Malaysian cocoa plantations, the standard being gamma-HCH. By 1984, resistance against this insecticide appeared not yet to have developed. The insecticides isoprocarb and propoxur are suitable, cost-effective alternatives for this cheap but persistant compound. Applications are commonly made with knapsack sprayers and mistblowers using low volume and ultra-low volume dosages. The cocoa industry has developed an effective scouting and early warning system for *Helopeltis* which allows the treatment of infested blocks without the need to engage in indiscriminate blanket applications. An effective method of biological control of *Helopeltis* spp. has not yet been found, although the cocoa black ant (*Dolichoderus thoracicus*) shows promise as a predator. Neither has a suitable cultural method nor a resistant cocoa variety been detected.

Fortunately, more options exist for controlling the cocoa pod borer (*Canopomorpha cramerella*) which was first detected in Sabah in 1980 and had within two years spread throughout Sabah. In 1986, the moth was also spotted in Peninsular Malaysia. The young larvae bore into young pods and caterpillars emerge from ripe pods before and after harvest. The caterpillar causes complete loss of production and thus it has been the object of intensive research which yielded the following integrated control approach:

* Selective spraying is directed at the underside of the horizontal, lower canopy branches - which form the daytime resting site of the moth - with pyrethroid or carbamate insecticides. Four to five applications at 10-14-day intervals in the period of low fruit-setting give good protection in the ensuing period of high fruit-setting. Care needs to be taken not to harm pollinating insects.

* The placing of polythene sleeves around pods just before they become attractive to female moths protects them against egg laying.

* Use of pheromones to monitor and trap large numbers of moths shows positive results, but the cost-effectiveness of this method is not yet fully known.

* Biological control using the indigenous egg parasite *Trichogrammatoidea bactrae fumata* gives encouraging results. The parasite is mass-reared and on an experimental scale released in large numbers, a technique known as inundation. Trials involving other, exotic parasites seem to have met with success.

* Frequent harvesting and immediate breaking to collect the beans and destruction of the pods through burying, drying or bagging is effective in preventing the emergence of borer larvae from ripe pods before and after harvest. If carried out consequently, this technique interrupts the life cycle of the cocoa pod borer and lowers the population density.

* Varietal resistance to cocoa pod borer infestation has been detected in Sabah. The resistance factor is based on a greater hardness of the pod shell; its usability for breeding purposes is under investigation.

A complete integration of all the above-mentioned control components would be desirable, but is considered to be too expensive. The majority of Malaysian cocoa growers practise selective spraying, frequent harvesting and bagging of the pods as the most convenient and cost-effective procedure. Scouting for pod borer occurrence is important.

COMPONENTS OF INTEGRATED PEST MANAGEMENT

The above examples clearly illustrate that procedures for controlling pests, diseases and weeds must correspond with the characteristics of the target organisms and those of the cropping environment. The latter not only includes aspects such as the kind of crops grown and methods of cultivation, but also the state of development of agriculture, the knowledge and economic resources of the farmers, and the expected cost-benefit ratio of the available control options. Some control procedures have a preventive character; others, a curative effect. Therefore, one needs to learn the advantages and disadvantages of all components which might be applied in an integrated control approach in order to make an appropriate choice.

The booming international trade in live plant material demands continuous vigilance against the introduction and spread of pests and pathogens. Quarantine personnel inspecting an incoming shipment of ornamental plants at the port of Amsterdam. (Photo courtesy of LNV, the Netherlands)

Preventive control procedures

Preventive measures have the objective of precluding the establishment of new pests or the limitation of the numbers occurring so that they will not become a problem. Examples of preventive measures are:

* Quarantine and regulatory control

The spread of pests and diseases to new areas from sources of infestation within or outside a country is often caused by the movement of people, commodities and equipment carrying the contaminant. Quarantine and containment laws or regulations and other legislation have been enacted and enforced in many countries to help overcome this problem. For example, plant materials to be imported into a country must be free from pests and diseases, and plants with the potential of becoming weeds are totally prohibited.

Countries with an efficient quarantine service require a rigorous inspection and fumigation of imported plant material at quarantine stations set up at ports of entry. Containment refers to limiting the spreading of new pests within permanent or ad-hoc in-country borders. Unfortunately, many countries still lack a disciplined and well-equipped quarantine service. Examples of recent introductions of unwanted pests are the greater grainborer (*Prostephanus truncatus*) in Tanzania, the potato cyst nematode (*Heterodera rostochiensis*) in various tropical regions, the thorny shrub (*Mimosa pigra*) in Thailand, and the coffee leafrust disease (*Hemileia vastatrix*) in Central America. The ever-increasing world-wide traffic and the weak quarantine infrastructure of many countries leave little hope for keeping such unwanted "guests" out.

Other examples of regulatory control are the prohibition to grow certain trees, plants or crops which are intermediate hosts to important pests or diseases of the economically most important crops of an area or the compulsory arrangement of crop rotation sequences. Governments are resposible for overseeing such regulatory control programmes which include eradication, containment and suppression. Eradication of a pest species is very difficult to accomplish, but it is possible in some cases to eliminate a pest from an isolated area. Containment programmes are used to limit the spread of pests that are likely to infest larger areas, e.g. the Mediterranean fruitfly in the United States, coffee leafrust disease in Central America and the larger grainborer in Eastern Africa. Suppression programmes are used when sudden outbreaks of pests, such as locusts, brown planthoppers and rats, occur over large areas and cannot be successfully dealt with by individual farmers alone.

* Pest-free propagation material

Infested or infected seed and plant material, such as tubers, rootstock, cuttings and grafts, can be the source of infestation or infection of an entire crop. The use of healthy propagation material, free of infectious material, is of particular importance if chemical control is inadequate or would be too costly as in the case of virus and bacterial diseases in low-value crops. Disinfecting propagation material with fungicides, fumigants, hot water, radiation etc., kills fungal spores and nematodes on the outside and in some cases also inside the seeds and vegetative material. Soaking seeds and plant parts in a fungicide dilution can also provide some limited protection against soil-borne pathogens, insects and nematodes. Cleaning seed from contaminating weed seeds is equally important. Plant protection authorities in developing countries should place more emphasis on the need for and possibilities of using healthy, cleaned seeds and planting material. Certification of seeds should be a trustworthy guarantee that the seeds do not contain dangerous diseases and noxious weed seeds.

Cultural control

Cultural control measures aim at modifying the cropping environment to the effect that it becomes less favourable for the development of pest populations, whilst maintaining the best possible conditions for high productivity. Some of the traditional cultivation practices were very effective in suppressing sources of infestation on or near the crop, and they are being revaluated for use in integrated control. These and newer cultural control methods are:

* Crop rotation

A traditional method of growing in the same field various crops alternately over succeeding years with or without leaving the land uncultivated (fallow) over intermittent periods of time. For instance, growing maize or sorghum for one or two seasons and planting vegetables or a root crop in the third season. The principle of a rotating cropping pattern is to interrupt the specific relationship between pests and host plants which favours or limits the development of these damaging organisms. For example, the golden potato cyst nematode, *Globodera rostochiensis*, is a parasite of potatoes and other solanaceous plants, but not of cereal crops. Crop rotation can be an effective measure against soil-borne pathogens, nematodes and soil insects provided the rotation cycle is sufficiently long. As different crops usually require different soil cultivation practices, such as shallow or deep ploughing, harrowing, ridging, broadcast sowing, crop rotation also plays a very important role in controlling weeds. It also interrupts the population build-up of above-ground pests. However, intensification of agricultural production may lead to growing only a few crops or to monoculture. Since modern intensive agriculture requires high investment, it has become a problem to find alternative crops which promise a sufficiently high economic return for the rotation cycle.

* Phytosanitation

Sanitation includes measures aimed at the removal and destruction of infected plant material as the potential source from which the pest or disease could spread. Collecting or burying crop debris and stubble from the field removes many diapausing larvae, eggs and pathogens. The cutting of weeds deprives many noxious organisms of their intermediate hosts, and rodents of food and shelter. Also, the eradication or prohibition of certain ornamental trees or cultivated crops from an area to protect a more important crop is a phytosanitary measure aimed at interrupting disease transmission. Furthermore, cleaning and disinfecting planting tools and machines, containers and soil, are necessary measures to prevent the spreading of pathogens, nematodes and weeds, especially in nurseries and protected cultures. The cleaning, disinfection and screening of stores and silos is a precondition in long-term food storage disinfection.

* Adequate cultivation practices

Techniques for growing crops should be effective in terms of promoting a healthy growth of the crop on the one hand and limiting the losses caused by abiotic and biotic factors on the other hand. Management of the water supply by irrigation and drainage, tilling of the soil for proper texture and aeration, fertilizer gifts, mechanical weeding, timing of planting dates, spacing and shading, are examples of cultural control factors. Rich gifts of nitrogen fertilizer induce a lush growth in a crop, but might also favour a stronger insect development. However, a balanced supply of nutrients also improves the health of a crop or tree so that it may better withstand parasitic diseases. Mulching with vegetative matter or polythene film (plastic) plays a very important role in reducing evaporation of soil humidity, furthering humification processes and beneficial soil organisms and in suppressing weed growth. Covering the soil with plastic sheets for the purpose of raising temperature and thereby killing noxious soil organisms is called solarization. Exclusion of pests by wire fences or screens is a common practice with high-value nurseries or small-area crops.

In trials at the Asian Institute of Technology, Thailand, the preventive effect of herbs against diamondback moth on cruciferous crops is studied. The beds planted with green cabbage in the front are surrounded by a hedge of basil Ocimum basilicum *and in the back with citronella grass. (Photo by the author)*

* Inter-cropping or mixed cropping

The growing of two or more crops on the same plot or field is a common practice in subsistance farming of many tropical countries. The mixed planting of food crops, vegetables and fruit trees often provides relatively high yields owing to the optimum use of soil, nutrients, water, space and other factors. The rich variety of tubers, fruits, vegetables and pharmaceutical herbs grown in home gardens offers a well-known example of mixed cropping. Losses from pests are generally low where no insecticides are used, which would disrupt natural enemy limitation, and the risk of yield loss is spread over the various crops available. To date, considerable research has been directed at investigating the advantages and disadvantages of inter-cropping for higher production and loss prevention.

A particular aspect is the interplanting of trap crops. These are cultivars or genotypes of crops which induce on a large scale the development of parasites without giving them a chance to complete their life cycle because of resistance, or because the trap crop concerned is destroyed by man to save the main crop. Trap crops are used to control nematodes and parasitic weeds such as *Striga* spp. and *Orobanche* spp.

Host-plant resistance

Centuries of natural selection in indigenous wild forms and cultivated forms of crops, grown under subsistence agriculture, have yielded "land races" with increased resistance to adverse enviromental factors such as drought, soil salinity and low temperature, as well as with a certain degree of durable resistance or tolerance against injurious organisms and viruses. These "land races", which have a distinct genetic composition, are morphologically recognizable; farmers have names for them and rely on their different properties concerning adaptation to soil type, time of seeding, date of maturity, nutritional value and taste etc. Most important, they are genetically diverse populations that are in equilibrium with both the environment and pathogens, and genetically variable. Land races have low productivity, but respond to selection pressures including selection for higher yields.

In previous decades, highly productive varieties have been developed by means of cross breeding and modern plant breeding technology; land races are being rapidly replaced. As they possess valuable genetic resistance to pests and diseases as well as other desirable traits, all wild forms and land races should be conserved. For instance, the only source of resistance against grassy stunt virus in rice was found in the wild species *Oryza nivara* collected from the State of Uttar Pradesh in India.

Resistance means the capability of a host to hinder the growth and activity of a parasite or phytophage and to limit the reproduction of a virus. Host-plant resistance can express itself in various ways; as the specific protection of plants against a fungus race, a virus strain or against a biotype of an insect or nematode.

Species of phytophagous or parasitic insects and mites, of fungi and bacteria etc., are continuously subjected to evolutionary and selective pressures which may favour certain mutations. For instance, the application of an insecticide may kill the vast majority of the target species, but a few individuals having a higher degree of resistance may survive. Resistance of an organism, e.g. an insect or bacterium, means its capacity to neutralize a harmful chemical substance such as an insecticide or bactericide applied against them. Under the pressure of repeated applications or long residual activity, a new population of the species concerned may develop, which is less sensitive to the pesticide. The new population differs from the original population only in a particular property; in this case, the acquired resistance to the insecticide is called a mutant strain. It is an important observation that, once the applications have been halted, the original wild strain can within a few generations get the upper hand again over the mutant strain.

Host-plant resistance with a race-specific expression is active against certain genotypes of the parasite (physiological races) and has no or little effect on others. This type of resistance is characterized by a genetical interaction between the genotypes of host and parasite. Often, the interaction is based on a gene-for-gene relationship between host and parasite or phytophage; this means that for every resistance gene in the host, one specific virulence gene can be found in the parasite. This type of resistance offers only a limited protection for the host plant, because it may not work against other races of that parasite and thus, the resistance in the plant may be broken when a new race develops. This weakness of race-specific resistance is another reason for restraining pesticide use as it furthers the development of new races of harmful organisms. For example, the fungus *Phytophthora infestans* has more than 20 races and the brown planthopper has at least three races (or biotypes). (The use of the term **race-specific resistance** has preference over the term vertical resistance, which is often

*New cultivars of radish (*Raphanus *spp.) demonstrate a great difference in resistance against artificial infection by* Fusarium oxysporum f. spp. conglutinans *in tests at the Plant Protection Research Institute (IPO) of the Netherlands. (Photo courtesy of IPO)*

used to describe the same property.)

Non-race-specific resistance of the host indicates a resistance that protects against all genotypes of the parasite or phytophage. In that case, there is no narrow genetical interaction between host and parasite. The resistance is usually not so strong as in the case of race-specific resistance. Just as the hosts differ in degree of resistance so do the genotypes of the parasites differ in aggressiveness. (The term **non-race-specific resistance** has preference over the used term horizontal resistance.)

Tolerance means the capacity of a host to keep the harmful effects of a parasite or phytophage or biotic factor at the lowest possible level. In cases of extreme tolerance, the host is carrier of the harmful agent, without showing any symptoms.

The growing of varieties of food and cash crops having a high degree of host-plant resistance is the major instrument in the prevention of crop losses, in particular in the case of diseases and viruses that can not be controlled with pesticides. For that reason, the international and national research institutes are paying much attention to the development of resistant, high-yielding varieties of such crops. An example is the development of IRRI-rice varieties with specific resistance against the Tungro-virus.

Unfortunately, no resistant varieties are as yet available for a number of urgent problems, in particular, virus-diseases. Furthermore, little effort has been spent on breeding for resistance in food crops of lesser economic value, such as *Pennisetum*, grain legumes and root crops. Subsistence farmers often have no opportunity to buy improved seed. Therefore, local Governments should take care of the multiplication, testing and distribution of seeds and planting material, and subsidize the introduction of new, resistant varieties as a component of the national integrated crop management strategy.

Biological control: means and methods

Biological control indicates the use of control agents in which the active principle is a living organism or virus (the so-called natural enemies) for regulating the incidence of pests or pathogens.

In an undisturbed ecosystem, the population density of a pest is kept within certain upper and lower limits over a period of time by naturally occurring parasites, predators or microorganisms. In a disturbed ecosystem, natural enemy limitation may, for a shorter or longer term, fail to prevent outbreaks of pests or epidemics of diseases. This may also happen when a new pest, pathogen or weed is introduced into a country or region, where its natural enemies are not available.

Theoretically, there are several ways in which insects might be controlled biologically; i.e. by

* Specific natural enemies, usually parasites;

* Non-specific natural enemies, usually predatory species;

* The sterile male method; and

* Micro-organisms and sex attractants.

* Natural enemies

Several cases of successful introduction of exotic parasites and predators for the control of a pest can be cited, the oldest being the introduction in 1888-1889 of the vedalia beetle (*Rodolia cardinalis*) from Australia into the citrus orchards of California that were being damaged by cottony cushion scales (*Icerya purchasi*). These predatory beetles kept the cottony cushion scales in check until, in 1947, they were destroyed by the spraying of DDT. As a result, the cottony cushion scale, thus freed from its regulating enemy, again became a serious pest in California.

A recent example is the control of the cassava mealybug in Africa by parasitic wasps (see box). Concerning the use of natural enemies for biological control, a distinction is made between the importation of exotic species (to control exotic pests rather than indigenous pests) and the conservation of indigenous natural enemies (to control both exotic and indigenous pests).

Some natural enemies become permanently established, but the number of those failing to establish themselves is much higher than the number of successful ones. The chance of establishment depends on various factors, such as the availability of large numbers of target insects, the type and availability of the crop, the presence of alternative prey and the weather conditions. In the case of the cassava mealybug, a few inoculative introductions of the wasp *E. lopezi* suffice for long-term control. But large-scale, inundative introductions of natural enemies in large enough numbers may be needed, if they provide only short-term control and survive for only a short time. If the inundative release has to be periodically repeated within a season, a reliable system of mass-production has to be set up, particularly if very large number of parasites are required. For instance, for the control of caterpillars on crops such as cabbage, sometimes a release of more than 100,000 *Trichogramma* wasps per hectare may be required to obtain the desired degree of control.

Inundative biological control aims at directly increasing mortality in a pest population that has already built up in a crop, whereby the released natural enemy is used as a biological insecticide. This method compares with the use of a chemical insecticide, and the same principle of assessing the cost-benefit ratio should be applied in chosing between these two alternatives. The cost-effectiveness of identifying, mass producing and releasing the suitable agent needs to be asserted in relation to the damage that is inflicted by the pest. The fact that already large numbers of target insects should be available for the parasites and predators to thrive upon

---

Successful control of mealybug

A famous example is the occurrence of the cassava mealybug (*Phenacoccus manihoti*) in Africa. This mealybug was for the first time reported in 1973 in the Congo, West Africa, and although being a wingless insect, it spread like a bush fire throughout 31 African countries within a period of 15 years. As the cassava mealybug can cause up to 80 per cent yield loss, it inflicted annually damage estimated at $US 2 billion to this very important food crop. Because of the immense threat to all cassava cultivation in Africa, the International Institute for Tropical Agriculture, Nigeria, started the "Africa-wide Biolocal Control Project" in 1980.

On the presumption that the cassava mealybug was brought in from South America on propagation material, a search for indiginous mealybugs and their natural enemies was launched in South America. After several years of searching, the same species of cassava mealybug was found in Paraguay together with two of its enemies, a ladybird beetle and a parasitizing wasp. The wasp, *Epidinocarsis lopezi*, proved to be an extremely effective parasite of the mealybug. It appears to be attracted by minute quantities of a compound that is exuded by plants which are attacked by mealybugs. Thus, these wasps immediately spot any mealybug even in cassava fields that are almost free of these insects.

Methods for mass-rearing have been developed and the wasp has meanwhile been released on many sites in more than 15 African countries. The wasps have become so well established that the pest might be brought under natural control. The predatory labybird beetle (*Diomus* spp.) initially performed well too after mass-release in Africa, but when at the end of the rainy season, the mealybugs became scarce, the ladybird beetles dispppeared. Behavioural studies incidate that ladybird beeltles have less well-developed sense of smelling their prey and require a greater nubmer of mealybugs as food for themsleves and their larvae.

The glasshouse whitefly, Trialeurodos vaporariorum, *is parasitized by the ichneumonid fly,* Encarsia formosa, *which lays its eggs in the larvae of the whitefly. Cardboard strips (EnStrips R) containing parasitized larvae are placed near tomato plants at regular intervals. Within several days, the* Encarsia *wasps hatch from the parasitized larvae and after maturation these will infest the next whitefly generation. (Photo courtesy of Koppert b.v, the Netherlands)*

and that they can not be released as a preventive measure means, in effect, that the farmer has to wait until much damage has been done before the pest can be eliminated. In seasonal crops such as vegetables, where unacceptable damage may be done in a very short time, this delay may create hesitance in using the natural enemy method of control.

The inoculative method of biological control is most successful in perennial crops, orchards and forests or those crops which are practically always available in the field, such as cassava and ratoon sugar-cane. Inundative releases are more applicable to seasonal crops, such as cotton and maize. For protected cultures, such as glasshouse crops, one seasonal inoculative release may give the required control during the whole growth period if other natural enemies, which as hyper-parasites or predators might attack the released species, can be kept outside. Thus, in glasshouses biological control of, for example, phytophagous mites by using predatory mite species may be cheaper than chemical control.

Many countries have gained experience in mass breeding of *Trichogrammas*, parasitic wasps belonging to the family *Trichochrammatidae*, which lay their eggs within the eggs of Lepidopterous moths. Commonly, eggs of the Angoumois grain moth, *Sitatroga cerealella*, or of the Mediterranean flour moth, *Ephestia kuehniella*, are used as alternative host-media for mass production of the *Trichogramma*. There are at least 100 species of these wasps, which differ in their selectiveness of hosts and usefulness as control agents (interspecific variation). A similar (intraspecific) variation is found to exist between various biotypes of particular *Trichogramma* species. In China and the Union of Soviet Socialist Republics (USSR), several million hectares of maize, cotton and other crops are being treated with *Trichogramma* spp.; the reason may be related to the need to compensate for a shortage of chemical insecticides rather than because of their proven biological and economic effectiveness (Van Lenteren, 1987). Recently, concern about the negative effects of pesticide use has stimulated new research which resulted in the commercially successful control of the corn

borer (*Ostrinia nubilalis*) in Europe by inundative releases of *T. maidis* and in control of various lepidopterous pests on cabbage by *T. evanescens* (Pak, 1988).

During a century of research on inoculative biological control, an estimated 3,000-5,000 introductions of natural enemies have been attempted world-wide, involving some 1,000 species of natural enemies against 200 pest species. One third of these introductions have resulted in the establishment of natural enemies. Of these, 58 per cent attained partial control and only 16 per cent provide complete control of the target pest (Hall *et al.*, 1980). Insects, mites and scales, having a sedentary mode of life, are an easier target for natural enemies than those species which are highly mobile. Thus, according to Greathead (1984), 40 per cent of the successful biological control programmes are directed against scales and mealybugs (the Coccidae family).

* Sterile male technique

A particular type of biological control involves the large-scale release of artificially sterilized male insects that belong to the same species as the target pest. The objective is to flood the habitat with sterile males which should compete with the fertile ones in mating with the females. If the inundative releases are repeated over a long period of time, the pest species may be suppressed or even eradicated owing to the absence of fertilized females. This technique is, however, really effective if the females mate only once. The best known example is the control of screwworm flies on sheep in the southern part of the United States and Mexico. A precondition for success is that the depleted population will not be continuously reinforced through in-migration of insects from outside the treated area.

* Microbial pest control agents

Biological control by means of entomopathogens and other microbial pest control agents involves the application of micro-organisms onto the crop for ingestion by insect pests or directly onto the noxious insects, fungus or weed with the objective of destroying them. Microbials include bacteria, fungi, viruses, nematodes and protozoaires. Probably the best known example is *Bacillus thuringiensis* against Lepidoptera larvae. Two varieties of *B.thuringiensis* are currently widely produced by various companies at competitive prices and marketed under trade names such as Bactospeine, Bactimos, Dipel and Thuricide. These formulations are sprayed onto crops and, after having been ingested by leaf-eating caterpillars, cause the latter's death. (For mode of action of *B.thuringiensis*, see Chapter 2.) Another bacterium, *B.penetrans* (nematoides), is effective against various nematode species.

Fungi have long been recognized as having the potential of killing insects. In the field one can often find dead insects covered with fungal mycelium and fruiting bodies. Identification and preparation of these fungi for use as a control agent, however, proves to be difficult and research on their entomopathogenicity, virulence, general

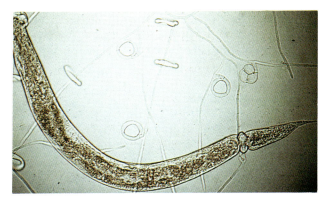

A soil nematode being caught in a loop formed by the parasitic soil-borne fungus Dactylella *spp. from which it cannot free itself.*

toxicity etc., is progressing slowly. The fungus *Beauveria bassiana* is reportedly used in the USSR for the control of Colorado beetles in potato fields and in China to control the corn borer, *Ostrinia nubilalis*, in maize. However, applications of pathogenic fungi in the open field are seldom successful as spores do not actively search for a host, and furthermore require specific conditions such as high humidity (dew) for establishing an infection.

Other applications of insect-killing fungi are *Verticillium lecanii* against whitefly in the glasshouse crops of Western Europe, and *Zoophthora radicans* against the two-spotted alfalfa aphid and blue-green aphid in Australia. Both aphid species are recent introductions and cause much damage to alfalfa; use of *Z. radicans* (itself brought in from a Mediterranean country) has reduced the frequency of sprays against aphids from five or more per season to zero, which means a saving of about $A 1.5 million per year.

Some fungal species are found to infect nematodes in the soil (see picture). Currently, research is carried out to indentify and cultivate pathogenic fungi that might be capable of controlling the golden cyst nematode, a major pest of potatoes. Some fungi can also successfully control weeds; for example, in Australia, skeleton weed in wheat fields is controlled by the combined action of a rust fungus and a mite. Fungi seem also able to kill other fungi, although their mode of action is not always clear. For instance, 17 different fungal species thriving upon other fungi were isolated and tested for their ability to control sporulation of cucumber powdery mildew, *Sphaerotheca fuliginea*. More than half of the isolated fungi reduced the number of healthy conidiophores (spore-carriers of the mildew fungus) to less than 10 per cent. It is not yet known if the effect is caused by fungiphagous (fungi-eating) behaviour or the result of fungicolous (living on fungi) behaviour; in the latter case the fungus attacked might be adversely affected by the toxic exudates produced by the attacking fungus. The production of such growth-limiting toxic exudates is well known from the *Pennicelium* fungus and can be demonstrated by growing *Pennicelium* spp. with other fungi on a laboratory medium.

Several viruses are extremely effective control agents of arthropod pests. Viruses have been isolated in

large numbers from insect and mite species and have been tested in practice for their suitability as selective pesticides. The group of baculoviruses, which includes the nuclear polyhedrosis viruses (NPV), are highly host-specific to the Class Insecta. Viruses are normally produced in larvae that are emulsified to prepare infectious sprayable formulations. Insects ingest the viruses when feeding on treated plants or on other insects. An example is the successful introduction into Western Samoa of *Baculovirus oryctes* from Malaysia in 1967 to control the rhinoceros beetle, *Oryctes rhinocerus*. (See also Chapter 2.)

Some nematodes have been identified as suitable entomopathogens. For instance, *Romanomermis* spp. are effective against insects of the Order Diptera (flies, gall midges, leaf miners) and *Neoplectana* spp. against caterpillars of the Order Lepidoptera and termites. Also, the protozoon *Nosema locustae* has been identified as a potential control agent of locusts.

Notwithstanding extensive research in all directions, by far the greatest number of successful biological agents and methods concern the control of arthropod pests. There are a few examples of effective control of weeds by plant-feeding insects, such as *Salvinia molesta*, an aqautic fern from South America and currently a pest in South-east Asia that is suppressed by the weevil *Cyrtobagous salviniae*. The noxious weed *Lantana camera* is being controlled in Hawaii with the help of several insects from South America. The control of weeds with fungi has been mentioned in a previous section. However, farms usually suffer from a more than one species of weed and thus, biolgical control of weeds does not offer a great prospect for economical weed control. Rather, it aims at removing particular problem weeds that are, for example, toxic to cattle or a pest in water reservoirs and forestry and right-of-way areas.

* Pheromones, hormones and growth regulators

Pheromones, juvenile hormones, natural plant regulators and insect growth regulators belong to the category of biochemical pest control agents, the classification of which is explained in Chapter 2.

The death symptoms from a virus infection are rather particular. Granulosis virus on a brown semilooper, Mocis frugalis *(F.) larva (left); Nuclear polyhedrosis virus (NPV) on a cutworm* Spodoptera litura *larva (right). (Photos courtesy of IRRI)*

Pheromones are substances emitted by insects and mites for communication between individuals of the same species. These pheromones modify the behaviour of the other insects to the benefit of at least one of the species involved. Sex attractants, for instance, are pheromones that enable adult male insects to find and recognize females of the same species. Most research has focused on these sex pheromones released by females prior to mating. In biological control, sex pheromones are used for trapping male adults and for communication disruption. Sex- pheromone-baited traps are used to:

- Monitor the distribution and abundance of an insect pest in order to achieve better timing of insecticide applications; and

- Catch a large number of male adults in order to reduce the population density. This technique is less effective at high densities and when the adults of the species are capable of multiple matings.

Other types of pheromones disturb the communication between insects or disrupt their sense of orientation. Usually, synthetic pheromones are used, which are modeled after the natural pheromones. There are many designs for pheromone dispensers and micro-encapsulation of pheromones has been developed. Sometimes relatively simple chemicals may have a similar attractiveness to insects as sex attractants; for instance, acetone can replace the sex pheremone as a lure for trapping tsetse flies.

Juvenile hormones are substances that regulate development in insects and that may prevent insects from reaching maturity. Similar synthetic compounds are being designed for use as biochemical control agents.

Growth regulators are natural chemical substances in plants and animals that control their processes of growth and that are very specific in their action. Obviously, any serious disturbance of growth may interfere with, for instance, vital processes of hatching, molting and pupation in insects or with processes of cell division and cell elongation in plants. Therefore, such natural compounds have been analysed and similar compounds synthesized. Some of these man-made chemicals possess a high efficacy either as insect growth regulator (IGR) or as plant growth regulator (PGR). Examples of IGRs are diflubenzuron, triflubenzuron, kinoprene and methoprene; these chemicals have been commercialized and are widely used for the control of caterpillars, bugs, mosquitoes, flies and ants. An example of a PGR are the auxins, which served as a model for the production of indoleacetic acid and herbicides such as 2,4-D and MCPA. (See also Chapter 2.) Because of their selectivity and relative harmlessness to mammals, the growth regulators fit well within the integrated pest management concept.

Biological control also includes the sowing of cover crops such as *Pueraria* and *Centrosoma* species, which suppress weed growth. Growing cover crops in combination with minimum soil tillage is practised in perennial tree crops to preserve soil moisture and to save the cost of weeding.

Biotechnology

A whole new field of research has been opened up by the development of molecular technologies. The concept of biotechnology was originally narrowly defined as "the integrated use of molecular biology, molecular genetics, micro-biology (cell biology) and process technology with the objective to arrive at a practical application of possibilities embodied by micro-organisms, cell-cultures or fragments of organisms or cells". This definition focused on industrial applications, such as the fermentation process in cheese-making, production of antibiotics by fungi, and the treatment of waste water with the help of bacteria.

Currently, a wider definition of the concept of biotechnology is being adopted that includes the application of molecular biology in the fields of agriculture, environment and health. This definition emphasizes the improvements of the genetical characteristics of the cell (of either plant, animal, bacterium or fungus) by exploiting recombinant deoxyribo-nucleic acid (DNA) and other molecular technologies in order to develop improved methods and modified organisms.

Biotechnology encompasses the following technological developments:

* Genetic manipulation
* *In vitro* regeneration
* Transformation systems
* Hybridoma technology

*Genetic manipulation refers to the technique of altering genetic material (DNA or genes) or transferring such material of different origin into one organism where it becomes functional. This can be achieved by means of "recombinant-DNA technology" or somatic cell fusion. Recombinant-DNA technology offers, in principle, the possibility of overcoming incompatibilities between species and to pave the way for the development of a genetic manipulation of micro-organisms, cells and eukaryotic tissues. The incorporation of resistance-genes in crops renders them less susceptible and vulnerable to injurious organisms and viruses or to the phytotoxic effect of pesticides. Example: the incorporation in high-value ornamental crops of the genes which are responsible for toxin-production in *Bacillus thuringiensis* and which prohibit the feeding by insects.

*In vitro **regeneration** techniques offer the possibility to grow complete plants from plant-cells, calluses and plant-tissue fragments in laboratory glassware on special media containing growth factors and hormones. The technique is used for the vegetative

In vitro *regeneration of plant tissue (tissue culture technique) on a growth medium in the laboratory enables rapid propagation of disease-free plant material; it is also an important aid for plant breeders.*

production of disease-free plants from plants infected by viruses, fungi or bacteria. For example, from a shoot-tip or from the auxilliary buds of a plant, the meristem dome and first pair of leaf primordia are excised and aseptically cultured on a medium. The meristem tissue may form further side shoots from dormant lateral buds which in turn are placed on a medium for rooting. By using an adequate medium and chemicals such as cytokinine, it is possible to obtain clones of virus-free plants from many food crops, fruit trees and ornamentals.

*Transformation systems exist for transferring alien DNA to cells of a different species, whereupon the arriving DNA is firmly integrated in the genetic constitution of the recipient cell. Such methods are commonly used for the transformation of plant cells and animal cells and to overcome barriers to interspecific hybridization in plant breeding.

Alien DNA can also be transferred into plant cells by viruses and particular bacteria such as *Agrobacterium tumefaciens* and *A. rhizogenes*.

*Hybridoma technology pertains to the production of monoclonal antibodies which can be used for the detection and identification of viruses in crops. The hybridoma technology, which was developed by Kohler and Millstein (1975), makes use of two properties: namely, that each lymphocyte (antibody-producing cell) produces an antibody which reacts specifically with one antigen and that transformed (tumor) cells can demonstrate unrestrained growth.

Antigens consist of alien material, such as toxins, pathogens and proteins foreign to the body, which penetrates the body of an animal organism and evokes therein the production of defensive antibodies. By fusing these two types of cells, Kohler and Millstein obtained a hybrid cell (hybridoma) that produces only one specific

antibody and this, in principle, in unlimited amounts. These so called monoclonal antibodies (MCAs) are very specific. Therefore, an MCA is perfectly suited for the detection of one single antigen with which it binds. World-wide, only a few specialized laboratories are producing selected monoclonal antibodies for use in immunospecific tests in which the binding of antigen to antibody can be made visible with the help of a colour reaction. An example is the "Enzyme Linked Immuno-Sorbent Assay" (ELISA), a testing method using antibodies for the detection and identification of viruses in cultivated plants, among others. The Netherlands has since had a leading position in the testing for virus infection in seed potatoes and flower bulbs, both of which are important, vegetatively propagated, export crops of that country.

Important in the context of integrated plant protection is the application of molecular technology with the aim of developing new methods and modified natural enemies. Besides techniques for detection of viruses and pathogens, biotechnology has already yielded herbicide-resistant, insect-resistant and virus-resistant transgenic plants. This development marks only the beginning of such breakthroughs for agricultural applications. Some genetically modified plants ("transgenics") are already available for cross-breeding into existing variaties of crops. Vlak (1989) cites, among others, the following examples of the practical application of biotechnology in plant protection:

Monoclonal antibodies as diagnostics
Diognostic tests have been developed on the basis of MCA technology for the testing of seed, planting material, cuttings and grafts for the presence of viruses and bacteria.

Plant material is crushed within a press to enable the extraction of plant sap for diagnostic tests on viruses by the standard ELISA method. (Photo courtesy of LNV, the Netherlands)

Transgenic plants with a build-in resistance gene against leaf-eating caterpillars (right) compared with a sensitive plant of the same cultivar.

In vitro regeneration

This technique makes use of the fact that each plant cell contains all the genetic information needed to regenerate to a complete plant. Meristemic tissue does not contain virus and is used in tissue-(or in vitro-)cultures to produce virus-free plants. The technique is also used for producing transgenic plants.

Herbicide-resistant plants

Incorporation of resistance against herbicides has been realized, e.g. by way of transferring a gene, that in a bacterium provides resistance against a herbicide, into a plant. (Comai *et al.*, 1985). Usually, the bacterium *Agrobacterium tumefaciens* is used as a vector for the resistance gene.

Transgenic plants resistant to virus infection

By incorporating into plants the "coat protein-gene" of six economically important viruses, such as tobacco mosaic virus (TMV) and potato X virus, several transgenic plants (tobacco, potato, tomato) have been obtained with a build-in protection against infection by viruses.

Transgenic plants possessing insect-resistance

By transferring a natural "insecticide-gene", originating from *Bacillus thuringiensis* into plants, the latter can be made resistant to certain insect species. Transgenic plants produce a protein (toxin) that, when ingested by a feeding caterpillar, for instance, permeates its intestine and kills it.

Insect-pathogenic symbionts of plants

If a similar gene governing the production of insect-toxins is introduced into the soil-borne bacterium *Pseudomonas,* which commonly lives in close association with plant roots (rhizosphere), the plant itself is influenced by the transgenic bacterium and becomes repulsive to soil-living insects which normally feed on its roots.

Hypervirulent baculoviruses

Through genetic manipulation, the virulency of insect-pathogenic baculoviruses can be increased in order to make them more effective insect control agents. Furthermore, these baculoviruses can be manupilated in such a way as to cause them to produce alien proteins in abundance, which can be used for therapeutic, diagnostic and prophylactic purposes.

Furthermore, research projects concern, among others, methods for:

* Investigating the molecular biology of key genes controlling insect development and reproduction;

* Studying molecular aspects of current biological and chemical insecticides with the aim of solving problems in their production and efficacy; and

* Studying gene-to-gene relationships by host-pathogen interactions.

In addition to these positive new options for improving agricultural production and reducing the use of pesticides, biotechnology may have some disadvantages, in particular for developing countries. Such drawbacks are the concentration of research capacity and funds on problems typical of large-scale commercial agriculture, and the development of resistance against certain herbicides in improved varieties of major food crops, which might influence their use by poor farmers.

Chemical control

Chemical control consists of the application of botanical or anorganic and organic synthetic compounds that have a killing, inhibiting or repulsive effect on injurious organisms threatening mankind, animals and plants. The active ingredient(s) causing such effect is contained in a formulation, the form of which enables efficient application and safe handling.

Pesticides can be applied either as a preventive treatment to protect crops or stored products against infection by diseases or infestations by animal pests, or as a curative treatment to destroy or limit population development of noxious organisms. The target of the application may be field crops, plantation crops, protected crops, seeds and plant material, soils, stored produce, buildings, surface water, right-of-way areas etc. Chapters 2 to 11 of this manual deal extensively with many technical aspects of the chemical nature, formulation, toxicity, safe handling and application of pesticides that are used for crop and storage protection.

* Pesticide use trends

The most recently available statistics estimate the value of the global pesticide market at over $US 20 billion in 1988: $8.9 billion on herbicides, $6.1 billion on insecticides, $4.2 billion on fungicides and $1.3 billion on other pesticides. The volume used is estimated to amount to 3.1 million metric tonnes (World Health Organization, 1989), of which about 20 per cent (equivalent to 600,000 tonnes annually) is used by the developing countries (Mowbury, 1988). Herbicides are most widely used in the United States, fungicides in Western Europe, and insecticides in the Asian and Pacific region.

In 1985, the Asian and Pacific region accounted for 16 per cent of global pesticide consumption, with average annual growth of 5-7 per cent. The greatest share (about 75 per cent) of the regional pesticide market is taken by insecticides which are used mainly on rice, cotton and vegetables. Fungicides are used mainly on vegetables, bananas and tobacco, whereas the bulk of herbicides is used in estate crops. Table 1.5 shows the percentages and values of the international market.

Most of the active ingredients are imported from the basic manufacturers in Europe, United States and Japan, but also India and the Republic of Korea are becoming important suppliers of technical grade chemicals.

The majority of the Asian countries have formulation and repacking facilities. However, the operational installations in some developing countries use outdated techniques and produce inexpensive organochlorine and organophosphorous insecticides no longer protected by patents, a number of which are either prohibited in Western countries or subjected to application restrictions.

These existing installations are sometimes a health hazard to the factory workers and their families as well as an environmental hazard because of unsufficient funds for safe waste disposal, water treatment and maintenance.

Table 1.5: The division of global pesticide use according to main groups (a), the percentage of regional share (b) and value (c) of pesticide consumption in the Asian and Pacific region

Class	Global percentage	Regional percentage	Regional value ($US million)
Herbicides	44.2	13.4	338.0
Insecticides	31.9	75.8	1,921.9
Fungicides	17.5	8.4	213.3
Others	6.4	2.4	61.3
Total	100.0	100.0	2,534.5

Source: Gaston, C.P.(1989); Environment and Agriculture Regional Office for Asia and the Pacific, FAO/RAPA.

Pesticides have played and will continue to play a great role in increasing agricultural production and in securing the supply of food and fibers needed by the people of the world. It is, however, essential that the reasons for using pesticides and the consequences of misusing them be carefully analysed in order to obtain maximum benefit from their application, while at the same time preventing and remedying their possible hazardous effects on non-target organisms and the environment. The proper and restrained use of pesticides must be a major objective of any integrated pest management strategy.

Advantages of chemical control

* Generally, pesticides are effective and reliable means for controlling pests and diseases and preventing losses in the field and in storage. They offer protection against many diseases, except for those inflicted by particular bacteria, viruses and mycoplasmas.

* Most pesticides demonstrate a fast activity, by which serious infections and outbreaks can be limited or possibly even controlled. Particularly when there is no time left for limitation by a build-up of natural enemies and the ripening produce is being threatened, then the use of pesticides is often the only remedy.

* Chemical control is effective under very diverse ecological conditions and it is less dependent on the scale of the operation than the various forms of cultural and biological control.

* No technical solutions, other than the use of pesticides, are as yet known with regard to a number of problems in intensive agriculture, horticulture, household pest and rodent control, and storage protection. Further, the proper preventive use of pesticides, such as the disinfection of soils and seeds, and the protective spraying of fungicides, are indispensable methods in modern plant protection.

* Pesticides are almost ubiquitously available in developing countries and, usually, at prices affordable even to poor farmers. Pesticide consumption is often stimulated through liberal subsidizing by national Governments and through credit schemes which include the supply of fertilizer and pesticides. Unfortunately, the choice is usually limited to older inexpensive products; the supply often does not correspond with the need and with safety requirements.

The application of pesticides, if measured in terms of harvested produce "saved" world-wide from damage by injurious organisms, is without doubt the most successful control method. Chemical control should therefore be seen as an essential component of integrated control. However, the current means and methods of chemical control must be further refined in order to delay resistance, to suit better the integrated approach and to avoid harmful interactions with biological control agents and the environment.

Disadvantages of chemical control

* Many pesticides cause the development of resistance in populations of organisms; many species of insects, mites, nematodes, micro-organisms and weeds have acquired a degree of resistance against one or several types or chemical groups of pesticides. These chemicals become, gradually, less effective, which necessitates the application of higher dosages, higher application frequencies and mixtures, and rapid replacement of those chemicals with new products. This is the so-called "pesticide treadmill".

* Pesticides, in particular broad-spectrum insecticides and miticides having a persistent activity, may destroy the natural enemy complex. Secondary pests, thus freed from their controlling factors, may then become major pests, or species that hitherto had no economic importance may suddenly assume such importance (e.g. cotton whitefly).

* Improper application of pesticides may cause unwanted or hazardous effects. With regard to the specific sensitivity of the target organisms and natural enemies, pesticidal selectivity and dosages need to be taken into account, whereas application technology must correspond with their behaviour and environmental conditions. Thus, a profound knowledge and skill is needed for the responsible use of chemical control, qualities which are often scarcely available among farmers and plant-protection technicians in developing countries.

* The cost of pesticides is rising steadily; the average wholesale prices for well-established insecticides rose about 20 per cent in the period 1970-1977. This trend has continued since then and is adversely affecting farmers as farm crop prices generally remained static at levels below the world market values. The increasing complexity of newer insecticides requiring a higher investment in research and manufacturing results in higher retail prices for the user.

* The inherent toxicity of pesticides and their often long persistence is a cause of many accidents involving mankind and non-target animals, and a grave source of environmental contamination.

Resistance against pesticides

Pesticides tend to induce in populations of organisms a capacity to resist the harmful effects of these chemicals. Pesticide resistance develops within a population by the selective pressure of the toxic chemical favouring those individuals which are capable of neutralizing the active ingredient. Resistance against chemicals is a variable entity, that may increase or decrease within a species. Resistance may also develop against more than one pesticide. **Cross resistance** enables resistant species to survive exposure to chemically related insecticides, e.g. the pairs DDT and methoxychlor, parathion and malathion, carbofuran and carbaryl, or

permethrin and fenvalerate. Cross resistance is often brought about by a common detoxification pathway. Far more serious is **multiple resistance**, which is found in more than 100 important pest species that display resistance to a variety of insecticide classes with differing modes of action and detoxification pathways (Sawicki, 1975). Cross resistance limits the choice of available insecticides, but multiple resistance reflects the past history of insecticide selection and precludes a return to those used previously.

An inventory has been made of the development of resistance since 1937 in animal pests and pathogens (Giorghiou, 1986). The total number of species of insects and mites resistant to insecticides or miticides stood at 447, of which 264 were agricultural pests, as compared with 168 pests of humans and animals. The number of plant pathogens resistant to fungicides amounted to 100, that of weeds 48, that of nematodes two and that of rodents five. The greatest cause of concern is the rapid increase in the number of insects showing multiple-resistance to insecticides. This development makes it ever more difficult to find effective compounds that can control these resistant insect populations. The supply of new pesticides entering the market, however, diminishes constantly owing to the very high development costs of new products (more than $US 30 million each on average). It is, therefore, evident that careful management of the currently available pesticides is necessary in order to prevent the development of resistance and to lengthen the time-span of their effectiveness. Examples of insecticides which have retained their usefulness for plant protection over more than four decades are malathion and monocrotophos, and the fungicide maneb.

The development of insecticide resistance (see table 1.6) can be prevented or delayed by limiting the frequency and amount of insecticide applications, which reduces selection pressure in the pest population. The use of specific insecticides should be discontinued whenever

Table 1.6: Development of resistance in insect and mite populations against insecticides and miticides

Number of species resistant to a) DDT and methoxychlor, b) HCH, lindane cyclodienes, toxaphene, c) organophosphates, d) carbamates and e) synthetic pyrethroids

Year	Total	a	b	c	d	e
1938	7	7	0	0	0	0
1948	14	13	1	0	0	0
1954	25	22	18	3	0	0
1969	224	155	42	23	4	0
1976	364	221	70	44	22	7
1980	428	245	95	53	25	10
1984	447	234	119	54	23	7

Source: *Giorghiou (1986) Plant protection and world crop production, Pflanzenschutznachrichten Bayer, Leverkusen.*

clear signs of a change in susceptibility of the pest are observed. If this method is neither practical nor economical, it may be useful to rotate the application of several conventional insecticides. Over the years, the most popular method of controlling resistant insect populations has been to use new insecticides with no cross resistance to the previously used chemical. This practice has, however, brought about the problem of induced pests such as whitefly in cotton. The only really effective way of preventing insecticide resistance appears to be a clever integration of insecticide applications with the use of resistant varieties and other practical methods of cultural and biological control.

Insecticides which have a specific (effective against certain species only) or systemic (absorbed into and translocated throughout the plant) action and a short residual activity, or which are stomach instead of contact poisons, are less harmful to natural enemies. Also, some formulations such as granules and baits, or methods of application such as spot treatments, adding pheromones or trunk injection, are relatively safe for natural enemies and thus fit better in an integrated control approach than, for instance, blanket spraying. However, these selective insecticides or methods may occasionally induce an outbreak of pests, suggesting that they have just tipped a delicate balance in favour of the pest.

POLITICAL VIEWS ON INTEGRATED PEST MANAGEMENT

Policy decisions on Integrated Pest Management

In many countries, agriculture is both a sustenance activity and a major driving force of the economy. Consequently, national agricultural policy is directed at strengthening farmers' productivity through increasing their income from production and providing the means for making production gains sustainable. One of the factors in furthering sustainability is the farmers' self-reliance in overcoming the adverse effects of pests in their crops. As many of the earlier official recommendations regarding intensive use of insecticides proved to be unnecesary or even harmful, several Governments are already redressing their pest control policies through regulatory measures and by withdrawing existing pesticide subsidies. The United Nations Development Programme (UNDP) together with the Food and Agriculture Organization of the United Nations have since 1975 initiated global programmes for the development and application of IPM programmes in rice, cotton, sorghum and millet. The objective is to improve overall food production and pest control capabilities in basic food and cash crops. In response to encouraging results achieved with biological and cultural control activities, more political and financial support is being given to the development of integrated pest management. Knowledge and experience are rapidly increasing and extension campaigns to train farmers in integrated pest

An efficient way of minimizing the negative effects of pesticide use is to apply the chemical by (ultra) low volume application to the target only. Accurate spraying of rows of plants can be achieved by a careful adjustment of the nozzles and the height of the tool bar of the tractor (Photo courtesy of LNV, the Netherlands)

control for major food and cash crops have been launched for quite some time in a number of countries.

The official standpoint of the industry regarding IPM concurs with such approaches:

"Industry supports IPM programmes when they are based on scientifically sound ecological and economic principles. Industry is involved in important research for the realization of IPM by developing appropriate biological, biotechnical and chemical compounds and measures" (GIFAP, 1984).

This supportive stance of the chemical industry is crucial to the development and implementation of IPM. Research and extension services in most countries are inadequate, overstretched and under-financed, whilst companies have the resources and often sophisticated sales promotion techniques, which in many countries appear to be the most effective channel for reaching the farmer. In cases where sales promotion is still used merely to sell a product without providing usable information on pesticide safety and responsible pest management, companies should be convinced of the long-term benefits of IPM by public-sector groups.

The 1986 FAO Code of Conduct, article 3.8, expresses clearly:

"Concerted efforts should be made by Governments and pesticide industries to develop and promote integrated pest management systems and the use of safe, efficient, cost-effective application methods. Public-sector groups and international organizations should actively support such activities."

References:

Bandong, J.P. and J.A. Litsinger, (1986). *Development of action control thresholds for major rice pests*. (in ref. Teng and Heong (eds.), 1988, pp. 95-102).

Brader, L. (1979). *Integrated Pest Control in the developing world*. Ann.Rev.Entemol, Vol. 24, pp. 225-254.

Brader, L. (1980). *Advances in applied entomology*. Ann. appl. Biol., Vol. 94, pp. 349-365.

Bennetzen, J.L. (1984). *Genetic engineering for improved crop disease resistance*. In: *Application of genetic engineering to crop improvement*. Martinus Nijhof Publishers, pp. 453-490.

Doeleman, J.A. (1989). *Biological control of Salvinia molesta in Sri Lanka: an assessment of costs and benefits*. Australian Centre for International Agricultural Research, ACIAR Technical Reports 12, Canberra, Australia.

Delucchi, V. (1987). *Integrated pest management: Quo vadis?* Parasitis, Vol. 86, Geneva, Switzerland.

Food and Agricultural Orginization of the United Nations, (1986). *International Code of conduct on the distribution and use of pesticides*.

Hall, R.W., L.E. Ehler and Bisabri-Ershadi B., (1980). *Rate of success in classical biological control of arthropods* Bull. Entemol. Soc. Am., Vol. 26, pp. 11-114.

Harrison, B.D.(ed) (1987). *Plant resistance to viruses*. Ciba Foundation Synposium 133. (J. Wiley & Sons, New York), pp. 215.

Ho, C.T. (1988). *Pest management on cacoa in Malaysia*, (in ref. Teng and Heong (eds.), 1988, pp. 193-204).

Joyce, R.J.V. *et al.* (1959). *Recent progress in entomological research in the Sudan Gezira*. Emp. Cotton Gr. Rev., vol. 36, No. 3.

International Rice Research Institute, Philippines (1984). *Judicious and efficient use of insecticides on rice*.

Kaske, R. *IPM activities in coconut in Southeast Asia*. (in ref. Teng and Heong (eds.), 1988, pp. 181-186).

Liau, S.S.; *Pest management on oil palm in Malaysia*. (in ref. Teng and Heong (eds.), 1988, pp. 187-192).

Mumford, J.D. (1986). *Control of the cocoa pod borer: a critical review*. In: *Cocoa and coconuts: Progress and outlook*. Incorporated Society of Planters, Malaysia, pp. 287-292.

Oudejans J.H., Meerman F. and Takken W. (1988). *Crop protection, vector control and pesticide use in developing countries*. Policy paper (background report) of the Ministry of Agriculture and Fisheries, the Netherlands.

Pak G.A. (1988). *Selection of Trichogramma for inundative biological control*. Agricultural University, Wageningen, Dept. of Entomology, the Netherlands.

Sawicki R.M. (1975). *Effects of sequential resistance on pest mangement*. Proc. 8th. British insecticide and fungicide conf., pp. 799-811.

Teng P.S. and Heong K.L., (1988). *Pesticide management and integrated pest management in South-east Asia*. Consortium for international crop protection, 4321 Hartwick Road, Suite 404, College Park, Maryland 20740, United States of America.

Van der Valk H.C.H. and Koeman J.H. (1988). *Ecological impact of pesticide use in developing countries*. Ministry of Housing, Physical Planning and Environment, the Netherlands.

Vlak J.M. (ed), (1989). *Biotechnology in de gewasbescherming*. Syllabus in Dutch language. (Biotechnology in crop protection). Agricultural University of Wageningen, Dept. of Virology, the Netherlands.

2

CLASSIFICATION OF PESTICIDES

Contents   Page

Classification of pesticides according to the type of pests they control 30

Classification by effect on pests 31

Classification of insecticides 31

- By chemical nature 31

- By mode of action 36

Classification of fungicides 37

- By chemical nature 37

- By mode of action 41

Classification of herbicides 41

- By chemical nature 41

- By mode of action 45

Classification of biological pest control agents 46

- Biochemical pest control agents 46

- Microbial pest control agents 48

References 52

Classification of Pesticides

By definition, a pesticide is a pest-killing agent (the Latin word *cida* means to cut or kill). The term usually refers to one or more materials developed and used to destroy a broad range of specific pests. In legal terminology, pesticides may be defined as "any substance used for controlling, preventing, destroying, repelling, or mitigating any pest". Hence, even chemicals that do not actually kill pests may, for practical and legal reasons, be considered pesticides. Included under the term are compounds used as repellents, attractants, anti-feedants etc.

A great many pesticides are available today for the control of unwanted organisms. The insecticides, fungicides, herbicides and microbial agents given below are classified by chemical nature, which stresses the relationship in chemical structure; this shows the principle involved for these well-known compounds. Some information is provided too on the type of activity, spectrum of pests to be controlled, persistency in the soil and toxicity. However, one should refer to the product label for details on each separate pesticide. Also shown are various ways of classifying these main groups of pesticides by their mode of action.

CLASSIFICATION OF PESTICIDES ACCORDING TO THE TYPE OF PESTS THEY CONTROL

(terms bearing the suffix -cide)

Acaricide .... mites, ticks and spiders
Adulticide .... adult insects
Algicide .... algae
Arboricide .... trees, brush, shrubs
Avicide .... birds
Bactericide .... bacteria
Fungicide .... fungi
Insecticide .... insects and sometimes related pest such as ticks and mites
Ixodicide .... ticks
Larvicide .... larvae
Miticide .... mites, ticks and spiders
Molluscicide ..... molluscs such as slugs and snails
Nematicide .... nematodes
Ovicide .... eggs
Piscicide .... fish
Predacide .... vertebrate pests
Rodenticide ...... rodents such as rats, mice, gerbils
Silvicide .... trees, brush, shrubs
Termiticide ...... termites, ants

For more detailed information on pesticides specifically suited to control certain kinds of animal pests such as rodents, snails and nematodes, and plant pests such as weeds and pathogens, or pesticides suited for

A pesticide product consists of an active material in a certain concentrated formulation, such as the above emulsifiable concentrate seen dispersing in water, designed to enable its safe and effective application.

treatment of seeds and fumigation, the reader should refer to the specialized chapters in the second half of this manual.

The newly generated term "biocide" is a nondescript word, commonly associated with the adverse effects of pesticides and frequently used as a synonym for pesticide. It has no exact scientific meaning, but is increasingly being used in writings on environmental issues to describe substances that kill living organisms.

CLASSIFICATION BY EFFECT ON PESTS

(terms not bearing the suffix -cide)

Anti-feedant . . . . inhibits feeding while insects remain on the treated plant; the insects eventually starve to death

Anti-transpirant . . . . . reduces transpiration

Attractant . . . . lures pests to treated location, e.g. sex attractants

Chemosterilant . . . . . destroys a pest's ability to reproduce

Defoliant . . . . removes unwanted plant growth without immediately killing the whole plant

Desiccant . . . . dries up plant parts and insects

Disinfectant . . . . destroys or inactivates harmful organisms

Feeding stimulant . . . causes insects to feed more vigorously

Growth regulator . . . . stops, speeds up or retards growth processes of plants or insects

Repellent . . . . drives pests away from treated object without killing them

Semiochemicals . . . . . pheromones, allomones and kairomones; substances emitted by plants or animals, which stimulate or inhibit certain behavioural activities of insects

Synergist . . . . enhances the effectiveness of an active agent

CLASSIFICATION OF INSECTICIDES

The term "insecticide" refers to chemicals which are capable of destroying, mitigating or repelling insects, but they usually also exert a certain efficacy against mites, ticks or even nematodes.

Classification by chemical nature

Most common insecticides can be classified into main chemical groups consisting of inorganic or organic, natural or synthesized compounds.

Inorganic compounds are derived from naturally occurring elements and do not contain carbon. They are stable, non-volatile chemicals frequently soluble in water. Many of them are persistent; those containing arsenic, cyanide, mercury and thallium are cumulative poisons.

Because of their residual persistence and high mammalian toxicity, the use of almost all inorganic chemicals has been discontinued in favour of less hazardous synthetic organic ones. Examples of inorganic chemicals are:

boric acid	sodium arsenite
copper hydroxide	sodium chlorate
copper oxychloride	sodium fluoride
copper sulphate	sodium fluoroacetate
mercuric oxide	thallium sulphate
mercurous chloride	silica aerogel
sodium fluoaluminate (cryolite)	

The elements copper and sulphur formulated as dusts or sprays are still widely used as fungicides (see below). Cryolite is regaining a place as a selective insecticide in integrated pest control. Boric acid and silica gels affect the waxes of an insect's cuticle, which becomes permeable to water from the insect's body. Consequently, the insect dies from dehydration. Both boric acid and silica gels are used in baits for controlling household insects such as cockroaches and ants.

Organic compounds are man-made or extracted pesticides consisting of carbon, hydrogen and one or more other elements such as chlorine, oxygen, sulphur, phosphorus and nitrogen. The main groups are described hereafter:

Botanicals, also known as "natural insecticides", are toxicants that have been derived from plant materials. These may be produced in the form of powder made from finely ground plant parts or extracts which are used in sprays.

Neem extracts obtained from the neem tree (*Azadirachta indica*) have an insecticidal, repellent, anti-feedant and growth inhibiting effect on many insects belonging to six taxonomic orders. Neem seed oil, admixed at 3-5 ppm concentration, protects stored beans against bruchid bean beetles for a period of up to six months. Watery extracts from neem leaves and fruit can be sprayed at weekly intervals against leaf-feeding caterpillars on vegetables. Neem is non-hazardous to man and other mammals, and there are no harmful side-effects on natural enemies.

Nicotine is an alkaloid extracted from tobacco leaves (*Nicotiana tabacum*) that has long been used as an insecticide, but since it is also a strong neuro-muscular poison in man and animals, its use is discouraged.

Pyrethrum is extracted from chrysanthemum flowers (*C. cinerariaefolium*). Pyrethrum is a powerful degradable contact insecticide causing a rapid knockdown action, but insects may recover. Insecticidal activity is markedly increased by the addition of synergists, e.g. piperonyl butoxide. Pyrethrum is very safe to man and mammals. As dusts or sprays, pyrethrum is used in animal husbandry, in stored food protection and in households.

Scilliroside is a very toxic glucoside with stomach action, which is extracted from ground bulbs of the red squill, *Urginea (Scilla) maritima*. Mixed with bait, it controls most species of rats and mice.

Rotenone is extracted from the roots of bean legumes, i.e. *Derris* spp. in South-east Asia and *Lonchocarpus* spp. in South America. It has long been used as an insecticide and is still used as a piscicide. Because of its toxicity to fish and its relatively short residual effect, it can be employed in fish farming to clean unwanted fish from ponds prior to restocking.

Ryania is obtained from the stem of *Ryania speciosa*, a shrub from South America. It has a high mammalian toxicity with symptoms resembling those evoked by strychnine, an intense mammalian poison extracted from *Strychnos* spp. seeds. Its use should be discouraged.

Sabadilla is an alkaloid extracted from seeds of *Schoenocaulon officinale* (Liliaceae). It too is a superseded compound.

Organochlorine compounds are synthetic organic insecticides that contain carbon, chlorine, hydrogen and sometimes oxygen. They may also be named "chlorinated hydrocarbons" or "chlorinated insecticides". They comprise a notorious group of insecticides, among which are DDT and BHC; both of these were synthesized more than a hundred years ago, but forgotten. It was only around 1940 that their insecticidal properties were recognized. This event gave birth to the era of synthetic organic pesticide manufacture.

DDT became famous for its efficacy against flies, mosquitoes, body lice and fleas, which as vectors transmit human diseases such as malaria, yellow fever, typhus and bubonic plague. DDT was also used on a large scale in agriculture and horticulture. Because of its low cost, it probably was the most economical insecticide ever sold. Four chemical classes of organochlorine compounds may be distinguished:

a. Diphenyl aliphatics

chlorbenside	DDT
chlorfenethol	dicofol
chlorobenzilate	metoxychlor

DDT has a long residual life or persistence in the soil and in aquatic environments. Also, because it accumulates in the tissues of animals and plants, it is banned for agricultural use. However, its use continues in health programmes for mosquito control. The other five compounds are very effective miticides, and do not have the long persistency of DDT.

DDT

1,1,1-trichloro-2,2-bis(*p*-chlorophenyl)ethane

b. Benzene derivatives

benzene hexachloride (BHC, HCH)
lindane (Gamma BHC)
pentachlorophenol

BHC is a chlorinated benzene consisting of a mixture of five isomeres, of which only the Gamma-isomer was found to have insecticidal properties; the other four being insecticidally inactive ingredients. A product containing 99 per cent Gamma-isomer, namely lindane, is highly effective, but too expensive for large-scale crop use. The four inert isomers have a musty flavour that persists in animal and plant tissues as well as in the soil. Hence, normal mixture BHC, containing only 12 per cent Gamma-isomer, affects the taste of root crops and vegetables. Because the Gamma-isomer is odorless, lindane was previously widely used in the control of storage pests.

HCH (BHC)

1,2,3,4,5,6-hexachlorocyclohexane

c. Cyclodienes

chlordane	endrin
chlordecone	heptachlor
endosulfan	mirex

Generally, cyclodienes are persistent soil insecticides, mainly used against coleopterous insects and termites; they also are generally photostable. Use of cyclodienes is progressively being cancelled or banned (aldrin and dieldrin) for reasons of resistance in insects, fish toxicity and serious environmental hazard. In Western countries, only endosulfan is still registered for use on field crops.

ENDOSULFAN

6,7,8,9,10,10-hexachloro-1,5,5a,6,9,9a-hexahydro-6,9-methano-2,4,3-benzodioxathiepin 3-oxide

d. Polychloroterpenes

camphechlor (Toxaphene)

For 25 years until 1980, Toxaphene had the greatest use of any single insecticide in agriculture. In cotton cultivation, it was widely used in mixtures initially with DDT and later with parathion-methyl. Although Toxaphene is persistent in the soil, it evaporates from plant surfaces within weeks. It has low mammalian toxicity, but is toxic to fish.

Organophosphorus insecticides are all esters of phosphoric acid (H_3PO_4) and are also called organophosphates (OPs) or phosphorus esters. They comprise a large group of compounds that have two characteristics: organophosphates are generally acutely toxic to man and vertebrate animals, and they are non-perstistent. Most of them have a short residual activity, which is desirable in keeping down residues on food crops, but is often a problem when longer protection is desired. Repeated applications may then be necessary. Organophosphates are nerve-poisons, acting through inhibition of the cholinesterase enzyme; their toxic action is explained in Chapter 4.

Phosphate:

Phosphorothioate:

Phosphorothiolate:

Phosphorodithioate:

Phosphonate:

Phosphoroamidate:

Phosphorus esters have distinctive combinations of oxygen, carbon, sulphur and nitrogen attached to the phosphorus atom. The names of the phosphor molecules above are part of much more complicated chemical names for the compounds that contain these building blocks (moieties).

There are three chemical classes of organophosphorus insecticides, namely:

* Aliphatic derivatives having a carbon chain structure.

* Phenyl derivatives containing a benzene ring with one hydrogen atom in the ring being replaced by the phosphorus moiety.

* Heterocyclic derivatives also containing a ring structure and the phosphorus group. In a heterocyclic carbon ring, however, one or more carbon atoms are displaced by oxygen, nitrogen or sulphur and the ring may consist of three, five or six atoms.

a. Aliphatic organophosphates

The oldest and best known insecticide of this class is malathion. Because of its high efficacy against many insect pests and its low mammalian toxicity, it has been used for almost 40 years on a large scale for agriculture, public health and the control of storage and household pests. Little resistance against malathion has developed among insect species and thus, it is still very useful.

MALATHION

O,O-dimethyl-*S*-1,2-di(carboethoxy)
ethyl phosphorodithioate

Naled is another safe OP-compound having a very short residual action and a fumigant quality. As is the case with malathion, it is widely used for the control of mosquitoes, flies and ecto-parasites of domestic animals.

Several aliphatic OPs, such as monocrotophos, demeton and dimethoate, have a systemic action and are particularly effective against sucking insects.

Some aliphatic OP-insecticides - although being highly toxic - have such a very short residual activity that they are preferred for use on vegetables. Examples are listed hereunder (generally superseded compounds are not included in the list of examples shown):

acephate	methamidophos
demeton	mevinphos
demeton-methyl	monocrotophos
dichlorvos	naled
dicrotophos	omethoate
dimethoate	oxydemeton-methyl
disulfoton	phorate
ethion	phosphamidon
formothion	thiometon
malathion	trichlorfon

33

b. Phenyl organophosphates

These compounds are generally more stable than aliphatic ones and more persistent. Parathion-ethyl was introduced in 1947; with widespread use, it came to be known as parathion. Because of its dangerously high toxicity, it has largely been replaced by the slightly less toxic and broader acting parathion-methyl. Several older

METHYL PARATHION

O,O-dimethyl *O*-*p*-nitrophenyl phosphorothioate

members of this class, such as fenitrothion and tetrachlorvinphos, are widely used in agriculture, horticulture and public health programmes. Included in this group are:

bromophos	parathion
carbophenothion	parathion-methyl
chlorfenvinphos	phenthoate
edifenphos	phosalone
fenamiphos	phoxim
fenitrothion	profenofos
fensulfothion	prothiofos
fenthion	temephos
fonofos	tetrachlorvinphos
isofenphos	triazophos

c. Heterocyclic organophosphates

Heterocyclic compounds have complicated molecular structures and are generally longer lasting than most of the aliphatic and phenyl derivatives. Upon degradation they may break down into many metabolites which may be difficult to analyse; some are very effective against mites and soil insects. Examples of this group are:

azinphos-ethyl	mephosfolan
chlorpyrifos	methidation
chlorpyrifos-methyl	phosmet
diazinon	pirimiphos-methyl
dioxathion	pyrazophos
etrimfos	quinalphos
isoxation	

PIRIMIPHOS-METHYL

O-[2-(diethylamino)-6-methyl-4-pyrimidinyl]
O,O-dimethylphosporothioate

TETRADIFON

p-chlorophenyl 2,4,5-trichlorophenyl sulfone

Organosulphur compounds contain a dominant sulphur atom and commonly two phenyl rings. Sulphur alone has a good efficacy against mites but, in combination with the two phenyl rings, its acaricidal or miticidal effect is enormously enhanced. Organosulphurs combine this high toxicity to mites with a remarkably low toxicity to insects; thus, they are very selective miticides and useful in integrated pest management. Examples of this group are:

ovex   propargite   tetradifon

Carbamates

These compounds resemble closely the organo-phosphates in biological activity and, like them, they

CARBARYL

1-naphthyl *N*-methylcarbamate

inhibit the cholinesterase enzyme required for nerve function in animals, (see Chapter 5). Carbamates are esters of carbamic acid. They generally have a short residual activity and a very broad-spectrum effectivity as insecticides, miticides, nematicides and molluscicides. The carbamate group also includes herbicides such as diallate and propham. (The groups of thiocarbamates and dithiocarbamates are considered under herbicides and fungicides, respectively). Among the carbamates, three classes of compounds may be distinguished:

a. Methyl carbamates with phenyl-ring structure

aminocarb	methiocarb
BPMC	metolcarb (MTMC)
carbaryl	mexacarb
isocarb	promecarb
isoprocarb (MICP)	propoxur

Carbaryl has become very popular in horticulture because of its low toxicity and broad spectrum efficacy. The carbamates propoxur, bendiocarb and dioxacarb are used to control mosquitoes, household and storage pests because of their rapid knock-down effect and long residual efficacy on wall surfaces.

CARBOFURAN

2,3-dihydro-2,2-dimethyl-7-benzofuranyl
methylcarbamate

b. **Methyl carbamates and dimethyl carbamates with heterocyclic structure.**

bendiocarb   dioxacarb
carbofuran   pirimicarb
dimetilan

c. **Methyl carbamates of oximes having a chain structure**

aldicarb   oxamyl
methomyl   thiodicarb

ALDICARB

2-methyl-2-(methylthio) propionaldehyde
O-(methylcarbamoyl) oxime

Thiodicarb is a relatively new compound with a double carbamate molecule linked with a sulphur atom; it has a long residual activity of up to 14 days.

Formamidines form a class of insecticide-acaricide having the characteristic nitrogen structure -N = CH-N. They are very effective against eggs and very young caterpillars as well as against most stages of mites and ticks. Formamidines are valuable for the control of insects resistant to OPs and carbamates. Examples are:

amitraz
chlordimeform (restricted use)
formetanate

Dinitrophenols are derivatives of phenol with two nitro (NO_2)-radicals attached. The group has broad spectrum toxicity as insecticides, ovicides, herbicides and fungicides. Below are outstanding miticides:

binapacryl   dinocap
dinobuton   dinoterbon

Organotins are tin-based organic compounds which generally exhibit miticidal as well as fungicidal activity. Cyhexatin and fenbutatin-oxide are very selective miticides with long residual effectivity.

cyhexatin   fenbutatin-oxide

Pyrethroids are synthetic compounds, the chemical structure of which is patterned after pyrethrins (extracted from the chrysanthemum flower), a mixture of six esters named cinerin I and II, jasmolin I and II and pyrethrin I and II. The synthethic pyrethroids mimick the broad efficacy of the botanical, but, as they contain only one of these esters, insect species tend to develop resistance to them. Compared with the botanical pyrethrum, which contains all six ester properties, no resistance has yet been observed as it becomes very difficult for insects to develop the requisite combination of alternative pathways to these six properties.

Generally, pyrethroids show low mammalian toxicity, but like pyrethrum, they are highly toxic to fish and bees. They are anoxic nerve poisons (see Chapter 4). Their residues in crops do not cause problems.

Pyrethrum and the older synthetic pyrethroids are very sensitive to sunlight, because their molecules split under ultra-violet light. Therefore, they are not suited to agricultural use, but some of them have good volatility and are very effective against indoor insect pests. For household use they are formulated as aerosols together with piperonyl butoxide for synergistic action. Examples are:

allethrin   phenothrin
bioallethrin   resmethrin
bioresmethrin   tetramethrin

From the early 1970s, pyrethroids have been produced which have better photo-stability with low volatility; therefore, they may be used for agricultural purposes. Examples are:

deltamethrin   permethrin
fenvalerate

The latest group of synthetic pyrethroids are photo-stable as well as extremely toxic to insects. Their efficacy is so good that a dose of only 10-40 g active ingredient per hectare is required. These new pyrethroids are not mixed with synergists. Examples are:

bifenthrin   fenpropathrin
cyfluthrin   flucythrinate
cyhalothrin   fluvalinate
cypermethrin   tralomethrin

FENPROPATHRIN

α-cyano-3-phenoxybenzyl 2,2,3,3-tetramethyl
cyclopropanecarboxylate

Fumigants are substances or a mixture of them which produce gas, vapour, fumes or smoke intended to kill insects, nematodes, bacteria or rodents. Usually, fumigants are volatile liquids or solids, or gaseous substances already consisting of small molecules, which often contain halogen radicals (Cl-, Br- or F-). They are used to disinfect buildings, stored produce or the soil. Toxicity varies; however, the substances are rapidly absorbed by the lungs and act through inducing narcosis (unconciousness which may lead to death). Examples are:

chloropicrin   methyl bromide
ethylene dibromide   naphtalene
formaldehyde   phosphine

Chloropicrin has a strong irritating "tear gas" effect and may be added to methyl bromide as a warning agent. (Soil fumigation, see Chapter 12; fumigation of stored products, see Chapter 15).

Petroleum oils, also referred to as mineral oils, have been used for many years as insecticides and miticides. Only light applications of refined paraffinic oils, formulated with an emulsifier, can be safely made to trees in leaf and plants. These so called "white oil" or "summer oil" treatments are made against aphids, mites and scale crawlers. "Winter washes" of semi-refined and more aggressive oils of lower viscosity have greater efficacy against the eggs of spider mites and aphids as well as against scale insects, but such treatments can be tolerated only by dormant or deciduous trees in winter. Furthermore, certain petroleum oils are herbicides by themselves or may be used as solvents or adjuvants to other pesticides. Examples are "Actipan" and "Fyzol".

Antibiotics are chemical substances, such as penicillin, produced by micro-organisms and having bactericidal and fungicidal activity. (See also the section below under fungicides). Abamectin, produced by a *Streptomyces* spp. fungus, is the only antibiotic with good insecticidal, miticidal and nematicidal efficacy. Because it possesses some systemic activity, it provides good control of leafminers and of various mites attacking citrus, even on the underside of the leaves.

ABAMECTIN

Plant protection oriented chemistry deals with increasingly complex compounds and is confronted with ever more demanding requirements regarding human and environmental safety. Thus, large capital-intensive investments are needed for research and for well-equipped analytical laboratories.

Classification of insecticides by mode of action

Insecticides may also be classified according to the various ways they enter or affect pests. Once on or inside a pest organism, an insecticide may act upon one or more of the life processes. The result may be death, illness or a change in the behaviour, growth, metabolism or reproductive capabilities of the pest. The following are some examples of this method of classification:

* Stomach toxicants generally enter a pest's body through the mouth during feeding (ingestion) and are absorbed through the digestive tract.

* Contact toxicants generally penetrate a pest's body as a result of contact with the legs or other external portions and treated surfaces such as sprayed leaves.

* Fumigants are volatile and enter a pest's body through the respiratory system and kill at lethal concentration.

* Systemic toxicants are characterized by high water solubility, by which property they can readily be taken into the plant through its roots, stems or leaves. Once inside the plant, systemics move through the plant's vascular system to other untreated parts from where insects acquire the translocated insecticide during feeding. Systemic insectices are particularly effective against sucking, boring and mining insects and nematodes and are also used to control pests of livestock.

* Suffocating materials, usually oils, clog the respiratory mechanism of pests.

CLASSIFICATION OF FUNGICIDES

The term "fungicide", strictly taken, refers to chemicals capable of preventing or eradicating diseases caused by fungi. Bacteria, the pathogens of bacterial plant diseases, may sometimes be controlled with bactericides or antibiotics. However, in the classification below, this distinction is not rigidly applied. Fungicides are generally not effective against virus diseases. However, because many virus diseases are transmitted by insects, mites or nematodes, their control is sometimes possible by using insecticides, miticides and nematicides to destroy these vectors.

Classification of fungicides by chemical nature

Fungicides can also be classified into main chemical groups consisting of inorganic or organic synthesized compounds.

Inorganic fungicides

Inorganic fungicides are derived from elemental sulphur or rather simple metal salts and do not contain carbon. Generally, they are stable, almost insoluble in water and persistent.

Sulphur, in its elemental form, is probably the oldest known cure against plant diseases. Originally, finely ground flowers of sulphur (the form of sulphur resulting from sublimation following the heating of crude sulphur and rapid cooling of the fumes) was applied as a fungicidal dust. It proved to be effective against powdery mildews and spider mites. Currently, elemental sulphur is mostly formulated as a wet paste or a "flowable" prepared from colloidal sulphur particles of 0.2 - 10 micrometers in diameter. The third form is a wettable sulphur powder which is mixed with a wetting agent during the milling process.

Previously, a liquid lime-sulphur mixture was widely used for orchard spraying against powdery mildews, anthracnose leafspot and brown rot diseases. This mixture is prepared by boiling sulphur and slacked lime in the ratio of 2:1 in five parts of water. Lime sulphur, however, cannot be sprayed in combination with organophosphates, copper compounds or other metal-containing pesticides.

The efficacy of sulphur against powdery mildews and mites is by direct contact and also by fumigant action; it has no systemic effect. For good fumigant action, the temperature should be above 20°C; however, above 32° C, sulphur vapour becomes phytotoxic. Because spores of powdery mildew germinate in the absence of a water film on the leaves, this fumigation effect is important for killing the fungus.

Sulphur is a cheap and safe fungicide, but outside temperatures need to be monitored during the time of application. Formulations of sulphur are Elosal, Kumulus S and Thiovit.

Copper fungicides

Another early discovery was the fungicidal effect of Bordeaux mixture (a watery solution or slurry of copper sulphate and hydrated lime). Bordeaux mixture, with 12 per cent copper, is a safe fungicide with low mammalian toxicity; it controls a wide range of diseases such as downy mildews and late blight of potatoes. The product has been largely replaced by "fixed coppers" in which a copper compound of low-solubility is packed in a form that is stable in storage and readily dispersable in water. The copper compound is usually a basic cupric salt with a bright red, blue, green or yellow colour. The solubility and the rate of release of the toxic copper ion (Cu-radical) are decisive for providing efficacy against the fungi and for preventing phytotoxicity to the host plant.

In general, protective fungicides have a low rate of ionization. However, if a film of water is present at the spot of infection where the fungal spore germinates, a portion of the fungicide residue may go into solution. The small quantity of Cu-ions absorbed by the germinating spore is then replaced in the solution from the residue. The spore accumulates the toxic Cu-ions and is killed in the process. Such compounds are less phytotoxic than Bordeaux mixture, but their fungicidal efficacy tends to be less. Copper oxychloride is the most widely used compound against *Phytophthora* diseases. Examples of fixed coppers are:

cupric carbonate   copper sulphate
copper hydroxide   cuprous oxide
copper oxychloride

Inorganic fungicides containing heavy metals, such as mercury, nickel, zinc and chromium ions, have strong fungicidal properties and have been in wide use over the past 35 years. These metal salts are generally very toxic to all forms of life and they are rather persistent in the soil and on treated surfaces. For these reasons, inorganic mercurial and other heavy metal fungicides have been banned in many countries. Manufacturing of such fungicides has been discontinued, except for some special purpose formulations, e.g. mercuric oxide is used as a paint against bark cankers and for sealing bark injuries, and insoluble mercurous chloride is still used for seed treatment.

Organic fungicides

Since the development in 1931 of the first organic sulphur compound, thiram, a wide range of synthetic organic fungicides have gradually replaced the less selective inorganic compounds. Currently, about 150 synthetic fungicides are available or under development. A common feature is their high efficacy at rather low dosages, their long residual activity and a greater safety to crops, animals and the environment. Generally, the organic fungicides are broken down by soil micro-organisms and leave no persistent toxic metabolites.

Dithiocarbamates are derivatives of the sulphur-containing dithiocarbamic acid. For half a century, these fungicides have been among the most effective and popular ones in use. In combination with metallic salts such as zinc salt (ziram), ferric salt (ferbam) and manganous salt (maneb), these organic sulphur compounds offer a particular effectivity, better stability and less phytotoxicity than elemental sulphur. They have no systemic activity. Since they provide a good cover, several dithiocarbamates are used as seed protectants against soil-borne diseases. Their toxic effect to fungi probably stems from the isothiocyanate-radical ($-N = C = S-$) which is formed as a break-down

THIRAM

bis(dimethylthio-carbamoyl)disulfide

component. Additionally, chelates are formed within the fungal cells when dithiocarbamates or heavy metal fungicides are applied. When an excess quantity of such chelates is present, they may interfere with enzymatic and metabolic processes within the fungal cells. Heavy metal dithiocarmates thus have great killing power. Examples are:

ferbam	propineb
maneb	thiram
metham-sodium	zineb
nabam	ziram

mancozeb: a complex of zinc (2-5% Zn) and maneb (20% Mn). This ready-to-use fungicide combines the benefits of both maneb and zineb.

Organometallic compounds

Mercury fungicides were popular because of their efficacy against fungi, their good disinfective and protective action, and often considerable volatility which properties made them very well suited for seed treatment. However, because of their high mammalian toxicity, all organic and inorganic mercurial fungicides have lost their registration and are no longer available for any purpose.

Organocopper compounds include organic salts of acetate, naphthenate, oleate and quinolinate. Copper acetate was first developed in 1889 and became the first factory-made basic copper fungicide. Copper compounds are not easily washed from leaves by rain, since they are relatively insoluble in water, and thus give longer protection against disease than do most of the organics. They are relatively safe to use and require no special precautions during spraying. The currently accepted theory for the mode of action of copper's fungistatic action is its nonspecific denaturation of protein. The Cu^{++} ion reacts with enzymes having reactive sulfhydryl groups -- which would explain its toxicity to all forms of plant life.

copper 8-quinolinate   cuprobam

Organotin compounds are triphenyl tin (fentin) salts, which are both toxic and phytotoxic. They demonstrate good control of blast diseases in rice, late blight in potatoes, coffee berry disease and brown spot disease in tobacco. The most widely used organotin fungicides are:

fentin acetate   fentin hydroxide
fentin chloride

Substituted aromatics are derivatives of the benzene or phenol ring with hydrogen atoms replaced by chlorine, nitrogen or oxygen atoms. Possessing fungicidal properties, most of them are suited for seed treatment and soil treatment to control soil-borne fungi such as smut and damping-off diseases of seedlings. Examples are:

chloroneb	hexachlorobenzene, HCB
chlorothalonil	pentachloronitrobenzene, PCNB
dicloran, DCNA	pentachlorophenol, PCP
etridiazole	tochlophosmethyl

Chlorothalonil and dicloran are used as protectants on foliage and fruits; PCNB and HCB for seed treatment; etridiazole and tochlophosmethyl for soil-treatment.

Dicarboximides constitute a group of fungicides which have the basic chemical structure shown in the diagram below; they contain a sulphur and nitrogen atom

at the central position. The group is also referred to as "sulfenimides". In general, they are considered to be among the safest pesticides and are used for seed treatment and for protectant sprays against *sclerotinia* diseases:

chlozolinate	procymidone
iprodione	vinchlozolin
metomeclan	

Phtalamides and related compounds are non-systemic fungicides with broad spectrum activity against many foliar diseases in fruit crops, vegetables and ornamentals. Captan is used on a large scale in tropical cultures; its use for food crops, however, is progressively being restricted:

captafol	folpet
captan	tolylfluanid
dichlofluanid	

Dinitrophenol fungicides with a non-systemic but specific action against powdery mildews are:

binapacryl   dinocap

Triazines comprise a group of compounds with herbicidal activity; they are reviewed below. Only anilazine is a fungicide; it is used for protectant application on vegetables.

Systemic fungicides

These compounds are absorbed by the plant by translocation of the active ingredient through the cuticle of leaves; hence they are taken into the vascular system. These toxicants are transported via the sapstream towards the foliage and the apical (growth) point. The direction of transport is generally towards the leaf tips, but not downwards to the stem base and the roots. Some systemics can be applied to the soil and are slowly absorbed through the roots to give prolonged disease control. Systemic fungicides give a much better protective effect than the non-systemic protectant fungicides which need to be sprayed evenly onto the plant surface where they remain, essentially where they are deposited. Most systemics also exert a curative (therapeutic) effect against pathogens that have already invaded the plant.

There are many systemic fungicides on the market, which may be classified in one of the groups below:

* Oxathiins

carboxin	methfuroxam
furmecyclox	oxycarboxin

Carboxin and related oxathiin fungicides control mainly pathogens of the Basidiomycetes, a class of fungi which includes such important pathogens as smuts and rusts of cereals, and the soil fungus *Rhizoctania solani*. Methfuroxam and furmecyclox control *R. solani* in cotton and potatoes.

* Benzimidazoles and thiophanates

benomyl	thiabendazole
carbendazim	thiophanate-methyl
fuberidazol	

This group contains highly effective, systemic, broad-spectrum fungicides which are widely used in many tropical crops. However, intensive and exclusive use has caused resistance in major crops. Fuberidazol is an important replacement for organomercury compounds as a seed dresser. Thiabendazol controls post-harvest diseases in fruits. These fungicides are not effective against Phycomycetes (*Phytophthora*, *Pythium* and *Perenospora* spp.)

* Pyrimidines

bupirimate	ethirimol
dimethirimol	fenarimol

* Acylalanines

furalaxyl   metalaxyl

* Ergosterol biosynthesis inhibitors (EBIs)

a. Imidazoles:

fenapanil	prochloraz
imazalil	triflumizole

b. Piperazine, pyridine- and pyrimidine compounds:

buthiobate	pyrifenox
fenarimol	triforine
nuarimol	

c. Morpholins:

aldimorph	tridemorph
dodemorph	trimorphamide
fenpropimorph	

The morpholine fungicides demonstrate a systemic, specific activity against powdery mildews of cereals, bananas, rubber and ornamental plants.

d. Triazoles:

bitertanol	myclobutranil
diclobutrazol	penconazol
etaconazole	propiconazol
fluotrimazol	triadimefon
flusilazol	triadimenol
flutriafol	triflumazol

The EBIs are a chemically heterogeneous group of systemic fungicides, grouped together because of a similar mode of action. They are also called sterol biosynthesis-inhibiting fungicides (SBIs) or demethylation inhibitors (DMIs). Most of these compounds have systemic, protective and curative properties, and are effective against powdery mildews and rust fungi which are important diseases of cereals and many other crops. They are also effective against many other leaf spot pathogens such as *Pyrenophora* spp., *Venturia* spp. and *Septoria* spp.

* Organophosphates

ESBP	pyrozophos
iprobenfos	tolclofos-methyl
edifenphos	trimiphos
ditalimfos	

These compounds display protective as well as curative activity because of their systemic property. ESBP (Inezin), iprobenfos (Kitazin P) and edifenphos (Hinosan) are applied against rice blast; the others, against powdery mildews. Toclofos-methyl controls soil-borne diseases, especially *Rhizoctonia solani*.

* Phenylamides and other fungicides against *Oomycetes*

Phenylamides:

benalaxyl	metalaxil
cyprofuram	ofurace
furalaxyl	oxadixyl

Others (including carbamates):

cymoxanil	propamocarb
fosetyl	prothiocarb
hymexazol	

This is a heterogeneous group of fungicides sharing the property of resistance against *Oomycetes*, the causal fungi of important root and foliar disesases such as *Perenospora*, *Plasmopara* and *Phytophthora*. Owing to their highly specific and systemic activity, resistance of this class of fungi to phenylamides may occur.

* 2-Aminopyrimidines

bupirimate	ethimirimol
dimethirimol	

A small group of systemic fungicides with high efficacy against powdery mildews (*Erysiphaceae*).

* Quinones

benodanil	futonil
chloranil	mepronil
dichlone	

Other organic compounds

There are important fungicides which do not belong to one of the chemical groups above; they comprise a random collection of fungicides with non-related chemical structures. Examples are:

anilazine	etridiazol, (echlomezol)
chlorfentezin	fenaminosulf
chlorothalonil	guazatine
dichlofluanid	pencycuron
dasomet	quinomethionate
dodine	sec butylamine
	thiocyclam

Most of these compounds are non-systemic protective fungicides. Pencycuron is specifically active against *Rhizoctonia solani*, sheath blight of rice, in particular. Dazomet is a soil fumigant with a broad spectrum activity, whereas etridiazole and fenaminosulf are soil fungicides with restricted use, mainly against Phycomycetes spp.

Guazatin and oxine-copper are applied in seed dressings, whereas the main use of sec-butylamine is for the control of post-harvest fruit-rotting fungi, for which purpose guazatin is also used.

Antibiotics

There are certain chemical substances produced by micro-organisms, such as bacteria and fungi (molds), that have the capacity, in very dilute concentrations, to inhibit the growth of or to kill bacteria and certain fungi that cause diseases in animals and plants.

To date, several hundred antibiotics have been discovered, but relatively few have been developed into commercial products. The Actinomycetales (an order of lower plants) constitute the largest source of fungicidal antibiotics. In particular, the family of Streptomycetaceae has become widely known for its members:

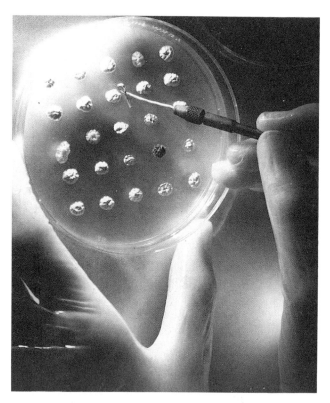

On a nutrient medium, clones of an antibioticum producing Streptomyces *species are grown. Its genetic properties have mutated following radiation with ultra-violet light.*

* *S. griseus* from which streptomycin and cycloheximide are produced that are used as bactericides for fruits, tobacco, maize and ornamentals. Their use is limited for toxicological reasons.

* *S. avermitilus*, from which abamectin is produced.

* *S. griseochromogenes*, which after fermentation yields the active substance of blasticidin-S.

* *S. kasugaenis*, from which kasugamycin is produced.

* *S. cacaoi*, which after extraction yields polyoxins.

Of the latter three Japanese antibiotics, the first two are effective against blast diseases of rice and the third against rice sheath blight. Abamectin has insecticidal, miticidal and nematicidal, but very little fungicidal, effectivity. Other antibiotics, such as penicillin, bacitracin and tetracycline, are used against bacterial diseases in man and livestock. Tetracyclines appear also to help control some mycoplasma-like plant diseases.

Cycloheximide, the least complicated antibiotic compound, affects the growth of yeast and fungi by inhibition of protein- and DNA-synthesis. Its growth regulating properties are used for promoting the abscission of fruits such as oranges and olives. Streptomycin specifically disturbs the build-up of the vital nucleic acids DNA and RNA. Cycloheximide (Acti-dione) is the most toxic fungicide (LD$_{50}$, oral, rats, is 2 mg/kg) and an extremely strong repellent to rats (it is a restricted use product).

Antibiotics having fungicidal and bactericidal activity which are used for agricultural purposes are:

cycloheximide   streptomycin

Antibiotics that have fungicidal action only are:

blasticidin-S   polyoxins
kasugamycin   validamycin

Classification of fungicides by mode of action

Fungicide sprays are of two types, protectants and eradicants.

Protectant fungicides. Several "classic" or conventional fungicides do not penetrate into plants. Instead, when applied as spray, dust or slurry, they provide a film of fungicide over the surface of plants and seeds that kills fungal spores upon germination. Plants are

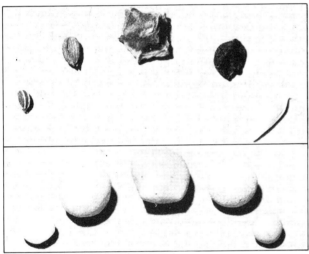

In the process of seed coating, a water-based solution of an insecticide and/or fungicide is applied around each seed and is fixed with a polymeric sticker. The film coat that is formed on the seed is not soluble in water, thus providing continued close protection when the seed germinates in the humid soil. Film coating reduces up to 90 per cent of the dosage needed for the control of soil pests. (Bare seed above and pelleted seed below.)

thus protected during stages when they are vulnerable to inoculation by pathogens, before there is any evidence of disease. As most fungicides have no systemic activity, it is necessary that a uniform coating be deposited on upper- and under-sides of the leaves and on other external surfaces. This is best achieved with high volume sprays with wetting agents and stickers admixed, or by dusting. For protection over longer periods, the application may have to be repeated. Protectant fungicides are effective in preventing leaf spots, mildews and rusts on foliage, blemishes on flowers and post-harvest rot of fruits and vegetables. Examples are copper fungicides, captan and maneb.

Curative or eradicant fungicides. To kill a fungus which has already invaded a plant, fungicides need to

exert a curative or chemotherapeutic effectivity. Generally, a fungal disease can be cured only in its early stage of infection. Once they are well established and symptoms have appeared, fungal diseases are difficult to control. Only a few fungicides have curative action; among them are benomyl, dichlone, dodine, liquid lime-sulphur, fenpropimorph, metalaxyl, propicinazol and triadimefon.

Systemic fungicides. Fungicides which after application are absorbed by the foliage or roots and are subsequently transported over long distances within the plant (from root to shoot or from one leaf to another) are called **systemic**. The highly effective systemic fungicides discovered so far are mostly taken along through the vessels and tracheids of the xylem by the transpiration sap stream; consequently, they are called xylem-systemics. An example is benomyl for the control of *Fusarium* diseases. Some systemics, such as fuberidazol and triadimenol, are used for treatment of seeds. Those fungicides which after having been absorbed show only a limited spread from the site of application into the tissues are termed **locosystemics.**

In practice, systemic fungicides are generally applied as foliar sprays and exert their disease-controlling activity mostly by direct contact between germinating fungal spores and the toxic deposit. Within the leaves, if penetration is sufficient, the translaminar transport of the systemic will stop the infection process in the unsprayed parts of the leaf in a curative action. Thus, systemic fungicides provide superior protection as compared with non-systemic protectants and they have a vital role to play in controlling fungi. This is particularly so if a crop is valued for its appearance, as is the case with vegetables, fruits and flowers. With such crops, fungicides must be applied preventively in advance of infection by damaging pathogens.

CLASSIFICATION OF HERBICIDES

A herbicide is a chemical intended for killing plants or interrupting their normal growth. They may either have a certain selectivity by chemical property by way of application in the control of weeds, grass or brush in cropland, rangeland or forests, or they may be "total herbicides" for use on industrial sites and rights-of-way locations.

Classification by chemical nature[*]

Inorganic herbicides

Most inorganic herbicides are salts; they have been used for decades. In particular, sodium arsenite solutions were very popular as herbicides and arsenic trioxyde as a soil sterilant until the 1960s, when they were banned for

---

[*] The book entitled *Weed Science in the Tropics: Principles and Practices* by I.O. Akobundu, published in 1987 by Wiley-Interscience, Chicester, Great Britain, gives more detailed descriptions of the various herbicide groups.

ecotoxicological reasons. Also iron- and copper sulphates as well as ammonium sulphamate were favoured herbicides for foliar application. Sulphuric acid was, atone time, used for selective weeding in cereal crops and pre-emergence in horticultural crops; however, it is very corrosive to spray machinery. These compounds are usually highly water-soluble and easily leach from the soil. They cause scorching of leaves by inducing plasmolysis and desiccation in the plant cells. Borates, i.e. various sodium salts of boric acid, are absorbed through the roots and are non-selective herbicides. Sodium chlorate can be used as a total herbicide at higher doses or for foliar application at lower ones.

ammonium sulphamate	sodium borate
sulphuric acid	sodium chlorate

Organic herbicides

Organic arsenicals are still being used, because they are much less toxic to mammals than the inorganic arsenic salts. They act by inhibiting metabolic processes in the plant as they compete with phosphate in essential chemical reactions. Examples are:

disodium methanearsonate (DSMA)
monosodium methanearsonate (MSMA)

Phenoxy aliphatic acid

This group comprises a series of compounds in which the phenoxy-nucleus is linked with acetic-, propionic- and butyric acids. Their solubility in water is generally high in comparison with other groups of herbicides. These herbicides have a selective, hormone-type effect on broadleaf weeds, which disturbs their normal growth, but grasses are very tolerant. Absorption takes place through the roots or shoots and the chemicals are readily translocated throughout the plants. 2,4-D and 2,4,5-T and MCPA have been immensely popular and

2,4-D

(2,4-dichlorophenoxy)acetic acid

were once regarded as being very safe. Around 1970, however, the 2,4,5-T compound was suspected of being contaminated with dioxin, an extremely toxic impurity. Therefore, its registration was cancelled. However, 2,4-D and MCPA have remained in wide use because of their low mammalian toxicity and their rapid degradation.

2,4-D	dichlorprop (2,4-DP)
2,4,5-T	fenoprop (2,4,5-TP)
MCPA	mecoprop (MCPP)
MCPB	

Substituted amides constitute a large group of organic nitrogenous herbicides which include two sub-groups:

a. Amides. The compounds in this sub-group act primarily in the soil against annual grass weeds, causing severe stunting of shoots. Amides are of low mammalian toxicity. Diphenamid is relatively persistent in the soil. Propyzamide has a selective and systemic effectivity against annual weeds. Quinonamid kills algae in lowland rice. Examples of amides are:

chlorthiamid	propyzamide
diphenamid	quinonamid

b. Anilides. Compounds of this numerous sub-group vary greatly in type of action and in the range of weeds they control. Propanil is a post-emergence herbicide, which controls annual grasses and broadleaf weeds through inhibition of photosynthesis, after absorption through the leaves. Other closely related compounds include propachlor, alachlor, butachlor, metazachlor and metolachlor. They are applied pre-emergence for control of germinating annual weeds, including grasses. Perfluidone has a similar pre-emergence effect, but in addition it may also induce strong retardation in the development of underground buds of *Cypereus rotundus* and *C. esculentus*. All the products in this group have a low mammalian toxicity. Persistence in the soil usually lasts not longer than a few weeks. Examples are:

alachlor	propachlor
butachlor	propanil
metolachlor	perfluidone
metazachlor	

Diphenylethers

These compounds possess two benzene rings joined through oxygen or a more complex chain of molecules. Nitrofen and fluorodifen are sprayed pre-

NITROFEN

2,4-dichlorophenyl *p*-nitrophenyl ether

emergence on the soil against germinating annual grasses and broadleaf weeds. Uptake is through foliage and by the roots. Bromofenoxim and lactofen are selective post-emergence herbicides for broadleaf weeds. Diphenylethers inhibit respiration and photosynthesis in plants, resulting in chlorosis, deformation and death of seedlings and the scorching of foliage. The mammalian toxicity of this group is fairly low. The compounds are relatively insoluble in water and do not easily leach from the topsoil. Persistency in the soil may last up to several months. Examples are:

aclfluorfen-sodium	fluorodifen
bromofenoxim	lactofen
diclofop-methyl	oxyfluorfen

Dinitroanalines are probably the most heavily used group of herbicides in agriculture. These herbicides, which have a similar dinitro-aniline-nucleus in common, are very effective against annual grasses and broadleaf weeds, when they are applied pre-emergence and incorporated into the soil. They even affect the underground buds and stolons of perennial weeds. These compounds and trifluralin and benfluralin, in particular,

TRIFLURALIN

α,α,α-trifluoro-2,6-dinitro-N,N-dipropyl-
p-toluidine

must be incorporated into the soil owing to the loss which occurs because of photo-decomposition and volatilization. Trifluralin is only slightly soluble in water, a property which minimizes leaching. Its efficacy stems from an inhibition of cell division in plants, which causes stunting of shoots and roots. Mammalian toxicity is low. The persistence in the soil ranges from a few weeks for low doses up to one year for high doses. Examples are:

benfluralin	oryzalin
fluchloralin	pendimethalin
nitralin	trifluralin.

Substituted ureas

Compounds of this group are based upon the simple nitrogen-containing molecule of urea, but have the

$$H_2N-\overset{\overset{\displaystyle O}{\|}}{C}-NH_2$$

hydrogen atoms at both sides replaced with various carbon-chain and ring structures. After the discovery around 1950 of monuron as a total herbicide, a large number of urea derivatives were developed as herbicides. The much used substituted ureas are primarily active as selective pre-emergence herbicides and are strongly adsorbed to the soil. After absorption by the roots, they show a good efficacy against the majority of annual weeds. Some compounds can also be sprayed post-emergence if surfactants are added to the spray mixture. Control of susceptible weeds results from inhibition of photosynthesis, which causes chlorosis of the leaves, stoppage of growth and finally death. The mammalian toxicity of urea derivatives is low. The persistence

generally lasts a few weeks in the soil, but efficacy is influenced by ecological factors (soil, rain, temperature). Examples are:

chlorbromuron	linuron
chloroxuron	methabenzthiazuron
chlortoluron	metobromuron
diuron	metoxuron
fenuron	monolinuron
fluometuron	tebuthiuron
isoproturon	thiazafluron

Carbamates

The ester structure of this group has been explained previously under carbamate insecticides. Since the discovery of the herbicidal activity of propham, many more products have been developed in the carbamate group. Phenmedipham and barban are post-emergence herbicides; asulam pre- and post-emergence. The other compounds in this group are soil-active and are to be applied pre-emergence. Uptake occurs through roots or foliage and translocation within the plant is good. Carbamates are effective in controlling annual grasses and some annual broadleaf weeds. They act through the inhibition of germination and cell division; phenmedipham inhibits photosynthesis. Carbamates are of low mammalian toxicity; persistence in the soil is short. Some examples are:

asulam	phenmedipham
barban	propham
chlorpropham	

Thiocarbamates

The thio- and dithiocarbamates belong to the group of carbamates which contain sulphur. They are selective herbicides for pre-emergence or pre-planting application. Because of their volatility, the compounds must be incorporated into the soil. Uptake through the roots is followed by retarded growth and failure of the plant to emerge properly. Control of annual grasses and broadleafs is good. Di-allate and tri-allate act specifically against wild oats (*Avena fatua*); molinate, against water-grass (*Echinochloa crusgalli*) in rice. The -allate compounds are persistent for many months in the soil; the other thiocarbamates persist for a much shorter period. Metham-sodium acts as a soil fumigant by being converted in the soil to methyl isothiocyanate. Examples in this group are:

EPTC	pebulate
di-allate	thiobencarb
metham-sodium	tri-allate
molinate	vernolate

Heterocyclic nitrogens are compounds with a ring structure in which carbon atoms have been replaced by nitrogen and occasionally by sulphur. The main groups are as follows:

a. Triazines have a ring structure consisting of six atoms of which three are nitrogen atoms (hence its name tri-azine) to which two amino groups are commonly attached. The compounds demonstrate strong efficacy

TRIAZINE NUCLEUS **SIMAZINE**

2-chloro-4,6-bis(ethylamino)-
s-triazine

against annual grasses and broadleaf weeds coupled with a remarkable selectivity. This selectivity depends on the ability of tolerant plant species to metabolize the triazine compound, whereas susceptible plants do not metabolize the compound. Triazines are applied to the soil, primarily being absorbed through the roots and translocated inside the plants. Germination and emergence of the crop and the weeds are not affected, but as soon as green leaves are formed, the activity causes severe inhibition of photosynthesis and chlorosis. Solubility in water is low, thus leaching from the soil is limited. The mammalian toxicity of triazine is generally low. Chlorotriazines and, in general, the "-azine compounds" such as atrazine, are very persistent in the soil and can be used for weed control in perennial crops and total weed control with higher doses. The methylthiotriazines, generally the "-tryn compounds" such as ametryn, exhibit a soil-persistence of only a few weeks and therefore can be used in crops having a short vegetation period and rapid rotation.

As of this writing however, it is expected that the use of atrazine in maize will be restricted in Western Europe, because serious contamination of groundwater (the source of drinking water) has resulted from intensive applications over the years. Examples of this group are:

atrazine	ametryn
cyanazine	dimethametryn
propazine	prometryn
simazine	simetryn
terbutylazin	terbutryn

b. Triazinones also have a six-member ring and are closely related. Hexazinon is used for weed control on non-cropland and industrial sites. Metribuzin is used for weed control in potatoes, tomatoes, tobacco and soybeans. Examples are:

hexazinon   metribuzin

c. Triazoles are compounds containing a five-member ring in which three nitrogen atoms have replaced the carbon atoms. They are effective against broadleaf weeds in many crops and have a contact and slight systemic activity. Examples are:

amitrole   bentazon   oxadiazon

d. Pyridines are chemicals derived from the pyridine molecule, a six-member ring with one carbon atom replaced by nitrogen. The best known is picloram,

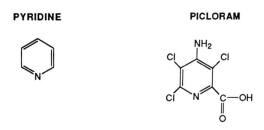

PYRIDINE   **PICLORAM**

4-amino-3,5,6-trichloropicolinic acid

which is very effective against deep-rooted herbaceous weeds and woody shrubs. Most grasses are resistant. Triclopyr is another related selective brush killer. Applied post-emergence, they may be taken up by either the roots or foliage. They act through stimulation of growth, resulting in elongation of cells, proliferation of tissues and formation of numerous rootlets. They are marketed for brush control in rangeland and utility rights-of-way. Examples are:

fluroxypyr   picloram   triclopyr

e. Uracils are derived from the uracil nucleus, another substituted six-member ring with two nitrogen atoms and a double bond. Primarily for pre-emergence application and uptake mainly via the roots, they are very effective in controlling annual grasses and broadleaf weeds over an extended period by inhibition of photosynthesis. Mammalian toxicity is low; at higher doses, they may remain in the soil for up to 12 months. Examples are:

bromacil   lenacil   terbacil

Bipyridiliums are composed of two pyridyl-rings and thus they are also termed pyridines. The two best known members of this group are diquat and paraquat. These herbicides have a strong contact action on above-

PARAQUAT

1,1'-dimethyl-4,4'-bipyridylium ion
(dichloride)

ground plant parts. Within the plant, the chemicals are reduced to free radicals, which destroy tissue under light. Wilting and drying out of the foliage follow within hours, making these herbicides useful pre-harvest desiccants for seed crops. Paraquat is particularly effective against grasses. Both are quickly deactivated in the soil due to strong adsorption to soil particles. However, because paraquat very slowly degrades in the soil, its use will soon be restricted in Western Europe. Both compounds are hazardous and need to be applied with great care, as no antidote is available for diquat and paraquat when swallowed. In many developing countries, paraquat is a major cause of fatal pesticide poisoning. It is much safer to use glyphosate (see sector under miscellaneous herbicides), a herbicide that is still more expensive, but which kills deep-rooted grasses as well as paraquat.

Aliphatic acids

Two heavily used herbicides of this group are chlorinated derivatives of acetic acid (TCA) and propionic acid (dalapon). They exhibit systemic action after having been absorbed by roots and leaves. They are effective against annual and perennial grasses. Used for non-cropland and forestry applications, they kill by causing precipitation of protein within the cells.

trichloroacetic acid (TCA)   dalapon

Phenol derivatives

Compounds of this group are highly toxic, selective foliar herbicides. There are two sub-groups:

a. Dinitrophenols. These are primarily contact herbicides, but have also some effect via the soil against annual broadleaf weeds. The dinitrophenols are not translocated within the plant and are used as selective pre- and post-emergence herbicides. They inhibit respiration and photosynthesis within the cells. Some other dinitrophenol compounds are discussed above in the paragraphs about insecticides and fungicides. Examples are:

dinoseb (DNBP)   dinoterb
dinoseb-acetate   DNOC

b. Chlorinated phenols. The only herbicide in this sub-group is pentachlorophenol (PCP or Penta), which is also a termiticide and a wood protectant. PCP is a restricted use product. The use of dinoseb and dinoseb-acetate is banned in the United States, and Western Europe.

Benzonitrilles or substituted nitrilles consist of a benzene ring containing a cyanide (C-N)-radical. They show a broad spectrum activity, and act by various processes of inhibition of growth and tissue disruption. Effective against annual grasses and broadleaf weeds. Examples are:

bromoxynil   dichlobenil   ioxynil

Miscellaneous herbicides

These are solitary compounds derived from rather unrelated chemical structures. Examples in wide use are:

alloxydim-sodium	glufosinate-ammonium
bensulide	glyphosate
bentazone	methyl bromide
dimethazone	quizalofop-ethyl
endothal sodium	pyridate
difenzoquat	sethoxydim
fluazifop-butyl	

Endothal sodium (Aquathol) is a preferred aquatic weed killer because of its low toxicity to fish, although it is toxic to mammals. Also, it is used as a systemic pre- and post-emergence herbicide against annual grasses and broadleaf weeds in field crops. Glyphosate is an important non-selective, non-residual, post-emergence herbicide for the control of deep-rooted grass and broadleaf weeds and woody shrubs. It is a foliar-applied systemic that is effective against all stages of weed growth. Action is through inhibition of protein and aromatic acid synthesis. Alloxydim sodium, sethoxydim, fluazifop-butyl and quizalofop-ethyl are new selective and systemic herbicides against annual and perennial grasses for post-emergence application.

Classification of herbicides by mode of action

Herbicides may also be classified according to the various ways they are being used or in which they act.

* Selective action. Compounds which kill weeds without harming crops.

* Non-selective action when all vegetation is killed.

* Contact action. Those plant parts are killed to which the herbicide is applied.

* Systemic action. The herbicide is taken up either by the roots or above-ground parts of plants and translocated throughout the tissues. They are particularly important for the control of perennial weeds, which continue growing from year to year.

* Pre-planting: Applied after the soil has been prepared but before seeding.

* Pre-emergence (contact): Non-residual dosages are used after seeding but before emergence of the crop seedlings.

* Pre-emergence (residual): Applied at the time of seeding or just prior to crop emergence; the compound kills weed seeds and germinating seedlings.

* Post-emergence: Application after emergence of the crop.

In scouting, pheromone traps are used to measure the rate of population build-up of certain pests in order to determine the kind and timeliness of chemical or other control methods needed when the economic threshold level is surpassed.

CLASSIFICATION OF BIOLOGICAL PEST CONTROL AGENTS

Biological pest control agents are naturally occurring or genetically modified agents that are distinguished from conventional pesticides by their unique modes of action, low use volume and target species-specificity. Two major catagories may be distinguished:

* Biochemical pest control agents, such as hormones and growth regulators.

* Microbial pest control agents, such as bacteria, viruses and fungi.

Biochemical pest control agents

A chemical must meet the following two criteria in order to be classified as a biotechnical pest control agent and meet the data requirements for this class of compounds:

* The chemical must exhibit a mode of action other than direct toxicity in the target pest (e.g. growth regulation, mating disruption). Botanical insecticides such as nicotine and pyrethrum kill and thus are not considered biochemical pest control agents.

* A biochemical must be naturally occurring, or if the

chemical is synthesized by man, then it must be structurally identical to a naturally occurring chemical. Identical in this sense means the molecular structure of the major component of the synthetic chemical must be the same as the molecular structure of the naturally occurring analog.

Minor differences among stereochemical isomers are acceptable unless an isomer is found to have significantly different toxicological properties than another isomer.

In cases of doubt, such as if the exact molecular structure of the naturally occurring compound is unknown, or if the mode of action is different in the target compared with non-target organisms, a country's regulatory authority should rule on a case-by-case basis whether such chemicals should be classified as a biochemical pest control agent or as a conventional pesticide.

Biochemical pest control agents fall into four general biologically functional classes:

a. Semiochemicals

These are chemicals emitted by plants or animals which modify the behaviour of receptor organisms of like or different kind. They include pheromones, allomones and kairomones.

* **Pheromones** are substances emitted by members of one species which modify the behaviour of others within the same species. Even at extremely low concentrations, these chemical messages induce insects to aggregate, help them to find a partner, provide an alarm signal or lead them to sources of food. The most common pheromones are sex attractants excreted by abdominal glands of the female to attract males for the purpose of mating; and the aggregation pheromones which are produced by one or both sexes of an insect species which bring both sexes together for feeding and reproduction. Sex hormones are common among moths and butterflies; the aggregation hormones, among beetles.

* **Allomones** are chemicals emitted by one species which modify the behaviour of different species to the benefit of the emitting species. Many plants produce secondary substances that repulse insects and prevent them from feeding; these are classified as allomones. The oil of *Citronella* grass has long been used by man as an insect-repellent applied to the skin.

* **Kairomones** are chemicals emitted by an animal organism which, at very low concentration, modify the behaviour of individuals of a different species to the disadvantage of the emitting and the benefit of the receptor species. For instance, an animal parasitoid may be guided by it in finding a host. Kairomones, like pheromones, can be used to attract insects to traps for the purpose of monitoring or catching them.

b. Hormones

Hormones are biochemical agents that are synthesized in one part of an organism and translocated to another part, where they have controlling, regulating or behavioural effects. Two major groups of hormones have been identified; they are:

* **Molting hormones** or **ecdysteroids**. These comprise a group of closely related water-soluble steroids in insects, of which several active analogs are found in plants. As yet, no effective control of insects through feeding or topical application of either natural or plant-derived ecdysteroid analogs could be obtained. Moreover, as their synthesis appears to be prohibitively expensive, the commercial production of ecdysteroids still remains in the research phase.

* **Juvenile hormones.** Four very similar juvenile hormones are produced by insects in the process of their immature development, which maintain the insects in their nymphal or larval form. These juvenile hormones, as well as many hundreds of synthesized active analogs -- the so-called **juvenoids** -- are able to affect insect development almost equally well by any mode of application. Conforming to their mode of action, these juvenile hormones and juvenoids are also designated as **insect growth regulators (IGRs).**

Table 2.1: Five juvenoid compounds that are currently registered for use against insect pests

Compound	Registered for use against:
diflubenzuron	*Lepidoptera, Diptera*, mosquitoes
fenoxycarb	fruit tree leafroller, fire ants
PRO-DRONE	fire ants
methoprene	flies, fleas, mosquitoes, pharao ants, stored food pests and tobacco pests
hydroprene	cockroaches
kinoprene	homopterous pests in greenhouses

Source: Staal (1987), Juvenoids and anti-juvenile hormone agents as insect growth regulators.

The natural hormones do not have enough environmental stability and are too costly to be synthesized; thus, they were not developed into commercial products. But the synthesis of their analogs, the juvenoids, turned out to be very productive and yielded many active compounds of often enhanced selectivity. To date, only five juvenoids have been registered for use against insect pests; they are listed in table 2.1. It appears that the listed juvenoids are used in protected environments; only fenoxycarb can be used in a field crop situation. This is one of several drawbacks of juvenoids; others can be summarized as follows:

- The instability of these compounds, when exposed to sunlight and wind as a foliar residue. Fenoxycarb is the most stable of the five compounds in field application.

- Except for kinoprene, the juvenoids do not cause direct mortality in the target insects.

- The juvenoids are able to affect the target insects only during two distinct but short periods of their life. These brief periods, the so-called windows of sensitivity, coincide with the last larval instar and with early embryonic development in the egg-phase. Application of juvenoids in the last larval instar inhibits the process of metamorphosis (see Chapter 12, the section on insect development). An excess of juvenile hormone or juvenoid results in the development of intermediate forms between nymphs and adults which, if they survive the molt, are incapable of reproduction. Application to the egg stadium causes an ovicidal effect that may result in abnormal embryo development, failure in hatching or death after emergence.

The advantages of juvenoids as chemical insect control agents are found in the following characteristics:

- Juvenoids demonstrate a rather extreme selectivity for insects, in which they affect receptor systems that do not exist in other forms of life and an often very high but selective activity on sensitive target species.

- These compounds have an extremely low toxicity to non-target organisms and mammals.

47

- Little cross-resistance is as yet expected to exist against these juvenoids.

Juvenoids which show neither a repelling nor direct killing action, as most conventional insecticides do, have proven to be very effective against social insects such as fire ants and pharao ants. Baiting with fenoxycarb and methoprene generally gives good control since juvenoids brought into the nest by returning workers affect the brood and the social interactions.

* **Diflubenzuron** is a synthetic compound with contact and stomach activity that interferes with chitin deposition at the time of molting and prevents the shedding of the old skin. This leads to the death of the larvae and pupae, or to the development of non-viable adults. It also prevents hatching of insect eggs. Diflubenzuron is effective against lepidopterous caterpillars, grasshoppers, fly and mosquito larvae and weevils. It has no systemic activity and thus sucking insects are in general not affected. This forms the basis of its selectivity in favour of many insect predators. The trade name is Dimilin. Another promising IGR showing similar activity is teflubenzuron (Nomolt).

DIFLUBENZURON (Dimilin®)

1-(4-chlorophenyl) 3-(2,6-difluorobenzoyl)urea

* **Kinoprene** is a juvenile hormone mimic that, when applied to larval stages, prevents their metamorphosis into viable adults. Uses include the control of stored product pests and indoor insect pests, but it is particularly effective against insects of the order Homoptera such as white fly, aphids, mealybugs and scale insects in greenhouses. Although this compound had proved highly effective for pest control on ornamentals in greenhouses, it appears to have been withdrawn from the market as its market share was too small to justify the cost of keeping it available.

* **Methoprene** is also a juvenile hormone mimic with similar activity; it was registered in 1975 as a commercial IGR for use against mosquito larvae. Other uses include control of storage pests, homopterous fleas, leafminers and pharao ants. Juvenoids such as hydroprene and methoprene are very useful in confined environments such as greenhouses, houses and storage buildings.

c. Natural plant regulators

Natural plant regulators are chemicals produced by plants that have inhibitory, stimulatory, or other modifying effects on the same or other species of plants. Some of these are termed "plant hormones" or "phytohormones".

Some examples of plant growth regulators (PGRs) are as follows:

* **Auxins-compounds** induce elongation in the cells of shoots. In 1934, it was discovered that auxins enhance the rooting of cuttings. Further uses of this PGR are for the thinning of fruits and the increasing of flowering. Auxins are formed within the plants from auxin precursors. Manufactured products with similar activity are indoleacetic acid, B-naphthaleneacetic acid and 2,4-D.

* **Gibberellins** or gibberellic acid derivatives are substances which stimulate cell division and cell elongation through the induction of increased enzyme production in the cell. They occur naturally in all plants. Gibberellins are used to increase stalk length, delay yellowing, break dormancy in seed potatoes, increase fruit set and yields etc.

* **Cytokinins** are naturally occurring or manufactured substances that induce cell division and the regulation of differentiation in plant parts. These compounds were discovered in 1955 and their usefulness lies in prolonging the storage life of green vegetables, cut flowers and mushrooms. Zeatin is a natural substance, whereas kinetin and adenine are synthetic cytokenins.

* **Inhibitors** are substances that interfere with a physiological process in plants regulating growth, germination of seeds, or the action of hormones, gibberellins and auxines. Inhibitors that occur naturally in plants are usually hormones. Substances with inhibiting or retarding action are used to prevent sprouting of bulbs and tubers, prevent sucker growth on tobacco plants and induce shorter stem growth in grains. Naturally occurring inhibitors are benzoic acid, gallic acid, cinnamic acid and abscisic acid. There are also synthetic inhibitors such as maleic hydrazide.

d. Enzymes

In this context, enzymes are defined as protein molecules that (1) are the instruments for the expression of gene action and that (2) catalyze chemical reactions.

Microbial pest control agents

The pesticides referred to as microbial pest control agents include formulations either of naturally occurring, infective micro-organisms, such as bacteria, fungi, protozoa and mycoplasmas, and viruses, or genetically modified micro-organisms that are used to achieve natural control of pests.

These pathogens and parasites are isolated and mass produced for use as a commercial pesticide. The basic general principle required for their registration is that the product should demonstrate effectiveness and not present unacceptable hazard to users, consumers of treated foods and the environment.

48

Microbial insecticides are preparations based on disease-inducing organisms (entemopathogens) or parasites which normally infect or poison an insect or mite and ultimately cause its death. Microbial insecticides are usually rather specific to certain orders of insects or mites and non-toxic to man, other mammals and plants.

Microbial insecticides have already been used for many years among a wide range of crops, forests and mosquito habitats where they are applied using conventional spraying techniques. They are often applied in combination with chemical insecticides and other means of control, within an integrated pest management approach.

Their main advantages are their environmental safety and highly specific action, but a major drawback is their lack of biological persistency under field conditions. Therefore, these microbial insecticides could not compete on the commercial market with more persistent and effective chemical pesticides, such as the successful synthetic pyrethroids. However, recent developments in the field of biotechnology, in particular in the areas of protein and genetic modification, have sparked a renewed interest in all groups of biological control agents.

Biotechnology appears to create new opportunities to overcome some of the long-standing disadvantages of microbial insecticides, such as their rapid breakdown, inconsistency of activity and costly production. In 1988, the world market for microbial insecticides was estimated at $US 70 million, or 1 per cent of the total insecticide market. The new developments seem to offer a possibility for slow expansion of the market share of microbially based products to an estimated $US 300 million by 1992 (Cannon, 1989). In addition to the microbial insecticides, other microbial pesticides have already been developed that demonstrate a fungicidal, nematicidal or herbicidal activity (see Chapter 1, the section on microbial pest control agents).

Bacteria are the most researched and most widely exploited pathogens for pest control, mainly because of the high efficacy of certain strains of the infectious bacterium *Bacillus thurigiensis* (Bt), against susceptible species of insects of the orders Lepidoptera, Diptera, Homoptera and (very recently) also Coleoptera.

The bacterium Bt forms within its cell an oval endospore and at the same time a proteinaceous inclusion commonly called a "crystal". This crystal, which represents up to 40 per cent of the dry weight of the bacterium cell at the time of sporulation, contains an endotoxin. Upon ingestion by an insect, this endotoxin may break down in an alkaline, mid-gut environment under the action of enzymes to release a smaller, active protoxin. This lethal protoxin damages the lining of the mid-gut wall leading to membrane rupture and lysis. The bacterial spores are thus released in the abdominal cavity of the insect where they multiply, eventually killing the insect. Furthermore, a cessation of feeding occurs usually within hours of ingestion owing to paralysis of the mouthparts.

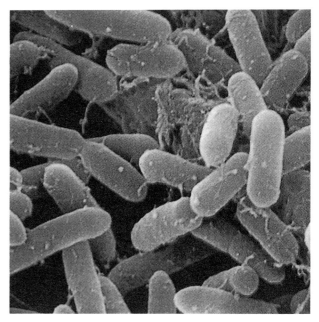

An electron-microscopic image of a rod-shaped bacillus, (magnified 12,000 times). The type of bacterium shown is Escherichia coli *which may affect the human intestines.*

The selectivity of Bt strains appears to depend on the acidity (pH-value) of the mid-gut content. Insects having an alkaline mid-gut content of pH 8-10 are susceptible (e.g. lepidopteran larvae). But the desert locust, *Schistocerca gregaria*, having a more acidic mid-gut content of pH 6.2-7.1 is non-susceptible as the known Bt endotoxins remain insoluble and hence inactive under such acidic conditions.

The microbial insecticides that have found widespread use against caterpillars and beetle grubs are produced from *B. thurigiensis* var. *kurstaki* under trade names such as Bactospeine, Thuricide and Dipel. Those microbials produced from *Bt* var. *israelensis* are most effective against larvae of mosquitoes and black fly; brand names are Bactimos and Teknar.

Vigorous research over recent years has yielded many new varieties of Bt which are effective against a number of different insect species. Some of the new discoveries are more than 10 times as effective as those used in traditional Bt formulations. A serious drawback of traditional Bt formulations has been their instability under field conditions; they break down quickly when exposed to ultra-violet light. By means of genetic manipulation, an ingenious solution to this problem of short persistency has been recently developed and patented under the tradename "Cellcap" (refer to box on the next page).

Bacillus sphaericus also produces a crystalline toxin, and several strains have proven to be effective against larvae of mosquitoes. No appearance of resistance in larvae against these toxins has as yet been observed either in the field or during laboratory tests at sub-lethal doses.

The Cellcap™ system for the production of microencapsulated Bt toxins

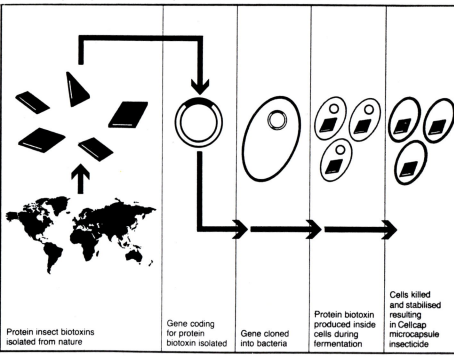

| Protein insect biotoxins isolated from nature | Gene coding for protein biotoxin isolated | Gene cloned into bacteria | Protein biotoxin produced inside cells during fermentation | Cells killed and stabilised resulting in Cellcap microcapsule insecticide |

In this process, the Bt gene which produces the toxic crystals is transferred into a Pseudomonas *bacterium. Having been killed and stabilized, this* Pseudomonas *bacterium then serves as a protective micro-capsule for the fragile Bt toxins. This microcapsulated Bt then remains stable in the field for much longer than traditional products, but once ingested by an insect, the cell walls break down rapidly and the toxins are released. Furthermore, no spores are produced by the transformed Pseudomonads, thereby reducing any possible risk of environmental contamination. (Source: Shell Agriculture: 5, 1989, p. 15)*

TM – Cellcap is a Mycogen Corporation trademark

Fungal microbial insecticides contain parasitic fungi which are capable of penetrating the insect cuticle directly. After piercing the cuticle -- often by means of enzymes that break down chitin and protein -- the fungus multiplies inside the insect, producing lethal metabolites. Peak mortality normally occurs 5-10 days after application, depending on dose and insect size.

Microbial insecticides are formulated, for instance, with the use of the following fungi: *Beauveria bassiana, Verticillium lecanii, Entemophaga asiatica, E. grylli, Metharizium anisopliae, Zoophthora radicans*. Parasitic fungi, generally, have host-specific strains and are not hazardous to non-target organisms. They are cheap to produce and may be formulated and applied in a way similar to chemical insecticides. Since these fungi broach directly through the insect cuticle, they have a contact action without having to await ingestion by the target insects. By this property, a fungal microbial avoids a prolonged exposure to ultra-violet light, low relative humidity and other adverse climatic conditions which may inactivate it. Thus, waiting for the pest to ingest the pathogen is a critical limitation for the use of microbials based on viruses, bacteria and protozoa.

Close observation, however, has revealed that under adverse conditions a pathogen such as *Beauveria* will kill a target insect only if it either hits the insect directly or contacts it quickly after application, but before it is inactivated. Past attempts to induce epidemics from single applications were not successful, as the disease did not automatically increase to epidemic levels. A high relative humidity is required for the production of spores. When low humidity prevails, the disease cannot spread and a high kill would only be obtained through a highly effective initial application. Constraints which apply to pathogen use are similar to those limiting effective chemical pesticide use; formulation and application technology are critical.

Recently, it has been found that the use of oil-based fungal pathogen formulations hold much greater promise than water-based formulations that evaporate very rapidly under dry conditions. In dose-response studies on a weevil pest, *Beauveria bassiana* was over 30 times more effective by topical application when formulated in a vegetable oil (Prior *et al.*, 1988). The suggested reason is that insect cuticles are lipophilic (fat-loving), so that the oil spreads out over the cuticle and penetrates into the thin membranes at the articulating surfaces of the joints, carrying the spores to the most vulnerable part of the cuticle. By contrast, water-based fungal formulations run off and the spores are lost. Thus, much interest exists in the development of oil-based fungal formulations for low

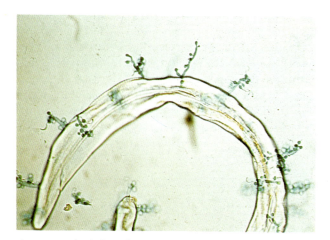

A nematode infected with and killed by a Verticillium lecanii *fungus (Photo courtesy of PD, the Netherlands)*

Table 2.2: Examples of commercially available microbial insecticides and their target pests

Product	Pathogen	Target	Company (country)
Dipel	Bacteria: Bt var.kurstaki	Lepidopterous larvae	Abbot Labs. (USA)
MVP	Bacteria: Bt var.kurstaki	Lepidopterous larvae	Mycogen Corp. (USA)
Skeetal	Bacteria: Bt var.israelensis	Mosquito and blackfly larvae	Novo BioKontrol (Denmark)
M-one	Bacteria: Bt var.san diego	Colorado potato beetle	Mycogen Corp. (USA)
Nemasys	Nematode/bacterium	Black vine weevil	Agricultural Genetics (UK)
Biosafe	Nematode *Steinernema* spp.	Soil insects	Biosys (USA)
Mamestrin	MbNPV[1]	*Heliothis* spp.	Calliope (France)
Spodopterin	SlNPV[2]	*Spodoptera* spp.	Calliope (France)
Virox	NsNPV[3]	*Neodiprion sertifer*	Oxford Virology (UK)
Nolo bait	Protozoan: *Nosema locustae*	Rangeland grasshoppers	Evans BioControl (USA)
Vertalec*	Fungus: *Erynia neoaphidis*	Aphids	Microbial Resources (UK)
Mycotal*	Fungus: *Verticillium lecanii*	Glasshouse whitefly	Microbial Resources (UK)
Development product	Fungus: *Neozygites*	Spider mites	Ecogen (USA)

Notes: [1] *Mamestra brassicae* nuclear polyhedrosis virus; [2] *S. littoralis* nuclear polyhedrosis virus; and [3] *Neodiprion sertifer* nuclear polyhedrosis virus. * Product withdrawn from the market.

Source: Cannon, R.; *Shell Agriculture*: 5,1989; pp. 13 - 15.

and ultra-low volume application. Further examples of fungal insecticides are:

Hirsutella thompsonii, a parasitic fungus from which a spore-formulation is produced that kills the citrus rust-mite and also infects spider mites. Formulations of *Verticillium lecanii* control aphids, scales, thrips and red spider mites in greenhouse crops (see also table 2.2). Furthermore, a number of other fungi are currently being investigated to determine their potential microbial value: e.g. *Nomuraea* spp. and *Paecilomyces* spp.

Viral microbial insecticides are formulated on the basis of nuclear polyhedrosis viruses.

A virus is a sub-microscopic infectious agent existing as a nucleo-protein entity that is composed of protein and a nucleic acid, RNA or DNA. The protein surrounds, as a crystalline shell (capsid), the nucleic acid component which transfers the genetic information. Owing to their lack of organ structures, viruses are able to replicate only within the living cells of a host. Since viruses contain only DNA or RNA and lack their own metabolism, they are not recognized as being organisms. (see also Chapter 13, section on viruses).

Nuclear polyhedrosis viruses infect the destructive *Heliothis* spp. caterpillars, such as the cotton bollworm and tobacco budworm, which damage many types of crops. Nuclear polyhedrosis viruses are very specific and highly virulent and can survive for years because of their protective capsid. Following ingestion by an insect, the protein capsid dissolves in the insect's alkaline mid-gut juices. The infectious, individual, mature virus particles, the virions, permeate the gut lining and eventually enter into the bloodstream where they multiply rapidly and take over control of susceptible cells. This results in the death of the insect within two to nine days, depending on the stage of development of the insect.

The baculoviruses include the group known as polyhedrosis viruses. Production of these viruses involves rearing them on living insects, which is a costly process.

Nematodes, which occur naturally in the soil and parasitize soil-insects, are excellent material for the formulation of microbial insecticides. Once released into the soil, they are able to disperse over short distances and they are attracted to their insect hosts. The infective juvenile stages of genera such as *Steinernema* and *Heterorhabditis* are usually ingested by the insects, but they may also enter via the anus or spiracles. From the gut of the insect they pass through the gut wall into the circulatory system, where they release bacteria. These released bacteria reproduce rapidly and cause the insect's death within one to two days.

Active nematodes need to be surrounded all the time by a water film; only some genera are able to survive drought conditions in a particular cyst-form. Thus, the use of nematodes as microbials has been mainly restricted to the soil environment, or protected habitats where they are able to survive. Above the soil surface they are easily exposed to ultra-violet light and desiccation, which prohibits their use against foliar-feeding insect pests.

Neoplectana carpocapsae is a nematode that acts as a vector in transmitting a pathogenic bacterium lethal to termites. Commercial microbial formulations are already on the market.

References:

Akobundo, I.O. 1987: *Weed Science in the Tropics: Principles and Practices,* Wiley-Interscience, Chicester, Great Britain.

Cannon, R.: Microbial insecticides: the opening of a new era? Shell Agriculture, nr. 5, 1989.

Food and Agriculture Organization of the United Nations, FAO Rome, (1988). *Guidelines on the registration of biological pest control agents.*

Heitefusz, Rudolf: *Pflanzenschutz: Grundlage d. prakt. Phytomedizin.* 2., neubearb. u. erw. Aufl. Stuttgart; NewYork: Thieme, 1987.

Lyr, H. (Ed.): *Modern, Selective Fungicides - Properties, applications, Mechanisms of Action.* Longman Group UK Ltd., London and VEB Gustav Fischer Verlag, Jena, 1987.

Meister Publishing Co. (1985). *Farm Chemicals Handbook, 71st edition.*

Prior, C., Jollands, P., and le Patourel, G. (1988): *Infectivity of oil and water formulation of B.bassiana to the cocoa weevil pest.* J Invertebrate Pathology, 52, 66-72 .

Staal, G.B.: *Juvenoids and anti juvenile hormone agents as insect growth regulators.* Contained in: Delucci.V.: *Integrated pest mangement*; *Quo vadis?* Parasitis, 1987.

The Royal Society of Chemistry, UK,(1985) *The Agrochemicals Handbook, update 4.*

United Nations. ESCAP Agriculture Division (1982) *ARSAP Pesticide Index.*

United Nations. ESCAP Agriculture Division (1987) *ARSAP/CIRAD, Regional Agro-pesticide Index Volume 1, Asia 1991 edition (revised).*

Ware, George W. (1986): *Fundamentals of Pesticide: a self-instruction guide. 2nd edition.*

Ware, George W.: *The pesticide book, 3rd edition.* Thomson publications Tresmo, Ca 93791, United States of America, ISBN 0-913702-52-8.

Worthing, C.R. and S.B. Walker (eds.): *The Pesticide Manual.* 8th Ed. British Crop Protection Council, Thornton Heath, UK, 1987.

3

PESTICIDE MANUFACTURE AND FORMULATION

Contents   Page

Manufacture and formulation 54

- The pesticide industry 54

- Synthetic pesticides 55

- Biological pesticides 56

- Formulation components 57

Pesticide formulations and their codes 60

- Concentrates for dilution with water 60

- Concentrates for dilution with organic solvents 62

- Formulations to be applied undiluted 63

- Miscellaneous formulations for special purposes 64

References 66

Pesticide Manufacture and Formulation

MANUFACTURE AND FORMULATION

The pesticide industry

The annual global consumption of more than 3 million metric tonnes of pesticides (World Health Organization, 1989) requires a wide range of structures on various levels of industry and commerce. On the level of technical ingredient manufacture, the industry is characterized by a wide range of processes. In contrast with the heavy chemical industry, which commonly produces one or a few basic chemicals in bulk, the pesticide industry produces a multiplicity of complex chemical products, mostly on a comparatively small scale. Owing to the research-intensive character of the pesticide industry, the complexity of the many modern compounds, and the elaborate legislation requiring expert investigation of environmental and health hazards, few chemical companies can afford to be in the pesticide manufacturing business alone.

Historically, the pesticide industry has often developed as an offspring of the manufacture of bulk chemicals by companies looking for secondary processes to utilize intermediate compounds from their primary production process. In particular, the association with the dyestuff industry appears to be a traditional one, since several of the leading industries in Europe have their roots in the production of vegetable dyes. In fact, such association of interests in the fields of dyestuffs, pharmaceuticals, plastics, pesticides and other related chemical fields which one can detect in giant chemical companies, has become a precondition to survival of the research capacity of the industry.

A pesticide manufacturing plant of Duphar b.v. in the Netherlands. The yellow building used to be a herbicide factory which, owing to a new and more efficient process technology, could be relocated to a smaller unit. Today the building is a diflubenzuron (insect juvenile hormone) factory. The entire plant must conform to strict environmental requirements regarding pollution control and reduction of chemical waste because of its location in the port area of Amsterdam.

Only by a close co-ordination of research in the synthesizing laboratories and through multi-site and multi-effect testing of the great variety of newly produced compounds can those compounds having any interest directly or as intermediates in any of these fields be detected. An efficient biological screening department is a necessity; it should be capable of testing compounds, quickly and all year round, for activity on a representative range of pests and pathogens of plants and animals.

Synthetic pesticides

Synthetic pesticides are manufactured as technical grade materials by specialized chemical factories. The manufacturing process yields the intended **active ingredient** (A.I.) together with associated impurities. **Technical grade material** (T.C.) may also contain small amounts of necessary additives. Technical grade materials are manufactured in various forms such as large crystals, lumps, flakes, viscous oils and so on.

However, if technical grade materials were used in such concentrated forms, it would be difficult to obtain an even distribution of the chemical over the area of application, which is essential for effective pest control. Therefore, raw pesticides must be processed into usable forms, usually called formulations, which can be distributed or sold to applicators.

Formulation is the processing of a compound by any method which will improve its properties of storage, handling, application, effectiveness and safety.

Technical material may be formulated by the basic manufacturer or sold to a formulator. The formulated material may be sold under the formulator's brand name or it may be custom-formulated for another company.

Because it takes only very small quantities of a highly concentrated active ingredient spread evenly over a target area to kill pests, technical grade materials are diluted to a formulation by blending with additives.

There are different types of pesticidal formulations to suit particular needs, but their number is more limited than that of the many types of manufacturing processes of technical grade material. Basically, the physical and chemical characteristics of the technical materials -- whether they are solids, liquids or gases as well as their volatility, stability to light and air, solubility in water and oil and so on -- largely determine which formulations are practical and which are not. These characteristics must be dealt with technically and economically since the final formulated product must meet regulatory specifications and standards of acceptibility to the user.

Furthermore, the requirements at the users' end, the biological efficiency and application means available may restrict the choice of formulation. Frequently, formulation is carried out locally in small units,

This pile of battered and corroded pesticide drums bear witness to the harsh transport and storage conditions in developing countries. As mechanical aids for lifting are rarely available, small-size packaging material should be employed wherever possible.

particularly if the pesticides need to be transported over great distances from the manufacturing sites. It may be more economical to obtain diluents from nearby sources, and a local formulator may be in a better position to supply according to customer preferences and local needs.

Another important task of the formulator is to choose the correct kind of packaging material and to use containers of the type and size which suit the requirements of the customers. Owing to the corrosiveness of many pesticides and their general hazardous nature, the selection of the container material and types should always be left to experts in this field. The difficult transport and storage conditions in many developing countries must also be taken into account.

Commercially, the industry is characterized by a highly sophisticated distribution and sales organization. This structure is dictated by the complicated and hazardous chemical nature of pesticides, the wide range of products offered, the worldwide pesticide market and its value of over $US 20 billion in 1988, which is expected to rise even further. The pesticide trade employs a highly trained sales force which must deal with a very large number of technical customers and the very varied complex of pests and diseases that threatens agricultural and animal production.

Biological pesticides

A new line in the pesticide industry is the production of **biological pesticides** for use in the integrated control of animal pests, pathogens and weeds.

The manufacture of biological pesticides applies the techniques of industrial microbiology used in the pharmaceutical (biotechnical) industry. For example, in the case of the bacterium *Bacillus thuringiensis* (see Chapter 2), the manufacturing process is aimed at the production of a "spore- crystal complex" which consitutes the active ingredient of the formulated biological insecticide.

Its manufacture entails the preparation of an inoculum with spores from a selected strain of absolute bacteriological purity. This inoculum is introduced into a sterilized fermentation medium in a seed-tank. The fermentation medium contains the essential proteins, sugars and mineral salts necessary for multiplication. Actual production takes place in large fermenter vessels under controlled conditions and passes through three stages: the spore germination stage, the exponential growth stage which is characterized by very rapid multiplication of the bacteria, and the sporulation stage in which the endospores and crystals are found.

The sporulation stage ends with the disintegration of the bacterial cells and the release of spores and crystals into the medium. With sophisticated recovery techniques, the spores and crystals are separated from the fermentation medium as the active ingredient, which then is further formulated to produce a wettable powder (WP) or flowable concentrate (FC).

The newest line in the production of pest control agents is the rearing of live natural enemies for the

The Spidex ^R *container holds live predatory mites for release in glasshouse cultures against noxious spider mites. Aerated packaging and rapid transport are essential for trade in live biological agents.*

incipient integrated pest control market. This activity is undertaken by a few newly formed, highly specialized enterprises without links to the pesticide industry or it may be a component of an internationally sponsored integrated control programme. Examples of natural enemies produced by such biological companies are predatory Trichogrammatid wasps, Ichneumonid flies and Tetranichid mites. Other biological products exist such as *Verticillium lecanii,* a spore-suspension of fungi. A particular feature of this trade in biological agents is that it deals with live organisms which lose their activity in a relative short time. Thus, formulation, packaging and speed of delivery to the applicator are of crucial importance.

Figure 3.1: Manufacture of a microbial insecticide

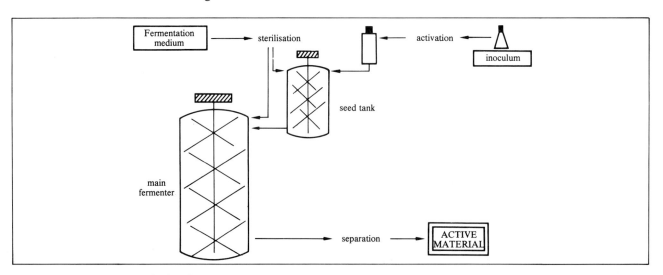

The manufacture of B. thuringiensis *includes: a. the seeding of B.t. spores in an inoculum; b. the preparation of a fermentation medium from proteins, sugars and salts; c. the implantation of the bacterial population in a seed tank; and d. the main fermenter tank in which sporulation takes place. The sporulation phase ends with the rupture of the bacterial cells and release of spores and toxic crystals into the medium. Finally, the spores and crystals are isolated from the medium and serve as the active ingredient base for different formulations to be used by growers.*

Formulation components

Formulations are prepared in a formulation factory, which is rather simple and much less complicated than a technical ingredient factory. Typically such a factory would have equipment for mixing powders and liquids as well as mills and sieves for screening. In addition, there may be granulating equipment consisting of a mixer, dryer and grader, as well as some kind of filling station.

Dilution of technical material to produce a formulated pesticide can be accomplished in several ways. Generally, the following ingredients are used in the formulation process:

* Technical material (active ingredient)
* Solvents
* Carriers
* Surface active agents
* Special additives

The active ingredient

The active ingredient (AI), as defined by its common name, is the most important ingredient in a formulation. The properties of this chemical have a profound influence on the choice of the other ingredients. Relevant properties are its physical phase, melting or boiling point, solubility, stability, toxicity and vapour pressure.

A technical material (TC), whether in solution or diluted with solid adjuvants to achieve an intermediate form for use only in the preparation of formulations, is termed a **technical concentrate (TK)**.

Solvents

The choice of a solvent for liquid pesticide formulations is based on the following considerations:

* Solubility of the active ingredient
* Phytotoxicity of the solvents used
* Toxicology and inflammability of the solvents (safety aspects)
* Volatility of the solvent (this determines the method of application)
* Cost of the solvent

An important property of solvents is their miscibility in water; they are divided into two categories accordingly:

* Solvents not miscible in water: these are used in emulsifiable concentrates. A good and safe solvent is xylene; it is the solvent of choice in many emulsifiable concentrates.

* Solvents miscible in water: these are used in solution concentrates. Examples are isopropanol and the glycolethers.

Carriers

The term carrier is generally used to denote the inert solid ingredients used to dilute the technical material, usually in dry formulations such as dusts, dispersible powders and granules. A typical carrier is pyrophyllite, an inert clay.

Surface active agents

Surface active agents (also called surfactants), for example soaps or detergents, are an important group of chemicals to which belong compounds such as emulsifying agents or emulsifiers, wetting agents or wetters, dispersing agents or dispersants, foaming agents and spreading agents.

A solution of surface active agents in water differs from pure water in several ways. Important for agricultural applications are the lower surface tension, the enhanced wetting power and the dispersing properties of these solutions.

Surface active agents have a characteristic structure which gives them unique properties. Their chain-like molecules have a part which is attracted to water (hydrophilic) and is water soluble; they also have a part that repulses water (hydrophobic) and is soluble only in organic solvents. In a watery solution, these molecules collect at the surface while pushing their hydrophobic tail-portion out of the water into the air. If the surfactant concentration is high, these molecules will cover the entire surface and influence its properties. Increasing the surface area of such a solution is much easier than

Figure 3.2: Surfactant molecules in water

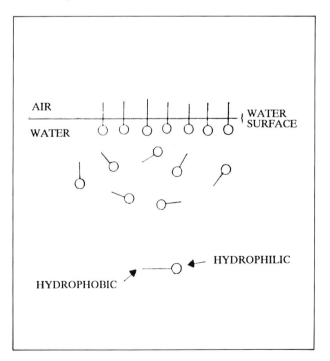

The apolar portion of each surfactant molecule forces it to the surface of a water solution thereby changing the characteristics of the surface.

57

inceasing that of pure water. In other words, its surface tension is lowered.

The hydrophobic tails at the surface of the solution are the hydrocarbon groups in the surfactant molecules. Owing to their presence, the surface of the solution resembles the surface of a hydrocarbon or oil. Because the surface of a leaf is often covered with a layer of wax to minimize transpiration, the wetting of a waxy leaf cuticle is easier to achieve with an oil than with pure water. This is the reason why a surfactant solution, with its oil-like surface, wets a leaf better than water does. Adding a surface active agent to a formulation thus improves its wetting power in solution.

These two aspects, lower surface tension and improved wetting power, play an important role in agricultural spraying. They have a profound influence on the size of the spray droplets and on the spreading, retention and run-off of the spray liquid on the target surface.

A third characteristic of a surfactant solution is its dispersing property, which is related to the same phenomenon that causes the improved wetting power. Because water and oil are immiscible, when these substances are mixed by stirring or shaking, the droplets of oil which float in the water quickly separate into a layer of oil and a layer of water. When oil and water are shaken together after an emulsifier has been added as a surfactant, the molecules of this surfactant will collect around the droplets of oil. The hydrophobic tails of these molecules then penetrate into the oil droplets, whereas their hydrophilic heads remain in the water portion of the solution.

Figure 3.3: Drawing of oil droplet in water

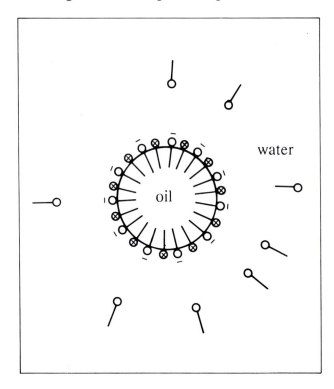

This congregation of surfactant molecules, in the so-called interface between water and oil, causes a more-or-less stable situation in which the oil droplets remain dispersed in the water giving the water a milky appearance. This is called a **dispersion** because the oil remains present as microscopic-sized droplets.

The dispersion of a liquid in a liquid is called an **emulsion**. An oil-in-water emulsion (EW) is the most common; fine globules of an organic solvent containing the technical ingredient are dispersed in water. Rare are water-in-oil emulsions (EO) in which fine water droplets containing the technical ingredient are dispersed in an organic liquid -- a so-called **invert emulsion.**

In a **solution,** no separate droplets can be found because the dissolving process is complete at the molecular level.

Wettable powders can be formulated as water dispersible granules (WG). A WG formulation offers the advantage that the granules can be mixed directly with water without any problem from dust or lumping of powder. (Photo courtesy of Ciba-Geigy)

A **suspension** is the dispersion of a wettable powder (WP), which is also called a water-dispersible powder, in a liquid. In this case, the hydrophobic tails of the surfactant molecules are adsorbed to the surface of the floating solid particles, while their hydrophilic heads "anchor" them to the surrounding water. In this way the solid particles remain dispersed in the water and sedimentation is prevented or considerably delayed.

Surface active agents are very important in pesticide formulations. Surfactants also have an influence on the biological activity of pesticides; the penetration and translocation of active ingredients can be enhanced by the presence of surface active molecules.

Special additives

These additives are used in a formulation or added as tank-mixes for special purposes. Examples:

* Stabilizers

Active sites on the surface of carriers may be responsible for the deactivation of the active ingredient. If it is not possible to change the carrier, then stabilizers or deactivators can be used to prevent deactivation. They are sprayed on the carrier to neutralize the adsorbtive spots before applying the active ingredient. Glycolethers and polyglycols are often used for this purpose.

In the case of a liquid formulation, the rate of deterioration can be slowed by adding catalytic amounts of metal ions. A chelating agent to catch and hold the ions may then be used as a stabilizer.

* Synergists

A synergist is a substance, such as a pesticide, that in combination with one or more other substances shows an increased activity.

Synergists, such as piperonyl butoxide, are commonly used with the pyrethrins and pyrethroids. Synergists block the detoxification mechanism in insects, thus enhancing the insecticidal activity of a formulation.

* Wetters

When a spray liquid does not have enough wetting power to wet crops with very waxy leaf cuticles such as cabbage, an additional wetting agent may be added. In some instances, liquid dishwashing detergent may be used for this purpose.

* Oils

Oils may sometimes be added to the spray liquid to enhance the biological activity of the pesticide. These oils are formulated as emulsifiable oils. Several types, from paraffinic to aromatic, are used. There may be different modes of action: inhibition of evaporation loss, inhibition of wash-off or assistance in penetration. Phytotoxicity may occur depending on the type and amount of oil as well as the prevailing weather conditions.

* Defoamers

In general, formulations should not foam. But the use of inappropriate spraying equipment or a highly concentrated spray liquid can cause excessive foaming. To prevent foaming, an anti-foam agent or defoamer may be added. These are usually silicone-based materials. As little as a few tenths of a per cent generally is sufficient to obtain the desired effect.

* Thickeners

These compounds, which are not frequently used, modify the viscosity of a spray liquid. They prevent the formation of very small droplets during spraying and the evaporation of small droplets afterwards. In this way they act as anti-drift agents. Another reason for their use is that their enhanced viscosity inhibits run-off of the spray from the leaves.

* Colouring agents

Colouring agents are used to reduce the possibility of accidents. Formulations used for seed dressing are often coloured with water-insoluble dyes to distinguish treated (coloured) seed from untreated seed.

Granules are sometimes coloured to make them visible on the soil. In this way the evenness of an application can be checked.

Coding system for pesticide formulations

The considerable increase in the worldwide manufacture, formulation and use of pesticides and the corresponding increase in the types of formulation being made available in the 1970s called for some sort of harmonized system for the designation of formulations. In 1978, a worldwide coding system was prepared under the co-ordination of the International Association of Pesticide Manufacturers (GIFAP); the system was last revised in 1989.

This coding system consists of two letters for each formulation type. Where possible, codes which were already widely used in some countries, for example EC for emulsifiable concentrate or WP for wettable powder, were maintained. Attempts were made to keep the coding system easy and self-evident. The system is a compromise between an expression of the physical status of a formulation and an indication of its intended use.

The coding system is meant to be internationally understood irrespective of the language in which it is being used and to replace the many different abbreviations being used in documentation. FAO and CIPAC (Collaborative International Pesticide Analytical

Figure 3.4: Effect of a wetting agent

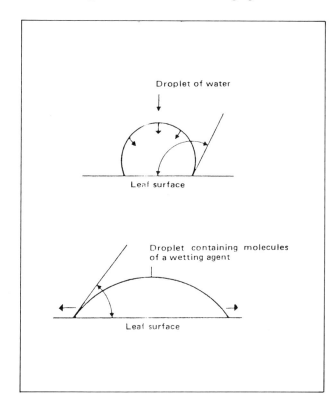

59

Council) have already adopted the two-letter code as have some national registration authorities. The code and the technical definitions given in GIFAP technical monograph No.2 (revised edition, 1989), are given in table 3.1. and further explained below.

Table 3.1: GIFAP's international coding system for some major pesticide formulation types

Group 1: Concentrates for dilution with water:
*DC - Dispersible concentrate
*EC - Emulsifiable concentrate
*SC - Suspension concentrate (flowable concentrate)
CG - Encapsulated granule
CS - Capsule suspension
*SL - Soluble concentrate
*SP - Water soluble powder
*SG - Water soluble granules
*WP -- Wettable powder
*WG-- Water dispersible granules

Group 2: Concentrates for dilution with organic solvents:
*OL - Oil miscible liquid
*OF - Oil miscible flowable concentrate (oil miscible suspension)
*OP - Oil dispersible powder

Group 3: Formulations to be applied undiluted:
*GR - Granules
*DP - Dustable powder (dusts)
*UL - Ultra low volume (ULV) liquid
*ED - Electrochargeable liquid

Group 4: Miscellaneous formulations for special purposes:
*RB - Bait (ready for use)
*GE - Gas generating product
*FU - Smoke generator
*HN - Hot fogging concentrate
*KN - Cold fogging concentrate
*AE - Aerosol dispenser
-- Controlled release formulations, such as briquett (BR), encapsulated granule (CG)

Note: * indicates basic formulation in each category. The table is not complete; a total of 71 codes have been assigned.

PESTICIDE FORMULATIONS AND THEIR CODES

Group 1: Concentrates for dilution with water

Emulsifiable concentrate (*EC)

An emulsifiable concentrate is a solution of active ingredient in non-water miscible solvents. It is a fluid, homogeneous formulation, usually containing 15 to 50 per cent of the toxic ingredient, that can be diluted with water to form a solution in the spray tank. Since the concentrate is not miscible in water, an emulsifier must be added. The concentrate should then emulsify spontaneously or with only little agitation. Then water is added to increase the spray volume needed for good coverage of the target area. The small droplets of solvent in the water emulsion contain the active ingredient in a dissolved state. These are the particles that remain as a residual deposit on the target once the water has evaporated.

Suspension concentrate (*SC)

This formulation, which is also termed a flowable concentrate, is a stable suspension of active ingredient in a fluid intended for dilution with water prior to its use. This formulation type has been ingeniously developed for active ingredients that are soluble in neither oil nor water. The technical material is blended with a solid carrier, such as inert clay, and a small quantity of water in a mixing mill to form a concentrated "pudding-like" suspension. Some wetting and dispersing agents and special additives may be added during the formulation process. Prior to application it has to be further diluted with water so that it can be sprayed.

There are advantages to using a suspension concentrate, which is a liquid formulation, over a solid, powdery formulation. Liquid formulations are easier to handle; for example, they can be poured, they are easy to measure and they give no dust problems.

Like all suspensions, sedimentation of the solid materials may occur in flowables. There is no problem as long as this sediment is easily redispersed; however, sometimes a sticky solid layer is formed. One way to prevent sedimentation is to use special additives, mostly swelling clays. They are used to build a skeleton-like structure, a three-dimensional network in the suspension, on which the particles rest. The suspension has the appearance of a stiff paste with a high viscosity that prevents sedimentation.

Agitation, in the form of stirring or shaking, destroys the structure and the suspension behaves like a liquid and is easy to pour. This phenomenon is reversible; after some time, the structure and viscosity will be restored.

When a container with suspension concentrate is not tightly closed, crusts may form on the walls or a solid skin may form on the surface. Again, this is no problem as long as the crusts are easily redispersed; otherwise, blockage of filters and nozzles may occur. Special additives have to be added to prevent crust formation.

Other forms of flowables are: the blending of finely ground technical grade material with molasses (carbaryl) or with latex (Sevin) as a binding or sticking agent which extends the residual activity of the pesticide.

A stable suspension of capsules in a fluid (normally intended for dilution with water before use) is the definition for a new formulation type developed in the early 1970s by Pennwalt from its patented micro-encapsulation process.

Encapsulated granule (CG)

A formulation type of granule with a protective and release-controlling coating.

Micro-encapsulation is a technique used to surround microscopic droplets of an active ingredient with a thin, cross-linked polymeric (plastic) skin that forms the capsule wall. The size of the capsule depends on the technical material to be encapsulated and on its use. In a typical formulation, the average particle size is about 20 to 30 microns. These capsules are mixed with wetting agents, thickeners, and a small amount of water to produce a flowable suspension of microcapsules in water. Such a capsule suspension is further diluted with water to a solution that can be used in conventional spray equipment.

After application, water surrounding the microcapsules, which initially inhibits the escape of the active ingredient, evaporates. The active ingredient contained inside the capsule then diffuses through the capsule wall covering the outer surface with a thin, concentrated film of pesticide. As this outer layer of pesticide dissipates, additional active ingredient diffuses through the wall to replace it. The result of this slow diffusion process is that the micro-encapsulation process extends the effective, residual life of the active ingredient typically two to three times that of a non-encapsulated formulation.

An additional advantage of the micro-encapsulation process is that the gradual release of the toxic material considerably decreases acute oral and dermal toxicity to mammals by means of diminished exposure. Similarly, a reduction in phytotoxicity is observed, as most of the active ingredient remains inside the capsules and the large amounts of aromatic solvents found in emulsifiable concentrates are eliminated. Micro-encapsulation also frequently reduces incompatibility problems found with conventional pesticide formulations.

Examples of this type of formulation are Penncap-M, a stable suspension of capsules containing

Because of their minuscule size, microcapsules may become lodged between body hairs of insects thus causing their continuous exposure to the toxicant. Social insects may carry this deadly load into their nest and poison fellow insects. (Photo courtesy of Pennwalt Corporation)

methylparathion in a fluid, intended for dilution with water before use in field crops. This new formulation type, which was developed by Pennwalt in the early 1970s from its patented micro-encapsulation process, is called capsule suspension and coded CS. Another Pennwalt capsule formulation is Knox out 2FM, a suspension of micro-capsules containing diazinon for household pest control.

Soluble concentrate (*SL)

After dilution in water, this liquid formulation is applied as a true solution of the active ingredient. The technical material may be water-miscible initially or it may be alcohol-miscible. The solvent of the concentrate dissolves in large amounts of water, leaving the solution clear and not milky as in the case of suspension concentrates. Wetting agents are added to the concentrate to give the spray liquid sufficient wetting power. Examples of such formulations are mono-crotophos and phosphamidon.

A small number of pesticides are soluble in water in such concentrations that it is possible to commercialize a concentrated aqueous solution as a solution concentrate. The active ingredient must be chemically

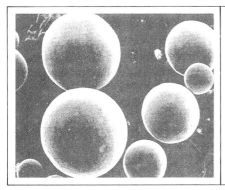 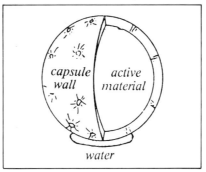 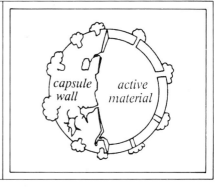

Microcapsules (magnified 100x) which contain the active ingredient are suspended in water. After application, the capsule sticks to the target surface and the water surrounding the microcapsule evaporates. The insecticide is then progressively released over a long period by filtration through the porous capsule wall made of nylon polymer. (Photo courtesy of Pennwalt Corporation)

stable in water. Examples of this type of formulation are the solutions of salts of the phenoxy herbicides.

Water soluble powder (*SP)
Water soluble granules (*SG)

A formulation consisting of either powder (SP) or granules (SG), to be applied as a true solution of the active ingredient after dissolving in water, may contain insoluble inert ingredients.

The technical material is a finely ground water-soluble solid and requires nothing else to assist its solution in water. These formulations dissolve immediately when they are added to a partially filled spray-tank; they neither settle out nor require frequent agitation as they are true solutions and form no precipitate.

These formulations are convenient to store and transport and usually contain a high concentration of the active ingredient. The development of bags made of water-soluble polymeric material allows for safer handling of pesticides because such dissolving bags containing water-soluble powder or granules may be dropped unopened into the spray-tank. Examples of this type of formulation are the herbicides dalapon (Dowpon) and amino triazole (amitrole) and the insecticide acephate (Orthene).

Wettable powder (*WP)

This type of formulation, also termed water-dispersible powder, is applied as a suspension after its dispersion in water. The technical ingredient is finely groundin a high-speed hammer mill, ball mill or "fluid energy mill". In the latter, the premixed, roughly ground material is passed through a small chamber in which air is circulated at very high velocity. The particles are shattered by the impact and centrifugal forces; this "micronizing" process yields a very fine powder. It is a widely used process.

Usually an inert diluent, such as finely ground talc or clay, is blended with the sometimes sticky technical material to prevent it from fusing together. The wetting and dispersing agents which it is necessary to add tend to increase the problem of stickiness. Thus, wettable powders are essentially dusts containing high concentrations -- 25 to 50 per cent -- of active ingredient with added wetting agents to facilitate the initial mixing of this fine powder with water before spraying.

When fine powders are put on a water surface, they are, as a rule, not easily wetted and will float on the surface. This is a general property of finely divided materials, but in the case of pesticides, it is often complicated by the hydrophobic nature of the active ingredient.

A good water-dispersible powder has a short wetting time and is wetted as individual particles and not as small lumps or flocks; it also would show no agglomeration or flocculation. However, as no suspension is stable, the high concentration of clay or talc particles cause rapid sedimentation. Sedimentation always occurs and it is only a matter of time before the solid particles separate from the liquid. Therefore, the spray liquid should be prepared just before application. In large spray-tanks, some agitation (pumping, stirring) is needed during the spraying operation.

In large sprayers with adequate mechanical or hydraulic agitation, the spray liquid is generally prepared by adding the water-dispersible powder to the partly filled spray-tank during the filling operation. A good quality wettable powder will give no problem, but a bad quality powder or an old, caked powder may contain small lumps which are difficult to break up.

In this case and always when using a backpack or compression sprayer with no mechanical agitation, it is advisable to stir the powder with a sufficient amount of water in a bucket, adding the water to the powder while stirring, to prepare a well-wetted paste (about the consistency of paint) without lumps. This concentrate is then poured into a partly filled tank.

Water-dispersible granules (*WG)

This type of formulation consists of granules and is applied after disintegration and dispersion in water. It is essentially a wettable powder granulated as small pellets, which offers several advantages. Such formulations are dustless and require no pre-mixing as they disintegrate rapidly and completely. They also do not form into lumps during storage.

Note: The water-dispersible granules formulation is sometimes referred to as a "dry flowable"; however, the use of this term is not recommended.

Dispersible concentrate (*DC)

This is a liquid homogeneous formulation applied as a solid dispersion after dilution in water.

Group 2: Concentrates for dilution with organic solvents

Oil concentrates of pesticides contain the technical grade material dissolved in an organic solvent; they are to be further diluted with oil before application. They may also be sold in a dilute, ready-to-use form. In either case, the compound is dissolved in oil and is applied as an oil spray, without emulsifier or wetting agent. Best known are the ready-to-use household insecticide sprays in small bottles or plastic containers equipped with a spray atomizer.

GIFAP's international coding system is used to described the oil-miscible formulations mentioned under this heading:

Oil miscible liquid (*OL)

A liquid formulation applied as a homogeneous liquid after dilution in an organic solvent.

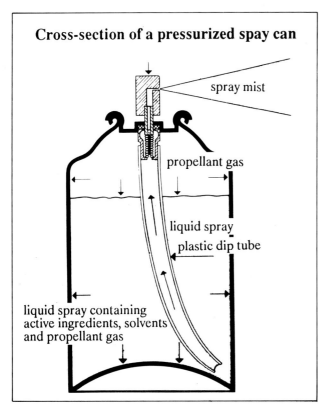

Cross-section of a pressurized spay can

spray mist

propellant gas

liquid spray

plastic dip tube

liquid spray containing active ingredients, solvents and propellant gas

The can should be kept in an upright position while spraying. If the can is kept upside down, the gas collects at the bottom of the can and escapes by way of the plastic tube and valve. (Photo courtesy of PD, the Netherlands)

Oil miscible flowable concentrate (*OF)
A stable suspension of active ingredients in an oil intended for dilution in an organic solvent before use.

Oil dispersible powder (*OP)
A powder formulation applied as a suspension after dispersion in an organic solvent.

Group 3: Formulations to be applied undiluted

Granules (*GR)
Granules are defined as a free flowing, ready-to-use, solid product of a defined granule size.

Granular formulations are usually made by spraying a liquid concentrate of the technical grade material onto small pellets of some absorptive material. Ground clay formed into granules by different techniques or uniform chips of broken bricks and coarse grains of sand are commonly used.

Granules are made and graded into various size ranges to suit particular purposes; standard ranges of size are:

* Macro granule (GG): particle size 2,000-6,000 micrometres

* Fine granule (FG): particle size 300 - 2,500 micrometres

* Micro granule (MG): particle size 100-600 micrometres

(1,000 micrometres = 1 millimetre)

Size ranges may also be expressed in "mesh", which refers to the number of grids per inch of screen through which they will pass. Only insecticides and herbicides are formulated as granules and the concentration of active ingredient generally does not surpass 10 per cent.

The particles of a granular formulation are small enough for a homogeneous distribution over a specific area, but large enough to prevent problems associated with drift and lodging in the vegetation. The granules themselves are too big for direct action; the active ingredient must be released from the granule by diffusion or by disintegration of the particle.

The application of granules is relatively safe and simple as long as no dust is present. Therefore, a good granular formulation is dust-free and resistant to attrition during transport and application. Good flowability of the granules and a constant bulk density are important factors if mechanical application equipment is used.

These types of formulation can be used without application equipment and at virtually any time of the day, since they can be applied even in strong winds without the problem of drift. They can also be drilled into the soil at planting time to protect roots against insects or to improve uptake of a systemic pesticide by the roots.

A disadvantage of the granular formulation is the high cost of manufacturing, handling, storing and transporting materials of low concentration.

Dustable powder (dust) (*DP)
This type of formulation is a ready-to-use, free-flowing powder, generally with a low concentration of active ingredient, and suitable for dusting.

Dustable powders are prepared by mixing and grinding the technical grade material together with an inert diluent such as pyrophyllite clay. If the technical grade material is a liquid, it has to be absorbed by the diluent before mixing and grinding.

The earliest pesticides were formulated either as undiluted dusts of sulfur for use in agriculture or of boric acid for the control of cockroaches. Only synthetic insecticides and fungicides are formulated as diluted dust, but not herbicides because of the hazards associated with drift.

A flo-dust (GP) is a very fine dustable powder for pneumatic application in greenhouses. Another type is aerosol dust, which contains ground silica as the diluent and a liquified gas as propellant. It is sold in aerosol cannisters for the control of indoor pests hiding in crevices.

During application, small particles are very susceptible to wind, which carries the particles away from the target area. This implies a loss of material and consequently a low rate of deposit on the target. Drift also increases the hazard of contaminating the applicator and adjacent areas.

These problems, as well as the low concentration of active ingredient and the concomitant high storage and transport costs, are the main disadvantages of this type of formulation.

Thus, despite their ease in handling and application, dusts are recognized as the least effective formulation and they are used only on a small scale.

Ultra low volume liquid (*UL)
A homogeneous liquid concentrate ready-to-use through ultra-low- volume (ULV) equipment.

This type of formulation consists usually of the technical grade product in its original liquid form, or if a solid, the technical material dissolved in a small quantity of solvent. Consequently, these formulations commonly have a very high concentration of technical grade material, i.e. 95 to 100 per cent.

An ultra-low-volume suspension (*SU) is a formulation of a technical material dispersed into a low-volatility organic solvent to form a suspension ready for use through ULV equipment.

Generally, they are applied by special aerial or ground spray equipment that generates a very fine spectrum of spray droplets, and limits the volume of application from 0.5 to 5.0 litres per hectare.

A critical property for determining whether or not organic solvents are suitable for the preparation of a UL formulation relates to their vapour pressure or volatility. Reduction of application volumes to rates below 5 litres per hectare, whilst attaining even and effective deposits, is achieved by breaking up the spray into ever smaller droplets. However, reduction of droplet size is accompanied by an exponential expansion of the droplet surface area at which evaporation takes place. ULV sprays, containing very many small droplets less than 50 um in diameter, are subject to severe loss from drift, even at low wind speeds.

Synthetic pyrethroid insecticides, which are applied at 20-50 grams of active ingredient per hectare, are also formulated as ultra-low-volume liquids at a much higher dilution rate to make up for the required application volume. Cottonseed oil or soybean oil may be used as a diluent.

Electrochargeable liquid (*ED)
This is a special liquid formulation for electrostatic (electrodynamic) spraying. Usually, these are oil-based formulations of very low volatility. Electrostatic spraying

has been developed for aerial and ground applications. The best known electrostatic atomizer is the Electrodyn sprayer, which produces droplets within a size range of 40-200 microns in diameter. The liquid pesticide is sold in a special bottle fitted with a particular nozzle. The combination of bottle and nozzle -- called bozzle -- is screwed directly onto the sprayer.

The electrodynamic sprayer, which uses electrostatic forces to create spray droplets, requires special oil-based ultra low volume formulations packed in a ready-to-use container known as a "bozzle". (Photo by the author)

Group 4: Miscellaneous formulations for special purposes

Bait (*RB)
Baits are formulations designed to attract and to be eaten by the target pest. Usually, the pesticide is first formulated as a solid or liquid bait concentrate (CB), intended for dilution with food or other edible substance attractive to the pest, and then mixed as a ready-to-use bait (RB) with an active ingredient concentration of less than 5 per cent. Such baits may be sold in special forms, such as granular bait (GB), plate bait (PB) and block bait (BB).

Another form of baiting involves the addition to a spray solution of insect-attracting chemicals. This practice enables the spraying of only a small portion of a crop instead of the entire leaf surfaces. For example, treating only one branch per tree of citrus and mango orchards reduces costs and labour enormously, while giving acceptable control of the Mediterranean fruit fly (*Ceratitis capitata*).

Gas generating products (*GE)
This is a type of formulation which generates gas by chemical reaction. Aluminium phosphide formulations in tablet form, which release the highly toxic phosphine

gas at moisture levels above 10 per cent, are a good example. The term fumigant is used for a chemical which, at the required temperature and pressure, can exist in the gaseous state in sufficient concentration to be lethal to a given pest organism. A gas (*GA), such as methyl bromide, is stored in pressure bottles or pressure tanks. Thus, the term fumigant can be used for a gas as well as for a gas-generating product.

Smoke generator (*FU)
This type of formulation is combustible: generally a solid, it releases the active ingredient(s) in the form of smoke upon ignition. Special forms of smoke generators are smoke tin (FO), smoke candle (FK), smoke cartridge (FP), smoke rodlets (FR), smoke tablet (FT) and smoke pellets (FW). They may be used for space treatment in greenhouses and warehouses against flying insects.

The terminology used for smoke generation, fumigation and fogging partly overlaps as in all these cases the toxic ingredients are released in extremely fine particles or molecular form, which for safety reasons should be handled as gases.

Hot fogging concentrate (*HN)
This term refers to a formulation suitable for application by hot fogging equipment; a similar concentrate for cold fogging (KN) has also been developed. Fogging is the process of disintegration of the pesticide liquid into very fine droplets by the impact of a high-speed airstream (see Chapter 10).

The pulse jet machine "Swingfog" breaks the chemical solution into small particles which it discharges as a dense fog. Certain types of fogger require special formulations. (Photo by the author)

Aerosol dispenser (*AE)
This is a container from which the active ingredient(s) is (are) dispersed, generally by a propellant gas upon the actuation of a valve, as fine droplets or particles. The technical grade material is dissolved in a volatile, petroleum solvent; it must remain soluble under pressure. The propellant gas used to be a fluorcarbon, which for ecological reasons is being replaced by carbon dioxide. When the formulation is atomized, the petroleum solvent evaporates rapidly, leaving the toxicant particles suspended in the air. Aerosols commonly produce droplets well below 10 micrometres in diameter. Because they are respirable (can be inhaled into the lungs), aerosols should not be breathed.

Usually, the valve of an aerosol container can be operated intermittently. Total release aerosols are available too, which discharge their entire contents in a single application when the valve is locked in a depressed position.

Controlled release formulations
In this type of formulation, the technical grade pesticide is incorporated within a permeable covering, through which the active ingredient slowly diffuses to the outer surface and dissipates. The first form appeared in 1963, i.e. the Shell Vapona Strip, a vapour-releasing product (*VP) that contains dichlorvos incorporated into panels of polychlorovinyl resin from which it is slowly released, thus providing a more or less constant, though low, concentration of dichlorvos vapour in enclosed spaces. The strips are suspendend in houses and storehouses to control flying insects.

A briquette (*BR) is a solid block designed for the controlled release of the active ingredient into water.

A newer slow-release formulation has been developed, i.e. the Pennwalt micro-encapsulation process, by which very fine droplets of active ingredient are covered with a very thin polymeric skin. The insecticide permeates through that skin to the outer surface where it volatizes. The diffusion and evaporation proceeds at a slow but constant pace, until the microcapsule has become empty. (The principle has been described previously in detail in the section on "encapsulated granule" formulations.) Since 1984, synthetic pyrethroids such as permethrin have also been formulated as microcapsules.

Microcapsules can be mixed with certain paints to provide residual efficacy against flies and mosquitos in houses and stables.

In controlled-release formulations, the pesticide is not offered all at once, but continuously. The pesticide is formulated in such a way that on the one hand there is a slow, continuous decomposition of the pesticide; on the other hand, there is a continuous release of pesticide. In an ideal situation, these counteracting mechanisms would balance the release of the active ingredient just above the minimum concentration required for control.

However, such a situation is very difficult to reach. Most controlled-release formulations show a curve like the one in the drawing above. Compared with the decomposition curve of a "normal" formulation, the curve of a slow-release formulation is much less steep. This means that controlled release formulations provide improved efficiency and longer residual action. However, controlled-release formulations are, in general, rather expensive.

Formulations for seed treatment

Formulations designed for seed treatment are designated by the use of an "S" as the second letter of the code. Examples:

DS - powder for dry seed treatment
ES - emulsion for seed treatment
SS - water soluble powder for seed treatment
WS - water dispersible powder for slurry treatment

Product combinations

In some cases products have been brought onto the market which contain two different formulations, packed in different containers in a common outer pack or in different compartments of one pack (combi-packs or twin-packs), which are intended for simultaneous application in a tank mix. In these cases, it is recommended that the codes for both formulations, separated by an oblique stroke, be used, e.g. EC/SC.

However, if a documentation sysytem does not allow five letters for the formulation code, the following two-letter codes should be used.

KK - combi-pack solid/liquid
KL - combi-pack liquid/liquid
KP - combi-pack solid/solid

References:

Green, M.B., G.S. Hartley and T.F. West, (1979). *Chemicals for crop protection and pest control,* Pergamon Press Ltd, Oxford.

International Group of National Associations of Manufacturers of Agrochemical Products (GIFAP) (1989). Technical monograph No. 2 (rev. edition): *Catalogue of pesticide formulation types and international coding system.* GIFAP, Ave Albert Lancaster 79a, 1180 Brussels, Belgium.

Pennwalt Corporation (1986). *Brochure: micro-encapsulation.* Philadelphia, PA 19102, USA.

Ware, George W. (1986). *Fundamentals of Pesticides; a self-instruction guide,* 2nd Edition.

4

PESTICIDE TOXICITY AND RESIDUES

Contents
Page

Pesticide toxicity 68

- Toxicity and hazard 68

- Determination of toxicity 68

- Testing requirements 71

- Tabulation of toxicity categories 72

- Toxicity of some chemical groups 72

- Toxicity of rodenticides 77

Restrictions for pesticide use 77

- Re-entry interval 78

- Hazards to bees, birds, wildlife and fish 78

Pesticide residues 79

- Residues in crops and produce 79

- Non-pesticide residues in food and feed 80

- Pesticide residue analysis 80

- Residues in soil and water 81

References 86

Pesticide Toxicity and Residues

PESTICIDE TOXICITY

Toxicity and hazard

Many pesticides are like two-edged swords; they are not only toxic to the intended target organism, but they are also toxic or hazardous to a certain degree to man, animals and the environment. One needs to differentiate clearly between toxicity and hazard.

Toxicity refers to the inherent toxicity of a compound to different forms of life. Hazard refers to the risk or danger of poisoning when a chemical is used or applied.

To the user of a pesticide, the use hazard or risk of exposure to a toxic chemical is of more immediate concern than its inherent toxicity. In order to establish a safe procedure for using a pesticide, one needs to pay as much attention to ways of minimizing exposure as to the actual toxicity of the pesticide and of its metabolites.

Ways and means of limiting exposure are discussed in the sections of this manual on formulation, safe handling, protective apparel, disposal and application.

The combined effect of toxicity and exposure may result in the poisoning of users, as well as animals and segments of the environment. Referring to man and test animals, a distinction is usually made between two types of poisoning: first, acute poisoning resulting from a single exposure or intake; and second, chronic poisoning following repeated or long-term exposure or absorption of smaller amounts of toxicants. The risk of acute poisoning is greatest for people engaged professionally in the manufacture, formulation, distribution and application of pesticides. Chronic risks, however, are a matter of much public concern in terms of long-term consumer and environmental safety.

In 1988, the global turn-over in pesticides amounted to $US 20 billion. Only 20 per cent of the total world pesticide consumption of approximately 3 million metric tonnes -- i.e. 600,000 tonnes of formulated product annually -- is used in developing countries (Mowburry, 1988). However, about half the poisoning cases and nearly three-quarters of the deaths are estimated to occur in developing countries. The International Organisation of Consumer Unions put the figure for 1986 at 375,000 human poisoning cases in developing countries of whom 10,000 died. A safe estimate is that about 20 per cent of all farmers in developing countries suffer from pesticide intoxication at least once in their working life (Kenmore, 1990). (See table 4.1.) Assuming an even higher number of intentional poisonings by suicide, these figures are a cause of grave concern. As a comparison, the statistics for 1983 of the National Poison Control Center in the United States record only 27 deaths from pesticide poisoning although the consumption of pesticides in that country is the highest in the world.

Such figures indicate that safety records for pesticide usage can be substantially improved by investment in regulatory control, education, skill training, better application techniques and promotion of integrated pest control methods. Much attention must also be directed towards strengthening the medical infrastructure, towards the teaching of symptom recognition and treatment, and towards properly equipping rural poisoning treatment centres.

Determination of toxicity

Pesticides differ considerably between each other in degree of toxicity to mammals and to other organisms. Where a small amount of a certain pesticide would already suffice to provoke severe illness or even death, larger quantities of another compound may cause hardly any effect at all, even if ingested. A pesticide may also demonstrate a selective toxicity, causing destruction of a group of organisms, whereas others would not be affected or perhaps much less affected by it.

An effective application requires skilled and well-protected spraymen, adequate equipment and timing as well as suitable formulations that spare natural enemies and the environment. (Courtesy of FAO IPC Rice Programme)

Focusing on mammalian toxicity, one can distinguish the following factors that influence the degree of poisoning:

* toxicity of the pesticide;
* the dose of the pesticide, particularly its concentration;
* the route of entry or absorption by the body; and
* the duration of exposure.

*Acute toxicity is determined by standardized tests, in which toxicologists administer accurately measured doses of the chemical to test animals under laboratory conditions. An important factor is the channel or route through which pesticides enter the body. They can enter the body through the mouth (oral route), through the skin (dermal), or through the lungs (respiratory, inhalation). Acute toxicity -- either oral or dermal -- is an indication of the hazards to humans and animals.

The units of measurement are milligrams (mg) of active ingredient (AI) per kilogram (kg) of body weight of the test animals (usually rats, rabbits, mice and guinea pigs). These animals are reared strictly under standardized conditions.

The accepted method of recording the relative toxicity of a pesticide is to give the median lethal dose (LD$_{50}$) value, which is the chemical dose needed to kill 50 per cent of a group of test animals of one species under stated conditions. Mortality counts are usually taken 24 and 48 hours after exposure to the pesticide concerned.

Because a dose or dosage indicates the quantity of an insecticide etc., applied per individual or per unit area, volume or weight, the median lethal dose is expressed as:

LD$_{50}$ value: milligrams of active ingredient per kilogram of body weight of the test animal (mg/kg).

Table 4.1: Percentage of pesticide users affected by unintentional poisoning from pesticides, estimated on the basis of a study in South-east Asia (Jeyaratnam *et al.*, 1987)

Country	Percentage of users ever poisoned	Yearly percentage of users poisoned
Indonesia	13.8	0.3
Malaysia	14.5	7.3
Sri Lanka	11.9	7.1
Thailand	19.4	n.a.

Table 4.2: Median lethal doses for test animals compared with probable fatal doses for humans (oral ingestion)

Oral LD$_{50}$ (in mg/kg)	Estimated fatal dose for an adult human
less than 5	a few drops of liquid or a grain of solid
5 - 50	a pinch to 1 teaspoon (5 ml)
50 - 500	1 teaspoon (5 ml) to 2 tablespoons
500 - 5000	30 ml to 470 ml (30 ml = 1 liquid ounce US) (1 ounce to 1 pint)
5000 - 15000	470 ml to 950 ml (1 pint to 1 quart US)

Adequate protection of workers is essential at ingredient blending stations of a formulation factory as well as at any other place where toxic dusts and vapours can be expected. (Courtesy of GIFAP)

The size of the dose is the most important single factor in determining the relative danger of a given chemical. The smaller the LD$_{50}$ value, the more toxic the compound.

Similarly, the relative toxity may also be expressed as the median lethal concentration (LC$_{50}$) value, which is the concentration of a toxicant required to kill 50 per cent of a large group of individuals of one species. It is an important concept for expressing the toxicity of a pesticide to fish when that pesticide is applied to water or pollutes surface water.

LC$_{50}$ value: milligrams of active ingredient per litre of water (mg/litre).

*The amount of toxicant required to kill a target organism depends on the weight, age and sex of the organism; also, physiological races of the same species may react very differently to the same treatment. For this reason, the LD$_{50}$ or LC$_{50}$ values developed for one species only begin to inspire confidence after numerous tests have been conducted by many researchers under varying conditions.

Because pesticides are selective in their action, different species of animals react differently. The LD$_{50}$ values for rats or beagle dogs may have little bearing on the value for humans. Nevertheless, LD$_{50}$ values are very useful in classifying pesticides according to their toxicity, so long as it is recognized that the values are not absolute and should be used only as a guide to relative toxicity. It is wise to assume that humans are at least as sensitive to these chemicals as the most sensitive animals used in the tests.

*The route of absorption is another important variable. The LD$_{50}$ is measured as oral, if the toxicant is fed to, or placed directly by tube into the stomach of, rats or dogs. When the pesticide is tested by applying it to the shaved skin of (usually) rabbits and guinea pigs, the result is referred to as the dermal LD$_{50}$. Inhalation studies are commonly performed on rats, rabbits and guinea pigs, and are referred to as respiratory LD$_{50}$.

Generally, oral ingestion is more toxic than respiratory inhalation (with the exception of lethal gases), which is more toxic than dermal absorption. With professional workers, however, contamination through dermal exposure occurs more often than ingestion. Therefore, dermal toxicity information is an important item in determining workers' safety.

*The duration of exposure is the fourth factor; it should always be kept as short as possible. Immediate cleaning of contaminated part(s) of the body by washing is one of the precautions to be taken, as is seeking urgent medical treatment.

Pesticide toxicity, as described in standard reference books, usually refers to the toxicity of the active ingredient (AI) or technical material (TC). However, for practical purposes, the pesticide user is more interested in knowing the toxicity of the particular formulation he is using.

Many pesticide manufacturers give the toxicity of the formulation as well as that of the active ingredient in technical leaflets. But where such information is not given, an indication of the approximate toxicity of the formulation can be obtained from applying the following formula, which is based on the concentration of active ingredient in the formulation:

Toxicity of formulation =

$$\frac{\text{LD}_{50} \text{ of active ingredient} \times 100}{\% \text{ active ingredient in formulation}}$$

Example: Technical grade carbofuran is a crystalline solid having an acute oral toxicity of 8 mg/kg (rats) and an acute dermal toxicity of 2,250 mg/kg (rabbits).

For carbofuran 5% granules (Furadan 5GR), the oral toxicity of the formulation would be:

$$\text{LD}_{50} \text{ (oral)} = \frac{8 \times 100}{5} = 160 \text{ mg/kg}$$

Table 4.3: EPA labelling toxicity categories by hazard indicator

Hazard indicator	Toxicity categories			
	I	II	III	IV
Oral LD$_{50}$ (mg/kg)	< 50	50-500	500-5,000	> 5,000
Dermal LD$_{50}$ (mg/kg)	< 200	200-2,000	2,000-20,000	> 20,000
Inhalation LD$_{50}$ (mg/kg)	< 0.2	0.2-2	2-20	> 20
Skin effects (at 72 hours)	corrosive	severe irritation	moderate irritation	slight irritation
Eye effects *				
- irritation:	corrosive	persisting	reversible	none
- corneal opacity:	not reversible	reversible	none	none
"Signal word"	DANGER/POISON	WARNING	CAUTION	CAUTION

* *Note*: Reversibility of corneal opacity and persistency of irriation are both measured over a period of seven days after contamination.

Thus, carbofuran 5% GR would be classified as very toxic (category II) and the technical material, extremely toxic (category I). (See table 4.3)

In determining toxicity, not only is the concentration of active ingredient in a formulation important, but also the type of formulation. Liquids are more toxic than solid chemicals because they enable more rapid absorption by the body.

For a given pesticide, powder and granular formulations are generally safer to use than spray formulations. By contrast, aerosols and fine powders are readily inhaled and can be very dangerous. This is why aerosols and dust formulations contain only a very low percentage of active ingredient.

For example, extremely toxic pesticides, such as aldicarb (LD$_{50}$ = 0.93 mg/kg, oral rats) and carbofuran (LD$_{50}$ = 8 mg/kg, oral rats), are best formulated as granules of low concentration.

Frequently, the hazard based on oral toxicity will be very different from a pesticide's hazard based on dermal toxicity. In the above example, even low concentration carbofuran 5% granules are still classified as very toxic in terms of oral toxicity (category II), but are practically non-toxic (category IV) in terms of dermal toxicity:

$$\text{Dermal LD}_{50} = \frac{2250 \times 100}{5} = 45,000 \text{ mg/kg}$$

As a general guide, the probable lethal dose for an adult human weighing 80 kg may be derived from the acute oral LD$_{50}$ (technical grade) for test animals.

Testing requirements

Before a pesticide formulation can be registered for use, an array of toxicity studies and hazard evaluations have to be carried out by the manufacturer. Requirements for **hazard studies** include the following general categories of tests on animals and in the environment:

Acute oral, dermal and inhalation toxicity

Eye and skin irritation

Neurotoxicity (effects on the nervous system)

Reproduction

Teratogenicity (birth defects)

Mutagenicity (genetic effects)

Dermal sensitization

Subchronic feeding (90 days -- two species)

Chronic feeding (two years -- carcinogenicity/cancer -- two rodent species)

Chronic feeding (one year -- one non-rodent species)

Effects on wildlife/fish

Fate of the environment (residual persistency).

Toxicity tests usually are even more demanding and require highly sophisticated scientific laboratory facilities. The very high cost of operating laboratories for such purposes or contracting outside ones contributes heavily to the development costs of new formulations. Toxicity studies required for obtaining registration of pesticides for use in food crops include:

- Acute effect studies evaluating toxicity effects from

single, short-duration exposure following oral, dermal and respiratory contamination; recording degree of eye and skin irritation, and dermal sensitization.

- Subchronic and chronic effect studies evaluating toxicity effects from multiple or continuous long-term (up to two years) exposure.

- Reproductive effect studies evaluating possible damage to reproduction and fetal development.

- Mutagenic effect studies evaluating potential structural or functional impairment of genetic material, such as DNA, and chromosome damage.

Tabulation of toxicity categories

Pesticides are grouped according to their LD_{50} values into various classes or categories of toxicity or hazard. Unfortunately, little uniformity can be detected in the literature regarding the way ratings and terms are combined with certain ranges of LD_{50} values.

The Environmental Protection Agency (EPA) of the United States has designed a table in which four toxicity categories are established for specific ranges of oral, dermal and respiratory $LD_{50}s$ and for symptoms of damage to the eye and skin (see table 4.3). Each category is also represented by a "signal word" indicating its potential hazard. The use of these signal words is obligatory on American pesticide labels. All labels must further bear the statement "Keep Out of Reach of Children".

A similar British table (O.D.N.R.I, 1980) quotes different LD_{50} values for solids and liquids, indicating that liquid technical materials are much more hazardous than solid ones by the oral as well as by the dermal routes of absorption.

A third tabulation gives six toxicity ratings and is more discriminating in the use of the terms "moderately"

and "slightly" toxic than others (see table 4.4). The values quoted for dermal LD_{50} are those obtained with rabbits as the test animal.

It is not clear if the EPA tabulation states dermal $LD_{50}s$ for rats or rabbits, although, usually, there is a great difference in dose response between these species.

Toxicity of some chemical groups

Organophosphates

The organophosphate group, which contains the largest number of insecticides, shows a diversity of biological activity and persistence. The group comprises insecticides with a broad range of toxicity, ranging from extremely toxic active ingredients such as disulfoton and parathion to the relatively safe malathion; from the rapidly degrading dichlorvos and mevinphos to the more persistent compounds such as diazinon and azinphos-methyl. Some have a systemic action on plants (dimethoate, mevinphos) and others are animal systemic insecticides effectively controlling internal and external parasites in mammals (e.g. coumaphos, fenchlorvos).

The organophosphates are nerve poisons which kill by inhibiting the action of certain enzymes in the nervous system. These enzymes are called choline-sterases. (See explanation at end of Chapter 5).

In general, the organophosphate compounds result in acute toxic reactions. Once the victim has been removed from further exposure to the compound, recovery is often rapid. There are usually no serious after-effects, provided that renewed exposure is avoided.

Because of their polar nature, organophosphates are not stored in body-fat, and they are readily excreted in the urine. Thus, mild exposures at infrequent intervals usually do not result in cumulative effects within the body. Nevertheless, there is always a danger from repeated light exposures to these types of pesticides, which may lower

Table 4.4: Combined tabulation of toxicity classes

| Toxicity rating | Commonly used term | Routes of absorption | |
		LD_{50} Single oral dose (rats mg/kg)	LD_{50} Single dermal dose (rabbits mg/kg)
6	Supertoxic	< 5	< 20
5	Extremely toxic	5-50 mg	20-200
4	Very toxic	50-500	200-1,000
3	Moderately toxic	500-5,000	1,000-2,000
2	Slightly toxic	5,000-15,000	2,000-20,000
1	Practically non-toxic	> 15,000	> 20,000

Source: Toxicity ratings modified from Gleason, Gosselin and Hodge. (1976). Clinical Toxicology of Commercial Products, 4th ed. (Baltimore, Williams and Wilkins Cy)

Notes: < means LD_{50} is less than the figure shown.
> means LD_{50} is higher than the figure shown.

the acetyl cholinesterase level without producing serious symptoms, or even any symptoms at all. This is believed to be a frequent cause of accidents.

LD50 values quoted here and elsewhere are taken from The Pesticide Manual, (6th ed.), 1980, The British Crop Protection Council.

In table 4.5 below, organophosphate (technical grade) pesticides are rated according to their oral toxicity in rats, and according to their dermal toxicity in rats, rabbits or mice. In many tests, female rats appear to be more sensitive than males, though not always. Consequently, test results normally form a range of LD50 values: in the tables below only the lower values are quoted.

Carbamates

Carbamate compounds are esters of carbamic acid. Carbamates resemble the organophosphates in biological action. They also inhibit the working of cholinesterase enzymes in the nervous system. However, because of their polar molecular structure, carbamates in

Table 4.5: Toxicity rating of various organophosphates according to their oral toxicity in rats and according to their dermal toxicity in (a) rats, (b) rabbits or (c) mice

	Oral LD_{50}	Dermal LD_{50}
Extremely hazardous		
(Category I: LD_{50} oral -- up to 50 mg/kg)		
azinphos-methyl (Guthion)	16	-
carbophenothion (Trithion)	32	1,270 b
demeton (Systox)	3	-
dicrotophos (Bidrin, Carbicron)	13	148 a
dioxathion (Delnav, Navadel)	23	63 a
disulfoton (Disyston)	3	20 a
ethion (Bovinox)	24	915 b
fensulfothion (Dasanit)	5	4 a
methamidiphos (Monitor, Tamaron)	30	50 a
mevinphos (Phosdrin)	3	1 a
monocrotophos (Azodrin, Nuvacron)	14	336 b
parathion (E-605, Folidol)	4	7 a
parathion-methyl (Folidol-M)	14	67 a
phorate (Thimet)	2	2 a
phosphamidon (Dimecron)	17	374 a
Highly hazardous		
(Category II: LD_{50} oral -- from 50 through 500 mg/kg)		
chlorpyrifos (Dursban)	135	2,000 b
demeton-S-methyl (Metasystox)	57	302 a
diazinon (Basudin)	300	2,150 a
dichlorvos (Dedevap, Vapona)	56	75 a
dimethoate (Cygon)	320	< 130 b
fenithrothion (Sumithion)	250	> 3,000 c
fenthion (Lebaycid, Queletox)	190	330 a
naled (Dibrom)	430	1,100 b
triazophos (Hostathion)	66	1,100 a
Moderately hazardous		
(Category III: LD_{50} oral -- from 500 through 5000 mg/kg)		
crufomate (Ruelene)	770	-
fenchlorphos = ronnel (Korlan)	1,740	2,000 a
malathion	2,800	4,100 b
tetrachlorvinphos (Gardona)	4,000	> 2,500 b

general are rapidly metabolized in the body and excreted in the urine. Like the organophosphates, carbamates do not accumulate in the fatty tissues of the body. Carbamates generally present less serious problems than the organophosphates. But repeated exposure to carbamates may lower the acetyl cholinesterase level to a point where renewed exposure may trigger acute symptoms of poisoning. Provided that renewed exposure is avoided, recovery from carbamate poisoning is usually rapid.

Carbamate insecticides, such as aldicarb and carbofuran, are persistent, systemic insecticides with high oral, dermal and inhalation toxicity. Because of their hazards, they are formulated only as a dust-free type of granule or as a flowable (suspension concentrate) formulation. In the class of carbamate chemicals, the insecticides are the most toxic; the herbicides, less toxic; and fungicides, the least toxic. The carbamates listed as examples in table 4.6 below have solid crystalline active ingredients, except the herbicides thiobencarb and tri-allate, which are liquids.

Organochlorines

Pesticides of the chlorinated hydrocarbon group generally and cyclodiens, such as aldrin and endosulfan, particularly have a high oral toxicity and long residual action. When dosages are light, even in the case of frequent or continuous exposure, these pesticides often do not cause visible symptoms of poisoning, but their residues accumulate in the fatty tissues of the body.

In the food chain, from lower forms of plant and animal life to large predators, this phenomenon may cause a serious build up in concentration, particularly in the case of DDT. The characteristics of persistency, low solubility in water and wide spectrum of insecticidal activity have already provoked legislation in many countries that restricts their use.

Some examples: Endosulfan and toxaphene should not be used on edible crops within 30 days of harvesting; endrin may be used on sugarcane provided a 45-day interval is observed before harvest. These pesticides are recommended more readily for use on cotton. Formulations of aldrin, chlordane and heptachlor are restricted or banned in many countries. During the 1985-1988 outbreak of the desert locust, *Schistocerca gregaria*, the use of dieldrin for their control was not acceptable to several donor countries. The use of DDT is often reserved for malaria control. Where its use is allowed for agriculture, it is in selected applications.

The organochlorines are axonic poisons (see the next section under pyrethroids). They generally act to destabilize the central nervous system. Although many of these insecticides or their degraded products can be

Table 4.6: Carbamates rated according to their oral toxicity (LD$_{50}$) in rats and dermal toxicity in (a) rats, (b) rabbits and (c) mice

Pesticide	Type* mg/kg	Oral LD$_{50}$ mg/kg	Dermal LD$_{50}$	Hazard/Category
aldicarb	I	0.7	5 b	
carbofuran	I	8	2,550 b	Extreme
methomyl	I	17	4,080 b	Category I
oxamyl	I	5	710 b	
promecarb	I	61	>1,000 a	
BPMC	I	410	4,200 c	
dioxacarb	I	60	1,950 b	High
isoprocarb (MICP)	I	403	>500 a	Category II
methiocarb	I	100	350 a	
propoxur	I	90	800 a	
carbaryl	I	850	>2,000 b	
asulam	H	>4,000	>1,200 a	Moderate
barban	H	1,300	>1,600 a	Category III
thiobencarb	H	1,300	2,900 a	
tri-allate	H	1,675	2,225 a	
chlorpropham	H	>5,000	no effect	
phenmedipham	H	>8,000	>4,000 a	Slight
carbendazim	F	>15,000	>2,000 a	Category IV
propamocarb	F	8,600	-	

Note: I = Insecticide, H = Herbicide, F = Fungicide

Table 4.7: Organochlorine insecticides rated according to their oral toxicity in (a) rats or (b) rabbits

Insecticides	Oral LD$_{50}$ (mg/kg)	Dermal LD$_{50}$ (mg/kg)	Hazard/ Category
endrin	7.5	15 a	Extreme/
dieldrin	46	10 a	Category I
aldrin	67	-	High/
comphechlor (toxaphene)	80	780 a	Category II
chlordane	457	-	
DDT	113	2,510 a	
endosulfan	80	359 b	
gamma-HCH (lindane)	88	900 a	
heptachlor	100	195 a	
pentachlorphenol	210	-	
lorobenzilate	700	-	Moderate
dicofol	668	1,870 b	Category III

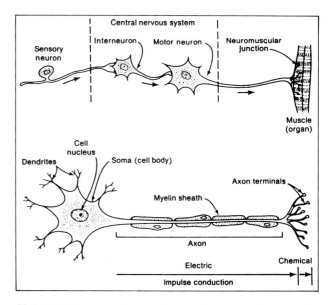

Uninhibited conduction of nerve impulses between the sensory and motoric neuron cells of the human nervous system is essential for proper functioning of the body; certain types of pesticides block the transmission of those impulses. (Courtesy of G. Ware)

stored in body-fat, they apparently cause no adverse effect so long as they remain in the fat. But if released as a consequence of fat depletion, they can be a source of poisoning, causing symptoms similar to those following heavy exposure. Rapid loss of weight is therefore a serious matter for persons contaminated by organochlorines. Deposits in the body-fat otherwise disappear only after a long period of time when further exposure has ceased. The organochlorines mentioned in table 4.7 all have solid crystalline or waxy active ingredients, except chlordane which is a viscous liquid.

Pyrethroids

These compounds have a synthetic structure resembling the natural pyrethrins, with which they share the advantages of high biological activity against insects and, generally, a lower mammalian toxicity than other insecticide groups.

Pyrethrum and synthetic pyrethroids are axonic poisons. The axon of a nerve cell is a long extension of the cell body and it is vital in the transmission of electrical nerve impulses from one nerve cell to another. Axonic chemicals affect the peripheral and central nervous system in such a way as to prevent normal transmission of nerve impulses in mammals and insects. Eventually, the neurons fire impulses spontaneously, causing the muscles to twitch, the repeated discharge results in muscle spasms and paralysis and finally death.

All pyrethroids share this mode of toxic action, which resembles that of organochlorine insecticides. Fortunately, the toxicity of pyrethroids is far more pronounced against insects than against man, but usual

safety precautions need to be taken. However, fish toxicity is generally very high; for example, the LC$_{50}$ of tralomethrin (96 hours of exposure) = 0.0016 mg TC/litre of water for rainbow trout; i.e. a concentration of 1.6 parts

Table 4.8: Pyrethroids rated according to their oral toxicity in rats and to their dermal toxicity in (a) rats, (b) rabbits or (c) mice

Insecticide	Oral LD$_{50}$ (mg/kg)	Dermal LD$_{50}$ (mg/kg)	Hazard/ Category
bifenthrin	55	>2,000	High
bioallethrin	425	1,545 b	Category I
cyhalothrin[1]	56	632 a	
cypermethrin	303	>2,400 b	
deltamethrin	135	2,000 b	
fenpropathrin	67	870 a	
fenvalerate	300	>2,000 b	
flucythrinate	81	>1,000 b	
fluvalinate	260	>20,000 a,b	
tralomethrin[2]	99-3,000	2,000 a	
allethrin	930	-	Moderate
cyfluthrin	500	>500 a	Category III
phenothrin	>500	-	
pyrethrins (botanical)	580	>1,500 a	
resmethrin	2,000	3,000 a	
tetramethrin	4,640	>15,000 c	
bioresmethrin	7,070	-	Slight Category IV

Notes: 1. The cis-isomere configuration of cyhalothrin "Karate";
2. The range in oral toxicity, i.e. 99 to 3,000 mg/kg (rats) for tralomethrin depends on the solvent or carrier used.

of technical tralomethrin per billion parts of water (1.6 ppb).

Because all synthetic pyrethroids are esters, they can be readily inactivated through ester cleavage. In mammals, they are generally eliminated rapidly by metabolic processes and excreted. In the soil, bio-degradation by soil micro-organisms diminishes residual action considerably.

The lipophylic character of these compounds causes strong adsorbtion to soil particles, precluding the possibility of leaching. Thus, this property and the fact that only low doses of active ingredient per hectare (10 to 25g AI/ha) are required for pest control in crops mean that, under normal field use, pyrethroids cause little harm to fish.

Pyrethrum and pyrethroids have a greater insecticidal effect when the temperature is lowered; this is called a **negative temperature coefficient**. DDT shows a similar effect.

Dithiocarbamates

Zinc, manganese and iron salts of dithio-carbamates are broad spectrum fungicides. They have been in great demand in agriculture for almost half a century. Nickel salts are used to control bacterial diseases in rice. Thiram, also named thiuram, is a protective fungicide and used for seed dressing.

Generally, this group is characterized by a low human and animal toxicity and its use poses little hazard. However, thiram (thiuram) and methan-sodium have a relatively high oral toxicity; thus these fungicides must be handled carefully.

Organotins

The group of triphenyl-tin compounds is very toxic on the whole, having an oral LD_{50} of less than 540 mg/kg (rats) and a dermal LD_{50} of less than 2,000 mg/kg (rabbits). The non-systemic fungicides fentin acetate, fentin chloride and fentin hydroxide and the insecticide cyhexatin belong to this group. The related insecticide fenbutatin-oxide is moderately toxic.

Table 4.9: Dithiocarbamate fungicides rated according to their oral toxicity in rats

Fungicide	Oral LD_{50} (mg/kg)	Category
ferbam (iron-salts)	17,000	IV
maneb (manganous salts)	7,990	IV
zineb (zinc salts)	5,200	IV
thiram (thiuram)	780	II
methan-sodium	820	II

Dinitrophenols

The pesticides in the dinitrophenol group often have multi- purpose uses as insecticides (I), miticides (M), fungicides (F) and herbicides (H). These are very toxic chemicals if swallowed. Similarly, absorption through the lungs and skin poses grave hazards, in particular for workers in hot environments. Like other phenols, they are toxic to the liver, kidneys and nervous system. Dinitro-phenols are likely to be banned in the near future.

Table 4.10: Dinitrophenols rated according to their oral toxicity in (a) rats, (b) rabbits or (c) guinea pigs

Pesticide	Type	Oral LD_{50} (mg/kg)	Dermal LD_{50} (mg/kg)
dinobuton	F, M	140 a	> 5,000 a
dinocap	F, M	980 a	-
dinoterp	H	25 a	150 c
DNOC	I, H	25 a	-

Note: See top of column for abbreviations

Chlorinated phenols

The pesticides in the chlorinated phenols and substituted aromatics groups comprise mostly protective fungicides (F), which may be used for seed dressing. These compounds are less hazardous than other pesticides and are generally rapidly metabolized in test rats and excreted in the urine. Pentachlorophenol (PCP) is a herbicide (H) and termiticide (T), with very high toxicity, that causes severe irritation of the mucous membranes and eyes.

Table 4.11: Chlorinated phenols rated according to their oral toxicity in rats (a) and rabbits (b)

Pesticide	Type	Oral LD_{50} (mg/kg)	Dermal LD_{50} (mg/kg)
chloroneb	F	> 5,000 a	> 5,000 b
chlorothalonil	F	> 10,000 a	> 10,000 b
dicloran	F	1,500 a	-
pentachlorophenol	H, T	210 a	-

Note: See top of column for abbreviations

Phenoxyalkanoic compounds

The chlorophenoxy compounds, which comprise acids, esters and salts, have a low toxicity for most individuals. The compounds are mildly to severely irritating to skin, eyes, respiratory and gastrointestinal linings. They are not stored in body-fat and are rapidly excreted. However, several cases of affliction of the nervous system have been reported in workers after seemingly minor exposure to 2,4-D.

Almost all are broad-leaf weed herbicides, which are widely used for selective weeding in rice and other grain crops. Well-known herbicides of this group are 2,4-D, MCPB, MCPA, 2,4,5-T and MCPA- thioethyl (Herbit). However, 2,4,5-T may contain traces of highly poisonous dioxine.

Bipyridyliums

The two important herbicides in this group are diquat (Reglone) and paraquat (Gramoxone). Both are desiccants which can be used for either complete weed control or as preharvest aids in some crops. They are very toxic to mammals when ingested and great care should be exercised in the handling and storage of the concentrates. No antidote is known for diquat and paraquat when swallowed. Both compounds are inactivated rapidly upon contact with the soil.

Organo mercury compounds

Organo mercury salts are powerful fungicides and desinfectants, which inhibit the growth of many bacteria and fungi. Therefore, they were used for certain purposes such as seed dressing notwithstanding their high acute oral and dermal toxicity to mammals, and accumulation in the environment. However, all organic and inorganic mecurial fungicides have lost their registrations and should no longer be used for any purpose.

Organic solvents

Petroleum solvents such as diesel fuel, deodorized kerosene and various petroleum distillates pose a respiratory risk. Upon inhalation they may provoke symptoms of pneumonia. Dermal toxicity is generally low. Aromatic solvents such as xylene and toluene offer much greater dermal hazard and are considered to be carcinogenic. Alcohol and esters such as methanol and isopropanol should likewise be considered poisons and handled with care.

Antibiotics

The antibiotic compounds, which are produced by bacteria and fungi, are to be approached with much care, as some are highly toxic substances. In particular, cyclohexamide (Acti-dione) and blasticidin-S are highly hazardous upon ingestion.

Toxicity of rodenticides

Anti-coagulant rodenticides

Anti-coagulants of the coumarin type (warfarin, coumafuryl etc.) and of the 1,3-indane-dione type (diphacinone, pindone, valone etc.) have the effect of inhibiting coagulation of the blood, causing rodents to die of internal bleeding. Also, the second generation anti-coagulants (brodifacoum, floccumafen etc.) have a

Table 4.12: Antibiotics rated according their oral and dermal toxicity in rats

Antibiotic	Oral LD$_{50}$ (mg/kg)	Dermal LD$_{50}$ (mg/kg)	Category
blasticidin-S	56	> 500	II
cyclohexamide	2	-	I
kasugamycin	22,000	-	IV
validamycin	> 20,000	-	IV

similar activity, but are generally more toxic. (See also Chapter 12, section on "using rat poisons".)

Anti-coagulant poisons are also hazardous to humans and non-target animals. Normally, these toxicants are applied in the form of low concentration baits to rodents. The risk of exposure and thus of accidental poisoning is thereby considerably lowered, but care in handling and storing should be exercised.

Rodenticides other than anti-coagulants

Arsenic trioxide and zinc phosphide are very poisonous inorganic compounds that kill rats and mice after ingestion of a single dose. Toxicity to man and domestic animals is high. The odour of zinc phosphide is repulsive to most animals but acceptable to rats.

New rodenticides such as alpha chloralose and bromethalin, affect the nervous system and should be regarded as being hazardous for all mammals, including humans.

Red squill, a botanical rodenticide, is less dangerous because it rapidly induces vomiting in all mammals except rats.

Fluoro acetamide (compound 1081) and sodium fluoro acetate (compound 1080) are extremely hazardous poisons for rats as well for other mammals. They should be applied by trained personnel only, wearing adequate safety apparel; they can be applied only in closed off areas such as sewers (see also Chapter 12).

RESTRICTIONS FOR PESTICIDE USE

In ever more countries lists of pesticides are being published officially, the heading of which may read:

"Pesticides that are banned or restricted for agricultural use within the country".

A pesticide is banned when it is not allowed for registration in a certain country or when its registration is officially withdrawn. Banning may be applied to a particular pesticide on grounds of human hazard, toxicity to beneficial insects and natural enemies, hazard to aquatic organisms, persistency in the environment etc.

The term "restricted" has a different meaning under various regulatory systems. Generally, it indicates a prohibition against use in areas of application or for purposes other then specified by the regulations. The pesticide may be restricted in some or in all of its formulation forms.

The U.S. Environmental Protection Agency (EPA), under its registration procedure, makes a distinction between two classes of pesticide: **restricted-use** and **unclassified**. Unclassified pesticides may be purchased and applied by any citizen. Restricted-use pesticides can be purchased and applied only by certified applicators, specially trained professionals who have been tested in the use, handling and application of pesticides.

In view of the highly hazardous nature of most pesticides, such far-reaching regulations and licensing procedures are very desirable. However, very few developing countries have a sufficiently well-developed infrastructure to enforce them. Fortunately, public awareness about the negative effects of pesticide usage is growing fast and should be promoted further by the free international exchange of information.

Re-entry interval

Some insecticides pose such a high hazard that a waiting interval is recommended for the sake of safety between the time of application and when workers are allowed to re-enter treated fields. Such an interval is designed to prevent contamination and dermal exposure. Recommendations concerning the duration of re-entry intervals may be issued by regulatory authorities or by the manufacturer. Examples of the prescribed waiting intervals for re-entry are given in table 4.13.

For all other insecticides, workers need only to wait

Table 4.13: Duration of waiting intervals for re-entry into treated fields prescribed for certain insecticides

48 hours	24 hours
carbo fenothion (Trithion)	acephate (Orthene)
chlorpyrifos (Lorsban)	azinphos-methyl
demeton (Systox)	ethion
dicrotophos (Bidrin)	EPN
endrin	fenvalerate (Pydrin)
monocrotophos (Azodrin)	methomyl (Lannate)
oxydemeton-methyl (Metasystox)	permethrin (Ambush)
parathion	phosalone (Zolone)
parathion-methyl	phosphamidon (Dimecron)
sulprofos (Bolstar)	

until sprays have dried or dusts have settled before entering treated fields again. If re-entry is necessary within the stated 24 or 48 hours, the workers must wear full protective clothing.

In the case of fumigation of food stocks or soil treatment with fumigants and gases, such as methyl bromide and phosgene, other intervals need to be observed before re-entering the premises. Generally, working with gaseous pesticides requires special training and the use of respiratory equipment.

Hazards to bees

Because most insecticides are very poisonous to bees, manufacturers recommend against spraying during the flowering period. If it is essential to do so, applications should be at a time of the day when bees are not very active. Some insecticides which are only slightly hazardous to bees are:

dicofol	endosulfan
prothoate	tetradifon
trichlorfon	white oil

Of course, there is no hazard to bees if the insecticide is applied as a soil treatment, seed dressing or bait.

Hazards to birds and wildlife

Large-scale abatement programmes for the control of desert locust, tse-tse flies, mosquitoes and other migrating pests commonly involve aerial applications over extensive areas. Such repeated applications may adversely affect wildlife, particularly in sensitive areas such as wetlands.

Intensive spraying in irrigation schemes and estates and also more localized pesticide use may likewise cause severe contamination of adjacent lands and surface water. A special hazard is created by burning of toxic waste and packaging materials and disposal of toxic waste. Various aspects of environmental hazard and contamination are discussed in other sections of this manual.

Hazards to fish

Many insecticides can be used safely on rainfed crops, but because of their high toxicity to fish, they should not be used in paddy fields. A good example is endosulfan (Thiodan) which is a very useful cotton insecticide, but is not normally recommended for rice pest control; LC_{50} (96 hours) = 0.002 mg/litre for golden orfe fish.

Synthetic pyrethroid insecticides are safer for fish in the fields, because of the much lower dose of active ingredient required, even if the LC_{50} is similar to that of endosulfan.

Fish cages are placed in flooded rice plots to test the relative fish toxicity of different formulations and dosage rates since fish cultures in paddy rice are important. (Courtesy of Shell Agriculture)

It should be noted that the emulsifier in the formulation may be as toxic or even more toxic to fish than the active ingredient itself. This is one reason why ULV sprays containing up to 100 per cent technical ingredient are preferred over conventional sprays with EC or WP formulations in areas with much surface water.

PESTICIDE RESIDUES

A pesticide residue is that fraction of a pesticide which, as a result of its practical use, has found its way into the produce or soil and is present there either in the form of a parent compound, or significant degradation product(s), or as "bound residue".

Thus, the term residue includes any derivatives of a pesticide, such as conversion products, metabolites and impurities considered to be of toxicological significance. Apart from those stemming from direct application to above-ground and under- ground parts of the crop, residues also include those from unknown or unavoidable sources such as from contaminated water and air pollution.

Most pesticides degrade fairly quickly on the crop and in the soil, but some may persist in one form or another for longer periods of time.

The word **persistent**, meaning remaining in existence, is generally used to describe pesticides that can be biologically or chemically detected long after the application has been made. The optimum duration of biological effects depends on agronomic needs and may vary from a few days to several months. For example, a protectant fungicide should prevent infection by fungi and a grain protectant should protect stored grain against storage pests over an as long as possible period. Further,

a contact herbicide should kill weeds rapidly and enable immediate planting, whereas a residual soil herbicide should control emerging weeds over a long period of time. Ideally, the biological activity should cease once the desired biological (pesticidal) effect has been achieved. When the biological activity continues beyond the desired duration and inflicts noxious effects on the succeeding crop or non-target organisms, this phenomenon is called **biological persistence**.

Residues in crops and produce

Pesticides may remain on a crop for some time after application. This is often desired in order to protect the crop against damaging organisms in the field, during transport or storage. However, it is not desirable if the produce is sold or consumed, as such residues would create a potential source of harm for consumers.

Fortunately, exposure to residues in food appears to be harmful only in extreme cases; if only minute quantities are consumed, there is no danger to health. As a measure of the risk, the concept of acceptable daily intake (ADI) is important. The definition is:

"Acceptable daily intake of a chemical is the daily intake which, during an entire lifetime, appears to be without appreciable risk to the health of the consumer on the basis of all known facts".

ADI is expressed in milligrams of the chemical per kilogram of bodyweight (mg/kg).

Since foods, both raw and processed, and animal feeds are agricultural commodities traded internationally, a need arose for their classification on the basis of their residue potential. Further, uniform procedures for sampling of traded lots, analysing and interpretation of the analysis results had to be developed. Thus, the Joint FAO/WHO Food Standards Programme was launched; it is being implemented by the Joint FAO/WHO Codex Alimentarius Commission to which more than 125 countries subscribe as members.

Its Codex Committee on Pesticide Residues (CCPR) establishes the maximum limits for pesticide residues and environmental contaminants in food and animal feedstuffs moving in international trade. It also monitors the methods of sampling and analysis, and proposes the definitions for relevant terms, such as the following:

Maximum residue limit (MRL) is the maximum concentration for a pesticide residue resulting from the use of a pesticide according to good agricultural practice that is recommended by the Codex Alimentarius Commission to be legally permitted or recognized as acceptable in or on a food, agricultural commodity, or animal feed. The concentration is expressed in milligrams of pesticide residue per kilogram (mg/kg) of the commodity.

MRL values for food or for a particular portion thereof, e.g. extracted vegetable oil, are usually very much lower than those that would be unacceptable from a health viewpoint. An MRL value is also called the **Residue Tolerance Level.**

Good Agricultural Practice in the Use of Pesticides (GAP) is the officially recommended or authorized usage of pesticides under practical conditions at any stage of production, storage, transport, distribution and processing of food, agricultural commodities and animal feed, bearing in mind the variations in requirements within and between regions, which takes into account the minimum quantities necessary to achieve adequate control, applied in a manner so as to leave a residue which is the smallest amount practicable and which is toxicologically acceptable.

In this definition, the "officially recommended or authorized usage of pesticides" is that which complies with the procedures, including formulation, dosage rates, frequency of application and pre-harvest intervals, approved by the national authorities.

Good agriculture practices in the use of pesticides with the aim of minimizing noxious effects on humans, animals and the environment include:

* Choosing the least toxic and least persistent pesticide that will effectively control pests in the field and in storage. As a general rule, persistent and/or cumulative pesticides should not be used on fodder crops.

* Similarly, choosing a formulation which combines maximum efficiency of the selected pesticide with minimum risk.

* Applying on the target area only the minimum amount of pesticide required while determining the number of treatments on a need basis in relation to actual pest infestation.

* Selecting the method of application which offers optimum control with minimum contamination of crops and the environment.

* Timing of the treatment in relation to vulnerable stages of a pest's development.

* The interval between last application and harvest (slaughter in the case of veterinary use) should be as long as possible in order to permit the greatest reduction in pesticide residues, bearing in mind the pest incidence, the degree of control required for maximum utilization of the commodity, and the vulnerability of the treated crop immediately prior to harvest. To this end, official pre-harvest intervals (PHIs) should be established and adhered to.

* Crop rotation should be adjusted in such a manner that residues in the edible parts of crops, as a result of previous treatments (carry-over effect), will be minimal.

Non-pesticide residues in food and feed

The use of pesticides is certainly not the only cause of dangerous substances or residues being present in food and animal feed. There are many harmful substances intrinsic to natural foods. Many plants and their various parts contain numerous glucosides which have pharmacological properties or are poisonous. For example, in a number of bean species (e.g. *Phaseolus lunatus* or lima bean) hydrolysis of a particular glycoside produces hydrocyanic acid (HCN), a well-known poison; this reaction can be prevented by prolonged cooking of the beans.

A very serious residue problem results from the infestation by moulds of agricultural produce which develop mycotoxins. Mycotoxins are poisonous metabolites produced by different varieties of fungi (see also Chapter 15). They bring about abnormal and damaging biological changes in man, animals, plants and bacteria. The acute and chronic health effects of the toxin lead to human ailments, such as liver diseases and to reduced animal performance. Mycotoxin contamination of food is thus a major health problem in warm humid countries where storage facilities are inadequate and insufficient opportunities exist for proper drying of produce.

Pesticide residue analysis

Pesticide residue analysis is a very specialized field of chemistry and physics and an extremely complicated science. This type of analysis requires highly trained and very experienced technicians and expensive laboratory facilities. In residue analysis, the chemists are searching for less than one milligram of an unknown pesticide in one tonne of material (or one part per billion). This is about the limit of detection of the smallest quantity that can currently be measured using gas chromatography or high performance liquid chromatography (HPLC) and the electron capture detector. The whole process of sample preparation and extraction is commodity-specific. The selection of solvents is largely determined by the chemical

This labourer drops with bare hands toxic insecticide granules into the leaf whorl of maize plants for borer control; training in safe handling is often not given to field workers. (Photo by the author)

properties of the pesticide, the clean-up procedure and concentration present, its distribution on or within plants, and its conjugation with glycosides, chlorophyl and other substances in plant systems.

Although the necessity for residue investigations is very obvious, the obstacles in the establishment of a national residue analysis laboratory are equally clear. A residue laboratory requires not only a huge investment in training and facilities, it also requires continuous effort to keep it operational and to maintain an effective network for the collection of samples. For many countries, the best answer for the time being is to direct most attention to keeping residues at a low and safe level through good agricultural practices.

Residues in soil and water

When pesticides are applied in the field, they can move into the soil, air, or water constituents of the environment. Generally, most pesticides will reach the soil in greater or lesser amounts either through application to the soil or indirectly through run-off, impacting spray droplets, or from decaying plant debris. Many pesticides degrade fairly quickly on the crop and in the soil. The active ingredients in pesticides are biologically inactivated in the soil primarily by two processes:

- adsorption to the soil through binding to clay and organic matter; and

- degradation by the countless organisms which digest and decompose organic matter in the soil.

The extent and speed of adsorption and degradation determine whether the pesticide and its residues remain beyond the time needed for controlling the pest organisms. When they remain as more or less intact chemicals for longer periods of time, this is called **chemical persistence**.

It is important to differentiate between the persistence of a biological effect (biological persistence) and the ability to detect the pesticide chemically after long periods (chemical persistence), because the chemical may be in a conjugated or bound form and quite unable to exert its original biological effects.

- Degradation mechanism

For millions of years natural organic substances of plant and animal origin have entered the soil in large quantities and have been degraded primarily by the soil microflora. Basic chemical components derived from decaying plant and animal organisms, which are incorporated into soil organic matter, include sugars, starches, proteins, pectins, celluloses, lignins, waxes, resins and tannins; the stability against degradation increases in this order.

Man-made organic chemicals are subjected to the same principle of transformation in soil as natural substances and are degraded also at different rates. The transformation of organic chemicals can occur by photochemical, chemical or microbiological action. Photochemical (photolytic) reactions occur mainly at or near the soil surface, whilst chemical reactions such as hydrolysis and oxidation are common phenomena in all layers of the soil.

Soil micro-organisms, which are able to degrade most organic chemicals, play the most important role by far in the degradation of plant protection substances. The degradation of these chemicals in soil, which usually involves one or more stages, results in extractable transformation products. These are further partly transformed to bound residues; degradation ultimately leads to complete mineralization of the products and residues.

A bound residue is a chemical that cannot be extracted from the soil with solvents commonly used in residue analysis. Its existence is demonstrated with the help of radioactive labelled pesticides containing for example ^{14}C isotopes. The radioactive material applied to the soil cannot be fully recovered either in gaseous form (e.g. as $^{14}CO_2$) or by extraction with solvents. The radioactivity which remains unaccounted for can usually be detected in the extracted soil only by total combustion of the organic matter or by destroying the chemical structure of the soil by digesting it with strong acids or bases. By comparison, non-ionised pesticides, which are only weakly adsorbed and have not undergone degradation, can be extracted from soil quantitatively with solvents under comparatively mild conditions.

In many cases, pesticides are metabolised to simple organic acids, which are commonly present in natural metabolic processes and can become fixed or "bound" into more complex, less easily degradable humic substances. These unextractable residues, originating either from natural substances such as ^{14}C-labelled cellulose or from a labelled pesticide, have been shown to remain bound to humic soil fractions for years. Final release appears to be in the form of $^{14}CO_2$ after complete mineralization. Thus, it is presumed that under field conditions the slow release of bound residues into the soil solution and their subsequent uptake by plant roots will be of little or no ecological concern.

- Half-life of a pesticide residue

Half-life, as an expression of chemical persistence in soil, refers to the time taken for the concentration of the parent compound to be reduced to half its original value. It thus gives a measure of the persistence of the chemical in that particular soil.

The ability of organic chemicals to be degraded by micro-organisms (biodegradability) or by chemical action varies according to the chemical structure, to the degree of its exposure to micro-organisms, and to environmental factors such as soil temperature and moisture content and climatic conditions.

Studies to evaluate the environmental fate of a pesticide

ENVIRONMENTAL FATE TESTS

DEVELOPMENT STAGE

SCREENING

OPTIMISATION

PHYSICOCHEMICAL PROPERTIES
- water solubility, hydrolysis, photochemical stability, partition coefficients, vapour pressure.

TRACE ANALYTICAL METHODS
- parent compound in soil and water.

DEVELOPMENT

DEGRADATION IN SOIL
- laboratory studies using radiolabelled pesticide.
- factors affecting degradation.

MOVEMENT IN SOIL
- laboratory study, leaching of parent pesticide (soil columns, thin layers).

DEGRADATION IN WATER
- laboratory study using radiolabelled pesticide.

TRACE ANALYTICAL METHODS
- degradation products in soil and water.

FATE IN FIELD SOIL
- field study long-term (three years or more), degradation and leaching.

FATE IN POND SYSTEMS
- fate in model pond or stream system.

COMMERCIAL TRIALS

ENVIRONMENTAL MONITORING
- soil, surface and/or ground water.

Spraying plots where soil persistence will be studied using a special type pressurized sprayer with tail boom for accurate dosaging.

Testing in progress to establish the fate of pesticides in rice paddy water.

(Courtesy of Shell Agriculture)

82

The rate of degradation ranges from rapid, if the chemical is usable immediately as a nutrient or energy source by the micro-organisms, to slow or delayed (lag-phase), if acclimatisation by micro-organisms or certain abiotic processes must take place first. Since most transformation processes take place in the soil-water solution, less strongly adsorbed compounds are more readily degraded than those which are almost completely adsorbed to soil fractions. Unfavourable environmental conditions, such as pH, soil temperature and low soil moisture, may impede the rate of transformation. So may unfavourable microbial conditions of the environment, which may for instance be negatively affected by soil sterilants and fungicides.

The half-life of a particular chemical in the soil will thus be shorter or longer depending on chemical and environmental conditions. Nevertheless, under well-defined conditions, approximate comparisons of half-lives of pesticides have been made and shown to be of value in predicting persistence in a wide range of circumstances, despite the limitations of the concept. Thus, the determination of half-life has become an important criterion in the pesticide registration procedures of many countries.

Experiments to determine the rate of degradation of pesticides in soil, the nature of degradation products and leaching potential are first conducted in the laboratory using agricultural soils taken freshly from the field. Checks should be run on the microbial activity of the tested soils since this activity decreases rapidly in the laboratory. Field studies are necessary to confirm the rate of degradation of longer lasting pesticides which might carry over into the next growing season. Moreover, leaching into the ground water can also be studied in field dissipation studies in which soil samples are analyzed which are taken from a depth of up to two metres. Experience has shown that the routes of degradation found in laboratory and field soils are identical and the same terminal residues are usually formed. However, degradation tends to proceed at a faster pace under field conditions than in laboratory soils.

In such studies, the transformation of a pesticide in soil is commonly measured by marking compounds with radioactive carbon-14 isotopes (^{14}C) and spraying the labelled and non-labelled chemicals onto the surface of field plots from which soil samples are taken from various depths at regular intervals. Degradation is monitored by radioactive assays of the soil and the evolved gases. Good results have been obtained in predicting the rate of transformation under field conditions by processing data from laboratory trials in computer simulation models.

It has been established by research over a long period of time that, by adaptation and rapid multiplication of the responsible soil micro-organisms, the bio-degradation of compounds may be accelerated. Therefore, even those compounds which have a half-life of a year or longer will not endlessly accumulate in the soil. Thus, although the rate of degradation may remain constant, as the residue of a pesticide slowly builds up with repeated application, the actual amount degraded will gradually increase in each successive year. A point will be reached when the amount being degraded per year will be roughly equal to the amount of each new annual application.

Thus, an equilibrium concentration between the amount applied and amount degraded, a so-called **plateau level**, will eventually be reached. The height of the plateau will be determined by the degradation rate of the pesticide and the dose and frequency of application.

- Pesticide persistence in tropical and sub-tropical soils

Within specific countries and climates, very large differences are measured in the degradation rates of pesticides in soil owing to different soil types, adsorptive properties and the extent and activity of the microbial populations. Comparative research on persistence in soils of temperate and tropical zones has revealed even wider variations in rates of degradation. Generally, appreciably greater pesticide loss has been recorded under tropical conditions. This can be explained by examining the various environmental factors which influence persistence. These factors are:

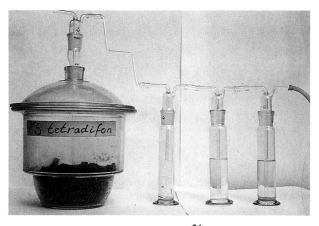

Vessel to study the metabolism of ^{24}C-labeled pesticides in soil. The side-vessel contains KOH-solution to trap possibly formed $^{14}CO_2$. (Photo courtesy of Solvay Duphar)

* Sunlight, particularly ultra-violet light, degrades pesticide deposits on leaf and soil surfaces by photochemical action (photodecomposition). This light is more intense in tropical and sub-tropical zones than in temperate zones and thus leads to greater loss.

* Temperature. The higher temperature in the tropics aggravates pesticide loss through volatilization and through increased microbial activity.

Pesticide disappearance brought about by evaporization, photo decomposition and chemical decomposition is more rapid on the surface of warm, moist agricultural soil than if the chemical is incorporated into the soil. For this reason, slow-action herbicides need to be incorporated into the soil to prevent much of their biological activity from being lost.

* Soil moisture directly affects the proliferation and activity of micro-organisms as well as the adsorption capacity of the soil fractions and the aeration of the soil.

Heavy rain and flooding (as in rice culture) favour the development of anaerobic micro-organisms. Several, if not all, chlorinated hydrocarbon pesticides, particularly the very persistent cyclodienes, are susceptible to anaerobic biodegradation. It is well documented that soil flooding can accelerate the degradation of some of the more persistent pesticides. By contrast, degradation involving chemical ring-cleavage reactions, occurs more readily under aerobic soil conditions.

In many tropical areas characterised by inter-mittent heavy rain and dry seasons, soils are subjected to alternate periods ranging from an arid to a flooded state, with frequent associated changes in the aerobic and anaerobic micro-organisms which predominate. Such alternate oxidation and reduction cycles provide a favourable environment for more destruction of the pesticide than in either one of the cycles alone.

* Microbial activity. The organic-matter content of a soil has a profound influence on the microbial activity and consequently on the persistence of pesticides in the soil.

Most tropical soils have a low organic-matter content, because at the prevailing high temperatures the organic-matter is rapidly mineralised. Application of organic materials, in the form of manure or mulch, increases the microbial activity of the soil and in turn accelerates the degradation of several pesticides.

- **Physical factors influencing persistency and mobility**

Physical properties of a pesticide such as vapour pressure, water solubility and adsorption/desorption characteristics are vital factors influencing its fate in the soil, i.e its mobility and distribution in the three constituents of the environment: soil, water and air. These properties dominate the leaching of pesticides in soil and their transfer from surface water to sediment and suspended matter. Low adsorption leaves more pesticide in the soil solution where it is more readily available either for chemical and microbial degradation or for uptake by target and non target-plants. By contrast, leaching of a substance, which is susceptible to microbial degradation, into deeper soil layers with low microbial activity can increase its persistence.

Concerning mobility of chemicals in the soil, a water solubility of 30 ppm is judged as a threshold enabling significant movement. Compounds with half-lives of more than six months also have, owing to their persistency, more opportunity for movement. This is the case with chlorotriazine herbicides, traces of which are showing up in deep layers of ground water in Western Europe.

- **Field conditions**

The main factors which determine whether or not a chemical will reach and contaminate ground water are the type of soil, and the geological and environmental conditions; these are collectivily called "field conditions".

Field conditions which are associated with ground-water contaminiation are:

* Soil profile. In soil which is well drained, percolation of water usually increases with a high sand content and low organic matter content as well as a high number of pores.

* Percolation and run-off. Leaching of water-soluble chemicals in sandy soils is enhanced by heavy rainfall and excessive irrigation required by soils with low water storage capacity during dry periods.

* Ground-water level. A high ground-water level, particularly if combined with low adsorptive soil, may lead to its contamination.

- **Concern about residues in soil and ground water**

It is important to differentiate between "free" pesticide residue and "bound" or non-extractable residue. As discussed previously, bound residues do not exert any immediate biological effect nor does the prospect of future release of the pesticide from the bound state present a practical problem. However, "free" residues which persist in the soil beyond the period of desired pesticidal activity may present the following problems:

* Free residues, especially those of herbicides, may cause undesirable biological effects such as phytotoxicity on the succeeding -- possibly more susceptible -- crop. Both the yield and quality of the crop may be adversely affected. This does not create a health problem, but should be prevented by providing adequate warnings on the pesticide label. Problems may occur if, after several years of monocropping, a field is planted with a new crop; for exemple, after long-time use of atrazin in

Well-protected spraymen are filling and readying their sprayers, but they are working too close to the water stream in the back that may easily be contaminated. (Photo courtesy of FAO-IPC Rice Programme)

maize, the new crop may suffer from herbicidal activity during the first two years after the last application of atrazin.

* Small quantities of free residues of a pesticide may be taken up from the soil by succeeding crops. This may not create a human or animal health problem, but it could result in illegal residues in a crop destined to be used as food or feed for which no tolerances have been set.

* Free residues may damage or destroy beneficial fauna living on or in the soil which have either predatory functions (beetles, spiders, ants, some species of nematode) or which improve soil structure and fertility (earthworms, bacteria, mycorrhiza). These may in turn be preyed upon by higher organisms and successively accumulate in the food chain. This phenomenon is a problem particularly associated with the more chemically stable and lipophylic insecticides such as aldrin, dieldrin and DDT.

* Long-time presence of chemical residues in the soil may inhibit the degradation of other pesticides. For example, in laboratory tests the transformation of the herbicide profam is slowed by residues of the nematicide ethoprofos.

* If the above-mentioned field conditions apply, free residues may, over a shorter or longer period, seep through and contaminate surface and ground water.

Therefore, if costly aquaculture projects for fish and prawn breeding are to be undertaken in areas of intensive pesticide usage, it is advisable to carry out

studies on the risk of pesticide leaching and contamination of surface and ground water. The risk of pesticide drift during application and direct run-off from sprayed and irrigated fields should also be taken into consideration. Similarly, much care needs to be taken in selecting and using pesticides in or near wetlands and estuaries. Wetlands frequently are a last resort for habitation by endangered species of birds and reptiles, whereas flood deltas and estuaries are the spawning grounds of many species of fish, shellfish and Crustaceae such as crabs, lobsters, shrimps and prawns.

Because there are differences in toxicological behaviour between pesticides of different chemical structure, toxicological risk assesments must be made on an individual case basis. Meanwhile, available evidence indicates that, in the overwhelming majority of situations, existing pesticide residue levels in water should not give rise to toxicological concerns for human.

The distinction needs to be made, of course, between the use of pesticides according to label recommendations and their grossly improper misuse or faulty disposal, which cannot be condoned. In those situations where pesticides are improperly used, or where spillage or improper disposal occurs, the likelihood that pesticides will reach surface or ground water is clearly greater.

Since 1989, dichlobenil herbicide has been manufactured in a new building at Solvay Duphar's factory in Amsterdam. Acquisition of advanced process technology improved the unit's output ratio of technical ingredient to chemical waste from 1:9 to 9:1 and minimized the use of solvents. This is an example of the industry's efforts to comply with strict state regulations concerning effluents.

References:

Guide to Codex recommendations concerning pesticide residues; part 1. Joint FAO/WHO Food Standards.

Jeyaratnam J., Lun K.E., and Phoon W.O. (1981). *Survey of acute pesticide poisoning among agricultural workers in four Asian countries*. Bull of World Health Organization, 65, pp. 521-527.

Overseas Development National Resources Institute (1980). *Training Supplements, ODNRI Pesticide Management Training Course.*

The British Crop Protection Council (1986): *The Pesticide Manual, a World Compendium*, eighth edition.

The persistence of pesticide residues in soil (1987). International Group of National Associations of Manufacturers of Agrichemical Products, GIFAP (BGJ/MS: 21.1.1987).

Ware George W. (1989): *The pesticide book*, 3rd edition, University of Arizona.

5

SYMPTOMS AND TREATMENT OF PESTICIDE POISONING

Contents   Page

How to recognize poisoning 88

Types of poisoning 89

First aid 90

Medical advice 91

Comprehensive advice for physicians 92

Cholinesterase enzyme poisoning 93

References 95

Symptoms and Treatment of Pesticide Poisoning

HOW TO RECOGNIZE POISONING

In many countries, an unacceptably large number of people fall ill or die because of pesticide-related poisoning. The symptoms of poisoning frequently are not recognized however, because of a lack of understanding of the causes and symptoms of poisoning; the result is morbidity and mortality.

Apart from intentional consumption of toxicants in suicide attempts, poisoning can be largely prevented if adequate safety precautions are taken. But because poisonings do occur, all people dealing professionally with pesticides should be prepared for any emergency and should also learn to recognize and treat cases of pesticide poisoning.

The symptoms of poisoning may appear almost immediately after exposure or they may build up over a period of several hours or longer. Different groups of pesticides, through their chemical properties and mode of action, may provoke very different types of poisoning, each requiring a specific method of treatment. The severity of the symptoms depends on the toxicity of the pesticide, the dose, the degree of contamination and the length of exposure:

* Symptoms and signs of *mild* poisoning are headache, a feeling of sickness (nausea), dizziness, fatigue, irritation of the skin, eyes, nose and throat, diarrhoea, perspiration and loss of appetite.

* Symptoms and signs of *moderate* poisoning are vomiting, blurred vision, stomach cramps, rapid pulse, difficulty in breathing, constricted pupils of the eyes, excessive perspiration, trembling and twitching of muscles, fatigue and nervous distress.

* Symptoms of *severe* poisoning are convulsions, respiratory failure, loss of consciousness and loss of pulse. In some instances, poisoning may be fatal.

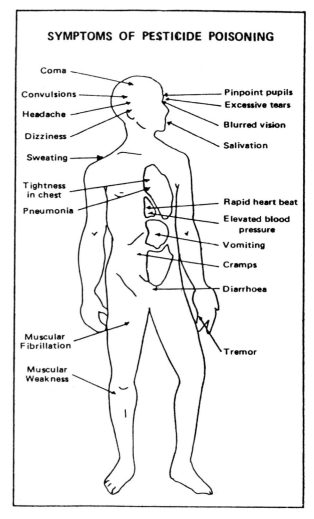

SYMPTOMS OF PESTICIDE POISONING

Coma — Convulsions — Headache — Dizziness — Sweating — Tightness in chest — Pneumonia — Muscular Fibrillation — Muscular Weakness

Pinpoint pupils — Excessive tears — Blurred vision — Salivation — Rapid heart beat — Elevated blood pressure — Vomiting — Cramps — Diarrhoea — Tremor

* Should an accident occur in which someone is doused with a pesticide and almost immediately goes into a coma, then there is no doubt about the cause. Proper treatment should be administered at once.

A more difficult problem occurs with mild, slow poisoning over a period of time; the symptoms are less severe and frequently difficult to distinguish from the effects of life's normal stresses.

88

If poisoning cannot be related directly to the use of pesticides, but exposure is suspected and symptoms appear, then medical advice should be sought. Pesticide poisoning may mimic other serious disorders such as heatstroke, pneumonia, asthma, low blood sugar (hypoglycemia), or intestinal infections; it requires a physician to make the correct diagnosis.

The symptoms of acute poisoning will occur generally within 12 hours of exposure. If symptoms occur later than this, the cause may be something else. But it is wise to check with a physician to be certain. The signs and symptoms of pesticide poisoning usually develop in a predictable order. If the signs or symptoms of poisoning appear, bring the patient to a physician quickly.

The physician should be told the name of the active ingredient in the pesticide, and given as much information as possible about the product, preferably by showing him or her the pesticide container with label and any other relevant technical literature or product sheet that may be available. Also, describe how the accident happened.

Physicians and hospital personnel should be familiar with procedures appropriate to the treatment of poisoning by important chemical groups of pesticides and frequently used products such as paraquat. Authorities and pesticide companies should co-operate in making such medical advice readily available to physicians and rural medical centres. Product sheets translated into the local language should contain essential advice that could assist physicians in providing first aid and the correct treatment, such as when to induce vomiting and when not to do so.

The effect of a pesticide on humans varies with the chemical group to which it belongs, the type of formulation, the route of absorption into the body and the quantity absorbed. Typical symptoms of poisoning are induced by the chemical type and are to be controlled by specific antidotes and medical treatment. While diagnosis and treatment ought to be left to trained medical personnel, emergency action prior to contacting a physician or while waiting for expert medical attention may help in saving lives.

TYPES OF POISONING

Pesticides are chemicals that are potentially toxic to all or in some cases to only some forms of life. This potential becomes harmful only when the organisms are actually exposed to the toxicant. The exposure, which may be sudden or prolonged, results in one of the following two types of poisoning:

Acute poisoning as a result of an accidental or intentional single or repeated exposure to a substantial dose of toxicant.

Chronic poisoning as a result of a prolonged or frequently repeated exposure to lower toxicant doses.

This refers to cumulative systemic damage over a relatively long period of time. Occupationally exposed workers may become victims of chronic poisoning, especially where familiarity breeds contempt in terms of safety precautions.

Both acute and chronic poisoning may result from intake by mouth, absorption through the skin or by respiration.

Oral ingestion (swallowing)

The mouth and alimentary tract provide a route by which a pesticide can be absorbed; upon entering the intestines, the poisonous substance is taken into the blood. Such poisoning may occur because of an accident, but is more frequently due to inexcusable carelessness.

Using the mouth to blow out plugged spray nozzles, smoking and eating without washing the hands after applying pesticides, or using a contaminated container for drinking water or for holding food are common examples of such carelessness.

Suicide attempts frequently involve pesticides, because these poisonous chemicals are easy to obtain.

Dermal contact

Absorption of a pesticide through the skin is the most important route of uptake when handling and applying pesticides. Arms and hands are the parts of the body most likely to come into contact with concentrated or diluted pesticides. Certain parts of the body (namely, the eyes, scalp, back of the neck, forehead and scrotal regions) have been found to absorp pesticides much faster than other parts. If applicators work without shirts or with bare legs, the area exposed is thus several times larger than if protected with clothes.

The danger of skin absorption is always greatest when the temperature of the working area is high and the skin is wet with perspiration. Under such conditions, the respiration of the skin is more intense, the pores are open and there is more blood in the skin tissues. Skin abrasions and cuts may further increase the risk of absorption.

Inhalation

Inhalation through the mouth and nose into the lungs is a route of usually lesser importance when handling and applying pesticides.

Aerosols, fog, smoke and very fine dust commonly contain particles well below 10 micrometres in diameter, which are respirable. This means that they will be absorbed by the alveolar tissue in the lungs rather than impinge on the bronchials as do larger droplets. When absorbed by alveolar tissue the toxicant enters the bloodstream more rapidly than via the other routes.

A particular danger exists with highly toxic substances which vaporize readily, such as parathion, TEPP, aluminium phosphide (Phostoxin), methyl bromide and cyanide compounds. Also, inhalation occurs readily where mixing or handling is being done in unventilated areas.

FIRST AID

First aid is extremely important; action taken during the first five minutes following the accident may determine whether the victim lives or dies. While first aid is being administered, someone should arrange for medical assistance.

The following course of action must be taken in case of poisoning:

1. Move the patient away from the source of contamination. Loosen his clothing to ease breathing.

2. Check to determine if the patient is breathing; if not, practise artificial respiration, while taking care not to contaminate yourself.

3. Remove contaminated clothing immediately and thoroughly wash the victim's body. The best way to do this is to wash the body with soap and water by rubbing with your hand or with a piece of cloth. Do not scrub with a brush because the outer protective layer of the skin may be abraded and thus induce a more rapid absorption of pesticide residues. Also, the hair on the patient's head and the nails of hands and feet should be thoroughly washed. Speed is essential.

4. If a pesticide is splashed into the eyes, immediately wash them with clean water for at least 15 minutes. If only one eye recieved the splash, be careful not to contaminate the other eye. Help the victim by keeping his eyelid open, which will have a tendency to shut because of pain and muscle twitching, and let the water run softly over the eyeball. Next, close the eye and cover it with a clean piece of cloth. Obtain medical attention as soon as possible.

5. If the poisonous material has been swallowed, DO NOT INDUCE VOMITING unless:

* It is *specifically* advised to do so on the label or information data sheet.

* The material (if a liquid) swallowed was one of the bipyridylium herbicides (paraquat, diquat).

* The material swallowed is known to be very toxic having an acute oral (rat) LD_{50} of less than 20 mg/kg.

If the victim is conscious and if induction of vomiting is advised, attempt to make the victim vomit by putting two fingers down towards the back of his throat, but only if the victim is conscious. Use the other hand to push the person's cheek between his teeth so that he cannot bite on your fingers.

First aid illustrations are helpful but not sufficient; first aid routines must be demonstrated and thoroughly exercised. (Courtesy of CIBA-GEIGY)

Do not use salt water as an emetic and NEVER GIVE ANYTHING BY MOUTH TO AN UNCONSCIOUS PERSON.

6. If the victim is conscious, but induction of vomiting is not called for, give activated medicinal charcoal (if available).

Absorption of poison through the gastro-intestinal tract can be prevented by the ingestion of activated charcoal which absorbs most kinds of poison except cyanide. Prepare an activated charcoal slurry for the conscious victim to drink by mixing three tablespoons of fine powder in half a glass of water. Repeat as often as possible.

7. If convulsions occur, prevent the patient from hurting himself. Usually, the patient inadvertently bites his tongue; this can be prevented by placing a tightly folded handkerchief or a piece of wood between his teeth. Beware of placing your fingers into the mouth of a patient having convulsions.

8. Keep the patient warm and quiet, preferably in a sheltered place.

IF UNCONSCIOUSNESS OCCURS:

* Clear the airway. Turn the patient on his side. Remove any vomit or food from his mouth with your finger, if necessary. Also, remove loose dentures.

* Ensure that the victim can breathe adequately, (you may need to pull the tongue forward to prevent blocking of the throat).

* Keep the victim positioned on his side with the head extended and lower than the stomach level. This position prevents obstruction of the airway and the inhalation of vomit into the lungs.

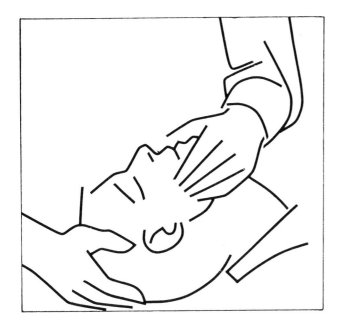

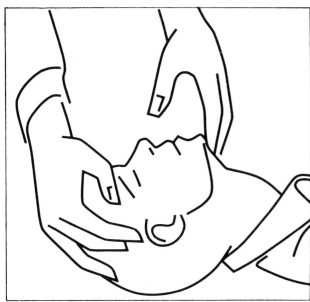

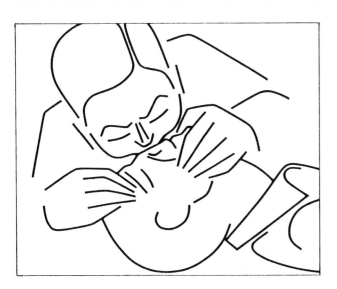

* Since pesticide poisoning requires hospital care, the victim should be brought immediately to a hospital. In instances of severe poisoning, the victim's breathing may be impaired. Therefore, after any serious exposure, the victim's breathing must be watched continuously.

IF BREATHING BECOMES DIFFICULT:

* Check to see if the person's mouth is clear. Wipe any obvious foreign matter from the mouth quickly. Use your finger wrapped in a cloth if possible.

* Put the victim on his back. Lift the neck and tilt the head backwards so that the victim's chin is pointing upward. Maintain the head in this position since it clears the airway moving the tongue away from the throat.

IF BREATHING STOPS, no time must be lost in getting fresh air into his lungs. Start resuscitation at once by following these steps:

* Act as described above. Commence with checking the mouth. Put the victim on his back and tilt his head backwards.

* Push the chin upwards with one hand while holding the head in position with the other hand.

* Take a deep breath, place your lips firmly around the victim's nose and slowly blow air into his lungs until you see his chest rise. While doing this, keep the victim's mouth closed.

* Remove your mouth and let the victim breathe out.

Repeat this procedure at the rate of 10 times a minute. Do not give up as it may take a long time to revive the victim.

If the victim starts breathing again, treat him for shock and stand by to provide further help in case it is needed.

MEDICAL ADVICE

Because treatment depends on the type of pesticide responsible for the poisoning, pesticide labels and product data sheets should always be required to carry an "in case of poisoning" section with advice on what to do and what antidote and other drugs to give. Also, the pesticide label or commercial container with label should be taken to the physician treating the patient.

An example of medical advice in a form suitable for product labels is the following on PARAQUAT and DIQUAT, both of which are highly toxic bipyridylium herbicides, often fatal if swallowed; contact with liquid products can cause severe damage to the skin or eyes.

Full texts for poisoning treatments

In view of the large number of insecticides which fall into categories of organophosphate, carbamate and organochlorine compounds, complete medical advice for physicians on dealing with these pesticides is given in the following paragraphs.

COMPREHENSIVE ADVICE FOR PHYSICIANS

* Full text for organophosphate (OP) poisoning

Organophosphate pesticides are cholinesterase inhibitors. Atropine is the emergency antidote.

In all cases of severe intoxication inject as early as possible, preferably intravenously (IV), atropine sulphate, 2-4 mg for adults (0.04 to 0.08 mg/kg body weight for children) every 3-10 minutes until adequately atropinised, as shown by dilated pupils, flushing of the skin and dry mouth. During the first hours of treatement, 20-80 mg or more of atropine sulphate may be required. If the intramuscular (IM) route has to be used to administer atropine sulphate, take care not to overdose when giving repeated injections, since distribution by the IM route takes longer than by the IV route.

While keeping the patient fully atropinised, administer also an oxime (cholinesterase reactivator) e.g. 2-PAM or P2S (CONTRATHION-Specia) 1000-2000 mg IM or IV for adults or 250 mg for children. Repeat, if necessary, after 30 minutes with half the dose. TOXOGONIN-Merck is a more recent cholinesterase reactivator. It can by given instead of 2-PAM or P2S at a dose of 250 mg IM or IV for adults. The dose for children is 4-8 mg/kg. If necessary, repeat after 1-2 hours.

Cholinesterase reactivators should be given within 36 hours after the onset of intoxication.

Convulsions can be treated with diazepam (VALIUM-Roche, STESOLID-Dumex) 5-10 mg IV respectively per rectiole. Repeat, if necessary, until convulsions have ceased.

Contraindicated are morphine or other opiates, phenothiazines, succinylcholine, xanthine derivatives, epinefrine and barbiturates.

Keep airway clear and watch respiration. Artificial respiration may be required. Observe patient in hospital for at least 24 hours. Guard against dehydration and acidosis. If necessary give intravenous infusion with dextrose or fructose solutions, electrolyte infusions, etc. as indicated by hematocrit, electrolyte spectrum, pH of the blood, diuresis, etc.

The diagnosis of OP intoxication should be confirmed as soon as possible by determination of the cholinesterase activity in venous blood.

* Full text for organochlorine (OC) poisoning

Organochlorine pesticides are central nervous system stimulants. Severe intoxication may cause convulsions without earlier symptoms. Convulsions may be delayed for 48 hours following exposure.

Treatment is symptomatic and should be aimed at controlling convulsions and restoring and maintaing tissue oxygenation.

Obtain and secure an unobstructed airway (if necessary by suction from the pharynx and trachea). Maintain adequate pulmonary ventilation, if necessary by artificial or controlled respiration and oxygenation.

Following *ingestion*, the stomach should be emptied as soon as possible by careful gastric lavage, avoiding aspiration into the lungs. This should be followed by intragastric administration of 3-4 tablespoons activated charcoal and 30 g magnesium or sodium sulphate in a 30% aqueous solution. No fats, oils or milk should be given by mouth since these promote absorption of OCs by the intestinal tract. In the case of ingestion of a solution or an emulsiable concentrate, the possibility of the occurrence of a chemical pneumonitis following aspiration of the solvent in the respiratory tract must be borne in mind.

Morphine or its derivatives, epinephrine and nor-adrenaline should never be given because of their depressive effects on the respiratory centre and because they may sensitise the myocardium and thus provoke serious cardiac arrhythmias.

Drug therapy of convulsive state

Rapid onset is required when tranquillizers or hypnotic drugs are used for controlling convulsions.

Table 5.1: Recommended dosage rate for benzodiazepines

Drug	Adults	Children		
		> 12 yr.	2-12 yr.	< 2 yr.
Diazepam IV	10 mg	10 mg	5 mg	2 mg
Diazepam rectiole	-	-	-	2 mg
Clonazepam IV	1 mg	1 mg	0.5 mg	0.2 mg

Treatments of choice are the anticonvulsant benzodiazepines clonazepam or diazepam. There is no major pharmacological difference between the two but clonazepam has a more selective anticonvulsant action in comparison with diazepam. They have in common that the central effects wane rapidly as a result of redistribution of the drug to other tissues. Repeated intravenous (IV) application of diazepam or clonazepam may therefore be required to maintain a blood level sufficiently high to control convulsions. (See table 5.1.)

Benzodiazepines are relatively safe and carry a much smaller risk of sedation and respiratory depression than other sedatives, e.g. barbiturates.

Very high daily dosages are known to be tolerated without unwanted side effects.

When convulsions are under control and do not recur and when the patient is able to swallow, it is recommended to continue treatment with phenobarbitone or other anti-convulsant drug for up to 2-4 weeks.

Barbiturates

If clonazepam or diazepam is not available, barbiturates may be used. Contrary to most barbiturates, phenobarbital has the advantage of a much higher anticonvulsant than sedative action and is therefore the barbiturate of choice. It may be administers only by slow IV and by intramuscular (IM) injection as phenobarbital sodium. It is important to wait for about 15 minutes after the initial dose (200 mg) for its effect to develop and to decide whether a repeat dose is necessary. (See table 5.2.)

Sedation is a major side-effect of treatment and respiratory depression is especially dangerous when breathing is obstructed.

Table 5.2: Recommended daily dosage of barbiturate

Drug	Adults	Children	
		> 12 years	< 12 years
Phenobarbital sodium IV:	-	-	-
IM:	1-5 mg/kg	1-5 mg/kg	3-6 mg/kg

When convulsions are under control and do not recur and when the patient is able to swallow, it is recommended to continue treatment with phenobarbitone or other oral anti-convulsant drug for up to 2-4 weeks.

* Full text for carbamate poisoning

Carbamates are cholinesterase inhibitors. Atropine is the emergency antidote.

In all cases of severe intoxication inject as early as possible, preferably intravenously (IV), atropine sulphate, 2-4 mg for adults (0.4 to 0.8 mg/kg body weight for children) and repeat at 2 mg for adults (0.04 mg/kg for children) every 3-10 minutes until adequately atropinised as shown by dilated pupils, flushing of the skin and dry mouth. During the first hours of treatment, 20-80 mg or more of atropine sulphate may be required. If the intramuscular (IM) route has to be used to administer atropine sulphate, take care not to overdose when giving repeated injections since distribution by the IM route takes longer than by the IV route.

Convulsions can be treated with diazepam (VALIUM-Roche, STESOLID-Dumex), 5-10 mg IV resp. per rectiole. Repeat if necessary, until convulsions have ceased.

Oximes, such as 2-PAM, P₂S, TOXOGONIN are contra-indicated. Also contra-indicated are morphine or the opiates, phenothiazines, succinylcholine, xanthine derivatives, epinephrine and barbiturates.

Keep airway clear and watch respiration. Artificial respiration may be required. Observe patient in hospital for at least 24 hours. Guard against dehydration and acidosis. If necessary give intravenous infusion with dextrose or fructose solutions, electrolyte infusions, etc. as indicated by hematocrit, electrolyte spectrum, pH of the blood, diuresis etc.

The diagnosis of carbamate intoxication should be confirmed as soon as possible by determination of the cholinesterase activity in venous blood. Restoration of cholinesterase activity is usually rapid compared with OPs.

(Source: Shell International Chemical Company Limited, Chemicals Health, Safety and Environment Division, Shell Centre, London SE1 7PG, U.K., 4 July 1989.)

CHOLINESTERASE ENZYME POISONING

The term cholinesterase refers to a number of enzymes in the body that have the ability to split, or hydrolize, acetylcholine. This is an important chemical mediator found throughout the central nervous system, at various junctions in the nervous system and at nerve endings in various glands and muscles throughout the body (such as in the eyes, in the gastrointestinal tract and in the lungs). When a signal goes from the brain to initiate movement of a muscle, it must pass through a number of nerve junctions. At reach junction the mediator, acetylcholine, is released at one nerve ending to stimulate the next nerve, and then is quickly destroyed by acetyl cholinesterase so that its stimulating action does not persist. More. acetylcholine is formed as needed for the next nerve impulse. Acetylcholine and cholinesterase enzyme are always in a delicate state of balance, interacting to modulate the complex nervous system which controls body functions. But if acetyl cholinesterase is not produced, acetylcholine builds up at the ends of the nerve fibres. Nerve impulse transmission then is not stopped and muscle movement races out of control.

This is what happens when cholinesterase enzymes are inactivated by organophosphate or carbamate poisoning. A variety of clinical signs and symptoms results depending on the severity of the poisoning and the system that is most affected.

However, the cholinesterase level may also be decreased in other types of tissue and in the blood serum, which therefore can be used as an index of poisoning. (The level in the blood can be lowered considerably without seriously affecting normal functions). At least two types of cholinesterase exist in the body. That which splits only acetylcholine is also found on the surface of red blood cells, and is known as "red cell cholinesterase". Another type, which splits other choline compounds as well, is known as "plasma cholinesterase". Because the activity of both enzyme types can be measured accurately with suitable laboratory procedures, the degree of exposure to organophosphate or carbamate pesticides can thus be estimated by measurement of cholinesterase activity in the blood.

The blood cholinesterase test is of value in two ways. Firstly, it is extremely valuable and reliable in diagnosing cases of poisoning by organophosphate or carbamate compounds. Secondly, regular tests are of value in monitoring the cholinesterase level of personnel using organophosphate or carbamate pesticides. Such testing forestalls illness in a worker who is unduly exposed day by day and who gradually approaches a state of clinical poisoning.

In order to be able to use the blood cholinesterase test (either in an emergency or on a routine check basis), it is essential to have the normal pre-season level of each person determined. Only then is it possible to know if the level has become depressed. There is no standard level of cholinesterase; it varies very widely among different people. The normal level for each individual must be known before the beginning of the pesticide application season.

A low cholinesterase reading may not necessarily be due to exposure to a pesticide, but critical poisoning in man is always accompanied by a drastic reduction in cholinesterase levels; and in every case, the normal level, the exposure history, the symptoms and the clinical findings have to be taken into consideration.

Slowness of recovery

In the absence of additional absorption or of medical treatment, the cholinesterase enzymes regenerate slowly. Red cell cholinesterase usually requires about 120 days to return from very low to normal values while plasma cholinesterase (which is manufactured by the liver) returns to normal in about 30

ACETYL CHOLINESTERASE REACTIONS IN THE BODY

ACETYLCHOLINE

ACETYLCHOLINE
CHOLINE
ACETATE ION
CHOLINESTERASE

ACETYLCHOLINE
CHOLINE
ACETATE ION
CHOLINESTERASE
ORGANOPHOSPHATE

Throughout the nervous system, there are electrochemical switching centres called synapses where the electrochemical nerve impulse is carried across gaps between a nerve cell called a neuron and a muscle, or from neuron to neuron. As the nerve impulse (the signal) reaches a synapse, it stimulates the release of acetylcholine. Acetylcholine acts as a sort of "bridge" over which the nerve impulse flows.

After the nerve impulse has been transmitted across the gap by acetylcholine, the enzyme cholinesterase (seen here as a wedge) breaks down the acetylcholine by hydrolyzing acetylcholine to choline and an acetate ion; the nerve impulse stops. These chemical reactions occur very rapidly.

(Courtesy of IRRI).

When organophosphate insecticides enter man or animals, they attach to the cholinesterase enzyme as seen here and inhibit the action of the cholinesterase. Because the cholinesterase cannot remove the acetylcholine, the nerve impulse flows continuously. The constant flow of the nerve impulse causes a rapid twitching of the muscles and finally leads to paralysis. When the muscles in the respiratory system no longer function, death occurs.

days. If additional exposure to organophosphates or carbamates occurs during the "recovery" period, the further inhibition of the enzymes will prolong the regeneration to normal levels. This has important implications for workers who may sustain a series of exposures during the application season.

After the recovery of a patient from poisoning, one should ensure that there are no signs of vomiting, temperature, thirst and convulsion, etc. The patient should be advised to take full rest if possible and keep away from pesticide stores and treated fields for at least 15 days.

If the poisoning was due to organophosphates and the patient has been treated for a severe case of cholinesterase poisoning, he should not be allowed to handle and spray pesticides for at least three months, the period it will take the body to recover sufficiently well. Returning to work earlier may trigger a second case of poisoning that may be more severe than the first one.

In general, because a carbamate-cholinesterase combination is less stable than organophosphate-cholinesterase, the effects of carbamate exposure tend to be not so long-lasting as the symptoms that follow organophosphate intoxication. With some carbamates, the cholinesterase level returns to normal very quickly and, because of the short duration of the depression, a blood cholinesterase test unless taken immediately may be of little value as an indication of poisoning.

The need for medical supervision and monitoring of pesticide residues and cholinesterase levels in body fluids cannot be over-emphasized in the case of workers engaged in intensive or regular spraying. It is highly desirable that a physician be readily contacted to advise personnel on medications, or to administer treatment if poisoning does occur. Pre-arrangement for emergency treatment is a must in every locality where large-scale or regular spraying operations are undertaken. Such operations may be emergency aerial or ground spraying in case of pest outbreaks, or routine rounds of spraying on estate crops such as oil palm or tea.

Records on state of health and frequency of exposure should be kept for pilots and air-strip personnel in aerial operations and members of regular sprayteams on estates and replanting schemes.

References:

Safety depends on you, 1988. CIBA-GEIGY, Agricultural Division AG 8.12, Basle.

Shell International Chemical Company Limited, Chemicals Health, Safety and Environment Division, Shell Centre, London SE1 7PG, U.K., 4 July 1989.

World Health Organization. *Multilevel Course on Safe Use of Pesticides and on the Diagnosis and Treatment of Pesticide Poisoning*, VBC/78.7 (inparticular sections VI and VII).

6

PESTICIDE REGULATION AND REGISTRATION AND THE PRODUCT LABEL

Contents   Page

Pesticide legislation 98

Pesticide registration 99

The product label 101

Pictograms or label symbols 106

References 106

Pesticide Regulation and Registration and the Product Label

PESTICIDE LEGISLATION

There can be no doubt about pesticides being of great benefit to mankind in effectively protecting agricultural and animal production against damaging organisms and in helping to control disease-transmitting vector-insects. However, these same chemicals, owing to their commonly toxic properties, pose potentially great hazards to users, livestock, wildlife and the environment as well as to consumers of food. Therefore, these hazards need to be minimized by the correct and safe use of pesticides.

The ever-increasing complexity, potency and volume of application of chemicals and growing public concern about their negative side-effects have made it necessary for Governments to set general standards and rules concerning proper use and to provide institutional support all along the line from their manufacture to their final use.

The need for regulatory control

The goal in regulating pesticides is to provide society with adequate protection from the harmful effects of improper pesticide use while giving society the opportunity to benefit from their effective use.

In developing rules and criteria to control the introduction of pesticide formulations and their proper handling and application, national authorities must have a good understanding of the need for using these cemicals in agriculture, animal husbandry, public health and industry. Also, they must understand the vulnerability of society and that all segments of society are inter-related to such an extent that pesticide use by any group unavoidably affects all others.

* **Users.** Farmers and other end-users are particularly interested in the effectiveness and efficiency of pesticides and any hazard associated with their use. They need instruction on doses, timing and methods of application, and on safety measures. They also need appropriate stimulus to observe the rules and precautions associated with that use. They should be provided with safe packaging and proper label instructions to prevent contamination and mistakes.

* **General public.** Children and the general public need to be protected against accidental exposure following spillage, disposal or careless handling. People who professionally handle pesticides during storage, transport or retail distribution equally need the protection of regulations.

* **Consumers.** The safety of food and animal feedstuffs is of paramount concern to public health. All treated food and feed should contain the least possible amount of chemical residues, if they contain any at all. Thus, residue-limits (tolerance values) need to be determined for such commodities.

* **Crop and plants.** Pesticide formulations may cause symptoms of phytotoxicty in crops, thus the effect of normal sprays, of overdosing and risk of damage from spray drift, run-off and carry-over effects in soils need to be evaluated and instruction given for the prevention of such undesirable effects.

* **Livestock and domestic animals.** Several species of animals are very susceptible to individual pesticides. Adquate precautions must be taken irrespective of whether the compounds are applied directly to the animals (e.g. against ecto-parasites), applied to feedstuffs or to the animals' environment.

* **Environment.** The potential adverse effects of pesticides on non-target insects, wildlife, soil and water organisms, flora and other ecological components are widely recognized. Harmful contamination of air, soil, surface and ground water should be avoided as much as possible.

* **Buyers.** Users who purchase pesticides need to be protected against the sale of poor quality products,

against misleading and unverified claims made on the label or in advertising and against adulterated products.

* **Manufacturers.** Manufacturers who develop, formulate, package and market pesticides need standard specifications by which the suitablity of products can be judged. Legislation should preferably also keep inferior, unsafe or untested products off the market and provide reasonable protection against patent infringement by unfair competitors.

* **Trade.** Local and overseas trade in pesticides requires the stabilizing influence of adequate legislation, reasonable standards, and the means of international enforcement. Effective products need to be protected against false and unfair claims from competing products. An essential aspect is the regulation of pesticide residues in traded agricultural commodities including food, animal produce and beverages to prevent their contamination with pesticide residues above acceptable limits.

Pesticide legislation deals with two fundamental aspects of pesticide use that are distinctly different. These are:

* **Use regulations** -- prescribing which pesticides, as well as how and when, they may be imported, marketed or used in view of their biological efficacy and hazardous side-effects, and

* **Tolerance regulations** -- establishing maximum limits for residues in food and feedstuffs.

In view of the many benefits derived from pesticide use, as well as their great inherent hazards, many countries have enacted specific pesticide legislation by which the distribution and use of any pesticide is prohibited within the country unless:

* It has been registered with the national registration authority; and

* It is covered by a provisional clearance.

Registration should be granted only by the national authority and for a limited period of time. A decision favouring registration is based on careful examination of the benefit/risk ratio. This usually involves submitting a vast number of data pertaining to the chemical and biological properties of the active ingredient and formulation, their efficacy and potential health and environmental hazards. Such information must be submitted along with the manufacturer's application.

According to common national legislation on pesticides, registration can be granted only if, on the basis of experimental data, it can be reasonably assumed that the product is effective for the purpose for which it is intended and that proper use of the product according to the instructions and recommendations does not entail unacceptable side-effects from either the pesticide or its degradation products.

On the one hand, registration enables the authorities to exercise control of quality, directions for use, claims of efficacy, labelling, packaging and advertising, thus ensuring that the interests of the final users and the general public are adequately protected. On the other hand, if every pesticide has to be registered, the public will know that the products on sale have satisfied the requirements of the law with regard to their effectiveness and safety when used according to the directions on the label. Ideally, a country should have the technical infrastructure and resources to guard against the misuse and faulty disposal of pesticides. Many countries, however, lack the capacity to enforce their pesticide laws.

PESTICIDE REGISTRATION

Before a product is imported or marketed it must be cleared by the national registration authority, which must be satisfied that the product fulfils its requirements. It is the responsibility of the manufacturer or importer to provide the scientific information on the following aspects according to national requirements:

* Chemical data concerning:

- Active ingredients: identity of physical and chemical properties, methods of analysis and manufacture;

- Impurities: same requirements as above;

- Formulation: composition, properties, methods of analysis and packaging.

* Toxicological data concerning:

- Acute, subacute and chronic toxicity (see Chapter 4);

- Special studies on metabolism, antidotes and neurotoxicity.

* Biological data:

- Spectrum of activity;

- Dose response studies for efficient pest control;

- Crop tolerances, phytotoxicity;

- Application techniques, timing and frequency;

- Crop rotation tolerances.

* Environmental data (see also Chapter 4):

- Fate and residues in plants;

- Fate and residues in animals;

- Fate and residues in soil, water and air;

- Wildlife studies.

It often takes many years to develop a new pesticide product and to perform all the studies needed to obtain the above-mentioned data (see table 6.1). The development costs are immense. Currently, they are estimated to be above $US 30 million for a single product, a long-term and high-risk investment which only major chemical companies can afford.

Table 6.1: The normal development path for a new pesticide

Stage	1	2	3	4	5	6
Duration	6 months - 1 year	6 months - 1 year	1-2 years	1-2 years	1-2 years	1-2 years
Biological tests	Initial tests in laboratory and greenhouse.	Further laboratory and greenhouse tests. Determination of range of efficacy and mode of action.	Field trials on small plots with varying soil types and climatic conditions. Drafting of provisional instructions for application. Decision on further field trials.	Farmer trials in all countries and crops for which the product is being developed. Drafting of provisional instruction for application. Decision on further development.	Field trials carried out by, or in cooperation with, official bodies. Investigation of secondary effects; e.g., compatibility with other compounds and assessment of varietal sensitivities. Final decision to go ahead with development.	Application for registration at home and abroad. Demonstration of the commercial product to farmers.
Toxicological trials and residue investigations		Range-finding trials (oral LD^{50} in rats and/or mice).	Initial work on analysis methods. Range-finding residue and breakdown investigations. Determination of acute toxicity - LD^{50} (oral and dermal) - for various animals, to decide what precautions are necessary during production and application.	Residue investigations. Start of breakdown investigations in soil, plants and animals. Ninety-day feeding trials using at least two different animal species to investigate active ingredient and metabolites, including the cumulative effect.	Continuation of residue and breakdown investigations; breakdown charts. Two-year feeding trial on at least two different animal species, and other special studies. Determination of "no effect level" and "acceptable daily intake".	Continuing investigations.
Effect on environment		Range-finding trials (toxicity to fish and bees).	Toxicity to fish, bees, birds and micro-organisms in soil and water.	Investigations in Stage 3 intensified and extended to other animal species.	Investigation of environmental risks	Continuing investigations.
Patents		Investigation of patentability. Application for patents.	Application for foreign patents.	Application for production patents.	Final survey of patents owned.	Granting of licences.
Production			Production of small trial quantities. Estimation of costs Start of storage trials.	Production of large trial quantities. Detailed calculation of costs	Study of large-scale production methods. Start of planning of production plant. Final decisions on formulation and packaging. Storage trials continue.	Full-scale production.
Marketing			Initial market studies.	Further market studies. Calculation of demand and acceptable price. Selection of trade name and agreement of common name with registration authorities.	Detailed market study Preparation for publicity campaign.	Preparation for sales. Sample marketing of small quantity if required. Association with other firms.

Source: Schering Agrochemicals Limited

Once a product is approved in principle for registration, the next important step is formulating the use instructions, which also provide legal status. These instructions describe the specific pests against and the specific crops on which the pesticide may be applied; they include dosage rates, modes of application, precautions and other details which must be included on the pesticide label (see below).

Equally important as the principles for granting pesticide registration are the principles for withdrawing that registration. Registration can be withdrawn either at the request of the registration holder or by enacting a pesticide law for that purpose. Usually, there are three basic criteria for withdrawing that registration. They are:

* **Ineffectiveness of the product.**

This may happen when certain pests develop such resistance to the pesticide that practical application becomes virtually useless.

* **Occurrence of unforeseen side-effects.**

This principle is of great importance because adverse side-effects may not be predictable at the stage of first evaluation, but are recognized only after a longer period of use.

* **Shift in the balance between benefit and risk.**

This shift would occur when a new pesticide is developed that is equally effective but considerably less hazardous than the pesticide in question, the side-effects of which had been considered previously barely acceptable.

Many countries, especially those with highly developed technologies, have long since set up an extensive regulatory system to control the trade in, and the production and use of pesticides. The majority of developing countries have not as yet obtained the degree of organization, technical infrastructure and expertise to implement and enforce such desirable regulatory procedures. Also, few of them are able to maintain sufficient facilities for monitoring pesticide residues in food and the environment.

However, developing countries need not introduce sophisticated regulatory facilities in order to control pesticides effectively. An enormous amount of basic laboratory work on chemistry, toxicology and analytical studies produced as a part of the regulatory procedures of technologically advanced countries is available and has gained worldwide credibility. Therefore, developing countries can use the results of such work without having to produce them independently.

Developing countries should design regulatory procedures suited to their specific needs, and not attempt to adopt all elements of regulatory schemes used in developed countries, which may have different agricultural practices and different kinds and species of pests, different climates and economic systems. The use of a pesticide should be permitted only if the benefits outweigh the risks involved and if no suitable alternative can be employed. For example, the risk/benefit analysis for a country plagued by vector-borne diseases, malnutrition and starvation may lead to a different decision than in countries without such problems. (See list of FAO guideline series contained in refereences.)

THE PRODUCT LABEL

The best source of information concerning the safe and effective use of a pesticide product is the label on the pesticide container. It is also a legal document that in most countries requires government approval. While requirements vary from country to country, almost all require labelling and registration of pesticide products. The model label shown in this section is not used in any one country, but the context and principles are applicable to every country. Every pesticide product label should contain the following types of information if the product is designed for use on crops:

1. Trade or brand name
2. Ingredient statement
3. Common or chemical name
4. Type of formulation
5. Net contents of the package
6. Name and address of manufacturer, distributor or formulator
7. Registration or license number
8. Warning or "signal words"
9. Hazards to humans and domestic animals
10. Environmental hazards
11. Physical and chemical hazards
12. Statement of practical first aid treatment
13. Re-entry statement
14. Storage and disposal directions
15. Warranty statement
16. Misuse statement
17. Use areas
18. Directions for use
19. Harvesting statement

1. Trade or brand name

Each company has brand names for its products. The brand name is the one usually used in advertisements. The brand name is printed plainly on the front panel of the label and is the most identifiable name for the product.

2. Ingredient statement

Every pesticide label must list the active ingredients present in the product. It is highly desirable that this information should appear on the front panel directly below the brand name. The amount of each *active ingredient* may be given in several forms: as a percentage by weight, as grams per litre or as pounds per gallon of concentrate. It may be listed by either the chemical name or the common name, but the use of the common name should be encouraged when one has been agreed upon. The *inert ingredients* need not be named, but the label must show what percentage of the contents they take up.

3. Common or chemical name

Many pesticides have complex chemical names. Some have been given another name to make them easier to identify. Such a name is called the *common name*. For instance, carbaryl is the common name for 1-naphthyl N-methylcarbamate. A chemical may be made by more than one company and thus be sold under several *brand names*, but the same *common name* or chemical name will appear on all the products.

4. Type of formulation

Different types of pesticide formulations (such as emulsifiable concentrates and dusts) require different methods of handling. The label must indicate what type of formulation the package contains as the same pesticide may be available in more than one formulation. The formulation type may be coded such as EC for emulsifiable concentrate etc. (See Chapter 3.)

5. Net contents of the package

The net contents number indicates how much is in the container. This can be expressed in litres, grams, gallons, pints, pounds, or other units of measure approved by the authorities concerned.

6. Name and address of manufacturer, distributor or formulator

The maker or distributor of a product is usually legally required to put the name and address of the company on the label.

7. Registration or license number

The registration number must appear on every pesticide label. It shows that the product has been registered with the Government. It should appear on the front panel of the label.

8. Warning or "signal words"

Every label should carry the words "KEEP OUT OF THE REACH OF CHILDREN", because most pesticide poisonings and deaths occur among children by accident. Labels usually have words printed in bold type such as CAUTION, WARNING, DANGER or POISON that indicate the relative toxicity of the product. The following system is often used:

Signal words	Toxicity
Danger, poison	Highly toxic
Warning	Moderately toxic
Caution	Slightly toxic
No signal word	Relatively non-toxic

Note: The language of the lable should be in the local language(s) to be understandable to users.

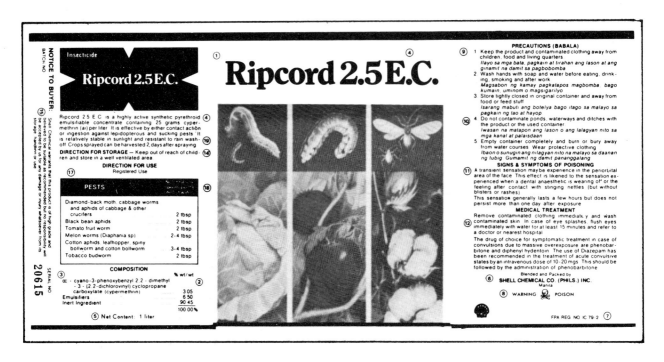

102

Symbols

One of the best ways of attracting attention is with graphic symbols. This is why a "skull and cross-bones" symbol is frequently used on all highly toxic materials along with the signal word DANGER and the word POISON. The symbol is a graphic reminder that the contents can kill.

9. Hazards to humans and domestic animals

This section indicates the ways in which the product may be poisonous to man and animals. It should also indicate any special steps needed to avoid poisoning, such as the type of protective equipment required.

10. Environmental hazards

Pesticides are useful tools; careless use could cause undesirable effects. To help avoid this, the label contains environmental precautions, such as:

"This product is highly toxic to bees exposed to direct treatment or residues on crops."

"This product is toxic to fish; do not contaminate bodies of water when cleaning equipment or when disposing of wastes."

The label may contain broader warnings as appropriate against harming birds, fish and wildlife.

11. Physical and chemical hazards

This section indicates any special fire, explosion or chemical hazards posed by the product.

12. Statement of practical first aid treatment

If the product poses a hazard from ingestion, inhalation and skin or eye contact, the label must indicate emergency first-aid measures. It also must indicate what types of exposure require medical attention.

13. Re-entry statement

If required, this section indicates how much time must pass before a pesticide-treated area is safe for entry by a person not wearing protective clothing. Local authorities should be consulted for special rules that may apply in specific areas.

14. Storage and disposal directions

Every pesticide should be stored and disposed of correctly. This section indicates how to store and dispose of the product as well as the containers once they have been emptied of their contents.

15. Warranty statement

Each product label specifies to what degree the manufacturer or distributor limits his warranty and liability.

16. Misuse statement

This section is a reminder that it is a violation of most national laws to use a product in a manner inconsistent with its labelling. Do not use a product on a crop or for a pest not listed on the label. Do not use it at more than the recommended rate.

17. Use areas

This section lists the crops, animals or other areas on which the product can be used. On many labels those targets of application are incorporated in the directions for use (Statement No. 18) and are not given separately.

18. Directions for use

The instructions on how to use the pesticide are an important part of the label. This section indicates:

- The pests for which the product is registered to control. (Labels often use common names for pests. Knowing these names helps in choosing the proper pesticide)

- The crop, animal or other areas on which the product can be used

- The form in which the product should be applied

- How much to use

- Where the material should be applied

- When it should be applied and how frequently

19. Harvesting statement

A harvesting statement is printed on some labels when there is the possibility or danger that the crop may be fed to animals or eaten or handled by humans. Because residues of the pesticide require a minimum number of days to break down, the Government may require a harvesting statement giving a specific number of days before the crop can be cut, harvested or consumed.

PICTOGRAMS FOR AGROCHEMICAL LABELS

The illustrations on this page may be used as artwork for the production of pictograms.

STORAGE Pictogram

Keep locked away and out of reach of children

ACTIVITY Pictograms

Handling liquid concentrate

Handling dry concentrate

Application

ADVICE Pictograms

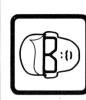

Wear gloves

Wear eye protection

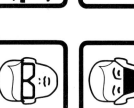

Wash after use

Wear boots

Wear protection over nose and mouth

Wear respirator

WARNING Pictograms

Dangerous/ harmful to animals

Dangerous/ harmful to fish — do not contaminate lakes, rivers, ponds or streams

(Pictograms courtesy of GIFAP)

PICTOGRAMS FOR AGROCHEMICAL LABELS

1 The activity pictogram for **handling the product** in its pack or container should appear to the left of the centre of the label, with associated advice pictograms grouped to the left of it. The activity and advice pictograms should be enclosed in a clearly-defined "box" to show they are linked. The group of pictograms shown here indicates that when measuring out a concentrated liquid product, gloves and eye protection should be worn.

2 The activity pictogram for the appropriate **method of application** should appear to the right of the centre of the label with associated advice pictograms grouped to the right of it, enclosed in a "box" to show they are linked.

This group of pictograms shows that gloves and boots should be worn when applying the product.

CENTRE

3 The pictogram showing that pesticides must be **locked away out of the reach of children** should be used on all labels and should appear to the left of the group of pictograms associated with handling the concentrate.

4 The pictogram showing that people must **wash after using pesticides** should be printed on all labels and should appear to the right of the group of pictograms associated with application of the product.

5 The **environmental warning** pictograms, when they are needed, should appear to the right of the "wash after use" pictogram.

TOXIC

If a hazard warning colour band (such as that recommended by FAO) is used on the label, the pictograms can appear within it. If one of the established hazard warning symbols, such as the

skull and crossbones, needs to be shown, it should be located at the centre of the hazard warning colour band, together with the accompanying warning statement.

The example shown here is a red band, used for a toxic or very toxic product. Other colour bands recommended by FAO are yellow (harmful), blue (caution) and green.

(Pictograms courtesy of GIFAP)

PICTOGRAMS OR LABEL SYMBOLS

Because of variations in the levels of literacy, training and supervision, quite a number of farmers may not be able to read or understand the warnings and advice on the labels of pesticides. Thus, they may not be aware of the need to take precautions for protection of themselves and others, and of the environment when storing and handling the concentrated chemicals, and during application.

It should also be acknowledged that labels sometimes do not contain all essential information. This may be the case when containers are of small size (e.g. 100 ml or less) and where the registration authority has not issued stringent requirements for labelling.

To help overcome these problems, International Group of National Associations of Manufacturers of Agrochemical Products (GIFAP) in co-operation with FAO has evolved a set of pictograms for use on the labels of agrochemicals.

A pictogram is a symbol or drawing which conveys a message without words. GIFAP's set of pictograms convey warnings and safe handling advice (see pictogrms on the previous two pages; the "skull and cross-bones" symbol in statement No. 8 above is another pictogram.)

However, there are two constraints in the use of pictograms on labels. Firstly, there is limited space on product labels and, therefore, only a few symbol messages can be printed on them. Secondly, the interpretation of symbols or drawings is largely dependent on cultural and social backgrounds. Thus, the selection of symbols which are understood by farmers all over the world remains a difficult problem and a matter of compromise. The GIFAP set of pictograms has been chosen following a worldwide survey of about 1,000 farmers and farm-workers in 42 countries. These pictograms were generally well understood.

The FAO International Code of Conduct on the Distribution and Use of Pesticides (see Chapter 7) recommends the use of these pictograms. Hence, Governments and industries are requested to promote their incorporation on agrochemical labels and to educate pesticide users about them. However, the following points should be kept in mind:

* Pictograms are intended to supplement the existing text on labels, not to replace it.

* A few pictograms should be used rather than many in order not to confuse the pesticide user.

* Only essential pictograms are selected by ranking the warning and advice messages in their order of importance.

* The use of pictograms should not conflict with national regulations.

How to use the pictograms

The pictograms should be featured on labels in a way best suited to the design of the particular label whilst conforming to national label regulations. The layout suggested here is consistent with the design of labels recommended in the FAO Guidelines on Good Labelling Practice for Pesticides (March 1985).

The pictograms should be printed in black and white and are most conveniently located at the bottom of the label. Their size should be appropriate for the label on which they are to appear. For labels on bottels of 1 to 5 litres' capacity, the preferred size is approxemately 15mm x 15mm, but they should never be smaller than 7mm x 7mm.

References:

Guidelines for environmental criteria for the registration of pesticides, FAO, Rome, 1985.

Guidelines for the registration and control of pesticides (including a model scheme for the establishment of national organizations), FAO, Rome, 1985.

Guidelines on efficacy data for the registration of pesticides for plant protection, FAO, Rome, 1985.

Guidelines on pesticide residue trials to provide data for the registration of pesticides and the establishment of maximum residue limits, FAO, Rome, 1986.

Pictogrammes for Agrochemical Labels (1988); International Group of National Associations of Manufacturers of Agrochemical Products (GIFAP).

Wilson, Dough, W., 1979, Pesticide Control Section, Ministry of the Environment, Ontario, Canada; consultant to ARSAP Training Project.

7

PLANT PROTECTION POLICIES
AND REGULATORY INFRASTRUCTURE

Contents   Page

Plant protection policies 109

- **Problems connected with pesticide use** 109

- **System-oriented plant protection policies** 110

- **Crop prices and subsidies** 113

- **Emission into the environment** 115

- **Non-chemical methods of crop protection** 117

- **Integrated farming systems** 117

Regulatory infrastructure 119

- **Responsibility of the Government** 119

- **Responsibility of the private sector** 120

- **Regulatory problems in third world countries** 120

- **International co-operation in crop protection, vector control
  and pesticide management** 121

- **International programmes on plant protection and pesticide
  management** 122

- **International guidelines on the use of pesticides** 123

- **The FAO International Code of Conduct on the Distribution
  and Use of Pesticides** 124

References 126

Plant Protection Policies and Regulatory Infrastructure

The splendor of nature will be preserved only through thoughtful and skilled exploitation of its potential. Integrated crop management with maximum use of natural pest control mechanisms as well as enforceable regulatory guidelines can contribute towards sustainability. (Photo by the author)

PLANT PROTECTION POLICIES

Problems connected with pesticide use

Worldwide more than a thousand different food and non-food crops are grown in a traditional or commercial way for consumption as food or feed, for industrial processing and for the flower and ornamental plant markets. All these crops are continuously threatened in their development by thousands of pathogens, pests and other causes of injury.

Very many species of pathogens, nematodes, insects or other organisms are soil-borne or remain for a greater part of their life-cycle in the soil where they injure roots, tubers and other underground plant parts. Other damaging organisms spend most of their life above ground where they attack other parts of the plant.

The likelihood that these organisms will cause a disease outbreak or physical damage depends on many factors such as climatic and area conditions, the frequency of planting similar crops in the rotation cycle, the sensitivity of the variety, the physical and biological soil structure and the range of phyto-sanitary measures taken. Considerable yield losses may be inflicted if no crop protection measures are taken. Physical losses of more than 30 per cent are quite common owing to competition from pests or weeds; usually the affected crop suffers an additional loss of quality.

The development of chemical pesticides has provided ways and means to control most pests and diseases. This is accomplished by preventive or curative treatments of seed and planting material, and crops in the pre-harvest (field) and produce in the post-harvest (storage) phases. It is also done by disinfection of the soil and production or storage buildings. Pesticides give good protection against a very great number of pests and diseases, except against those caused by viruses, viroids, certain bacteria and mycoplasmas. They are also relatively cheap, fast working, and reliable means of control and rather easy to apply.

For these reasons, farmers by and large got used to countering any risk of pest or disease outbreaks by immediately applying chemical pesticides. In this way, chemical pest control became an essential component of modern intensive agriculture. (For example, fig. 7.1 shows the increase in total value of the agro-pesticides imported by Thailand over the period 1976-1985). However, it is a well-known phenomenon that, after prolonged application of only one or a few formulations in frequently or continuously grown crops, one or more diseases, pests or weeds may no longer be effectively controlled. This happens when such organisms have become resistant to a pesticide or class of pesticides that should have protected the crop. Diseases in particular can spread rapidly and over large areas if they are not recognized and controlled early, or if the crops cannot be protected preventively. The danger of a rapid spread of a disease is even greater if a new race of a pathogen has broken the resistance of a cultivar. The resulting loss of yield may be considerable and the crop may even fail completely.

In surveying the problems that exist in connection with the use of chemical pesticides, one can distinguish two inter-related major groups of problems.

* The first group includes all types of pesticide resistance in plant and animal organisms as well as the resurgence of pests and epidemics resulting from pesticide use.

* The second group includes the problems arising from contamination of man, domestic and wild animals, foodstuffs and other agricultural produce, surface and ground water, and of other components of the environment.

These groups of problems have a common cause, namely the very high structural dependency of the agricultural production system on chemical methods of pest and disease control.

In developing countries, however, pesticides are often used in situations of serious food shortages and because such countries must produce for export in order to earn foreign currency for payment of essential imports such as oil. Pesticides are also used to control vector insects that transmit human and animal diseases, or to stop epidemics such as of desert locusts, army worms and plant diseases.

Against a background of perilous food shortages and diseases or plagues in developing countries, a weighing of the disadvantages of pesticide use against its advantages may lead to very different decisions than would be the case in well-developed countries. In the latter countries, public awareness of the negative effects of excessive pesticide use in intensive farming is rapidly

Figure 7.1: Total value of imported agrochemical products, 1976-1985

Millions of baht (noted at top of bars)

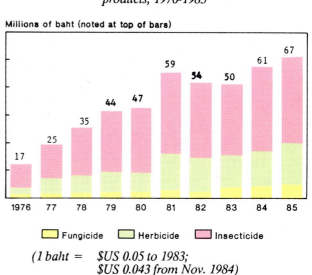

(1 baht = $US 0.05 to 1983; $US 0.043 from Nov. 1984)

Source: Ministry of Agriculture and Cooperatives of Thailand (MOA)

109

increasing. People realise that intensive farming has become the victim of both its success (higher productivity) and its defects (excessive reliance on inputs).

Consequently, while objecting to the ever-increasing contamination of their food, drinking water and environment, the general public demands that corrective action be taken. Forced by increasing political pressure, Governments of some developed countries have begun to revise their agricultural policy. Girardin (1990), however, warns them to remain realistic and to recognise that in many places considerable increases in productivity are still possible and large areas thus will continue to switch to intensive production. In the case of growing gramineous cereals, for example, the unit cost of nitrogenous fertilizer and herbicides is so low that there is no incentive for farmers to practise moderation. Therefore, a more rational use of agro-chemicals (which is not yet possible for all crops) can be achieved mainly by training farmers in integrated pest management and, perhaps, also through legislation.

The main objective of the new plant protection policies is to reduce overdependency on agro-chemicals in sectors of intensive agriculture. When reviewing current plant protection practices and problems, one cannot but come to the conclusion that chemical control must be largely reduced and, if possible, completely eliminated. This means breaking the current trend in agriculture, in which cropping systems are geared to maximizing physical production, by encouraging those systems that primarily promote a competitive, safe and sustainable agriculture. In sustainable agriculture, the emphasis is put on obtaining optimum quality rather than on obtaining maximum yields. From this point of view, such factors as an unpolluted environment, safe labour conditions and public health, are regarded as integral aspects of quality-oriented production. However, to reverse a trend that has taken about half a century to develop will require radical changes in current agricultural production systems and, thus, a great effort from all the people involved.

Agricultural society itself has a great interest in an early realisation of the needed changes. A "wait-and-see" attitude will allow new problems to arise and existing ones to become more serious and, in the end, it will make any change more complex and costly. One has only to think, for instance, of the accumulation of persistent degradation products in soils and surface water, which will make the purification of drinking water ever more costly. Therefore, government administrators, pesticide distributors and large-scale agricultural producers in developing countries should inform themselves about the consequences of excessive pesticide use. They should try to understand the objectives of new plant protection policies, and strive to identify those policy components which lend themselves to implementation nationally.

In the Netherlands, for example, a country which has a highly developed and intensive agriculture, a new strategy towards safe and sustainable agriculture is being plotted along three main lines of action. (Meerjarenplan, 1990). These lines are:

* Decreasing the dependency upon chemical methods of crop protection;

* Decreasing the volume of chemical pesticides used; and

* Decreasing the emission of chemical pesticides into the environment.

In the Netherlands, the cultivation of flower bulbs, seed- and consumption potatoes and onions in open fields and of vegetables and ornamental plants in glass-houses are important activities. Since nematodes and soil-borne diseases represent a major threat to these cultures, soil treatments with nematicides and fumigants account for more than 50 per cent of the country's pesticide use. But soil treatments cause large-scale emission of contaminants into the ground water and environment. Consequently, in order to decrease emission into the environment, a new regulation is being proposed that will make permits for soil treatment mandatory and allow the sale of such chemicals under certain conditions only. In glass-house cultures, the use of soil as a growth medium has already been largely replaced by sterile substrate material, such as stone wool, that can be recycled for reuse. In this way, the volume of growth medium to be treated against soil-borne pests and diseases is enormously reduced and thus the quantity of pesticides too.

A detailed review of all aspects of crop production and protection, and of vector control in the public- and animal-health sectors could reveal comparable possibilities for decreasing pesticide use. If highly hazardous pesticides are used in cases for which alternative means and methods are available, these should be replaced. If developing countries lack the necessary resources, they should be helped through the transfer of knowledge and technical assistance from international institutions and donors, and from the agrochemical industry, which all would seem to have a moral obligation in this respect.

System-oriented plant protection policies

In attempting to decrease the structural dependency of a country on chemical pesticides, the following aspects must be taken into account:

- The structure of agriculture and horticulture of the country concerned; and

- The types of production systems being operated within the national structure, particularly regarding crop protection.

The agricultural structure of a country may be characterized by, for instance, the state of its agricultural development, the kind and scale of its crop production

In modern glasshouse cultures, a sterile substrate such as stone wool replaces soil as the growth medium. The use of substrate material reduces the need for chemical soil sterilization. Furthermore, surplus fertilizer solution, which is drip-fed to individual plants (hydroponics), can be drained and disinfected for re-use. Used substrate material can be recycled at high temperatures for re-use. (Courtesy of LNV, the Netherlands)

activities, the degree of specialization of those activities, the local climatic conditions etc. Furthermore, a national agricultural structure may consist of several sub-structures with specific production systems. For instance, a tropical country may contain an arid zone cultivated by small-scale producers growing a few traditional food crops for self-sufficiency, as well as a humid, more fertile zone with advanced, large-scale agricultural or horticultural production systems growing crops for the export market.

Evidently, there is a close relationship between the state of agricultural development in a country and the use of chemical inputs. Therefore, it is useful first to distinguish three main phases of agricultural development in order to get an impression of the need for changing the present situation:

* Traditional agriculture. This is typically a poor man's activity on undeveloped land, such as shifting cultivation in bushland or forests or, more common-ly, on infertile soils and under marginal climatic conditions, such as rain-dependent highlands. Generally, these are small-scale farmers growing some grain- or root crops as food for their family and for exchange with neighbours. The size of their plots is usually determined by the availability of family labour for land preparation and hand-weeding.

Almost all traditional farmers are illiterate and never receive any formal agricultural training. Crop protection measures, such as weeding, are aimed at securing the food supply for the family. In traditional agriculture, diseases and pests are usually endemic in nature. This means that harmful organisms are present all the time, but cause little damage because of their low population density. They are effectively regulated by natural enemies, host plant resistance and weather conditions. Traditionally grown crops usually consist of low-yielding land races in which, through a process of long-time natural selection, the genetic properties determining the degree of host-plant resistance match the harmful properties of the pest- and disease organisms. Therefore, there is little need for insecticides and fungicides in traditional agriculture.

However, because of their extreme poverty and often damaging land use (e.g. slash-and-burn land preparation which leads to soil erosion), traditional farmers may become a target group for development efforts by their Government or foreign donors. Among the measures taken, the farmers may be supplied with some subsidized agricultural inputs, including pesticides.

* Agriculture in a state of transition. Generally, the development of some local market structure induces

farmers to grow more food and fiber crops for the purpose of selling a part of the yield. The introduction of money in a farmer's household usually leads to gradual expansion of its activities with the help of hired labour and draught animals, whilst production is raised by applying purchased inputs such as seed of improved cultivars, fertilizers and some pesticides.

A feature of agricultural progress is a gradual intensification of the cultivation practice and a decrease in the number of crops grown in the rotation cycle. New and more profitable crops are introduced, traditional land races are replaced by modern cultivars, crops are grown in monocultures instead of in varied cropping systems, and water management is improved through irrigation or drainage works.

In traditional agriculture, many farmers keep seed of several cultivars which require a different number of days to maturation. This enables early or late sowing depending on the onset of the rainy season. Availability of improved seed and irrigation or drainage facilities enables the replacement of such low yielding, but genetically very diversified varieties (land races) by a few higher yielding, but genetically uniform, cultivars. The planting of such cultivars, which are selected on the basis of their high-yield potential rather than on pest and disease resistance, and in monocultures rather than in varied rotation, creates problems with regard to the development of many kinds of pests and diseases which previously remained rather unnoticed. The high dosages of fertilizer which improved cultivars require stimulate a proliferous development of weeds, pathogens and pests.

Examples of transition agriculture include flooded rice cultivation, intercropping of maize and soybeans, the planting of coffee, rubber or coconut trees as cash crops, and the growing of vegetables and fruits in addition to food crops.

Normally, a whole range of transitional forms ranging from traditional to market-oriented agriculture can be observed in third world countries. The so-called "Green Revolution", that led to a general adoption of high-yielding cultivars of important commercial crops, has proved to be a mixed blessing. On the one hand, it contributed to much larger global agricultural production, but on the other hand it made agriculture more dependent on inputs (table 7.1 gives an example for Bangladesh). The "Green Revolution" has thereby widened the gap between the "haves" and "have-nots" since poor traditional farmers generally have gained neither access to nor knowledge about the rational and safe use of agrochemicals, irrigation, and improved seed and planting materials. Research to improve their traditional crops, such as dry-land and deep-water rice, sorghum and millet, sweet potato, yams and other minor root crops, are only recently receiving more attention from international institutes.

* Large-scale commercial agriculture. A few selected crops with high market value are grown in monoculture or with little rotation on estates or large-size farms. The efficacy of biological regulation of pests and pathogens is drastically reduced mostly through the planting of genetically homogenous, high-yielding cultivars with low natural host-plant resistance, and through intensive use of fertilizers and pesticides. When application kills their natural enemies, pests get a chance to recover rapidly and to reinfest the crop. Also secondary pests, i.e. damaging insects and other organisms that normally are effectively controlled by natural enemies, reach economically significant population densities.

Important commercial production systems include irrigated field crops such as cotton and sugar-cane; estate crops such as bananas, oil-palm and coffee; horticultural fruit- and vegetable growing; and protected crops such as ornamental plants and flowers. (Protected crops are grown under artificial climatic conditions; by growing plants under glass or plastic, the temperature, humidity of the air and evaporation can be regulated, while the plants are being protected against wind and rain.)

Usually, estate companies and large-scale commercial farmers hire crop protection experts or they themselves may have considerable experience in chemical plant protection. If market prices of food and industrial crops are set by the Government in advance, they can calculate how much they can afford to invest in crop protection. Usually, a farmer's dependency on chemical fertilizer and pesticides increases proportionally to his financial investment and to the degree of specialization of his farm.

Table 7.1: Quantity and costs of pesticides purchased by farmers in Bangladesh for use on rice

Year	Quantity used (in tons)	Value (farmer's price) (million taka)	Value (million $US)
1983/84	2 676.43	269.88	10.30
1984/85	2 565.64	250.33	8.66
1985/86	2 603.73	340.00	11.37
1986/87	3 244.01	375.70	12.28
1987/88	3 825.00	685.10	22.10

Source: FAO (against official taka exchange rates).

Table 7.2: Pesticide subsidies in various countries

Region and country	Estimated average rates of pesticide subsidy: rate of subsidy in% of full retail costs	Value of annual subsidy in $US	
		Total ($US million)	Per capita of total population ($US)
Africa			
Senegal	89	4	0.7
Egypt	83	207	4.7
Ghana	67	20	1.7
Latin America			
Honduras	29 (35)[a]	12	3.0
Colombia	44	69	2.5
Ecuador	41 (55)[a]	14	1.7
Asia			
Indonesia			
food & other annual crops	82	128	0.8
Pakistan	negl. (19)[a]	negl.	-
China	19	285	0.3

Note: a Includes subsidies captured by private distributors as higher margins between their costs and selling prices, but not passed along as lower retail prices.

Sources: - World Resource Institute (Data derived from detailed country tables and country reports).
          - Repetto, 1985.

When traditional farmers turn to cultivating new crops, such as certain types of vegetables, ornamental plants, or fruits, they often have no experience with associated pest- and disease problems. Since these farmers seldom receive any guidance from agricultural services, they depend on the sales-agents of pesticide companies for advice and may never hear about the disadvantages of indiscriminate pesticide use. Sometimes, farmer co-operatives supply pesticides and spray equipment at reduced prices and also provide guidance on crop protection matters.

Where the cultivation of certain crops such as fruits, vegetables and ornamentals becomes a speciality of certain areas, the chance for contamination between production units and farms increases. Diseases, pests and weeds are easily carried from one farm to another by the exchange of plant material and seed, by equipment and transport. In areas of specialization, the need for phyto-sanitary measures increases as does the need for chemical control.

Furthermore, in the international exchange of plant material and agricultural produce, the import of foreign pests, weeds and diseases has become un-avoidable, against which the local agro-ecosystems have no defense. In spite of quarantine measures and costly elimination campaigns, many organisms gain a permanent foot-hold. The necessity to keep such intruders under control leads to an ever-higher use of pesticides. Similarly, in the international trade of agricultural propagation material and ornamental plots, that often must be completely free of any contaminating organism, the use of pesticides tends to be higher than would be required for pest control in field crops.

Crop prices and subsidies

It is evident that a whole range of different policies and measures will be needed to deal effectivily with such system-specific plant protection problems. The size of his investment and the availabity of reliable non-chemical means and methods will be decisive factors in persuading a producer to minimize the use of pesticides on his crops. He will certainly be reluctant or unable to afford major risks. Any possibility for changing current control practices in field crops depends on economic factors and autonomous developments in the various sectors. These factors are for one part dictated by international commodity prices and trade negotiations, but for another part also by national conditions such as high prices for arable land, high costs of labour, credit facilities, subsidies and poor infrastructure.

In many third world countries, the production and sale of fertilizers and pesticides are subsidized as a means of increasing agricultural production. Government programmes, such as Masagana 99 in the Philippines and Bimas in Indonesia, promoted the practice of monoculture by subsidizing the growing of rice, maize and a few other major grain crops in particular. Repetto (1985) (see table 7.2) analysed the subsidizing of pesticides in nine countries, three each in Africa, Latin-America and Asia, all having very different agricultural and economic systems. In Senegal, Egypt and Indonesia, total subsidies on pesticides over the period 1982-83 amounted to more than 80 per cent of the cost. The average rate of subsidy in the nine countries was 44 per cent. It appeared that, in some countries, a part of the subsidy was not calculated in the sales price, but given to

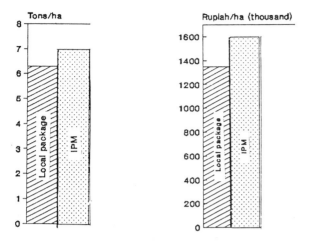

Figure 7.2(a): Average yield of IPM and local management in CISADANE rice cultivar, Indonesia, wet season 1989/90

Figure 7.2(b): Average yield value minus pesticides for IPM and local management in CISADANE rice cultivar, Indonesia, wet season 1989/90

Source: FAO IPC Rice Programme.

the trade. A calculation shows that the annual per capita expenditure for pesticides in Indonesia amounted to $US 0.80 and in Egypt to $US 4.70 per capita. In the early 1980s, the subsidy base for rice alone in Indonesia was put at $US 120 million.

Subsidies are given in various forms such as premiums on investment, tax exemptions, cheap credits to farmers and by retail selling at a reduced or below-cost price. Generally, subsidies tend to make fertilizers and pesticides so cheaply available that commercial farmers see little advantage in trying alternative non-chemical means and methods of crop protection. The latter may even look expensive in comparison as they often result in some more damage to the harvested produce. The

advantage to a farmer of using biological control often is related more to its cost-saving aspects than to increased yields [compare figs. 7.2(a) and 7.2 (b)]. Cost savings through a reduction in pesticide use becomes ever more attractive to a farmer when the actual cost of formulated products is charged at the retail outlet. Thus, a political decision to abolish subsidies on pesticides creates an opportunity for promoting integrated pest management.

The impact that such policy decisions can make is demonstrated by the Indonesian example. In November 1986, the Government of Indonesia declared, through Presidential Instruction (Inpres 3/86), that Integrated Pest Management would be the national pest control strategy for rice. In order to quickly limit the excessive use of insecticides on rice -- a practice recognized as the main cause of the alarming rice brown planthopper outbreaks -- the decree banned the use of 57 registered brands of broad spectrum insecticides on rice, about 20 of which had been widely used by farmers and, up to that point, heavily subsidized by the Government. (Refer to fig. 7.3, which shows the relationship between outbreaks of the rice brown planthopper and pesticide subsidy in Indonesia.) Within three years, the number of applications and, thus, the volume of pesticide use decreased dramatically and so did the expenditure of farmers on pesticides and of the Government on pesticide subsidies (see fig.7.4).

The Indonesian Government in fact shifted funds to purchase adequate stocks of a single, narrow spectrum insecticide affecting particularly brown planthoppers and no predator insects. It also recruited 1,500 new pest observers and provided senior pest observers with motorcycles to improve surveillance and impart training to farmers. The large-scale promotion campaign for integrated pest management in rice, that started in 1986

Figure: 7.3: Area infested by brown planthopper and relative subsidy on pesticide, Indonesia, 1974-1990

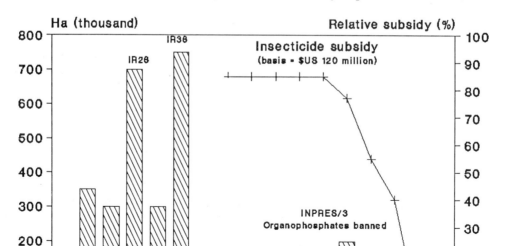

Rice brown planthopper outbreaks may result from intensive use of pesticides killing their natural enemies. Planthoppers feed at the base of rice plants near the water table (left). A heavy plant hopper infestation causes rice plants to wither and dry out thereby giving the field a burned appearance, called hopperburn (centre). Grubs of lady bird beetles prey on planthopper nymphs and adults (right). (Photos courtesy of IRRI)

and is still going on, has helped to prevent outbreaks of rice brown planthopper and other major rice pests. Subsidy funds can also be re-allocated to the financing of research on and popularization of integrated pest control. Subsidies on fertilizers also have some unwanted effects, and should also be reviewed. For example, high nitrogen gifts appear to make high-yielding short-straw cultivars of rice and wheat more sensitive to many pests and diseases which used to be of little importance previously on traditional cultivars.

Prices of agricultural commodities and thus farmers' income are often kept low for political resaons. For several years, donor agencies have exerted pressure on developing countries to raise farm-gate prices and gradually to decrease subsidies on agricultural inputs. Raising prices for major field crops creates an economic opportunity for a less intensive use of arable land and for crop diversification. These are two important factors in limiting the proliferous growth of soil-borne pathogens and other injurious organisms, and thus in reducing the need for pesticide applications. A 1987 World Bank report on Indonsia gave much attention to decreasing subsidies as a way for lowering public expenditure. It suggested that increasing farmers' income and charging normal cost prices for agricultural inputs and supply of irrigation water would help to re-organize Indonesian agriculture. In 1989, Indonesia complied by withdrawing all subsidies on pesticides by Presidential Instruction (Impres 6/89).

Figure 7.4: Pesticide production, rice production and pesticide subsidy, Indonesia, 1984-1990

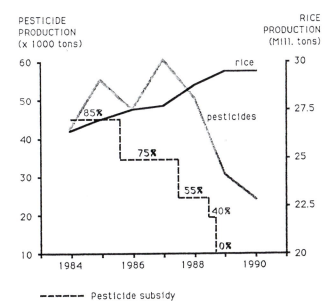

Source: Ministry of Finance, Indonesia, 1990

Emission into the environment

The intensive use of chemical pesticides leaves a high level of residues in the soil and ground water. Organic activity in the soil, an essential component of humification and of other processes which improve the soil structure, is thereby impaired. Slowly degrading compounds cause a gradual build-up of residues in the soil that may only after many years be recognized as a threat to underground deposits of drinking water; this happened in the case of the triazine herbicides. Similarly, the degradation products of, for instance, organo-chlorine insecticides tend to accumulate in the biological food chain of predators.

In seeking to prevent environmental contamination, any likely source of emission needs to be considered. Firstly, a part of environmental contamination stems from faulty application practices in plant protection. Leaking or badly adjusted spray apparatus, in particular that of larger units on aircraft and tractors, deliver an inefficient deposit pattern, and cause spilling and spray drift. A wrongly adjusted spreader may pulverise granules that are blown away as dust. Because insufficient training or discipline of the workers are also very common causes, several countries have made it mandatory that pesticide applicators be licensed.

Secondly, the method of application can usually be improved. Examples of saving on active ingredient are: applying insecticides by repeated spraying of low doses at short intervals (a technique called "incremental spraying") rather than to apply a high dose of active ingredient in a single treatment. Also, spraying liquid or dispersing solid formulations in field crops to rows of plants or individual plants only, instead of applying them to the entire soil surface. Similarly, combining mechanical weeding in-between rows with herbicide spraying within the rows is another practical way of minimizing product use. Chosing a formulation type that suits best the nature of the target to be treated is equally important: for instance, the use of insecticide granules for transplanted rice, a herbicide flowable for pre-emergence soil application, and a slow-release encapsulated insecticide for foliar application. Selecting proper spray nozzles and adjusting the rotational speed of atomizers in order to obtain a range of droplet sizes which can give an effective spray deposit, or adding a wetting agent to the spray solution in order to minimize loss from runoff, are two further examples of how environmental contamination can be avoided. Measures such as coating of seeds with insecticides and fungicides reduce the need for applying pesticides to the soil.

Thirdly, a policy of restricting or prohibiting the use of pesticides which are most harmful to the environment would help to further reduce environmental damage. So should the use of endosulfan be avoided in an aquatic environment, owing to its high toxicity to fish. Fertilizers too cause emission of residues into the groundwater and environment and should therefore also be applied with moderation.

An experimental orchard sprayer designed by IMAG, the Netherlands, enables a large reduction in pesticide emission. While moving along the rows of low-stem fruit trees, the machine completely surrounds and sprays individual trees. Spray particles which are not deposited on the trees are collected and pumped back into the spray tank through a filter system. This design holds promise for reducing the spray volume in other row-planted crops too. (Courtesy of IMAG, the Netherlands)

Box 7.1: Representation of steps to be taken from chemically dependent to
integrated agricultural production

Chemical control
    +   = **Rational chemical**
                     control
* Economic   +   = **Integrated control**
  thresholds   Biological control   +   = **Integrated protection**   = **Integrated**
* Damage   * biological method   *crop protection   +   **agricultural**
  assessment   - entomophagous   measures   * rational production   **production**
* Resistant   insects   - sanitation   techniques
  varieties   - entomoparasitic   - weeding   - green manure
                   nematodes   - barriers   - rotation
                 - entomopathogens   - quarantine   - mixed cropping
               * biotechnical methods - soil tillage   - fallow etc.
                 - pheromones   - early/late sowing
                 - sterile males
                 - growth regulators
                 - genetic engeneering

Source: Girardin, 1990.

Non-chemical methods of crop protection

A prerequisite to reducing the structural dependency on chemicals in crop protection is the availability of non-chemical methods and means of control, and their application in as many situations as possible. The most important non-chemical approaches are biological and integrated control of soil-borne and above-ground pathogens, pests and weeds. Since cost-effective methods of biological- and integrated control are as yet available for a few crops only -- with most successes booked for plantation crops -- the research for biological control agents and integrated crop protection techniques should be intensified. Researchers should, however, be aware that traditional farmers take some measures that help to prevent and control pests and diseases.

In several developed countries, a trend exists towards biological agriculture in which separate "ecological" and "biological-dynamic" currents can be distinguished. Biological agriculture strives, in principle, for a soil-bound crop production system on mixed farms, in which soil and crops and cattle function as a living entity (ecosystem) and as components of a closed nutrient cycle. The use of animal manures and of nitrogen-fixing crops for green manuring stimulates organic soil activity and keeps soil-borne pathogens in check. Biological agriculture rejects, on principle, the use of synthetic-chemical pesticides. Diseases, pests and noxious weeds are controlled as well as possible by preventive measures. These include the use of resistant cultivars, the promotion of a stable and active soil ecosystem, and a varied rotation cycle. If diseases and pests occur nevertheless, they are controlled by natural, non-synthetic products. Biological agriculture is labour-intensive and capital extensive; the land is extensively used. In general, yields of biological farms are more than 20 per cent lower than those in intensive commercial agriculture. Consequently,

biological producers have to ask a higher price for their produce. Moreover, the marketing of biologically grown produce requires much attention in terms of keeping transport and retail activities separated from the regular grain, vegetable and fruit trade.

Integrated farming systems

The term "integrated" indicates farming systems in which the emphasis is shifted from maximizing the physical production to cost-saving and quality improvement of both the product and production process. In integrated systems, a farmer does not merely grow field crops which give high yields and a high financial return in the short term, but he searches for a sound rotation schedule with satisfactory, sustainable financial results. Sustainability can be pursued by including crops in the rotation sequence which improve the soil structure and tend to suppress diseases, pests and noxious weeds. A beneficial rotation schedule could, for instance, contain

In southern Viet Nam, marigold flowers, Tagetes *spp., are grown for decoration at ceremonies. Incidentally, these plants have a suppressing effect on the population build-up of soil-nematodes. Thus, planting marigolds could provide a double advantage for vegetable growers. (Photo by M. Gorlugh)*

Table 7.3: Crop-wise use of chemical pesticides (in kg of active ingredient) in a chemically intensive (CI) and an integrated production (IP) system of the research farm OBS at Nagele, the Netherlands (1985-1987)

	Herbicide		Fungicide		Insecticide		Nematicide		Crop total	
	CI	IP	CI	IP	CI	IP	CI	IP	CI	IP
Seed-/consumption potato	3.1	0.4	14.7	7.6	0.4	0.0	170.6	-	188.8	8.0
Sugarbeet	4.7	1.7	0.0	0.0	0.4	0.2	-	-	5.1	1.9
Winter wheat[b]	4.8	1.5	2.1	1.5	0.1	0.0	-	-	7.0	3.0
Green peas	4.7	4.3	1.6	1.2	0.2	0.2	-	-	6.5	5.7
Seed onions[b]	9.7	2.7	9.0	4.1	0.1	0.0	-	-	18.8	6.8
Winter carrot	2.9	1.6	0.7	0.3	4.8	1.0	-	-	8.4	2.9
Average of rotation cycle[a]	4.5	1.7	5.0	2.7	0.6	0.1	42.7	-	53.1	4.6

Notes: a Rotation cycle: potato - various crops(1/2 peas, 1/4 onion, 1/4 carrots) - sugarbeet - wheat.

b Growth regulators are used on seed onions at 2.3 kg in CI and IP both, and on winter wheat at 0.5 kg in CI only.

Source: P. Vereyken; Landbouwkundig Tijdschrift, 101 (1989), 1.

two years of tuber- or root crops followed by two years of cereals and one year of nitrogen-fixing crops. (Refer to table 7.3). Furthermore, a farmer practising integration does not apply fertilizers as a matter of routine nor out of fear of nutrient shortages, but he tries to avoid unnecessary costs to his farm and to the environment by using a combination of animal manure, fertilizer and leguminious plants as green manure. Current fertilization advice is based on maximum production levels, but it is hardly recognized that high fertilizer doses attribute to a rapid population build-up of soil-born pathogens and a destabilisation of the soil-ecosystem.

In integrated agriculture, the use of chemical pesticides is not rejected on principle. Rather, the most harmful compounds in terms of persistency and leaching hazard as well as soil sterilants are not applied. Pesticides are used rationally, which means that a farmer determines when such compounds are needed by monitoring the incidence of major pests and diseases and by applying the economic threshold principle, and determining also how they can be applied with the least negative effects to people and the environment. The farmer makes optimum

Sweep-nets are frequently used in scouting to determine insect numbers present. However, sweep-nets collect insects only from the upper plant parts and, thus, may present a biased sample. The lower plant parts should be checked also for the occurrence of planthoppers, stemborers and beneficial insects. (Courtesy of FAO-IPC Rice Programme)

use of preventive control methods which include the use of resistant cultivars, healthy seed and planting material, coated seeds and phyto-sanitary measures. By keeping a beneficial (micro-)flora and fauna in and around the cultivated plots intact, he creates a more stable agro-ecosystem for the control of diseases, pests and weeds.

Although the application of biological means and methods of pest and disease control is a very important aspect of integrated agriculture, the availability of sufficient selective pesticides thus remains a pre-condition for its economic viability. But these selective pesticides may not be harmful to the natural enemies (biological control agents). For example, in the case of late-season planting of crops, certain pests may multiply so rapidly that their natural enemies fail to control them. A suitable chemical pesticide with selective action will then be needed to prevent high crop losses.

The most important social effects expected to result from large-scale application of integrated farming systems are:

* The maintainance of the production potential of the soil (the sustainability of agricultural production);

* Less contamination of agricultural produce, meat and fish from pesticide residues;

* A considerable reduction of environmental contamination;

* A better safeguarding of public health; and

* A net return on investment in integrated systems that is comparable to the net return on investment obtained in the current intensive systems (see fig. 7.5).

Another precondition to achieving widespread acceptance of integrated crop production and protection is to train the masses of farmers. In planning extension programmes on integrated control, one should commence by making a baseline study of farmers' current practices and their knowledge about non-chemical control

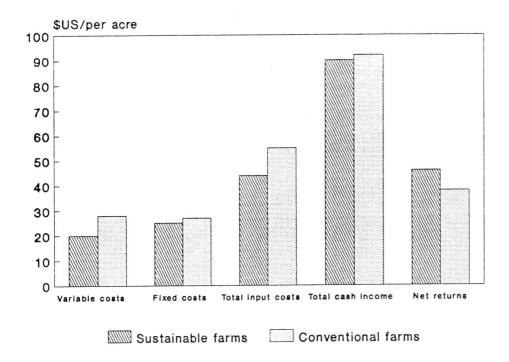

$US/per acre

Figure 7.5: Profit

Profits from sustainable farms can exceed those of conventional farms, according to Steven L. Kraten, formerly of Washington State University, United States. The cash incomes per acre for the two types of farms were comparable over two years, but because the input costs of sustainable agriculture are lower, its net returns are 22.4 percent higher. Variable costs include those for fuel, machinery maintenance, seed, fertilizer, pesticide and labour. Among the fixed costs are property taxes and interest on loans.

methods. Such a baseline study can provide clues for designing the technical guidance (extension messages and training content) that will help to solve problems which farmers recognize as realistic ones. The study should also identify ways by which farmers can be reached effectively. Suitable methods include display of posters and the broadcasting of special radio programmes for farmers. An example of the latter is the programme "IPM on Air" produced by the Philippine-German Crop Protection Programme. However, radio programmes and posters need to be updated from time to time in to order to retain their appeal. This requires a frequent monitoring and evaluation of farmers' acceptance and response. As in all other sectors of agriculture, providing technical guidance on integrated crop protection demands a well-trained and highly motivated extension service with a sizeable long-term budget.

REGULATORY INFRASTRUCTURE

Responsibility of the Government

The use of pesticides and the risks involved demand a well-developed social infrastructure which is able to respond adequately and efficiently to prevailing problems and which is alert to the appearance of new problems. Not only are economic and personal interests of individuals at stake in the production and distribution sector, but also those of many others, including consumers of agricultural products and users of the environment, of which the production system forms an inseparable component. This fact calls for a strongly developed sense of responsibility on the part of all parties concerned, and a strong government presence.

It is the responsibility of national authorities and international organizations to create the conditions for sustainable crop protection as well as for rational and safe pesticide use. Consequently, all developed countries have taken action to introduce the necessary legislation for the regulation, including registration, of pesticides. Such legislation directs that pesticides can be registered for domestic use only after the testing of their chemical and toxicological properties and environmental hazards. The registration procedures include an evaluation of product efficacy and use areas, and the establishment of permissible residue levels and possible use restrictions. The registration authority should from time to time review the pesticides to be marketed in the country and may decide, on the basis of new test data, either to completely ban or to withdraw admission of active ingredients and products or to impose restrictions on their use in certain areas and by unlicensed applicators. (See table 7.4).

Table 7.4: Summary of data on chemical pesticides registered for use in agriculture in the Netherlands in 1989

Category of admitted chemicals (year 1989)	Pesticide products			Active ingredients		
	a	b	c	d	e	f
Fungicides	479	14	41	76	2	5
Insecticides	287	32	25	60	4	1
Herbicides	491	37	44	85	7	1
Other pesticides	246	27	16	73	4	0

Source: Kwartaalbericht milieu (CBS) 90/2.

Note: Number of admitted pesticide products: total (a), withdrawn (b) and new products (c), number of admitted active ingredients: total (d), withdrawn (e) and new active ingredients (f). Periodical reviews of admitted active ingredients and products are a powerful instrument in promoting integrated and safe plant protection technology.

Registration provisions further demand that products be adequately packed and that labels contain full information on the contents, use directions and safety warnings. A pesticide may be applied only according to official prescriptions and in approved use areas. In order to ensure compliance with the regulations, Governments should also enforce them. Other government tasks include the monitoring of the import and export of pesticides, quarantine, testing of pesticide quality and analysing residue levels in food and the environment and, finally, education of users concerning safe pesticide use and integrated pest management.

Responsibility of the private sector

The private sector is composed of several categories which can be roughly divided into a supplier group and a user group. The suppliers are primarily responsible for the development of appropriate pesticide products and application equipment, for their effective distribution and for the availabilty of adequate advice on the safe management and effective use of those products and apparatus. The second group includes agriculturists and horticulturists, livestock farmers, professional applicators, farmers' organizations, processing industry and individual households.

Another important group consists of highly critical non-governmental organizations (NGOs) which give voice to the growing public discontent concerning the negative effects of excessive pesticide use, the hidden costs of which are not yet sufficiently well taken into account by agricultural policy makers and the pesticide industry. Their continuous pressure on Governments and industry has led to an increasing willingness to review current legislation, to develop more specific pesticide formulations having less broad-spectrum toxic activity, and to develop and apply integrated pest management methods.

Regulatory problems in third world countries

Generally, developing countries lack the required infrastructure and technical capabilities to ensure the safe manufacturing, marketing and use of pesticides. Short-comings in management capacity, in knowledge and skill, and in the execution of tasks can easily be pointed out at many organizational and executive levels in government institutions and industry.

On the supply side, many hazardous situations and practices exist in the manufacture, storage, transport and retail distribution of pesticides which are not rectified or even not recognized as being dangerous. In general, since no special license or diploma is required for trading or selling pesticides, very few retailers care about informing themselves and their clients about product properties, suitable application methods and use restrictions. Sometimes, even illegal practices such as product adulteration and misleading advertising are observed.

On the regulatory side, government authorities are barely or not at all equal to their task because of a shortage of funds and lack of trained administrators, scientists and executive staff to operate the services and laboratories involved in pesticide management and crop protection. The legislative basis is usually incomplete and the regulatory system lacks the means to enforce regulations concerning the trade, distribution and rational use of pesticides. Socio-economic conditions in developing countries often hinder communication between researchers on the one side and field staff and farmers on the other side, which fact might render results from research largely irrelevant to farmers' actual problems. The unattractive employment situation of agricultural staff and resistance to working outside urban areas accounts for their often low motivation and inefficiency.

On the users' side, there are very great differences,

Educating farmers concerning integrated pest management involves extensive demonstrations and participatory training at the field level, whereby farmers get ample opportunity to exercise and rehearse their skill in recognizing pests and natural enemies and taking control decisions. (Courtesy of FAO-IPC Rice Programme)

Research and extension staff should talk directly with farmers about their plant production and protection problems in order to understand their needs as well as knowledge, attitude and practice levels. The author (second from left) and Dr. J. Mumford, Silwood Centre for Pest Management, U.K., interview farmers in the Philippines. (Photo supplied by the author, 1989)

in terms of knowledge and skill levels and of access to products and equipment, between large-scale commercial enterprises and small-scale farmers. The first either possess sufficient knowledge and means or have the resources to hire them. The latter have little knowledge either about pest- and disease problems, or about selecting the right product, dosage rate and timing for controlling individual pests or diseases, or about safe handling -- includes the need for protective clothing -- and efficient application equipment and methods. Legislation is usually uniform and docs not take into account specific hazards or needs of either of these groups. Because of the generally high rate of illiteracy among small-scale farmers, special methods and materials for informing them have to be developed that should suit local conditions. Agricultural extension services, if in existence, often lack the organization and funding to provide guidance to poor farmers. Therefore, the latter are mainly dependent on pesticide sales agents from whom they may get biased advice but rarely instructions on the safe

handling of pesticides and the potential hazard they pose to individuals and the environment.

International co-operation in crop protection, vector control and pesticide management

It is evident that technical and financial aid from donors will continue to be needed for a considerable time, especially to support national services in carrying out the above tasks. A number of international organizations are active in the field of crop protection, vector control and the use of pesticides. They are usually concerned with co-ordination, research, education and extension, legislation and regulation, quarantine and the control of migratory pests and epidemics. Among them are seven United Nations organizations, of which FAO, WHO and UNEP are the most influential (see box 7.2). UNDP co-ordinates the multilateral co-operation between United Nations organizations and organizes funding of programmes. IAEA undertakes research on genetic modification of pests and pathogens.

Box 7.2: Organizations of the United Nations active in the field of crop protection

Organization	Field			Tasks				
Food and Agriculture Organization (FAO)	C P V			r	e	a	o	l
World Health Organization (WHO)	P V			r	e	a	o	l
United Nations Environmental Programme (UNEP)	P	E		r			o	l
International Atomic Energy Agency (IAEA)	C	V		r				
United Nations Industrial Development Organization (UNIDO)	P						o	l
International Labour Organisation (ILO)	P				e		o	l
United Nations Development Programme (UNDP)	C P V E						o	

Note: Crop Protection (C), Manufacture and use of pesticides (P), Control of disease transmitting vectors (V), and Environmental protection (E). Tasks of these organizations include: research (r), education and extension (e), actual control (a), organization and co-ordination (o), legislation (l).

In 1951, several international agencies including FAO, founded the International Plant Protection Convention (IPPC) with the objective of promoting and co-ordinating inter-governmental activities concerning the prevention and control of diseases and pests of international importance. Under the aegis of the International Plant Protection Convention, seven regional organizations were set up. (See box 7.3.) Also the World Bank, the International Fund for Agricultural Development (IFAD) and the regional development banks for Asia (ADB), Africa (AFDB), Inter-America (IADB) and the Arabian Gulf Fund contribute funds for plant protection programmes.

International programmes on plant protection and pesticide management

The above-mentioned global agencies, in collaboration with national authorities and international institutes, execute several international assistance programmes in the field of plant protection. These programmes can be divided into three categories, namely:

* Programmes aimed at the control of epidemics and plagues: examples are the FAO emergency control programmes for desert locust (1985-89) and screw worm in Libya (completed in 1991). Under the FAO Technical Co-operation Programme also medium-term programmes for the control of cassava mealy bug and cassava mite, and long-term programmes, such as the

Desert Locust Surveillance and Forecasting Service, are being maintained.

* Programmes aimed at the development and reinforcement of the plant protection infrastructure, inclusive of regulatory systems, education and research, executive services, inspection and enforcement, distribution and extension. These types of programmes require a medium- to long-term commitment from donors. Examples are: the FAO Global Programme for the Implementation of the International Code of Conduct on the Distribution and Use of Pesticides, with regional sub-programmes in Asia, West Africa and Central America; the FAO-UNDP (proposed) programme Support to Control of Migratory Pests and Plant Protection in Africa; and, the bilateral CILSS-Netherlands Education in Plant Protection project for Sahelian Countries in Niger, Africa. Other supportive programmes are the Cooperative Action Programme for Improved Plant Health which organized 194 training courses in the period 1980-1985, and the on-going FAO Fellowships Programme, which granted 474 fellowships to plant protection students in the period 1977-1985.

* Programmes directed at the development of new control methods. Examples are: the FAO Global Programme for the Development and Application of Integrated Pest Control in Agriculture, and the International Programme on Horizontal Resistance (a new approach to identifying sources of sustainable resistance).

In 1975, FAO and UNDP jointly launched the Global Programme for the Development and Application of Integrated Pest Control in Agriculture with the objective of researching and introducing an ecologically sustainable plant protection technology. The FAO/UNEP Panel of Experts on Integrated Pest Control advises on policy making, planning and execution of projects in this field. Examples are the current integrated pest management projets in cotton in Africa and Latin America, in rice in South-east Asia and East Asia, in cassava in Africa, and in vegetables in the Asian and the Caribbean regions.

Integrated pest management requires prolonged field studies to identify possible natural control agents and to understand their interaction with major pests. Cages made of gauze fabric provide a niche for their undisturbed development. (Courtesy of FAO-IPC Rice Programme)

An international programme for data collection and pesticide management, called the Agricultural Requisites Scheme for Asia and the Pacific (ARSAP), was initiated in the 1960s by the United Nations Economic and Social Commission for Asia and the Pacific (ESCAP), Bangkok, with financial assistance from the Netherlands. The ARSAP/agro-pesticide project, which became operational in 1978, analysed data on pesticide supply, distribution and use in 13 Asian countries. Furthermore, ARSAP organized training programmes for government staff and retailers in 10 countries. Since 1986, the French International Co-operation Centre of Agricultural Research for Development (CIRAD) has been supporting ESCAP/ARSAP in the publication of the Regional Agro-pesticide Index. The 1991 Index contains data from 15 Asian countries on 700 active ingredients and 8,000 products.

In 1982, the UNIDO Regional Network for the Production, Marketing and Control of Pesticides in Asia and the Far East (RENPAF) was founded at Manila. This network, while taking over some of ARSAP's data collection and information tasks, focuses on standardization of quality control and analytical methodologies, and on harmonization of registration procedures and trade tariffs in Asia.

In 1972, UNEP commenced operation of a programme called the International Register for Potentially Toxic Chemicals (IRPTC), at Geneva. Its objective is to evaluate industrial chemicals, inclusive of pesticides, and to provide full information on their chemical properties and hazards. The programme forms an important contribution to the data exchange on restricted chemicals and to the organization of pesticide management and legislation in the participating countries. In a joint project, UNEP and FAO are establishing a global IRPTC-network in support of implementing the International Code of Conduct.

The complementary International Programme on Chemical Safety (IPCS) is a joint undertaking of UNEP, WHO and ILO. Its objective is to evaluate potential hazards of chemical compounds for the purpose of establishing Environmental Health Criteria. A second objective is the improvement of the procedures for measuring the negative effects of chemical compounds on health and the environment.

The Joint Food Contamination Monitoring Programme (JFCMP) is an international programme of UNEP, WHO and FAO, and a component of the UNEP Global Environmental Monitoring System (GEMS). Its objective is to analyse data on chemical residues in food. The programme component GEMS/Food is associated with the Joint FAO/WHO Codex Alimentarius Commission, an organization in charge of elaborating proposals for residue-tolerance levels in food. A total of 20 developed and 14 developing countries participate in the GEMS/Food programme.

Crop protection research largely takes place in the major International Agricultural Research Centres (IARCs) and the International Centre for Insect Physiology and Ecology (ICEPE), at Nairobi. These are situated in developing areas and are co-operating closely with national research centres.

International guidelines on the use of pesticides

Already in the 1950s, scientists became aware of the negative effects of pesticide use and United Nations organizations soon took up the issue. In 1969, FAO, in collaboration with WHO, formulated guidelines for legislation concerning registration for the sale and marketing of pesticides. The organizations continued by presenting in 1970 a "model scheme for founding national organizations for official control of pesticides". At the World Conference on the Human Environment at Stockholm in 1972, three leading international agencies, FAO, WHO and World Bank, took the initiative in developing a Code of Conduct to help minimize the unwanted side-effects of pesticide use in the developing countries. The formal decision to develop the Code was taken at the FAO Second Government Consultation on International Harmonization of Pesticide Registration Requirements at Rome in October 1982.

In 1975, WHO proposed for the first time a classification of pesticide hazards. In 1978, UNEP took the decision to request Governments of exporting countries to halt the export of pesticides until the importing countries could take a decision on the admission or refusal of pesticide imports based on the submitted product information. This principle of "prior informed consent" (PIC) was further discussed in three resolutions of the General Assembly of the United Nations in the period 1979-1982. The PIC-issue was then also taken up in GATT (General Agreement on Tariffs and Trade) discussions and by the OECD (Organization for Economic Cooperation and Development). The OAS (Organization of American States) decided that a list should be made of all toxic chemicals which were banned in the United States or in other countries. In 1983, the European Parliament also accepted a resolution seeking an amendment of the existing European guidelines on the export of pesticides. Since then, the following international guidelines have been drawn up:

* FAO-International Code of Conduct on the Distribution and Safe Use of Pesticides; underwritten by the Environment Liaison Centre (ELC) and the International Association of Pesticide Manufacturers (GIFAP). (Adopted by the FAO Conference at its twenty-third session in 1985 by way of resolution 10/85, as amended by resolution 6/89 of the twenty-fifth session of 1989.)

* World Bank: internal policy guidelines with reference to crop protection programmes and delivery and use of pesticides under Bank-financed projects.

* EC-regulations with regard to the export from and import into the European Community of certain hazardous chemical substances, including nine pesticides.

* UNEP-guidelines for the Exchange of Information on Chemicals in International Trade.

* The OECD recommendation concerning Information Exchange Related to Export of Banned and Severely Restricted Chemicals.

* The London Guidelines for the Exchange of Information on Chemicals in International Trade. (Decision 15/30 of the Governing Council of UNEP of 25 May 1989)

The FAO International Code of Conduct on the Distribution and Use of Pesticides

The Code was born out of concern on the side of Governments and public sector organizations about the propriety of supplying pesticides to countries which do not have infrastructures to register pesticides and thereby to ensure their safe and effective use. The objective of the Code is to set forth the responsibilities and to establish voluntary standards of conduct for all parties involved in the manufacture, distribution, marketing and use of pesticides. The Code is based on the necessity of voluntary collaboration between exporting and importing countries in the realization that anyone dealing with these toxic chemicals must also share responsibility. The articles of the Code contain numerous technical guidelines and intend to encourage responsible and acceptable marketing practices. They provide guidance to countries that have not yet developed adequate legislation and control, and where knowledge of chemical pest control matters is generally very insufficient. In this way, they serve to facilitate the safe and rational use of indispensable pesticides whilst minimizing hazardous side-effects to people and the environment.

In promoting the implementation of the Code, FAO collaborates with 156 member Governments, each of them having its own regulatory system. FAO helps national authorities in introducing the necessary legislation, including registration of pesticides, and in the training of personnel. Harmonization of regulations and registration procedures is being pursued at the regional level.

Under an overall global programme, regional projects for the implementation of the provisions of the FAO Code of Conduct are undertaken in collaboration with donor countries. Examples are the Japanese-funded project for Asia, the Netherlands-funded project for Central America and a UNDP-funded project for West Africa.

The Groupement International des Associations des Fabricants des Produits Agrochemiques (GIFAP) stimulates and co-ordinates the implementation of the Code by means of publications, training, symposiums and by special instructions to associated national associations of manufacturers and distributors. These national associations are supposed to adhere to the principles of the Code.

The following paragraphs give a brief description of the provisions laid down in the articles of the Code of Conduct. The complete text of the Code is attached at the end of this chapter.

The Code is subdivided into 12 articles, of which the main points for attention are:

* Article 1: Objectives of the code

The objectives of the Code are to identify the responsibilities concerning agro-pesticides of the various segments of society and to lay down standards for attitudes and behaviour which all public and private parties involved should adopt or demonstrate voluntarily. To improve chances for adherence to these standards the Code encourages countries to develop the necessary legal infrastructure and to strengthen international co-operation between Governements, industry and public sector organizations.

* Article 2: Definitions (text revised in 1989)

This article explains the meaning of concepts and important words used in the code in order to make the legal text more understandable for the reader.

* Article 3: Pesticide management

This article places overall responsibility for efficient pesticide management with Governments and stresses that only they have the power of enforcement. This responsibility does not stop at the border. By allowing the export of toxic chemicals from the country, a Government automatically accepts an obligation to assist the recipient country in managing the imported chemicals. By underwriting voluntarily the principles of the International Code of Conduct, all 156 member states of FAO have de facto assumed this moral responsibility. In a similar way, all manufacturing and trading companies should try to ensure that their pesticides reach the users everywhere in a condition of good quality, packaged in proper containers and accompanied by the necessary information (on labels and in promotional messages) for their safe and effective use. Since individual Governments and companies do not have the resources or influence to remedy the many cases of mismanagement taking place in developing and underdeveloped countries, they should collaborate through national and international organizations to ensure that pesticide management improves all over the world.

The article draws special attention to the rapid development of pesticide resistance in organisms, which limits the availability of valuable pesticides.

* Article 4: Testing of pesticides

Each pesticide product, brought onto the market in whatever formulation form, must be tested by reliable standard procedures so that all its properties can be known in terms of control potential, toxicological hazards, application requirements and pathways of chemical degradation under the conditions of the country where the product will be actually used. All countries in which the product will be offered for sale should be provided with evaluated test reports by the manufacturer for evaluation by qualified registration experts. The manufacturer should enable countries that do not have the facilities to check the numerous technical data by providing analysis methods and encouraging the training of laboratory staff.

The country itself should develop quality control facilities or have access to analytical laboratories which are sponsored by international or private sector organizations. Product control should be continued after a formulation has been registered (product reviews) and the fate of the chemical in the environment after its application should be surveyed.

* Article 5: Reducing health hazards

On the basis of numerous studies of intoxication from pesticides in Asia, Central America and Africa, it is generally assumed that 20 per cent of all farmers and professional applicators in developing countries suffer from pesticide poisoning once or several times during their working life. Of these incidents, 10 to 15 per cent are fatal. These alarming figures call for guidelines on safe marketing, handling and application that aim to, on one side, reduce toxicity of the products and, on the other (particularly important) side, minimize the chance of exposure. This requires that the toxic properties and hazards of each pesticide product are well-known and clearly indicated on safe-to-use containers. It also requires that Governments control the range of toxic pesticides on the market, enforce their safe management in storage, transport and distribution, and encourage safe disposal of used containers and waste. Governments should not hesitate to ban or restrict the use or to recall products if safe use proves to be impossible. The article furthermore calls for the establishment of national poison information and control centres which would assist medical staff in improving the diagnosis and treatment of intoxications through the organization of tele-communication, training and antidote distribution.

* Article 6: Regulatory and technical requirements

This article invites Governments to develop an adequate regulatory system along FAO guidelines and a registration scheme. It should, however, warrant property rights on products and data files provided by manufacturers. Governments should also encourage data collection on supply, distribution and use of pesticides as instruments for policy making in the fields of agriculture, public health and the environment.

The remaining sub-articles state that the industry should provide objective statements about the purity and quality of the pesticide products marketed in correspondence with international WHO and FAO specifications. The industry should itself correct mistakes or help Governments to find solutions to problems.

* Article 7: Availibility and use

This article addresses the Governments and intends to protect the users of pesticides. Governments should allow only products which can be safely managed by users, thus which are compatible with their level of training and expertise, and the application equipment available to them. Similarly, the packaging and labelling should be adapted to the target group, particularly if the users are illiterate.

* Article 8: Distribution and trade

The industry has the responsibility that its pesticide products are well tested under the conditions of the country where they will be used. Further, that they conform to the same high standards of quality and packaging as required in the country of manufacture and as laid down in international specifications and guidelines. Because of the global character of pesticide trade, parent companies are required to ensure that their daughter companies or agencies collaborate in avoiding malpractices, and sell through well-trained and reliable retailers. In case the imported or locally formulated pesticide represents an unacceptable hazard, the parent company should stimulate its recall from the market. The last sub-article refers to the international code regarding residues in food (Codex Alimentarius) and recommends the drafting of national residue legislation that will enable the rational use of pesticides in agriculture.

* Article 9: Information exchange (text revised in 1989)

This article deals with the exchange of information between countries and international organizations concerning any restriction in the use of pesticides imposed by any country. In particular, no Government should permit the export of banned or seriously restricted chemicals before it has officially informed the importing country of these restrictions and has received notification of the importing country's decision. This procedure is called the "Principle of Prior Informed Consent" (PIC). It offers countries, which do not possess the technical capabilities of evaluating hazardous chemicals, protection against the import of pesticides that pose a threat to health and the environment. It should also stimulate decision-making capability and development of regulatory mechanisms in importing countries.

After many years of discussions, this PIC-procedure was finally accepted by the twenty-fifth session of the FAO Conference in 1989 with agreement of the industry. Its acceptance was disputed owing to the difficulty of establishing in individual developing countries a designated national authority capable of handling and evaluating the PIC-data. The industry feared delays in obtaining the requested import permission and a negative effect on international trade in pesticides. To date, only the Netherlands has practical experience with the working of the PIC-procedure. Following the enactment of a bill on the management of environmentally hazardous chemicals (Wet Milieugevaarlijke stoffen) in 1985, the Netherlands Government and industry entered upon a voluntary agreement concerning the export of banned and restricted pesticides. On the basis of the Netherlands experience, FAO and UNEP are pursuing the global implementation of the PIC-procudure in a joint project through a network of national or regional authorities.

* Article 10: Labelling, packaging, storage and disposal

This article consists of guidelines on label design and the information it should contain. The use of local languages, pictogrammes and colour indications should enable the transmission of warnings and use instructions to uneducated users. Governments and industry share a responsibity in preventing illigal and unsafe repackaging and the wrong use of containers.

* Article 11: Advertising

This article gives voice to public concern about promotional activities that are based on misleading or exaggerated claims or that appear to have a scientific basis which they do not possess. Promotional material or advertisements should contain complete information on use restrictions and hazardousness, but should not depict unsafe practices. Statements on safety should always contain the phrase "when used as directed" and should not compare one product with another. Pesticides which are restricted to use by licenced operators should not be advertised other than in professional journals with mention being made of said restriction.

* Article 12: Monitoring the observance of the Code

This article stresses the importance of a world-wide publication of the Code through Governments, United Nations organizations and individual parties in order to give it maximum impact. It calls for observance of all relevant legal rules already in existence. Finally, Governments are invited to monitor the observance of the Code and to report on progress made in order to enable the international bodies to review periodically the relevance of its articles.

References:

Crop protection, vector control and pesticide use in developing countries. Report prepared by a working group of the Ministry of Agriculture, Nature Management and Fisheries of the Netherlands, Wageningen, July 1990.

Girardin P. (1990); **Willing to pay more?,** Institut National de la Recherche Agronomique. BP 507, F-6804 Colmar, France.

Maerjarenplan Gewasbescherming (1990); A 10-year plan for plant protection in the Netherlands, 1990-2000. Ministry of Agriculture, Nature Management and Fisheries, the Netherlands.

Niessen, H.J. (1986); **Pesticide regulatory analysis from an industrial viewpoint.** Paper given at the 6th International Congress of Pesticide Chemistry. GIFAP.

Reganold, J.P., Papendick R.I. and Parr J.F. (1990). **Sustainable agriculture.** Scientific American, June 1990.

Report of the Conference of FAO, twenty-fifth session, Rome, 11-29 November 1989. Food and Agriculture Oraganization of the United Nations (C 89/Rep).

Report of the expert consultation on the introduction of "Prior informed consent" in article 9 of the International Code Of Conduct on the distribution and use of pesticides. Rome, 23-25 March 1988. Food and Agriculture Organization of the United Nations.

Tweede Kamer der Staten-Generaal, vergaderjaar 1986-1987, stuk 19 310 nr 4. Uitvoer van gevaarlijke stoffen en preparaten.

International Code of Conduct on the Distribution and Use of Pesticides

(Amended to include Prior Informed Consent in Article 9 as adopted by the 25th Session of the FAO Conference in November 1989)

**FOOD AND AGRICULTURE ORGANIZATION
OF THE UNITED NATIONS
Rome, 1990**

Preface

The action by FAO to develop, in consultation with appropriate United Nations agencies and ohter organizations, an International Code of Conduct on the Distribution and Use of Pesticides follows and accompanies many other events, some going back 25 years. All these events were designed to benefit the international community and to serve to increase international confidence in the availability, regulation, marketing and use of pesticides for the improvement of agriculture, public health and personal comfort.

One of the basic functions of the Code, which is voluntary in nature, is to serve as a point of reference, particularly until such time as countries have established adequate regulatory infrastructures for pesticides.

The Director-General of FAO in 1981 suggested that such a Code could help to overcome a number of difficulties associated with pesticides. The FAO Panel of Experts on Pesticide Specifications, Registration Requirements and Application Standards, at its meeting in 1982, agreed that activities involving the export and import of pesticides, and thereby their safe use, might be best dealt with through the adoption of a Code of Conduct. To that end a working paper was prepared for the FAO Second Government Consultaiton on International Harmonization of Pesticide Registration Require- ments, Rome, 1-5 October 1982. The formatl decision to develop the Code was taken at that Consultation., which recommended that FAO, in consultation with the appropriate United Nations organizations and bodies and international organizations outside the United Nations system, should draft a Code (1).

The Code itself was adopted by the FAO Conference at its Twenty-third Session in 1985 by way of Resolution 10/85, which appears as an Annex to the present publication.

A number of governments and organizations have expressed concern about the propriety of supplying pesticides to countries which do not have infra- structures to registered pesticides and thereby to ensure their safe and effective use. It should be noted that the development of national regulatory programmes is the first priority of FAO activities in this field. There has also been concern over the possibility that residues of certain pesticides, not needed or not permitted in particular countries, are present in imported agricultural commodities produced in other countries where the use of such pesticides is not restricted. While recognizing that it is impossible to eliminate all such occurrences, because of diverging pest control needs, it is none the less essential that every effort be made to apply pesticides only in accordance with good and recognized practices. It is at the same time important for industrially developed countries to recognize, in their regulatory activities concerning residues, the pest control needs of developing countries, particularly the needs of countries in tropical regions.

In the absence of an effective pesticide registration process and of a governmental infrastructure for controlling the availability of pesticides, some countries importing pesticides must heavily rely on the pesticide industry to promote the safe tnad proper distribution and use of pesticides. In these circumstances foreign manufacturers, exporters and importers, as well as local formulators, distributors, repackers, advisers and users, must accept a share of the responsibility for safety and efficiency in distribution and use.

The role of the exporting country needs to be considered. Much emphasis has been given recently to the desirability of regulating the export of pesticides from producing countries. It is generally accepted that no company should trade in pesticides without a proper and thorough evaluation of the pesticide, including any risks. However, the fact that a product is not used or registered in a particular experting country is not necessarily a valid reason for prohibiting the expoert of that pesticide. Developing countries are mostly situated in tropical and semitropical regions. Their climatic, ecological, agronomic, social, economic and environmental conditions and therefore their pest problems are usually quite different from those prevailing in countries in which pesticides are manufactured and exported. The government of the exporting country, therefore, is in no position to judge the suitability, efficacy, safety or fate of the pesticide under the conditions in the coufntry where it may ultimately be used. Such a judgement must, therefore, be made by the responsible authority in the importing coufntry in consultation with industry and other government authorities in the light of the scientific evaluation that has been made and a detailed knowledge of the conditions prevailing in the coufntry of proposed use.

The export to developing countries of epsticides which have been banned in one or more other countreis or whose use has been severely restricted in some industrialized countries has been a subject of public concern which has led to intensive discussions. In addressing this issue, the FAO Conference at its Twenty-fifth Session in 1989 agreed to introduce provisions for Prior Informed Consent (PIC) procedures. These procedures are described in the revised Article 9 on Information Exchange and Prior Informed Consent.

While a Code of Conduct may not solve all problems, nevertheless it should go a long way toward defining and clarifying the responsibilities of the various parties involved in the development, distribution and use of pesticides, and it should be of particular value in countries which do not yet have control procedures. Where there is a pesticide regulatory process in a country, the need for a Code of Conduct will obviously be less than where there is no such scheme in operation.

The Code of Conduct is not a short or simple document, mainly because the nature, properties, uses and effects of pesticides are diverse and therefore require comprehensive consideration. Furthermore, the strong public pressure for banning or restricting the use of some effective and much needed pesticides often stems from a lack of understanding of the many important issues involved. This document is designed, therefore, also to provide the general public with some basic guidance on these issues.

Edouard Saouma
Director-General
FAO

Text of the Code

Article 1.   Objectives of the Code

1.1   The objectives of this Code are to set forth responsibilities and establish voluntary standards of conduct for all public and private entities engaged in or affecting the distribution and use of pesticides, particularly where there is no or an inadequate national law to regulate pesticides.

1.2   The Code describes the shared responsibility of many segments of society, including governments, individually or in regional group ings, industry, trade and international institutions, to work together so that the benefits to be derived from the necessary and acceptable use of pesticides are achieved without significant adverse effects on people or the environment. To this end, all references in this Code to a government or governments shall be deemed to apply equally to regional groupings of governments for matters falling within their areas of competence.

1.3   The Code addresses the need for a cooperative effort between governments of exporting and importing countries to promote practices which ensure efficient and safe use while minimizing health and environmental concerns due to improper handling or use.

1.4   The entities which are addressed by this Code include international organizations; governments of exporting and importing countries; industry, including manufacturers, trade associations, formulators and distributors; users; and public-sector organizations such as environmental groups, consumer groups and trade unions.

1.5   The standards of conduct set forth by this Code:

1.5.1   encourage responsible and generally accepted trade practices;

1.5.2   assist countries which have not yet established controls designed to regulate the quality and suitability of pesticide products needed in that country and to address the safe handling and use of such products;

1.5.3   promote practices which encourage the safe and efficient use of pesticides, including minimizing adverse effects on humans and the environment and preventing accidental poisoning from improper handling;

1.5.4   ensure that pesticides are used effectively for the improve ment of agricultural production and of human, animal and plant health.

1.6   The Code is designed to be used, within the context of national law, as a basis whereby government authorities, pesticide manufacturers, those engaged in trade and any citizens concerned may judge whether their proposed actions and the actions of others constitute acceptable practices.

Article 2.   Definitions

For the purpose of this Code:

Active ingredient means the biologically active part of the pesticide present in a formulation.

Advertising means the promotion of the sale and use of pesticides by print and electronic media, signs, displays, gift, demonstration or word of mouth.

Banned means a pesticide for which all registered uses have been prohibited by final government regulatory action, or for which all requests for registration or equivalent action for all uses have, for health or environmental reasons, not been granted.

Common name means the name assigned to a pesticide active ingredient by the International Standards Organization or adopted by national standards authorities to be used as a generic or non propri etary name for that particular active ingredient only.

Distinguishing name means the name under which the pesticide is labelled, registered and promoted by the manufacturer and which, if protected under national legislation, can be used exclusively by the manufacturer to distinguish the product from other pesticides containing the same active ingredient.

Distribution means the process by which pesticides are supplied through trade channels on local or international markets.

Environment means surroundings, including water, air, soil and their interrelationship as well as all relationships between them and any living organisms.

Extension service means those entities in the country concerned responsible for the transfer of information and advice to farmers regarding the improvement of agricultural practices, including production, handling, storage and marketing.

Formulation means the combination of various ingredients designed to render the product useful and effective for the purpose claimed; the form of the pesticide as purchased by users.

Hazard means the likelihood that a pesticide will cause an adverse effect (injury) under the conditions in which it is used.

Integrated pest management means a pest management system that, in the context of the associated environment and the population dynamics of the pest species, utilizes all suitable techniques and methods in as compatible a manner as possible and maintains the pest populations at levels below those causing economically unacceptable damage or loss.

Label means the written, printed or graphic matter on, or attached to, the pesticide; or the immediate container thereof and the outside container or wrapper of the retail package of the pesticide.

Manufacturer means a corporation or other entity in the public or private sector or any individual engaged in the business or function (whether directly or through an agent or through an entity controlled by or under contract with it) of manufacturing a pesticide active ingredient or preparing its formulation or product.

Marketing means the overall process of product promotion, in cluding advertising, product public relations and information services as well as distribution and selling on local or international markets.

Maximum residue limit (MRL) means the maximum concentration of a residue that is legally permitted or recognized as acceptable in or on a food, agricultural commodity or animal feedstuff.

Packaging means the container together with the protective wrapping used to carry pesticide products via wholesale or retail distribution to users.

Pesticide means any substance or mixture of substances intended for preventing, destroying or controlling any pest, including vectors of human or animal disease, unwanted species of plants or animals causing harm during or otherwise interfering with the production, processing, storage, transport, or marketing of food, agricultural commodities, wood and wood products or animal feedstuffs. or which may be administered to animals for the control of insects, arachnids or other pests in or on their bodies. The term includes substances intended for use as a plant growth regulator, defoliant, desiccant, or agent for thinning fruit or preventing the premature fall of fruit, and substances applied to crops either before or after harvest to protect the commodity from deterioration during storage and transport.

Pesticide industry means all those organizations and individuals engaged in manufacturing, formulating or marketing pesticides and pesticide products.

Pesticide legislation means any laws or regulations introduced to regulate the manufacture, marketing, storage, labelling, packaging and use of pesticides in their qualitative, quantitative and environmental aspects.

Poison means a substance that can cause disturbance of structure or function, leading to injury or death when absorbed in relatively small amounts by human beings, plants or animals.

Poisoning means occurrence of damage or disturbance caused by a poison, and includes intoxication.

Prior Informed Consent (PIC) refers to the principle that international shipment of a pesticide that is banned or severely restricted in order to protect human health or the environment should not proceed without the agreement, where such agreement exists, or contrary to the decision of the designated national authority in the participating importing country.

Prior Informed Consent Procedure (PIC procedure) means the procedure for formally obtaining and disseminating the decisions of importing countries as to whether they wish to receive future shipments of pesticides that have been banned or severely restricted. A specific procedure was established for selecting pesticides for initial implementation of the PIC procedures. These include pesticides that have been previously banned or severely restricted as well as certain pesticide formulations that are acutely toxic. This procedure is described in the Guidelines on the Operation of Prior Informed Consent. (15).

Product means the pesticide in the form in which it is packaged and sold; it usually contains an active ingredient plus adjuvants and may require dilution prior to use.

Protective clothing means any clothes, materials or devices that are designed to provide protection from pesticides when they are handled or applied.

Public sector groups means (but is not limited to) scientific associations; farmer groups; citizens' organizations; environmental, consumer and health organizations; and labour unions.

Registration means the process whereby the responsible national government authority approves the sale and use of a pesticide following the evaluation of comprehensive scientific data demonstrating that the product is effective for the purposes intended and not unduly hazardous to human or animal health or the environment.

Repackaging means the transfer of pesticide from any commercial package into any other, usually smaller, container for subsequent sale.

Residue means any specified substances in food, agricultural commodities, or animal feed resulting from the use of a pesticide. The term includes any derivatives of a pesticide, such as conversion products, metabolites, reaction products, and impurities considered to be of toxicological significance. The term "pesticide residue" includes residues from unknown or unavoidable sources (e.g. environmental) as well as known uses of the chemical.

Responsible authority means the government agency or agencies responsible for regulating the manufacture, distribution or use of pesticides and more generally for implementing pesticide legislation.

Risk means the expected frequency of undesirable effects of exposure to the pesticide.

Severely restricted – a limited ban – means a pesticide for which virtually all registered uses have been prohibited by final government regulatory action but certain specific registered use or uses remain authorized.

Toxicity means a physiological or biological property which determines the capacity of a chemical to do harm or produce injury to a living organism by other than mechanical means.

Trader means anyone engaged in trade, including export, import, formulation and domestic distribution.

Use pattern embodies the combination of all factors involved in the use of a pesticide, including the concentration of active ingredient in the preparation being applied, rate of application, time of treatment, number of treatments, use of adjuvants and methods and sites of application which determine the quantity applied, timing of treatment and interval before harvest, etc.

Article 3.   Pesticide management

3.1 Governments have the overall responsibility and should take the specific powers to regulate the distribution and use of pesticides in their countries.

3.2 The pesticide industry should adhere to the provisions of this Code as a standard for the manufacture, distribution and advertising of pesticides, particularly in countries lacking appropriate legislation and advisory services.

3.3 Governments of exporting countries should help to the extent possible, directly or through their pesticide industries, to:

3.3.1 provide technical assistance to other countries, especially those with shortages of technical expertise, in the assessment of the relevant data on pesticides, including those provided by industry (see also Article 4);

3.3.2 ensure that good trading practices are followed in the export of pesticides, especially to those countries with no or limited regulatory schemes (see also Articles 8 and 9).

3.4 Manufacturers and traders should observe the following practices in pesticide management, especially in countries without legislation or means of implementing regulations:

3.4.1 supply only pesticides of adequate quality, packaged and labelled as appropriate for each specific market;

3.4.2 pay special attention to formulations, presentation, packaging and labelling in order to reduce hazard to users, to the maximum extent possible consistent with the effective functioning of the pesticide in the particular circumstances in which it is to be used;

3.4.3 provide, with each package of pesticide, information and instructions in a form and language adequate to ensure safe and effective use;

3.4.4 retain an active interest in following their products to the ultimate consumer, keeping track of major uses and the occurrence of any problems arising in the actual use of their products as a basis for determining the need for changes in labelling, directions for use, packaging, formulation or product availability.

3.5 Pesticides whose handling and application require the use of uncomfortable and expensive protective clothing and equipment should be avoided, especially in the case of small scale users in tropical climates.

3.6 National and international organizations, governments, and pesticide industries should take action in coordinated efforts to disseminate educational materials of all types to pesticide users, farmers, farmers' organizations, agricultural workers, unions and other interested parties. Similarly, affected parties should seek and understand educational materials before using pesticides and should follow proper procedures.

3.7 Governments should allocate high priority and adequate resources to the task of effectively managing the availability, distribution and use of pesticides in their countries.

3.8 Concerted efforts should be made by governments and pesticide industries to develop and promote integrated pest management systems and the use of safe, efficient, cost-effective application methods. Public-sector groups and international organizations should actively support such activities.

3.9 International organizations should provide information on specific pesticides and give guidance on methods of analysis through the provision of criteria documents, fact sheets, training sessions, etc.

3.10 It is recognized that the development of resistance of pests to pesticides can be a major problem. Therefore, governments, industry, national institutions, international organizations and public sector groups should collaborate in developing strategies which will prolong the useful life of valuable pesticides and reduce the adverse effects of the development of resistant species.

Article 4.   Testing of pesticides

4.1   Pesticide manufacturers are expected to:

4.1.1   ensure that each pesticide and pesticide product is adequately and effectively tested by well recognized procedures and test methods so as to fully evaluate its safety, efficacy (2) and fate (3) with regard to the various anticipated conditions in regions or countries of use;

4.1.2.   ensure that such tests are conducted in accordance with sound scientific procedures and good laboratory practice (4) – the data produced by such tests, when evaluated by competent experts, must be capable of showing whether the product can be handled and used safely without unacceptable hazard to human health, plants, animals, wildlife and the environment (3);

4.1.3   make available copies or summaries of the original reports of such tests for assessment by responsible government authorities in all countries where the pesticide is to be offered for sale. Evaluation of the data should be referred to qualified experts;

4.1.4   take care to see that the proposed use pattern, label claims and directions, packages, technical literature and advertising truly reflect the outcome of these scientific tests and assessments;

4.1.5   provide, at the request of a country, advice on methods for the analysis of any active ingredient of formulation that they manufacture, and provide the necessary analytical standards;

4.1.6   provide advice and assistance for training technical staff in relevant analytical work. Formulators should actively support this effort;

4.1.7   conduct residue trials prior to marketing in accordance with FAO guidelines on good analytical practice (5) and on crop residue data (6, 7) in order to provide a basis for establishing appropriate maximum residue limits (MRLs).

4.2   Each country should possess or have access to facilities to verify and exercise control over the quality of pesticides offered for sale, to establish the quantity of the active ingredient or ingredients and the suitability of their formulation (8).

4.3   International organizations and other interested bodies should, within available resources, consider assisting in the establishment of analytical laboratories in pesticide importing countries, either on a country or on a multilateral regional basis; these laboratories should be capable of carrying out product and residue analysis and should have adequate supplies of analytical standards, solvents and reagents.

4.4   Exporting governments and international organizations must play an active role in assisting developing countries in training personnel in the interpretation and evaluation of test data.

4.5   Industry and governments should collaborate in conducting post-registration surveillance or monitoring studies to determine the fate and environmental effect of pesticides under field conditions (3).

Article 5.   Reducing health hazards

5.1   Governments which have not already done so should:

5.1.1   implement a pesticide registration and control scheme along the lines set out in Article 6;

5.1.2   decide, and from time to time review, the pesticides to be marketed in their country, their acceptable uses and their availability to each segment of the public;

5.1.3   provide guidance and instructions for the treatment of suspected pesticide poisoning for their basic health workers, physicians and hospital staff;

5.1.4   establish national or regional poisoning information and control centres at strategic locations to provide immediate guidance on first aid and medical treatment, accessible at all times by tele phone or radio. Governments should collect reliable information about the health aspects of pesticides. Suitably trained people with adequate resources must be made available to ensure that accurate information is collected;

5.1.5   keep extension and advisory services, as well as farmers' organizations, adequately informed about the range of pesticide products available for use in each area;

5.1.6   ensure, with the cooperation of industry, that where pesticides are available through outlets which also deal in food, medicines, other products for internal consumption or topical application, or clothing, they are physically segregated from other merchandise, so as to avoid any possibility of contamination or of mistaken identity. Where appropriate, they should be clearly marked as hazardous materials. Every effort should be made to publicize the dangers of storing foodstuffs and pesticides together.

5.2   Even where a control scheme is in operation, industry should:

5.2.1   cooperate in the periodic reassessment of the pesticides which are marketed and in providing the poison control centres and other medical practitioners with information about hazards;

5.2.2   make every reasonable effort to reduce hazard by:

5.2.2.1   making less toxic formulations available;

5.2.2.2   introducing products in ready-to-use packages and other wise developing safer and more efficient methods of application;

5.2.2.3   using containers that are not attractive for subsequent reuse and promoting programmes to discourage their reuse;

5.2.2.4   using containers that are safe (e.g. not attractive to or easily opened by children), particularly for the more toxic home use products;

5.2.2.5   using clear and concise labelling;

5.2.3   halt sale, and recall products, when safe use does not seem possible under any use directions or restrictions.

5.3   Government and industry should further reduce hazards by making provision for safe storage and disposal of pesticides and containers at both warehouse and farm level, and through proper siting and control of wastes from formulating plants.

5.4   To avoid unjustified confusion and alarm among the public, public-sector groups should consider all available facts and try to distinguish between major differences in levels of risk among pesticides and uses.

5.5   In establishing production facilities in devoloping countries, manufacturers and governments shoud cooperate to:

5.5.1   adopt engineering standards and safe operating practices appropriate to the nature of the manufacturing operations and the hazards involved;

5.5.2   take all necessary precautions to protect the health and safety of operatives, bystanders and the environment;

5.5.3   maintain quality-assurance procedures to ensure that the products manufactured comply to the relevant standards of purity, performance, stability and safety.

Article 6.   Regulatory and technical requirements

6.1   Governments should:

6.1.1   take action to introduce the necessary legislation for the regulation, including registration, of pesticides and make provisions for its effective enforcement, including the establishment of appro priate educational, advisory, extension and health-care services; the FAO guidelines for the registration and control of pesticides (9) should be followed, as far as possible, taking full account of local needs, social and economic conditions, levels of literacy, climatic conditions and availability of pesticide application equipment;

6.1.2   strive to establish pesticide registration schemes and infrastructures under which products can be registered prior to domestic use and, accordingly, ensure that each pesticide product is registered under the laws or regulations of the country of use before it can be made available there:

6.1.3   protect the proprietary rights to use of data;

6.1.4   collect and record data on the actual import, formulation and use of pesticides in each country in order to assess the extent of any possible effects on human health or the environment, and to follow trends in use levels for economic and other purposes.

6.2 The pesticides industry should:

6.2.1 provide an objective appraisal together with the necessary supporting data on each product;

6.2.2 ensure that the active ingredient and other ingredients of pesticide preparations marketed correspond in identity, quality, purity and composition to the substances tested, evaluated and cleared for toxicological and environmental acceptability;

6.2.3 ensure that active ingredients and formulated products for pesticides for which international specifications have been developed conform with the specifications of FAO (8), where intended for use in agriculture; and with WHO pesticide specifications (10), where intended for use in public health;

6.2.4 verify the quality and purity of the pesticides offered for sale;

6.2.5 when problems occur, voluntarily take corrective action, and when requested by governments, help find solutions to difficulties.

Article 7. Availability and use

7.1 Responsible authorities should give special attention to drafting rules and regulations on the availability of pesticides. These should be compatible with existing levels of training and expertise in handling pesticides on the part of the intended users. The parameters on which such decisions are based vary widely and must be left to the discretion of each government, bearing in mind the situation prevailing in the country.

7.2 In addition, governments should take note of and, where appropriate, follow the WHO classifications of pesticides by hazard (11) and associate the hazard class with well-recognized hazard symbols as the basis for their own regulatory measures. In any event, the type of formulation and method of application should be taken into account in determining the risk and degree of restriction appropriate to the product.

7.3 Two methods of restricting availability can be exercised by the responsible authority: not registering a product; or, as a condition of registration, restricting the availability to certain groups of users in accordance with national assessments of hazards involved in the use of the product in the particular country.

7.4 All pesticides made available to the general public should be packaged and labelled in a manner which is consistent with the FAO guidelines on packaging (12) and labelling (13) and with appropriate national regulations.

7.5 Prohibition of the importation, sale and purchase of an extremely toxic product may be desirable if control measures or good marketing practices are insufficient to ensure that the product can be used safely. However, this is a matter for decision in the light of national circumstances.

Article 8. Distribution and trade

8.1 Industry should:

8.1.1 test all pesticide products to evaluate safety with regard to human health and the environment prior to marketing, as provided for in Article 4, and ensure that all pesticide products are likewise adequately tested for efficacy and stability and crop tolerance, under procedures that will predict performance under the conditions prevailing in the region where the product is to be used, before they are offered there for sale;

8.1.2 submit the results of all such tests to the local responsible authority for independent evaluation and approval before the products enter trade channels in that country;

8.1.3 take all necessary steps to ensure that pesticides entering international trade conform to relevant FAO (8) WHO (10) or equivalent specifications for composition and quality (where such specifications have been developed) and to the principles embodied in pertinent FAO guidelines, and in rules and regulations on classification and packaging, marketing, labelling and documentation laid down by international

organizations concerned with modes of transport (ICAO, IMO, RID and IATA in particular);[1]

8.1.4 undertake to see that pesticides which are manufactured for export are subject to the same quality requirements and standards as those applied by the manufacturer to comparable domestic products;

8.1.5 ensure that pesticides manufactured or formulated by a subsidiary company meet appropriate quality requirements and standards which should be consistent with the requirements of the host country and of the parent company;

8.1.6 encourage importing agencies, national or regional formulators and their respective trade organizations to cooperate in order to achieve fair practices and safe marketing and distribution practices and to collaborate with authorities in stamping out any malpractices within the industry;

8.1.7 recognize that the recall of a pesticide by a manufacturer and distributor may be desirable when faced with a pesticide which represents an unacceptable hazard to human and animal health and the environment when used as recommended, and cooperate accordingly;

8.1.8 endeavour to ensure that pesticides are traded by and purchased from reputable traders, who should preferably be members of a recognized trade organization;

8.1.9 see that persons involved in the sale of any pesticide are trained adequately to ensure that they are capable of providing the buyer with advice on safe and efficient use;

8.1.10 provide a range of pack sizes and types which are appropriate for the needs of small-scale farmers and other local users to avoid handling hazards and the risk that resellers will repackage products into unlabelled or inappropriate containers.

8.2 Governments and responsible authorities should take the necessary regulatory measures to prohibit the repackaging, decanting or dispensing of any pesticide in food or beverage containers and should rigidly enforce punitive measures that effectively deter such practices.

8.3 Governments of countries importing food and agricultural commodities should recognize good agricultural practices in countries with which they trade and, in accordance with recommendations of the Codex Alimentarius Commission, should establish a legal basis for the acceptance of pesticide residues resulting from such good agricultural practices (7, 14).

Article 9. Information exchange and prior informed consent

9.1 The government of any country that takes action to ban or severely restrict the use of handling of a pesticide in order to protect health or the environment should notify FAO as soon as possible of the action it has taken. FAO will notify the designated national authorities in other countries of the action of the notifying government (15).

9.2 The purpose of notification regarding control action is to give competent authorities in other countries the opportunity to assess the risks associated with the pesticides, and to make timely and informed decisions as to the importation and use of the pesticides concerned, after taking into account local, public health, economic, environmental and administrative conditions. The minimum information to be provided for this purpose should be:

9.2.1 the identity (common name, distinguishing name and chemical name);

9.2.2 a summary of the control action taken and of the reasons for it – if the control action bans or restricts certain uses but allows other uses, such information should be included:

---

[1] ICAO: International Civil Aviation Organization
IMO: International Maritime Organization
RID: International regulations concerning the carriage of dangerous goods by rail
IATA: International Air Transport Association

9.2.3 an indication of the additional information that is available, and the name and address of the contact point in the country to which a request for further information should be addressed.

Information exchange among countries

9.3 If export of a pesticide banned or severely restricted in the country of export occurs, the country of export should ensure that necessary steps are taken to provide the designated national authority of the country of import with relevant information.

9.4 The purpose of information regarding exports is to remind the country of import of the original notification regarding control action and to alert it to the fact that an export is expected or is about to occur. The minimum information to be provided for this purpose should be:

9.4.1 a copy of, or reference to, the information provided at the time of the notification of control action;

9.4.2 indication that an export of the chemical concerned is expected or is about to occur.

9.5 Provision of information regarding exports should take place at the time of the first export following the control action, and should recur in the case of any significant development of new information or condition surrounding the control action. It is the intention that the information should be provided prior to export.

9.6 The provision to individual countries of any additional information on the reasons for control actions taken by any country must take into account protection of any proprietary data from unauthorized use.

Prior Informed Consent

9.7 Pesticides that are banned or severely restricted for reasons of health or the environment are subject to the Prior Informed Consent procedure. No pesticide in these categories should be exported to an importing country participating in the PIC procedure contrary to that country's decision made in accordance with the FAO operational procedures for PIC.

9.8 FAO will:

9.8.1 review notifications of control actions to ensure conformity with definitions in Article 2 of the Code, and will develop the relevant guidance documents;

9.8.2 in cooperation with UNEP, develop and maintain a data base of control actions and decisions taken by all Member Governments;

9.8.3 inform all designated national authorities and relevant international organizations of, and publicize in such form as may be appropriate, notifications received under Article 9.1 and decisions communicated to it regarding the use and importation of a pesticide that has been included in the PIC procedure;

9.8.4 FAO will seek advice at regular intervals and review the criteria for inclusion of pesticides in the Prior Informed Consent procedure and the operation of the Prior Informed Consent scheme and will report to Member Governments on its findings.

9.9 Governments of importing countries should establish internal procedures and designate the appropriate authority for the receipt and handling of information.

9.10 Governments of importing countries participating in the PIC procedure, when advised by FAO of control action within this procedure, should:

9.10.1 decide on future acceptability of that pesticide in their country and advise FAO as soon as that decision has been made;

9.10.2 ensure that governmental measures or actions taken with regard to an imported pesticide for which information has been received are not more restrictive than those applied to the same pesticide produced domestically or imported from a country other than the one that supplied the information;

9.10.3 ensure that such a decision is not used inconsistently with the provisions of the General Agreement on Tariffs and Trade (GATT).

9.11 Governments of pesticide exporting countries should:

9.11.1 advise their pesticide exporters and industry of the decisions of participating importing countries;

9.11.2 take appropriate measures, within their authority and legislative competence, designed to ensure that exports do not occur contrary to the decision of participating importing countries.

Article 10.   Labelling, packaging, storage and disposal

10.1 All pesticide containers should be clearly labelled in accordance with applicable international guidelines, such as the FAO guidelines on good labelling practice (13).

10.2 Industry should use labels that:

nclude recommendations consistent with those of the recognized research and advisory agencies in the country of sale;

10.2.2 include appropriate symbols and pictograms whenever possible, in addition to written instructions, warnings and precautions;

10.2.3 in international trade, clearly show appropriate WHO hazard classification of the contents (11) or, if this is inappropriate or inconsistent with national regulations, use the relevant classification;

10.2.4 include, in the appropriate language or languages, a warning against the reuse of containers, and instructions for the safe disposal or decontamination of empty containers;

10.2.5 identify each lot or batch of the product in numbers or letters that can be read, transcribed and communicated by anyone without the need for codes or other means of deciphering;

10.2.6 are marked with the date (month and year) of formulation of the lot or batch and with relevant information on the storage stability of the product.

10.3 Industry should ensure that:

10.3.1 packaging, storage and disposal of pesticides conform in principle to the FAO guidelines for packaging and storage (12), the FAO guidelines for the disposal of waste pesticides and containers (16), and WHO specifications for pesticides used in public health (10);

10.3.2 in cooperation with governments, packaging or repackaging is carried out only on licensed premises where the responsible authority is convinced that staff are adequately protected against toxic hazards, that the resulting product will be properly packaged and labelled, and that the content will conform to the relevant quality standards.

10.4 Governments should take the necessary regulatory measures to prohibit the repacking, decanting or dispensing of any pesticide into food or beverage containers in trade channels and rigidly enforce punitive measures that effectively deter such practices.

Article 11.   Advertising

11.1 Industry should ensure that:

11.1.1 all statements used in advertising are capable of technical substantiation;

11.1.2 advertisements do not contain any statement or visual presentation which, directly or by implication, omission, ambiguity or exaggerated claim, is likely to mislead the buyer, in particular with regard to the safety of the product, its nature, composition, or suitability for use, or official recognition or approval;

11.1.3 pesticides which are legally restricted to use by trained or registered operators are not publicly advertised through journals other than those catering for such operations, unless the restricted availability is clearly and prominently shown;

11.1.4 no firm or individual in any one country simultaneously markets different pesticide active ingredients or combinations of ingredients under a single distinguishing name;

11.1.5 advertising does not encourage uses other than those specified on the approval label;

11.1.6 promotional material does not include use recommendations at variance with those of the recognized research and advisory agencies;

11.1.7 advertisements do not misuse research results or quotations from technical and scientific literature; and scientific jargon and irrelevances are not used to make claims appear to have a scientific basis they do not possess;

11.1.8 claims as to safety, including statements such as "safe", "non-poisonous", "harmless", "non-toxic", are not made, with or without a qualifying phrase such as "when used as directed";

11.1.9 statements comparing the safety of different products are not made;

11.1.10 misleading statements are not made concerning the effectiveness of the product;

11.1.11 no guarantees or implied guarantees — e.g. "more profits with...", "guarantees high yields" — are given unless definite evidence to substantiate such claims is available;

11.1.12 advertisements do not contain any visual representation of potentially dangerous practices, such as mixing or application without sufficient protective clothing, use near food, or use by or near children;

11.1.13 advertising or promotional material draws attention to the appropriate warning phrases and symbols as laid down in the labelling guidelines (13);

11.1.14 technical literature provides adequate information on correct practices, including the observance of recommended rates, frequency of applications, and safe pre-harvest intervals;

11.1.15 false or misleading comparisons with other pesticides are not made;

11.1.16 all staff involved in sales promotion are adequately trained and possess sufficient technical knowledge to present complete, accurate and valid information on the products sold;

11.1.17 advertisements encourage purchasers and users to read the label carefully, or have the label read to them if they cannot read.

11.2 International organizations and public-sector groups should call attention to departures from this Article.

11.3 Governments are encouraged to work with manufacturers to take advantage of their marketing skills and infrastructure, in order to provide public-service advertising regarding the safe and effective use of pesticides. This advertising could focus on such factors as proper maintenance and use of equipment, special precautions for children and pregnant women, the danger of reusing containers, and the importance of following label directions.

Article 12. Monitoring the observance of the Code

12.1 The Code should be published and should be observed through collaborative action on the part of governments, individually or in regional groupings, appropriate organizations and bodies of the United Nations system, international governmental organizations and the pesticide industry.

12.2 The Code should be brought to the attention of all concerned in the manufacture, marketing and use of pesticides and in the control of such activities, so that governments, individually or in regional groupings, industry and international institutions understand their shared responsibilities in working together to ensure that the objectives of the Code are achieved.

12.3 All parties addressed by this Code should observe this Code and should promote the principles and ethics expressed by the Code, irrespective of other parties' ability to observe the Code. The pesticide industry should cooperate fully in the observance of the Code and promote the principles and ethics expressed by the Code, irrespective of a government's ability to observe the Code.

12.4 Independently of any measures taken with respect to the observance of this Code, all relevant legal rules, whether legislative, administrative, judical or customary, dealing with liability, consumer protection, conservation, pollution control and other related subjects should be strictly applied.

12.5 FAO and other competent international organizations should give full support to the observance of the Code, as adopted.

12.6 Governments should monitor the observance of the Code and report on progress made to the Director-General of FAO.

12.7 Governing Bodies should periodically review the relevance and effectiveness of the Code. The Code should be considered a dynamic text which must be brought up to date as required, taking into account technical, economic and social progress.

Annex

FAO Conference Resolution 10/85:
International Code of Conduct
on the Distribution and Use of Pesticides

THE CONFERENCE,

Recognizing that increased food production is a high priority need in many parts of the world and that this need cannot be met without the use of indispensable agricultural inputs such as pesticides,

Noting that FAO's study entitled *Agriculture: toward 2000* foresees a steady increase in the worldwide use of pesticides,

Convinced that such growth in pesticide use is likely to take place in spite of necessary intensive parallel efforts to introduce biological and integrated pest control systems,

Acknowledging that pesticides can be hazardous to humans and the environment and that immediate action must be taken by all concerned, including governments, manufactures, traders and users, to eliminate, as far as possible and within the scope of their responsibility, unreasonable risks, not only in the country of origin but also in the countries to which pesticides may be exported,

Being aware that the requirements for the safe and proper use of pesticides in some developed countries have led to the adoption of complex systems of regulations and of enforcement mechanisms, but that many other countries have neither such mechanisms nor the necessary legislation, regulations or infrastructures to control the import, availability, sale or use of pesticides,

Convinced that additional efforts are needed to enable such countries to control pesticides more effectively and to assess the hazards which could result from their use or misuse,

Recognizing that a voluntary International Code of Conduct, based on internationally agreed technical guidelines, would provide a practical framework for the control of pesticides, especially in countries that do not have adequate pesticide registration and control schemes,

Noting that such a draft Code was reviewed by the Committee on Agriculture at its Eighth Session, and endorsed by the Council at its Eighty-eighth Session,

Having further noted the conclusions and recommendations of these bodies,

1. *Hereby adopts* a voluntary International Code of Conduct on the Distribution and Use of Pesticides as given in the annex to this Resolution;

2. *Recommends* that all FAO Member Nations promote the use of this Code in the interests of safer and more efficient use of pesticides and of increased food production;

3. *Requests* governments to monitor the observance of the Code in collaboration with the Director-General who will report periodically to the Committee on Agriculture;

4. *Invites* other United Nations agencies and other international organizations to collaborate in this endeavour within their respective spheres of competence.

(Adopted 28 November 1985)

References

1. *Report of Second Government Consultation on International Harmonization of Pesticide Registration Requirements,* FAO, Rome, 11-15 October 1982.

2. *Guidelines on efficacy data for the registration of pesticides for plant protection.* Rome, FAO. 1985.

3. *Guidelines on environmental criteria for the registration of pesticides.* Rome, FAO. 1985.

4. *Good laboratory practice.* Paris, Organisation for Economic Co-operation and Development. 1981.

5. *Codex guidelines on good practice in pesticide residue analysis.* Rome, FAO. 1984.

6. *Guidelines on crop residue data.* Rome, FAO. 1985.

7. *Codex recommended national regulatory practices to facilitate acceptance and use of Codex maximum limits for pesticide residues in foods.* Rome, FAO. 1985.

8. *The use of FAO specifications for plant protection products.* Rome. 1986. FAO Plant Production and Protection Paper.

9. *Guidelines for the registration and control of pesticides (including a model scheme for the establishment of national organizations).* Rome, FAO. 1985.

10. *Spécifications for pesticides used in the public health,* 6th ed. Geneva, World Health Organization. 1985.

11. *The WHO recommended classification of pesticides by hazard and guidelines to classification 1986-87.* Geneva, World Health Organization. 1986.

12. *Guidelines for the packaging and storage of pesticides.* Rome, FAO. 1985.

13. *Guidelines on good labelling practice for pesticides.* Rome, FAO. 1985.

14. *Guidelines on good agricultural practice in the use of pesticides: guide to Codex recommendations concerning pesticide residues.* Rome, FAO. 1984.

15. *Guidelines on the operation or Prior Informed Consent (PIC).* Rome, FAO. 1990.

16. *Guidelines for the disposal of waste pesticides and pesticide containers on the farm.* Rome, FAO. 1985.

Additional technical guidelines adopted after the FAO Conference of 1985 and published up to May 1990.

Pictograms for pesticide labels. Rome, FAO.

Guidelines on pesticide residue trials to provide data for the registration of pesticides and the establishment of maximum residue limits. Rome, FAO. 1986.

Addenda to the guidelines for the registration and control of pesticides. Rome, FAO. 1988.

Guidelines on post-registration surveillance and other activities in the field of pesticides. Rome, FAO. 1988.

Guidelines on the registration of biological pest control agents. Rome, FAO. 1988.

Guidelines for retail distribution of pesticides with particular reference to storage and handling at the point of supply to users in developing countries. Rome, FAO. 1988.

Guidelines for legislation on the control of pesticides. Rome, FAO. 1989.

Guidelines on personal protection when using pesticides in hot climates. Rome, FAO. 1990.

8

SAFE HANDLING, STORAGE
AND MARKETING OF PESTICIDES

Contents   Page

Safe handling of pesticides 139

- Protective clothing 139

- Types of respiratory equipment 141

- Personal protection in application 143

Storage of pesticides 144

- Pesticide shelf-life 144

- Recording system for stored pesticides 145

- Safety requirements for storage personnel 146

- Safe storage systems 146

- Storage buildings 147

Marketing of pesticides 150

- Responsible marketing 150

- Recommendations for safety in shops 152

References 152

Safe Handling, Storage and Marketing of Pesticides

SAFE HANDLING OF PESTICIDES

Regularly, incidents are reported of accidental poisoning of pesticide applicators, bystanders and animals, or of serious environmental contamination. They paint an alarming picture of ignorance and carelessness in the handling of pesticides. This situation calls for stricter regulatory control and effective training of those handling pesticides.

Pesticides are designed to be toxic and can be dangerous if they are misused or handled wrongly. Therefore, the safe and effective use of pesticides should be the concern of everyone.

The immediate hazard and burden of good practice in the use of pesticides falls on the manufacturers, distributors and professional applicators. These groups of people, who are in frequent contact with pesticides because of their occupation, should understand and follow safety recommendations as carefully as possible. They should also, by advice, demonstration and their own behaviour, teach the occasional pesticide user about safety.

A toxic pesticide poses a potential danger to man and the environment. But it is the combination of toxicant and EXPOSURE that creates the real hazard.

Safe handling of pesticides involves seriously trying to minimize always and everywhere the risk of exposing oneself and the environment to the unwanted effects of these poisons.

Protection against unwanted exposure requires the use of protective clothing, proper application techniques, good application equipment, correct storage and transport etc. It also involves knowing what action must be taken in case something goes wrong such as accidents and spills. These aspects and remedies are described in detail in this chapter.

Protective clothing

All pesticides are toxic, but toxicity and hazards differ for each chemical. Even formulations with the same technical ingredient differ in degrees of hazardousness. Some formulations can be mixed and applied with the only protection required being very good hygiene. Other formulations require the use of protective clothing or even special breathing equipment. The degree of protection needed should be indicated on the pesticide label.

◄ Apparel for working with hazardous powder formulations protects against inhalation of dust and vapour. This feature is provided by the half-face mask of the worker and the exhaust pipe above the bucket which draws out dust from the work area. (Courtesy of LNV, the Netherlands)

If a label instructs you to use a specific piece of protective equipment, it is because the product may kill or injure you if you do not have that protection.

In a hot climate, it is uncomfortable to wear heavy protective clothing. Rubber boots and gloves, face masks and respirators soon become extremely irritating to wear, especially if the humidity is high. Applicators are generally very reluctant to wear such protective equipment. If they wear such apparel, the risk of them shedding some of it is great when they become hot and tired. One should, therefore, select protective clothing and equipment that is most comfortable to wear and not use more of it than is needed to do the job safely.

If the required protective equipment is not available, it is better to study labels of other available pesticides to determine whether there is any other product that can be used to control the pest, but does not require the use of such equipment.

People handling pesticides should:

* Understand the reasons for wearing protective clothing;

* Know the various kinds of protective clothing available;

* Know which types of pesticide and which tasks in handling and application require the use of each protective item;

* Read the safety instructions on the label;

* Realize that dirty, contaminated or defective protective equipment is a source of contamination in itself;

* Clean or wash all safety equipment after each day of use.

Protective clothing for tropical conditions

As a general principle for tropical conditions, materials should be as light as possible and provide maximum respiration in correspondence with the hazards of the job. The most suitable minimum protection for persons working with pesticides is a pair of light, durable, cotton overalls.

If overalls are not available, shirts and trousers should be worn that cover the full length of the arms and legs and fasten at the wrist and neck. These should be washed immediately after use and kept separate from other clothing.

Professional applicators, store and transport personnel, and labourers working in manufacturing and formulation factories may need even better protection than the spray-man in the field. Thus, employers should supply all workers with overalls and demand that they change into their own clothes before going home. Also, they should provide gloves and boots and other pieces of protective equipment according to the hazard created by the toxicity of the pesticide and the degree of exposure involved with the particular task to be done. Different types of protective equipment are described below:

Overalls

These should be made of tightly woven cotton fabric, that fasten at the neck, with sleeves long enough to overlap with the gloves, and preferably having no pockets. White overalls have the advantage of showing up contamination. Overalls offer good minimum protection if washed regularly and stored away from pesticides. Clean overalls should be a standard requirement for employees of stores and formulation factories.

Coveralls

New, light-weight, disposable coveralls are available that resist splashes and can be laundered and reused several times if handled carefully.

Aprons

Ideally, these should be made of neoprene rubber. They should reach from the top of the chest to below the tops of the boots and wrap around to cover the sides of the legs. The apron does not protect shoulders, arms or back but is adequate for mixing and loading jobs not involving highly toxic, concentrated pesticides. Aprons made locally from tarpaulin fabric should be carefully decontaminated after spills or splashes.

Spray suit

For greater protection, you can choose from among several good quality spray suits. They must be liquid-proof, and resist tears and the solvents commonly used in liquid pesticides. Heavy-weight and light-weight neoprene or plastic spray suits will last for years if they are given proper care. These spray suits can be easily cleaned.

Raincoat

A liquid-proof raincoat can be used when mixing and handling highly toxic, concentrated pesticides. Unlike an apron, a raincoat will protect the applicator's arms, back and shoulders as well as the front of the body. Choose light-weight material for use in hot weather.

Protective gloves

Gloves are the most important items of safety equipment required in any pesticide store or during mixing and handling of concentrated pesticides. They should be worn wherever there is a risk of skin contamination.

The type of glove is important and sufficient sets of gloves should be available to meet any emergency. They should always be long enough to be worn over the sleeves of overalls (see picture on opposite page); short gloves ending at the wrist should not be used. Shirt sleeves should fasten at the wrist and be tucked inside the gloves. For efficient handling, gloves should neither be too thick nor too thin.

A storekeeper at a farmers' co-operative repackages an insecticidal powder. Although he wears a dust-mask, his forearms, hands and feet remain unprotected and are badly contaminated. (Photo by the author)

Damaged gloves or those made of materials which absorb pesticides (such as cotton or leather) must not be worn when handling liquid formulations. To wear such gloves could be more dangerous than wearing no gloves at all. Newly purchased gloves and used gloves should be tested periodically for flaws.

It should be remembered that certain solvents used in pesticide formulations, such as xylene, can penetrate quickly through even the thickest gloves. Therefore, spills on gloves should be treated as spills on the hands and washed off quickly and thoroughly using plenty of detergent and water. Discarded gloves must be torn or cut and burned -- not just thrown away -- to prevent further use by someone else.

Types of protective gloves

Nitrile rubber, polyvinylchloride (PVC) and neoprene are materials used for manufacturing rubber gloves. Nitrile and neoprene are more resistant to the corrosive action of many solvents used in pesticide formulations than, for instance, latex (natural rubber) and PVC. The material thickness should not be less than 0.4 mm and the glove length should be at least 30 cm in order to cover the hands and wrists. The inner glove surface may be reinforced with a high-density cotton liner which facilitates glove removal and absorption of perspiration. As this liner may also retain pesticide residues more than a smooth rubber surface would, it should be washed very thoroughly after use.

Cotton gloves may be used for handling granular formulations of insecticides, herbicides and fungicides. However, beware that the cotton gloves do not become wet during handling, since this can enhance penetration of the pesticide into the skin. Cotton gloves should be washed thoroughly with water and detergent after each use.

Cleaning gloves after use

Immediately after use, gloves should be washed well with soap or detergent and water BEFORE being removed from the hands. After removal they should be turned inside out and again thoroughly washed, rinsed, rubbed dry and allowed to dry completely in a well-ventilated place before being stored again in a locker for safety equipment. Sprinkling of talcum powder on the inside makes it easier to put them on.

If organophosphorous pesticides have been handled, gloves should be soaked in an alkaline solution such as a dilution of soda lye for at least eight hours, rinsed and dried before re-use.

Note: Long and frequent contact with organo-phosphorous pesticides can result in the rubber becoming impregnated with the pesticide. When there is a risk of this happening, the gloves should be destroyed.

Goggles

Goggles offer the best protection to keep your eyes from absorbing toxic pesticide fumes. However, goggles will not protect your face or mouth from pesticide spills, splashes and drift. Goggles are comfortable and can be used when you wear a cartridge type respirator; wear shatterproof goggles that will not fog. Make sure they fit well and are adjusted for maximum eye protection. They can be worn over spectacles.

Face shield

A face shield will protect your eyes against spills and splashes. It offers the eyes little protection from irritant pesticide vapours, spray mists and fine dusts. The face shield has the advantage of being cooler in hot

Only because the trousers were worn outside the boot was the applicator saved from the adverse consequences of a pesticide spill.

climates and it does not "mist up" as easily as goggles. It can be easily attached to a plastic hard hat.

Boots

Boots should be inspected regularly for damage and possible leaking; also, they should be washed regularly and allowed to dry upside down. Boots should be removed from the feet only after the outer surface has first been cleaned. Boots should have thick soles and preferably an inner lining, which makes boots more comfortable to wear and easier to take off.

Head covering

Because fine spray and dust particles are effectively caught in the hair, and the scalp is very sensitive to contamination, it is advisable to cover the hair at all times when handling pesticides. Wearing an easily washable cotton cap or scarf is recommended. A wide-brimmed, waterproof hat without a sweatband gives additional protection of the neck. Plastic "hard hats" with plastic sweatbands can be easily cleaned. They are waterproof and are comfortable to wear in hot weather.

Types of respiratory equipment

Simple dust mask

These are pads made of paper, cellulose, synthetic material or gauze. They cover the mouth and nose only. Dust masks protect against dust particles and to a limited degree against spray droplets and aerosols. They do not provide protection against vapours. Dust masks must be discarded after a single use.

Half-face masks (cartridge type)

This type of repirator covers the nose, mouth and chin, but not the eyes. For eye protection, the half-face mask should always be worn in combination with goggles or face shield (see photo at beginning of this Chapter). Air is inhaled through both a filter pad and a cartridge which contains an absorbent material such as activated charcoal. It is not suitable for wearing during fumigation or when the oxygen level is low, as it would be in a silo or sewer, for example.

A face-shield and overalls provide comfortable and safe protection against contamination by many pesticides; however, the instructions on the label of the pesticide package should be followed as it may indicate the need for more complete precautions. (Courtesy by GIFAP)

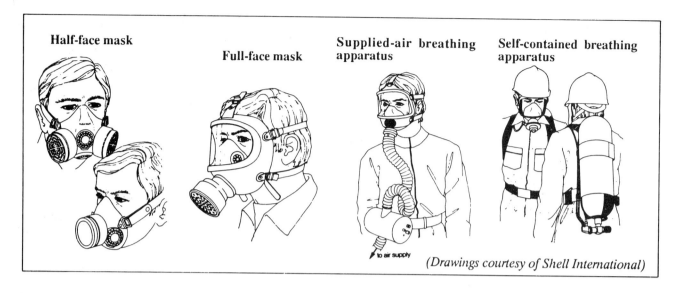

Half-face mask

Full-face mask

Supplied-air breathing apparatus

Self-contained breathing apparatus

to air supply

(Drawings courtesy of Shell International)

Full-face masks (canister type)

The canister or chemical cartridge respirator (gas mask) should be worn when one is exposed to a continuous concentration of a toxic pesticide. It protects the lungs as well as the eyes, nose and mouth against vapours, fumes, mists and fine dusts. It has a canister on the face-piece or one that is attached to the face-piece by a flexible hose. The part covering the eyes can be a one-piece visor or separate eye-pieces resembling goggles. Full-face respirators restrict vision, are heavier than half-face masks and perspiration tends to accumulate inside the visor.

The cartridges differ in the range of gases or vapours they absorb and must be chosen carefully. For the handling of organophosphorous and carbamate compounds, gas filters marked "against organic vapours/gases" must be selected. Situations where full-face respirators must be worn are, for instance, hot or cold aerosol, smoke generators and dust treatments in warehouses and green-houses, or whenever preparing a spray mixture with a volatile formulation in rooms without ventilation.

The canister type respirator contains more absorbing material and longer-life filters than the cartridge type. It affords limited protection during fumigation or when the oxygen supply is low.

Supplied-air breathing apparatus

This type of apparatus consists of a full-face mask with a flexible air line (up to 30 feet in length) to draw fresh air from an outside source. A pump may be supplied with some types; others may be connected to a pressurized air cylinder or air line of a building or installation. Only clean, fresh air should be used for this type of apparatus; read carefully and follow the instructions for using it.

The supplied-air respirator should be used while filling, mixing or applying pesticides under the following conditions:

* When the oxygen supply is low, as it may be in a grain

elevator building or sewer system; and

* When one is exposed to high concentrations of highly toxic pesticides in enclosed areas such as in a fumigation chamber.

Self-contained breathing apparatus

This apparatus serves the same function as the supplied-air respirator, but allows greater mobility as the air supply is always with the operator. The mask is connected to a compressed air cylinder, usually carried on the operator's back. This allows the operator to move freely and over a wider area than would be possible with a supplied-air respirator. It can be readily used for work in grain elevators and fumigation chambers and is usually part of normal fire-fighting equipment.

There should be at least two serviceable, self-contained respirators in or near all stores where the more poisonous pesticides are kept. They should be stored in individual containers in a specially marked locker.

Those employees who are likely to deal with spills, leaks, fires etc., of the more poisonous products should be instructed on how to wear protective respiratory equipment so that they can use it properly if the need arises.

Replacing cartridges or canisters

Cartridges should be replaced after exposure to pesticides following the manufacture's recommendations. Many makers recommend that cartridges be replaced after a total of about 30 minutes to eight hours of use in the presence of pesticides. The maker's specific recommendations on this matter should ALWAYS be followed. Regardless of the length of time that a cartridge or canister has been used, it should be changed immediately if the wearer can detect even a slight smell of the pesticide. It should be noted that, in the presence of a lot of dust, the filters may become clogged thus making breathing difficult. If this occurs, the wearer should move outside the area and change the cartridge.

Use of respiratory protective equipment

It must be recognized that in warm climates, the use of respirators is very uncomfortable owing to heavy perspiration around the face-piece and because of condensed moisture accumulating inside the face-piece. The user is therefore constantly tempted to move the respirator up and down, often with a contaminated glove. Taking rests out of the danger area may be a sound precaution.

In order to function effectively, the respirator should fit closely to the face to prevent any air from entering except through the cartridge or canister. However, it should not fit so tightly that it causes excessive discomfort to the wearer. If a worker is bearded or wears a large moustache, it may be very difficult to fit the half- and full-face masks and breathing apparatus to give an airtight seal.

Note 1: It may be highly dangerous for workers to wear respirators if the cartridges (or canisters) have exceeded their service life. If there is any doubt at all, new cartridges or canisters should be fitted.

Note 2: It may be highly dangerous to rely on cartridge/canister respirators in enclosed spaces deficient of oxygen. Under such circumstances, a supplied-air or self-contained breathing apparatus should be used.

Maintenance of respirators

After use, remove the cartridge (and filter if separate) and carefully wash the face-piece inside and outside with soap and water and rinse well to remove all traces of soap. Dry with a clean cloth and leave in a well-ventilated place to dry completely. If the filter still can be used, it should be put into a clean plastic bag, sealed and stored away from any pesticide. Filters and cartridges cannot be washed.

Gas and vapour filters are effective for only a limited period of time. If the filter, or cartridge, or canister has reached the end of its service life, it should be destroyed immediately so as to prevent any accidental re-use.

Note 3: If the maintenance, storage and change of filters and cartridges is likely to be poor, the respirator may do more harm than good, whilst becoming a serious source of secondary exposure.

Personal protection in application

An applicator is exposed to a pesticide during the time that he is mixing and preparing the spray solution, filling the sprayer or duster and actually applying the chemical to the target. Until the moment he has finished cleaning his equipment and disposing of the empty container, the applicator should wear appropriate protective clothing.

While mixing, the applicator should be aware that he is handling a concentrated pesticide, which is more dangerous than the spray solution; that the risk of exposure is highest during measuring, mixing and filling. Clean buckets, stirring stick and measuring cups should be used in preparing the spray solution. Liquids should be poured carefully without splashing; powders should not be allowed to puff up into the face. This can be done by standing upwind and by keeping the package close to the mixing bucket while pouring.

Empty a liquid pesticide container into the bucket or spray tank and drain it for about one minute. Carefully fill the empty pesticide container about one-quarter full with water. Close the pesticide container and shake well. Drain the rinse water into the spray tank or mixture; repeat three times. This reduces the amount of residue in the container and makes it less hazardous. Do not leave the empty container laying around. If the hands or other parts of the body get contaminated, wash immediately with water and soap; keep some clean water and soap ready for this purpose. Care must be taken that, while the applicator is working in the field, children and animals do not come into contact with the pesticide containers used for mixing.

Before entering the field, the spray-man should check once more for leaking spray hoses, hose and lance attachments and the functioning of the cut-off valve. Leaking and dirty sprayers as well as contaminated

Blocked nozzles can usually be cleaned with a soft brush. This professional applicator is wearing heavy-duty protection for use in a temperate climate. The colour coding of the fitted flat-fan nozzles indicates the size of the orifice.

shoulder straps are a common source of exposure to pesticides. Blocked spray nozzles should never be blown out with the mouth.

In the field, an applicator should spray with the wind and avoid walking in the spray drift or dust and in crops already sprayed. Spray sideways and not in front of the body. Using a tail-boom sprayer, which emits the spray backwards, helps in avoiding contamination of the legs.

Often the spray-man walks barefoot in the fields. Although this may be unavoidable in wet paddy fields, this procedure should be discouraged in spraying dry-land crops; some type of protective footwear should be worn. If a proper dust mask is not available, covering the mouth and nose with a clean cloth offers some protection against spray and dust particles. A hat or piece of cloth covering the head will protect the operator from pesticide contamination and from strong sun. The hat and cloth should both be washed after use. The use of sunglasses offers cheap, additional protection for the eyes while spraying if other types of eye protection are not available.

The product label should state whether special safety equipment such as a respirator and liquid-proof clothing should be worn. Always read the label carefully for safety instructions before you start mixing and spraying the material. If a respirator is required, one must use it to protect his life, even if it becomes hot and very uncomfortable to wear. The requirement for protective clothing or a respirator may be a factor in deciding which pesticide to purchase and what method of application to use. The applicator should use the least toxic pesticides that will do the job, thus lessening the toxic effect on himself and his surroundings. Routine fumigation tasks in food stores and ship holds should be left to adequately trained professional applicators.

STORAGE OF PESTICIDES

At the site of manufacture and all along the distribution line down to the village retail outlet or farm level, pesticides must be stored in bulk or in smaller quantities for certain periods of time. Larger companies usually operate with well-equipped facilities, but very often, storage facilities elsewhere are poor. Nevertheless, the necessity for maintaining high standards of safety is present everywhere. Hence, attention must be paid to those safety requirements that can be met regardless of the facilities available. Such requirements may be divided systematically into catagories pertaining to workers' behaviour, storage systems, storage facilities and display for sale.

Pesticide shelf-life

Good storage of pesticides involves the control of all factors that may cause deterioration of the stored products. These factors may be physical (heat, humidity), biological (moulds, insects), chemical (acidity, corrosion) or mechanical (pressure packing). Even under optimum storage conditions, some products may spoil after a period of time. Foodstuffs, vegetable oils and chemicals are examples of substances that may become unusable after a period of time because of internal metabolic processes.

The period of time over which products of this type may be safely stored, before deterioration becomes sufficiently serious to affect their use, is known as shelf-life.

Nearly all pesticides have a limited shelf life. Even pesticides in sealed containers may change over a period of time with one of two principal results.

* The active ingredient may change as a result of chemical reactions and thereby produce breakdown products; thus the concentration of the original active ingredient may drop. One or more of the breakdown products may even be more toxic, more volatile or more persistent than the active ingredient itself.

* The formulation of the pesticide may break down, making it impossible to mix or effectively apply the pesticide; e.g. crystallization of liquids, and caking of powders and granules.

Technical material of DDT and other organo-chlorines are chemically very stable (persistent) over many years, but their formulations may break down more rapidly. However, organo-phosphorous and carbamate compounds are much less stable and consequently have a much shorter shelf-life than organochlorines. Dust and wettable powder formulations in lined paper bags and cardboard boxes are more susceptible to breakdown due to high temperature, high humidity, strong sunlight and pressure packing than are liquids in sealed containers.

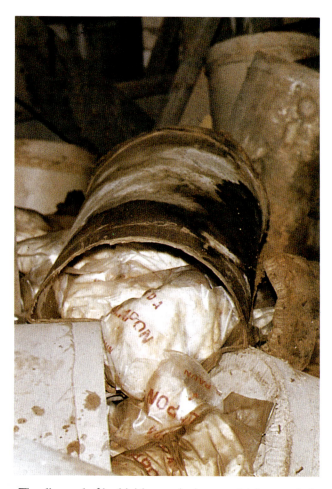

The disposal of herbicide powder in completely corroded cardboard containers in a government store is impaired because store bookkeeping regulations do not allow the writing-off of old stock. (Photo by the author)

The advanced status of formulation and packaging technology has prolonged the shelf-life of most pesticides beyond at least two years from the time of manufacture. However, if insufficient attention is given to the selection of appropriate formulation ingredients and packaging materials, the shelf-life may be significantly shortened.

Shelf-life rapidly declines once containers have been opened, especially if they are left partially empty. The turnover of stock should be managed so that the contents of containers, once opened, should be used as quickly as possible. Unsealed containers of dusts and wettable powders should not be kept for longer than one year.

Containers are not only subject to deterioration caused by external factors (climatic, biological and mechanical), they can become corroded internally through the action of the pesticides they contain. EC formulations, particularly, cause corrosion of weak spots, especially along seams or where there are imperfections in the internal coating of the container. Some pesticides increase in acidity during storage; this increases their capacity to corrode containers from within. Signs of this type of corrosion should be looked for during stock inspections.

Storekeeping in relation to shelf-life

Ordering

When ordering new supplies of pesticides, the storekeeper should not order a larger quantity than can be used during the stated period of shelf-life. He or she should also take into account the often considerable time pesticides will be in transit between the formulation factory and arrival at the store.

Inspection

Stocks should be inspected regularly for signs of corrosion, leakage and deterioration of the containers. Check also for missing lids. Look for caking of powder, pulverization of granules, sedimentation or jelling of liquids, change in colour owing to oxidation and dampness of packages.

Rotation

At the time of inspection and arrival of new stocks, older stock should be moved to the front row on the shelves or stacks. Strictly following the rule "first in - first out" will prevent accumulation of out-dated products.

Write-off

Store bookkeeping should always allow for write-off of old stocks. If such provision is not made and a storekeeper is held personally accountable, he or she will not be able to dispose of old stock in time. This often leads to hazardous situations which sometimes remain unchanged for years. (See photograph at the left.)

Recording system for stored pesticides

A major cause of problems in pesticide stores is the lack of a proper system for keeping records of stocks received, held and issued. Otherwise, if such a system exists, the store manager may not be trained to use it or may fail to keep it up because he or she is not under supervision.

If stock records are not kept accurately, it will soon become impossible to keep track of older products. Customers always want "fresh pesticides" and workers are inclined to take from the easiest accessible stacks. However, the shortness of shelf-life makes it imperative that older stocks be issued first.

The basic principle in any storage operation is rotation of stocks on a FIRST-IN-FIRST-OUT basis.

Everything, from store layout to ways of stacking and supervising, should be aimed at preventing out-of-date stocks ending up in the darker recesses of the store. Not only do such stocks represent a financial loss to the owner, they also present a health hazard to those working in the store and an environmental hazard when, eventually, they must be disposed of in some way.

Although the type of record system adopted will depend on the size and function of the store, there are two practices that MUST be followed in ALL pesticide stores, no matter what their size. These are:

* Write the date of purchase or the date of receipt on each container immediately upon its arrival in the store.

* Ensure that containers are properly labelled; that the labels remain on the containers, and that they are clean and readable. Labels in poor condition should be replaced.

In any retail store, a record system needs to be kept either in the form of a card index or as sheets on a clip-board for placement on file later. For each new consignment, a new card or sheet should be completely filled out and all movements of the units of that consignment should be recorded on it. In case the pesticide containers need to be repacked or out-of-date stock needs to be disposed of, these actions are also to be considered as "movements" and thus should be recorded. The record system should be kept separate from the pesticide stock, so that it will not be destroyed in case of fire or major accidents. (Store owners should also consider making a copy of the master record sheet.)

Out-dated stocks

Information on shelf-life should be requested from the distributor at the time the pesticide is purchased. Generally, pesticide labels contain a batch number, but no date of manufacture. However, a manufacturer can trace a batch's date of manufacture using that number.

Out-dated stocks may still be usable if the formulation has not broken down and this can be ascertained by mixing a small quantity according to label instructions. If an emulsion or a suspension can be achieved and there are no large particles, sediments or other matter which might block a sprayer, then the pesticide may be tried against the pest it is desired to control. If control is achieved at the recommended rate, it may be assumed that, although the pesticide has exceeded its shelf-life, it may still be used. If there is any reason to suspect that the concentration of active ingredient has dropped, it may be necessary to increase the concentration of the spray to allow for this. Many tons of otherwise usable pesticides are wasted, because this "try and see" approach is not followed with old stocks. However, where old stocks are being applied to crops, a watch should be kept for signs of phytotoxicity.

Safety requirements for storage personnel

* All personnel who work in a pesticide store should receive proper instruction and on-the-job training before they are sent to work in the storage area. Knowledge and skill levels need to be tested periodically.

* Appropriate protective clothing must be worn; a respirator must be used whenever recommended. Protective apparel and instruction in its proper use must be provided by the employer.

* Do not work alone when handling very dangerous pesticides.

* Do not permit smoking, eating or drinking in pesticide warehouses and storage areas.

* Practise good personal hygiene; wash work clothing frequently. (Clean water and soap should always be made available for this purpose.)

* Inspect pesticide containers for leaks first before handling them; avoid leaving containers open.

* Should a leak or spill occur, keep people and animals away from the area and give priority to thorough decontamination of the area.

* Always keep material on hand for dealing with spillage and decontamination. For larger warehouses, have this equipment available at strategic points.

* In order to prevent damage, never handle containers roughly or carelessly.

* Have available throughout the storage areas dry-powder fire extinguishers, or substitutes such as sand buckets, for fighting small fires. Further, train all personnel in dealing with small fires.

Safe storage systems

Storage systems should have the objective of not only ensuring a safe working environment, but also

A large, well-organized pesticide store, which features wide gangways and metal shelves, facilitates the movement of palleted stocks by fork-lift truck. (Courtesy of ICI)

enabling a rapid turn-over of stocks and helping to deal with spillage and fire. Frequently, pesticides are stored in the same place as fertilizer, seeds and other agricultural inputs. However, pesticides may require special storage conditions according to their toxicity, corrosiveness or viability. Fertilizers and pesticides are hardly compatible; poisonous gases may develop if nitrogen fertilizer is contaminated with organophosphates. Obviously, storage of pesticides requires strict organization and adherence to some kind of system. Aspects to be taken into consideration:

* Storage systems should be flexible in view of changing stock quantities; they should also be adaptable to the storage structure.

* The system should provide for orderly stacking and shelving with sufficiently wide gangways in order to (a) enable access to and easy movement of stocks, and (b) minimize the risk of contamination from handling pesticide containers.

* Stock should be arranged so that the old stock will be used first.

* Keep floor space clear and uncluttered, with gangways between stacks and shelves for easy inspection and good ventilation.

* Keep doorways and fire exits free from obstacles; also keep access to the washroom and fire station free.

* Do not store pesticides and contaminated empty pesticide containers near food and animal feedstuffs, seeds and other plant material, beverages, clothing and other such articles.

* Separate pesticides into product types such as herbicides, insecticides and fungicides, and allocate separate stacking areas for each type.

Reason: To avoid cross contamination and to eliminate errors such as dispatching the wrong product.

* Segregate the more toxic products and the combustale ones in separate sections of the warehouse, preferably separated by firebrick walls.

Reason: In the event of spillage and leaking containers, more stringent measures need to taken for personal protection and decontamination when highly poisonous chemicals are involved. Further, in the event of a fire, high-toxicity products create a contamination problem if water is used to fight the fire. Thus, if it is necessary to use water, it should be kept to the smallest amount possible.

* Stack solid products separately from liquid products.

Reason: In the event of a spillage or a fire, liquid products require different safety procedures than solids. Moreover, solids are commonly packed in cardboard boxes or lined paper bags which fall apart when soaked by liquids.

* Never place containers of pesticides directly on the floor. Use dunnage -- bricks, pallets or timber -- under the containers so that leaks may be discovered easily. Dampness from the floor or leaking pesticides will lead to corrosion of containers and more leakage.

* Larger containers, such as drums, bags and boxes, may be stacked in such a manner that the stacks are stable and sufficiently low to enable easy handling.

* Dust, granule and wettable powder formulations may cake under pressure. Store and stack these formulations in their original cardboard or fibre containers to prevent compression, If such formulations are supplied in bags, they should not be stacked high.

* Small containers may be stored on shelves which should not exceed two meters (six feet) in height; this precludes the use of ladders.

* It should never be necessary to climb on pesticide containers to reach other stocks.

Reason: there is danger that lids may give way under a person's weight which would cause gross contamination of the person with pesticide

* Mark all warehouse sections and stacks of products with portable waterproof signs indicating the contents.

* Keep accurate account of all stock movements and inspection reports.

Pallets

Where fork-lift trucks are available, the use of pallets will improve the ease and safety of stacking pesticide containers. Mechanical handling minimizes the risk of contamination. Common dimensions of pallets are 80 x 120 cm, or 120 x 180 cm, the latter size pallet being most frequently used in container transport. The number of packages on a pallet will obviously depend on their size and nature, but normally each loaded pallet should not exceed one metre (3.3 ft.) in height. Care should be taken about the method of stacking packages on pallets so that maximum stability can be achieved (use ties if necessary).

Storage buildings

Usually pesticides are stored in existing ware-houses or buildings that were not specially designed for the purpose of storing pesticides. Many buildings, however, can be made more suitable for pesticide storage. A manager of a pesticide store should make a critical appraisal of the existing situation with the help of a checklist of features that should be incorporated in a well-designed pesticide storage building. It is a good practice to seek the advice of experts on construction, industrial hygiene and occupational health from outside the company. Necessary improvements in an existing store can often be made by studying the main criteria for designing special-purpose pesticide stores. Important criteria pertaining to site selection and the design and construction of the building are as follows:

Site selection

Much improvement can be achieved (at relatively low cost) if early and careful consideration is given to site selection. Once a store is built, it is much more difficult -- if at all possible -- to rectify problems, and extremely costly to do so. With the growing concern about environmental safety and greater public awareness, regulations concerning site selection and precautionary measures are bound to become stricter in the future. The following are a few guidelines:

* A pesticide store should be a dry and secure building; thus, it should be constructed on well-drained, high land not affected by seasonal flooding.

* Locate the store at a safe distance from residential buildings, and from stockyards or warehouses used for storing foodstuffs, fuel and fertilizer.

* Do not construct a pesticide storage building near wells, water tanks or ponds or other sources of drinking and domestic water.

* In case the store has been or has to be constructed near waterways or inland ports, extra care must be taken to prevent contamination from major spills and run-off water in the event of a fire or similar emergency.

* The site should offer good access, preferably from more than one side, for delivery trucks and fire-fighting trucks.

* In certain temporary situations, pesticides may have to be stored in the open if lack of warehouse space does

Owing to the siting of this toxaphene (camphechlor) factory at the border of the Lake of Managua, Nicaragua, the water of this vast lake has been very seriously polluted by the discharge of untreated effluent from the factory over many years. (Photo by the author)

not permit covered storage. In such cases, the storage yard should have an impervious (concrete) floor and an appropriate drainage system terminating in an interceptor pit. Because high temperatures may destabilize some pesticides, full containers should be shaded.

* The site and the storage building should be fenced off and warning placards should ALWAYS be prominently displayed on the outside of ALL stores and shops:

"DANGER PESTICIDES, AUTHORIZED PERSONS ONLY!"

"NO SMOKING" signs should also be displayed on the inside and outside of stores and pesticide shops.

Building design

The design of pesticide storage buildings will depend on function, quantity and types of pesticide to be stored as well as cost. The following points should be borne in mind:

* The storage facility should be designed to consist of several warehouses or warehouse sections in order to enable segregation of products according to type, toxicity, flammability etc.

* The building should be cool, well ventilated, and inaccessible to children, unauthorized persons and animals.

* The store should be designed so that the need for handling pesticide containers is minimized, yet access to older stocks is not impeded.

* The store should have direct access to the outside (not through some other store or building). There should be sufficient emergency exits in relation to the size and layout of the store.

* If the pesticides are to be dispensed from bulk containers, repacked or otherwise handled, then a well-lit, well-ventilated working area should be provided, but not near the doorway of the store.

* Provision should be made for washing facilities and the separate storage of protective clothing and respiratory masks.

* Emergency stations for fire-fighting and decontamination equipment should be positioned at strategic, well-lit points and clearly indicated.

* If the store is to be continuously occupied, proper office accomodation for the storekeeper should be provided separately from the main storage area.

* Sufficient space should be allowed for storing out-of-date stocks and empty containers awaiting disposal.

Building construction

Construction details will depend on local practices, materials, functions and costs. However, all stores should be secure -- and kept secure -- and dry. Doors should have locks. Windows (if any) should be barred, as it is essential to keep out unauthorized persons, especially children. Rising dampness from the soil as well as rain must also be kept out. Construction materials should be, as far as possible, fireproof, and the building should be secure against rodent attack.

Floors should be of smooth concrete or other impervious material so that pesticide spills will not be absorbed, but may easily be cleaned up. Unsuitable materials for floors in pesticide stores include beaten earth, unglazed brick or tile, bitumen, rubber tiles and wood.

Warehouse floors should slope slightly (gradient not steeper than 1:100) so that accumulated liquids drain into a concrete-lined sump or tank from where they can be pumped out for recovery or disposal. Even where floors do not slope, sumps should be constructed and then covered over with slats for greater safety.

Bundwalls and gutters. Stores should be surrounded by an outside bundwall 15 to 20 cm high or lined to the same height (including thresholds) with impervious material so that, in the event of fire or catastrophy, the contents of the store can be contained within the store area, with no risk of gross environmental contamination. Access is via ramps. Alternatively, the floors should be recessed by 15 to 20 cm and should drain into a sump.

Storage areas inside the warehouses should be surrounded by a gutter or gully which drains into a tank or sump, so that spills and leaks can be neutralized and removed.

Gullies, tank, sumps and drains should never be connected to the public sewer system nor to site drainage. Where such gullies, sumps and drains are not sheltered by roofs, they should be covered to prevent entry of rain which could complicate disposal work. With storage in the open, drainage of rain cannot be avoided. The drainage system should then terminate in an interceptor pit. The contents of the pit should be released only after ensuring that the effluent is uncontaminated with pesticides; otherwise it must be neutralized first.

Walls. Internal walls should be smooth, free from cracks and ledges and easy to clean. If walls do not reach up to the roof, strong wire netting should be installed to prevent unauthorized entry. However, broken glass bottles should not be placed on top of these walls to keep intruders outside.

Windows. If alternative means of ventilation and lighting are provided, it is better not to have windows at all. If they are present, they should be barred against unauthorized entry and shaded, because strong sunlight can degrade some pesticide formulations.

Roofs should be of light material, such as corrugated iron sheet or fibre cement, which will collapse in the event of fire, enabling smoke and fumes to escape. In a blazing fire, the unrestrained supply of oxygen will give a more complete combustion of the chemicals, which would result in less toxic residues in the smoke.

Ventilation. This is one of the most important factors in the design of any chemical store for three main reasons:

* To prevent the build-up of toxic vapours which would endanger the health of those working in the store;

* To prevent the build-up of inflammable vapours which would present a fire risk; and

* To keep the store as cool as possible since high temperatures can destabilize many pesticides. Also, a high temperature within the warehouse is tiresome for the workers and creates laxity concerning adherence to safety precautions.

Ideally, the size of the ventilators should be equivalent to 1/150th of the floor area. Where this is not possible, doors to the outside should be open for several hours at least two days per week. Large stores should be equipped with exhaust fans, preferably operated by a time switch. Ventilators are needed to extract light fumes and hot air at levels high off the floor; adequate ventilation must also be provided near floor level to help disperse heavy vapours.

Lighting and electrical fittings. Stores must be well lit throughout by electrical lights (200 lux is recommended) to make it easy to read labels on containers and to facilitate periodic inspection of the stock for signs of leaks and other damage.

Sparks from switches and electrical equipment are a major cause of fires in pesticide stores and should be guarded against. Electrical cable should be housed in conduit for protection, or be of the mineral-insulated or armoured type. Fittings should be flame- and dust-proof.

Water supply. Adequate supplies of water, either static or piped, should be available for the washing room and for decontaminating the store in the event of spills.

Also, water should be provided near the loading platform for decontamination of the delivery trucks. The washings should be drained into the collector pit or sump.

MARKETING OF PESTICIDES

The typical retail outlets in villages throughout developing countries are small, open-fronted shops either selling only pesticides and other agricultural inputs or offering a broad array of foodstuffs, textiles and household supplies. Sometimes pesticides are even displayed together with baby food and pharmaceuticals in the same area of the shop. Generally, safety standards tend to be lower and knowledge about pesticides less among those shopkeepers who supply pesticides as a sideline only. Among these unsafe practices are the arrangement of shelving from floor-level to arm-reach height with the chemicals also piled up on the counter and floor. Larger stocks are frequently stored in the back of the store, partly blocking access to the kitchen and living quarters.

The danger of accidents is always present in a pesticide shop where careless customers and children may enter at any moment. Hazards are even more evident if the shopkeeper is ignorant about the very poisonous nature of specific pesticides; in such situations his precautions would be minimal. Training-cum- licensing of shopkeepers and enforcement of safety rules by plant protection inspectors would help to improve the currently undesirable situation.

An orderly display of properly labelled pesticide bottles and products should be protected to prevent accidents. Highly hazardous pesticides should be kept in a cupboard or behind glass. (Photo by the author)

Responsible marketing

Much of the technical information presented in this manual is designed to help people engaged in crop protection to better understand the advantages and disadvantages of chemical pest control and to pay more attention to the safe management and application of pesticides. It should, however, always be remembered that such information will have little impact on the daily practice of crop protection, unless it is well understood and accepted by farmers and other pesticide users. What the farmer considers important is how he percieves what is being offered. Thus, essential technical information on product performance and application, that has been put together by manufacturers and researchers, needs to be passed on to the users in a clear and convincing manner.

Pesticides reach the farmers largely through the efforts of those professionals who make their living from the distribution of goods and services. These people are the distributors, storekeepers and retailers or shopkeepers of the public and private sectors, to whom farmers come to purchase pesticide formulations and application equipment and to get advice about the effective use of these items for controlling worrisome pest problems.

A farmer cannot be misled in any way as to what the product or method will do for him or what benefits he can derive from them. If a sales agent or retailer misleads the farmers into believing that either a product will do more than it actually can or that a method will work without the right conditions actually being set, these farmers will become dissatisfied customers and may neither purchase the product again nor be easily convinced about such information the next time.

Pesticide distributors and retailers form a crucial link in providing reliable information to farmers and making needed products and equipment available. This is not always sufficiently recognized by governement authorities. In many developing countries, too few facilities exist for the proper training and licensing of retailers and distributors in both the public and private sectors. Frequently, the national structures for enforcement of the regulations and quality control of the formulations on sale are also underdeveloped. This situation might expose customers to unnecessary hazards or even to malpractice.

The agro-pesticide market is a difficult field for any distributor or retailer as they are expected to advise their customers objectively on the effectivity of available pesticide formulations for controlling pest and disease problems in their crops and on the most suitable method of application. "Marketing" involves a great deal more than selling (which term covers only the process of offering products and services for sale at a certain price.) Marketing comprises all activities involved in the flow of goods and services from producer to consumer. It includes elements of forecasting the need for and deciding on ordering the products, providing the place and means for their distribution, deciding on product pricing and promotion etc. All these activities are viewed from the position of satisfying the farmer's demand, since without him there is little need for the product or service.

Marketing advocates consumer orientation; it grows from attempting to meet the needs of the potential consumer. The ultimate goal of marketing is an efficient organization doing the things it was designed to do and satisfying customers.

When dealing with private business, it involves making a profit both at the supplier's end and the consumer's end. Successful marketing should be achieved by:

* Offering a product or service that satisfies needs felt by the customer;

* Offering this product or service at a price which is considered fair by all parties;

* Offering the product and service at a place which is convenient to the customer; and

* Promoting the product or service by advertising its availability and providing reliable assistance in its proper use; and

* Showing responsibility by recommending pesticide use on need-base only and avoiding highly hazardous formulations.

When discussing these five keywords -- product, price, place, promotion and responsibility -- in terms of the marketing mix of an average pesticide retail distributor, one could think of the following aspects:

Product

The term "product" includes individual pesticide formulations, as well as the product range, the assortment the shop is selling and the packaging of the product. Formulations on sale should be of good quality and free from contaminating matter. Formulations should be stocked and sold in their original good-quality packings and contain the quantity that is indicated on the label. The containers should be properly labelled (see also Chapter 6). The size of the containers should correspond with the common application pattern, i.e. small-holders usually require bottles or tins containing 100 to 500 cc of concentrated liquid pesticide or packages up to 500 grams of wettable powder. Left-over products are a major source of hazard to a farmer's family owing to poor storage facilities at the farm.

The formulations on sale should neither be too old, nor degraded, nor adulterated in any way. It does not pay in the long-run to sell sub-standard products. A product must never be sold in a leaking container; any necessary repacking should be done at the retailer's store with proper precautionary measures being taken (see Chapter 9). Because of the limited shelf-life of pesticides under tropical conditions, a retailer should neither order nor stock larger quantities than he or she can sell within one season. On the other hand, the retailer should not run out of stock of much wanted pesticides. Nothing is more frustrating for a farmer than to find the shop empty whilst pests are ravaging his crops. Rather, a retailer should make a flexible arrangement with his or her supplier to take back redundant and out-of-date stock as well as contaminated empty containers. Larger companies have the moral obligation and the resources to invest in facilities for flexible distribution and safe disposal of chemical waste as a contribution to minimizing pesticide hazards to the environment.

The "product" should also include a choice of suitable sprayers, dusters and other application equipment. Good service consists of providing a selection of spare parts and repair facilties etc. Providing an opportunity for treating larger quantities of seed with protectant fungicides -- seed dressing -- and selling treated seed of improved varieties are other services a retailer might consider offering his customers.

It might be considered bad service if a retailer includes an item in his product range just because a salesman is pushing it with incentives, but the item itself does not fit the need of the farmers. It may appear very attractive to a retailer to include a particular formulation in his product range if it has a very high profit margin. But he should be rather suspicious about the origin and quality of such a product and first arrange to have its efficacy checked in the field before placing a large order.

A service which customers obviously expect to get at the retail outlet is the retailer's help in identifying pests and diseases occuring in their crops and his knowledgeable advice on suitable control methods. If the retailer has not got that specific knowledge, he ought to employ an assistant with training in crop protection. At least he should obtain from the local agricultural extension officer or the distributor some list of recommended pesticides for major crops grown in the area and pictures of common pests and diseases.

The price

The selling price at the retail level is determined by many factors beyond the control of the retailer. These factors include wholesale prices, duties and tax, government subsidy rates and price setting by the competition. However, some elements, such as offering credit or discounts, are within the reach of an avarage retailer. The retailer should not try to charge his customers a "little extra" unless he is able to show a fair justification for it. Farmers will be inclined to go to another shop or settle for a less suitable product, if they feel unfairly over-charged. It should be normal agricultural policy to regulate the trade in agrochemical inputs in order to guarantee the supply of effective products and services against fair prices for distributors and users of pesticides both.

A street-side retail outlet needs careful arrangement of hazardous products since anyone, including children, may enter it. (Photo by the author)

The place

The best place for an agrochemical retail shop is determined by the attitude of the customers as well as by the local conditions. Are the farmers accustomed and willing to travel to distant markets? Are there many or few villages in the area and do the farmers have transportation?

The location should meet the needs of the farmers, but at the same time it should also comply with the demand for public and environmental safety (see the section under storage buildings). The location should also enable the retailer to display the products in a safe and attractive fashion.

Promotion and responsibility

Promotion of pesticide use may be based on different motives. The farmer is best served by objective information about the means and methods available for the control of pests and diseases in his crops. The most probable place from which he may expect such objective information is the agricultural extension service. It is the function of the extension service to inform and train farmers in the integrated use of pesticides and to demonstrate suitable methods of plant protection. In particular, if a Government decides to accept integrated plant protection as the leading policy in this regard, it must first educate the plant protection and extension officers accordingly, and further provide them with the means of convincing the farmers. An agricultural extension service is an indispensable link between research and practice as direct communication between researchers and farmers hardly ever takes place.

The promotion of pesticide use by retailers is profit oriented. It is the task of both the pesticide industry and plant protection authority to guard against misleading information and malpractice at the retail level. Apart from the Government installing a regulatory infrastructure and reliable inspection, the best method for achieving this aim appears to be the organization of retailer training. Furthermore, an ample supply of high quality instructional leaflets and wall posters, designed to be understood and read by farmers, is a powerful instrument for promoting sound plant protection practices. A retailer should also be able to demonstrate the use of products and durable sprayers to farmers in the field.

Recommendations for safety in shops

* Because pesticide shops are often located near the centre of towns, attention should always be given to fire prevention.

* Highly poisonous pesticides should be stored behind glass doors on shelves or in glass compartments or cupboards, preferably under lock and key.

* Do not display pesticides near foodstuffs, pharmaceuticals and other consumer goods; always try to achieve the best segregation possible.

* Display a "DANGER, POISON" notice on the outside and on the inside of the shop so that it can be seen immediately.

* All containers, packs and bottles should carry a complete label, preferably in the major national language(s) which can be understood at least by the shop manager.

* Pesticide containers and bottles should not be piled upon the sales-counter where they can be easily knocked over.

* Do not keep too many containers of the same kind in the shop area. Keep your stocks in a separate store.

* Customers should not be allowed to open containers to smell the contents - because they might be poisoned by inhalation.

* Ensure that pesticides taken by customers from the shop to the farm are wrapped adequately and carried separately from food, drinks and others consumer articles. Do not sell pesticides in leaking containers.

* Do not hand over dangerous pesticides to children sent by parents or others to collect such products.

* Store adequate supplies of water, soap and towels ready for use by customers in case of contamination.

* Do not allow customers to use a pesticide shop as a place for lengthy conversations or friendly gatherings.

References:

Oudejans, J.H.: *Agro-pesticides: their management and application.* United Nations Economic and Social Commission for Asia and the Pacific, Bangkok, 3rd edition, 1985. ISBN 974-8606-08-2.

Overseas National Resources Development Institute (1980). *Training Supplement*; Pesticide Management Training Course

Shell International Chemical Company Ltd., (1984). *The Shell Guide to Pesticide Safety.*

9

TRANSPORT, SPILLS AND DISPOSAL

Contents   Page

Transport 154

Spills and decontamination 155

- Leaks 157

Disposal of pesticide wastes 158

Preventing waste always pays 158

Major emergencies, fires and floods 161

References 162

Transport, Spills and Disposal

TRANSPORT

All year-round large quantities of pesticides are moved by air, sea, rail and road from the manufacturing sites to large distribution centres. Strict rules for the transport of chemicals apply internationally and are usually reasonably well-observed because bulk transport takes place under the responsibility of experienced shipping companies. However, major accidents do occur and can lead to massive contamination of road systems and waterways. Efforts to contain or clean up such contamination are very expensive. Chemical companies and national authorities, therefore, need to research ways and means for dealing with such environmental threats and should have emergency plans ready.

The downstream distribution of pesticides from larger depots to retail stock-points, shops and, finally, applicators is frequently carried out by insufficiently trained people and under little or no supervision. Furthermore, in many developing countries, because the means of transport are often hardly adequate and road conditions harsh, extra attention should be given to training in handling and safe transport procedures.

Only when a driver is well aware of the hazards associated with transporting these poisons will he take the extra care required in loading and driving. Hazard data and emergency instructions on dealing with accidents during the transport of chemicals should be given to transport personnel. For instance, drivers should be provided with a transport emergency (TREM) card or its equivalent. Kept in the driver's cab, this card would provide essential data, even if the driver would be badly hurt. Such data would include the following:

* The name of the dispatching company, including its address and emergency telephone number.

* Type of products being carried.

* Basic hazards posed by those products and the safety precautions to be followed in case of an accident.

Moreover, all containers must be properly labelled and the pesticides packed in the manufacturer's original containers. Warning placards for hazardous materials must be placed on the four sides of the vehicle used to transport pesticides if they are of a corrosive, flammable or highly toxic nature. A suitable fire-extinguisher, a basic clean-up kit, brooms and shovels and protective gear should be available for use by the driver.

Before releasing the stock for transport, the shipping company should always check to ensure that the correct product is being dispatched and that the cargo is labelled correctly. Do not ship large containers containing, for example, 200 litres (45 gallons), if neither a fork-lift truck nor a loading platform is available at the receiving end. Such large containers weigh up to 300 kg and cannot safely be lifted by men unaided by specialized lifting equipment. Quite often, a very dangerous practice is followed -- barrels containing pesticides are pushed off a truck onto old tyres which are supposed to act as a cushion; however, the container might burst in the process.

Special care should be taken during the loading and unloading of pesticides to prevent damage to or breakage of containers. Boxes should be placed with the proper side up according to signs on the outside. Never transport open or leaky containers. Always inspect the cargo by reading the labels to determine if protective clothes (aprons, gloves, boots etc.) should be worn while handling the shipment. Do not carry more than one large box or container at a time to avoid stumbling with it. Keep an eye on workers so that discipline is maintained during loading and unloading.

Load carefully to prevent containers from falling off the transport vehicle. Large shipments should be secured carefully to prevent shifting of the load or to prevent drums from striking each other. Use braces and dunnage to prevent punctured containers, loosened lids or caps and damaged labels. Pesticides should never be transported with food, animal feed, beverages or clothing

A Sudanese spray team departs for a desert locust control mission on an open pick-up truck loaded with people, sprayers, insecticidal dusts, food and drinking water, and fuel. (Photo by the author)

because of the danger of contamination. Also, drivers should not take passengers.

If it becomes necessary to transport small quantities of pesticides in a van, station wagon or in the boot of a car, it is essential that the load be secured and the vehicle kept well ventilated. At no time should pesticides be in the driver's cab. Extra attention should also be given to parking a vehicle, which should never be left unattended.

When pesticides have been unloaded from a vehicle, inspect the body of the vehicle, the tarpaulins and the rest of the cargo for evidence of leaks or spills. Check all newly arrived goods for leaks, loose lids, loose labels and re-pack immediately if necessary before storage. Replace torn or unreadable labels without delay before one forgets to which lot the unlabelled container belongs. If spills have occurred, immediately decontaminate the vehicle and do NOT let it depart before it has been cleaned completely.

The greatest chance of exposure or damage from pesticides in transport is in the event of an accident. The driver must know what he is carrying and be informed of the emergency measures which he should take in the event of an accident. In a crash, fire or spillage, the driver should act rapidly by:

* Switching off the engine; and not lighting a cigarette.
* Sending someone to call the police and the pesticide company, and warning other traffic to keep away.
* Trying to contain the spill by covering it with an absorbtive material or earth while being careful about his own and others' safety; putting on protective gear and attempting to avoid walking in splashes and fumes.

* Staying with the vehicle, but upwind of the spilled chemical and keeping people away.

* Collecting the absorbed spill, broken containers and all contaminated waste for disposal in a safe place.

There is always a great danger that toxic liquids or powders will spill out of broken drums, bottles and bags, thereby contaminating crops, pastures, vegetation and waterways. In many countries, cattle and goats frequently graze along roadsides. Also, water from canals and roadside streams is often used for consumption by people and animals. Immediate containment of a spill, surveillance and thorough decontamination is, therefore, essential. If the roadside is contaminated, warning placards should be displayed to warn farmers about the hazards to grazing cattle.

For decontamination of spills, see the next section. For cleaning up large spills and decontaminating the area affected and the vehicle, advice should be sought from the pesticide company, the local plant-protection or agricultural extension officer and the local police force. Environmental protection agencies could also be involved.

If there is any possibility that food, animal feedstuffs, clothing or general consumer goods have been contaminated, these goods must be destroyed by burning under supervised conditions. Many people have died because of eating poisoned food that had been transported together with pesticides.

SPILLS AND DECONTAMINATION

Even in the best run stores, spills occasionally occur, especially where concentrates are decanted and repacked.

SPILLS MUST BE CLEANED UP IMMEDIATELY.

If spills are not cleaned up quickly, they may be absorbed by the floor and other containers. Floors should be made of non-absorbent material such as polished concrete. The only way to decontaminate brick, earth or wood floors may be to remove and replace the contaminated part. Otherwise, the spill may be trodden in, thus contaminating personnel. Such spills could corrode adjacent pesticide containers, or worse, produce toxic or inflammable fumes.

Protective gear and cleaning equipment should always be kept ready in manufacturing plants, warehouses, stock yards and shops for dealing rapidly and effectively with spillages and leaking containers. Immediate action helps to minimize damage and loss. The following list of equipment should be sufficient to enable someone to deal with common incidents of spillage and leakage:

* Two pairs of neoprene or PVC gloves (elbow-length)

* Two pairs of rubber boots

* Two neoprene or PVC aprons; 1 metre wide and 1.5 metres long

* Two respirators, canister type (see Chapter 7)

* Two face shields, helmet type covering eyes and face or at least two pairs of well-fitting goggles

* Fifty kg of powdered lime†, or sawdust, or sand, or other absorptive material

* Five litres of Teepol, or washing soda (sodium carbonate) or strong soap

* Two yard-brooms

* Two shovels

* Some empty decontaminated drums of various sizes for waste disposal and re-packing

* Paper labels and glue for marking drums and boxes

* Funnels, buckets, drum spanners and earthing wire

† **Note:** Lime is particularly suitable as a general purpose absorbent for liquid pesticides, especially in view of its alkaline reaction which assists the degradation of many of the more toxic pesticides, such as organo-phosphates and carbamates.

The storekeeper deals with a spill by absorbing the liquid with soda ash after first having donned safety overalls, boots and neoprene gloves. (Courtesy of ICI)

DO NOT HOSE DOWN SPILLS as this merely disperses the pesticide over a wider area. Rather, first absorb as much of the spill as possible and then, follow up with a thorough decontamination of all affected surfaces and objects. Remember that in case liquid fumigants, such as the volatile nematicides D-D or methyl bromide, have been spilled in a building with poor ventilation, self-contained breathing apperatus has to be donned first before any steps are taken to deal with the contamination.

Spills of **liquid formulations** should be soaked up with an absorbent material (sawdust, lime, sand, earth), supplies of which should always be kept in the store. After the spill has been absorbed, it should be swept up and placed in marked, metal containers (or strong plastic bags) for disposal.

Spills of **solid formulations** (dusts, wettable powders and granules) should be swept up carefully to avoid creating dust and placed in marked containers for disposal. Sprinkling of damp sand or sawdust onto spilled dry powder helps in preventing dust.

After sweeping (repeated more than once if necessary), the area affected by the spill should be scrubbed with water, and strong soap or detergent. Absorb the first washing too if no sump is available; do not allow the wash water to run off into public sewers or surface water. Repeat the scrubbing.

Water, soap and detergent are usually most readily available, but sometimes other chemicals sold for domestic or common commercial purposes may also be useful.

Organophosporous compounds, carbamates and pyrethroids can be broken down by hydrolysis, usually by mixing with sodium hypochlorite (bleach) or a 10 per cent solution of sodium carbonate (washing soda, Na_2CO_3) or 5 per cent sodium hydroxide (caustic soda, $NaOH$). These chemicals are useful for decontamination and can be applied following initial scrubbing with soap and water. Beware, however, because caustic soda is corrosive to skin and eyes and should be handled with care.

Organochlorine compounds. These are more difficult to deal with. Although household ammonia and washing soda can be used for decontamination, the main method is tho scrub with water and detergent.

Bipyridylium herbicides (paraquat and diquat). Scrubbing with washing soda or strong soap should be used for decontamination.

If other impermeable containers, such as drums, or handling equipment, or the delivery vehicle have been contaminated, particularly by the more toxic products, all surfaces should be well scrubbed with a concentrated detergent solution or strong soap and then rinsed with water. If powders or granules are spilled onto other goods, the outer surfaces of such contaminatd containers should be well brushed. Check thoroughly to ensure that no spilled product, in particular herbicides, has penetrated into the content of other containers.

Decontamination of personnel (skin)

Successful decontamination requires:
- working with plenty of soap and water
- extreme thoroughness in using the soap and water
- great speed in washing with soap and water

A person contaminated with a pesticide should remove his clothing to allow the affected part of the body to be quickly and thoroughly washed with soap and water. Do not use a hard brush, since this may damage the outer layer of the skin and thus promote penetration of the poison. After washing, rinse thoroughly, particularly when a very toxic concentrate was the cause of the contamination. Finally, wash the contaminated area with methylated spirits or alcohol. The last action is all the more important if there has been a delay in beginning the decontamination process.

Protective clothing should be thoroughly washed with industrial-grade detergent followed by several rinsings. Such clothing should be washed in hot water if available. If replacement clothing is available, destroy heavily affected clothing by burning. Also, decontaminate the protective gloves and boots and the equipment used, such as shovels.

Leaks

Leakage is a major problem in the storage and transport of pesticides. With good housekeeping procedures and careful handling, containers normally would remain intact for long periods.

Rough handling, which dents drums, weakens or splits seams and weakens closures (lids, caps, stoppers), causes initial damage and the containers may later begin to leak as a result.

Other causes of mechanical damage are puncturing or abrasion during transport when packages and containers rub against each other or against the sides of the truck. The use of hooks for unloading bags at inland ports may damage the bags, from which poisonous granules may flow.

Leaks also result from corrosion of the container, which may be accelerated as a result of mechanical damage (dents may rupture linings, for example).

Corrosion may start internally, when the pesticide itself (or its breakdown products) is the primary cause. Or it may begin externally, owing to rusting in damp storage conditions or contamination of nearby containers. Many EC formulations are very corrosive and some, including monocrotophos, dicrotophos, dichlorvos and phosphamidon, are incompatible with steel and should be packed in plastic or aluminium containers.

Containers may leak for other reasons; for example, strong sunlight can degrade some plastic containers, including plastic bottles and plastic sacks. Rodents may damage paperboard or fibre containers.

Stores should be inspected regularly (at least every two months) and any leaking container should be dealt with IMMEDIATELY.

Old, rotting and leaking containers are extremely difficult to move; hence the need to deal with leaks as soon

For almost 10 years, these badly corroded and leaking drums containing MCPA herbicides have been causing serious contamination in the centre of a large government store; for lack of a writing-off procedure for redundant stocks, the storekeeper could not help to correct this dreadful situation. (Photo by the author)

they are noticed. But, **always**, put on protective clothing first and isolate the area.

Usually the only way to deal with a leak is to repack the material in a strong container. Put small leaking containers into large open-top drums. For this reason, it is useful to save a few empty containers of various types and sizes. If old containers are saved for this purpose, the old labels must be completely removed and the containers thoroughly decontaminated. Wash the containers with a 10 per cent sodium carbonate solution, added at the rate of at least 1 litre per 20-litre drum. Swirl the mixture around to rinse the walls; empty and add the rinsings to the absorbents used to soak up the leak.

Repack pesticides in containers made of the same materials as the original containers (as previously explained, some pesticides are not compatible with steel). Ensure that the new labels which are written out contain all the information on the old label; the new labels should be fastened securely to the new container. Record the date of repacking (and the date of original receipt by writing this information on the replacement container and in the store record). Finally, ensure that the repacked material is used first.

If in neglected stores old abandoned drums are found to be too badly corroded to move, one may proceed by emptying them through the bunghole by means of a handpump. If no handpump is available, one could try to dig a hole at the side in the floor and fill it with absorbent materials. Puncture the rotten drum stepwise from top to bottom while replacing the saturated absorbent material each time with fresh material. This is not an ideal situation so proceed with much care.

DISPOSAL OF PESTICIDE WASTES

Pesticide wastes include everything from redundant out-dated formulations with high pesticide content to contaminated containers, left-over spray-solutions and rinse water. The sheer volume of pesticide waste has become a matter of great concern. Considerable research capacity is aimed at finding acceptable solutions for this problem. This is proving to be a very difficult task even for countries with an advanced infrastructure. For a number of developing countries, it is as yet hardly possible to find safe means of disposal for the mounting stocks of pesticide waste. Some such countries are looking for international assistence in disposing of obsolete, high-content pesticides, which are frequently the result of poor transport practices, and inadequate storage facilities and store husbandry. Solving the problem of disposal needs always to be accompanied by a programme of training, aimed at preventing the creation of waste from the factory to the end-user. The avoidance of waste build-up starts with the initial order. Store-keepers should order only as much pesticide as can be sold within the shelf-life of the consignment. Inspect the store regularly to spot leaks; work carefully to avoid spills. Further, farmers or users should mix only the quantity of chemical that can be applied in one day.

PREVENTING WASTE ALWAYS PAYS

There are three recommended procedures for disposal of pesticide wastes:
* Incineration
* Hydrolysis (for certain pesticides)
* Burying or dumping in an approved landfill

Incineration is an effective and safe means of disposing of the majority of pesticide waste materials and is currently recommended as the generally preferred method. However, some pesticides, such as hormone weed-killers, should not be destroyed by incineration, because the combustion gases may contain hazardous compounds. If larger quantities of pesticides are to be disposed of, advice should be sought from major pesticide companies or the Association of Agrochemical Manufacturers (GIFAP), or agencies such as the Food and Agriculture Organization of the United Nations (FAO), the United States Agricultural International Development Agency (USAID) and the Overseas Development National Resources Institute (ODNRI) of the United Kingdom.

The following technical advice is taken from the Safety Guides of Shell International Chemical Company Ltd., and of CIBA-GEIGY Ltd.

The temperature and time required for effective destruction of pesticide materials by incineration is specific to the type of material and therefore the type of incinerator required and method of firing.

1. **Solid waste with a high pesticide content such as obsolete technical pesticides or wettable powders.** If these materials are packed in steel containers, they must first be removed and may have to be broken into lumps before charging the incinerator. Such wastes require auxilliary burners and fuel for ignition and maintenance of combustion. Up to 30 minutes at 900°C is required; continuous and controlled feeding of the waste and continuous agitation are necessary to achieve complete combustion and destruction. The rotary kiln furnace is preferred for such duty.

A scrubber is required for removing acid fumes from the flue gas before dispersion into the atmosphere.

Only major manufacturing sites produce enough waste of the above category to consider the investment expenditure and expertise required for operating such an incinerator. In some countries, liquid waste is incinerated in rotary cement kilns, which also reach high temperatures.

2. **General solid waste, lightly contaminated with pesticides.** Lightly means about 1 per cent w/w of active pesticide material. Most formulation plants have this mixed waste, consisting of sawdust, sand, empty cartons, fibre kegs, old gloves, cotton waste etc.

Such waste may be destroyed in a simple incinerator but with controlled temperatures. If the waste contains a great deal of fibrous matter such as paperbox packing material, the combustion temperature may rise to 2000°C. At such high temperatures dangerous emission gases may be formed also; combustion temperatures should remain within the range of 900° - 1600°C. Ask expert advice if flue gas scrubber and stack is necessary. If enough waste has accumulated, the incinerator is fired, say once a week, or once a month.

If occasionally there are larger amounts of concentrated pesticide waste for destruction, the waste should be mixed with additional solid fuel to remain within the capabilities of the "simple" types of incinerator.

Empty metal, glass or non-combustible containers cannot be destroyed by incineration except under special conditions. However, these containers can be effectively decontaminated by being passed through an incinerator, if the containers are open- top or suitably punched with, say, 2-inch holes.

If no incinerator is available, glass containers

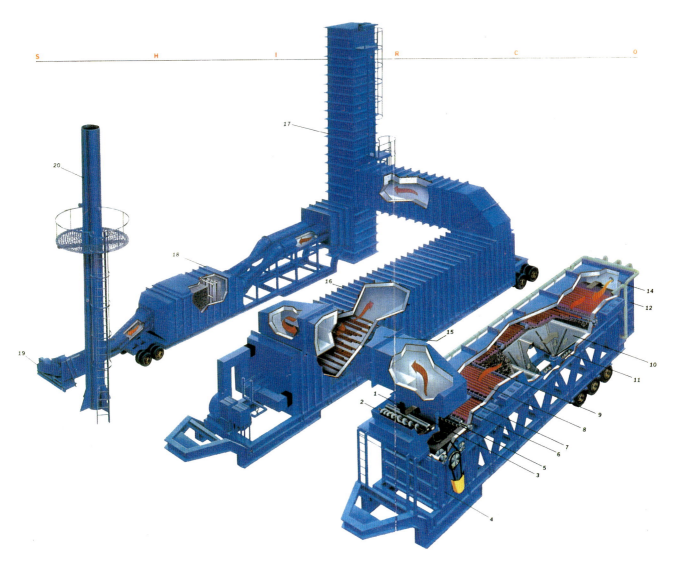

should first be crushed and small metal containers punctured to render them unusable for other purposes. Diesel fuel should then be poured over the fragments and ignited to burn any remaining pesticide residue. Next, the remaining waste should be collected and buried. On the farm, burning of empty paper packages and plastic bottles in the open air may be considered. But burning must take place away from housing and people, as it may release toxic vapours. Burn in a hot fire, standing upwind. The ashes should be buried.

3. **Liquid waste.** Such waste can be finely dispersed by pumping through an atomizing burner while adding auxilliary liquid fuel as necessary. Homogenized liquid wastes can be burned in a rotary kiln. When finishing off the spray job, small quantities of left-ver spray solution and rinsate from containers should be sprayed over the target crop, whenever possible.

Some major chemical manufacturing plants and larger cities in industrialized countries have built rotary drum furnaces (kilns) or combustion rooms with ignition grate system for large-scale waste incineration. Several specialized companies are involved in designing mobile incinerators for use in third world countries (see picture above). Some donor countries, such as the United States

A complete incineration of pesticides and chemical waste that does not leave residual toxic gases requires high temperature furnaces and extensive processing of exhaust gases. A mobile incinerator mounted in shipable containers (patented) is being tested for use in developing countries.

Process Description: *A Feed System (not shown) (1) delivers preconditioned waste material to the primary furnace through a rotary airlock into the Spreading Section (2). There the material is spread and leveled on the Metering Conveyor (3) before entering the Feed Module (4). The Furnace Conveyor (5) moves the waste material through Fiber Blanket (6) insulated Powered Modules (7) where it is brought to combustion temperature by infrared Heating Elements (8) and gently turned by Rotary Rakes (9). If residual material falls through the belt, it is collected in Hoppers (10) where it is removed by the center mounted Collector Screw (11) for analysis. Processed material (or ash) passes from the Discharge Module (12) into the Ash Discharge System (13) to an external receptacle.*

The Combustion Air System (14) delivers air from a blower through manifolds to side ports. Exhaust gases are drafted through the furnace to the Exhaust Duct (15) and to the Secondary Process Chamber (16) to incinerate any remain- ing combustibles. Gases exit the secondary chamber and go through the base of the Emergency Bypass Stack (17) into the Emissions Control System (18) where they are cooled and cleaned before being exhausted by an induced Draft Blower (19) through the Exhaust Stack (20).

of America, Germany and the Netherlands, implement projects to assist developing countries in investigating their chemical waste problems and to find solutions for their safe disposal. For instance, in 1991 the Netherlands shipped back obsolete stock of dieldrin from Niger, Africa, for destruction at the incineration plant of Rotterdam.

Hydrolysis

Hydrolysis of many pesticides of the organophosphate carbamate and pyrethroid classes can be achieved, usually by mixing with a 10 per cent solution of sodium carbonate (Na_2CO_3) or 5 per cent solution of sodium hydroxide (NaOH, caustic soda). The time required for complete hydrolysis is computed on the basis of the half life of a particular pesticide, which is the time needed for the chemical break-down into metabolites of the pesticide compound to 50 per cent of its original concentration at a certain pH and temperature. The required hydrolysis period can be determined by taking twice the half-life time plus some reserve. Before release of the effluent into a sump or sewer, it should be checked by laboratory analyses to determine whether the pesticide toxicity has been removed.

Burial

Where facilities for incineration are not available and hydrolysis is not practical, burial may remain the most practical method of disposal. However, this is not associated with safeguarding public health and the environment. The main problem is to keep the buried pesticide in place, to prevent leaking that would contaminate surface run-off water and groundwater.

Local or area authorities, or waste disposal contractors, may be able to advise on suitable dumping sites, e.g. land-fills. Pesticide wastes should never be dumped on public rubbish-tips, nor in areas subject to flooding.

This procedure is the least satisfactory of the three options for disposal of pesticide material, but is suitable for disposing of empty, decontaminated containers. Burial should preferably be on privately owned land and in any event, in an isolated place away from ponds, water-courses and boreholes. The area should be marked and a record kept of the site's position and its burial load. Wastes should be buried to a depth of at least one metre (about 40 inches).

Contaminated absorbents and surplus products should be thoroughly mixed with crystals of sodium carbonate (washing soda) to help neutralize the waste before burial. Adding powdered lime is also helpful.

Some organochlorine pesticides may remain in the soil for many years; other groups break down more rapidly. Pesticides containing heavy metal ions will never

Old damaged drums are crushed to prevent re-use while cleaned sound drums are neatly stacked before being refilled at the Shell Factory near Kuala Lumpur, Malaysia. (Photo by the author)

disappear completely and the metal will remain in the soil indefinitely.

Disposal of pesticide containers

A very great problem is the fact that in developing countries empty pesticide containers represent a great value. Containers of polythene or metal are durable in comparison with traditional materials of wood or clay. Since many products, such as cooking oil and cereals, are sold loose, there is always a great demand for containers, even those of dubious origin.

Nevertheless, empty pesticide containers should never be used to hold water or food, because pesticide containers can never be cleaned out completely. Studies conducted at the Tropical Development and Research Institute, U.K., have shown that even after thorough washing of laquered steel containers, aluminium containers and polythene containers, small residues of pesticides remained absorbed on the inner surfaces and coatings, which slowly keep leaching out into fillings of water, despite further rinses and daily water changes. For example, in washed polythene containers that had contained lindane concentrate, a sustained release of this compound was observed despite weeks of daily water changes.

Therefore, all empty smaller containers should, after rinsing three times be rendered impossible to reuse by puncturing, crushing, breaking or burning and subsequently be disposed of at least by burial if not another safer method.

The value of empty drums and large plastic containers is in some areas so high that reuse cannot be prevented. In principle, such empty containers should not be reused except for special purposes and then only after thorough decontamination with caustic soda, for example.

These special puposes include refilling with the same pesticide or with another formulation of the same active ingredient of reuse for cleaning up spills in storage areas. Another use could include storing fuel.

The best solution for large-container disposal is for them to be taken in by the distributor or formulator for reconditioning and re-use. Such containers should be thoroughly drained and triple rinsed, washed on the outside and tightly closed before being handed over to the transporter. While awaiting transport for disposal, the empty containers should be stored in a special secure area to ensure that they are not stolen and used for other purposes.

MAJOR EMERGENCIES, FIRES AND FLOODS

Fire

AN OUNCE OF PREVENTION IS BETTER THAN A POUND OF CURE. One of primary objectives of the design and management of pesticide stores is the reduction of fire risks.

Pesticides, especially those formulated as liquids, present major fire hazards because the solvents (oils and petroleum distillates) used in such formulations have low flash points and readily vapourize at normal temperatures.

In poorly ventilated stores, heavy vapours may accumulate near the floor if drums are left open or if leaks and spills are not cleared up (as mentioned previously, stores need low-level as well as high-level ventilation). An electric spark, naked flame, or even the sun's rays concentrated by a glass container may cause an explosion followed by fire. Therefore, keep glass containers away from windows and out of the sun. Smoking and the use of naked flames must be forbidden in pesticide formulation plants and stores.

Some wettable powders are suspected of starting fires through spontaneous combustion. Sodium chlorate (used as a herbicide, defoliant, dessicant and soil sterilant) is a powerful oxidising agent which easily catches fire. Once sodium chlorate containers have been opened, their entire contents should be used at once.

Notices should be placed on the outside of pesticide stores stating "DANGER PESTICIDES: AUTHORIZED PERSONS ONLY". Notices and symbols stating "NO NAKED FLAME" should be prominently displayed as well. To be effective, these rules must be strictly enforced by storekeepers.

Have emergency plans ready for dealing with fires and hold fire emergency exercises regularly. Know what products are stored and where they are situated.

Also, the fire department and local police should be informed about the whereabouts and quantities of stored pesticides. In turn, the fire department should visit the premises and learn in advance what special measures should be taken to control the hazards associated with fires in chemical stores.

Since pesticide shops are often located in densely populated areas, evacuation of the neighbourhood may sometimes be necessary.

Fire extinguishers should be available in the store at strategic places and should be regularly checked to ensure that they are properly charged. Static or running water should be available as well. Buckets of sand or earth are very useful for putting out small fires and for absorbing any spills or leaks of liquid pesticides.

Action in the event of fire

Fighting large out-of-control fires should be left to professional fire-fighters. The advice given here is chiefly with regard to "first aid" fire fighting, that is, extinguishing small fires and most important, preventing a fire from developing into a conflagration.

As a rule, water should not be used on fires involving pesticides because of the problems associated with toxic run-off. However, water as a fine spray can be used for cooling nearby metal or non-absorbant drums. For dealing with small fires or those involving stacks of paper or fibreboard containers, dry powder or carbon dioxide extinguishers should be used. Foam is also suitable, but it should be of the alcohol-resistant type.

What to do when a fire is discovered:

* Sound the alarm.
* If you think you can deal with the fire yourself, use the nearest available extinguisher. Afterwards, make sure the area is properly cleaned and decontaminated and replace the used extinguisher. Report the incident and use of the extinguisher so that it can be replaced or recharged.
* If assistance is required, DO NOT DELAY. Clear the area of all personnel except those dealing with the fire.
* Call the fire fighting service and inform them of the hazards involved.
* Ensure that the firemen are adequately protected. They should wear entire body protective clothing and, if exposed to smoke or fumes, the use of a self-contained breathing apparatus is essential. Do not risk the life of non-professional personnel.
* Protect employees and nearby residents. Warn residents downwind of the fire to close all windows and be prepared to evacuate.

When fire gets out of control

If the fire is in a building containing toxic materials and there is little chance of the fire spreading to other buildings or property, the safest procedure, once a large fire is burning, may be to let it burn under supervision. This decision should be taken on the advice of the chief fire officer. Remember that incomplete oxidation of toxic materials may occur in a smouldering fire during certain phases of fire-fighting. Under such conditions, toxic fumes and residues can be formed. The likelihood of this is smaller in a blazing fire. Ventilation will aid more complete combustion and reduce the amount of toxic smoke. A light-weight roof designed to collapse easily will at least permit the fumes to be carried upward away from the firemen. Attempts to extinguish such a fire with water would prove counter-productive; in addition to toxic effluent, it would cool the plume, thereby restricting efficient dispersion of the fumes.

Therefore, the use of large volumes of water should be avoided in order to keep toxic run-off to a minimum. If possible, construct a barrier of earth to contain run-off and burning liquid.

Once the fire is under control or extinguished, immediate attention should be given to the environmental implications posed by the nature of the fire. Work should be started to clean up and decontaminate the site and surrounding area with the advice of a major pesticide company.

Floods

In some countries, flooding is a common event. Special hazards follow when pesticide stores are flooded.

Many pesticides are packed in cardboard or paper containers and these lose strength and may leak or burst open when wet. Other containers, especially drums of liquids, may be carried away in a flood. The result of such events is potential environmental contamination over a wide area, pollution of water supplies and potential hazards to the public who may find the pesticide containers. The danger of floods and fires emphasizes the value of keeping stock records in a place where they will be safe in these types of emergencies. They will prove invaluable in subsequent efforts to trace missing containers as well as to assess the monetary losses caused by the emergency.

References:

CIBA-GEIGY Ltd., *Application Advisory Service (1988) Operator Guide: Safety depends on you.*

Overseas Development National Resources Institute (ODNRI). *Training supplements,* CORP Pesticide Management Training Course. 1980

The Shell Guide to Pesticide Safety (1984) Shell International Chemical Company Ltd., London

10

SPRAY EQUIPMENT AND MAINTENANCE

Contents   Page

Common types of application equipment 164

- Hand-operated sprayers with hydraulic pumps 164

- Air carrier sprayers 171

- Fogging equipment 173

- Hand-operated rotary dusters and granule spreaders 174

- Sprayers with centrifugal-energy nozzles 175

- Electrostatic spraying equipment 177

- Rope-wick herbicide applicators 178

Spray delivery assembly and types of nozzle 179

- Hydraulic-energy type nozzles 180

Maintenance of application equipment 182

- Problems with motorized sprayers 184

References 184

Spray Equipment and Maintenance

COMMON TYPES OF APPLICATION EQUIPMENT

Insecticides and fungicides are usually applied as foliar sprays, and herbicides are mostly sprayed either onto the foliage or the soil. Thus spraying of liquid and wettable powder formulations is the common method of application and consequently a wide variety of hand-operated and power-driven spray apparatus has been developed over the years. Also, other formulation types are marketed, such as granules, dusts and fumigants, which require different equipment for their application, or none at all. For example, granules can either be applied by mechanical spreaders or broadcast by hand. However, granules and other non-sprayable formulations have a limited market and are not available everywhere.

In Asian countries, most pesticides are applied with small, hand-operated, hydraulic sprayers or hand-held spinning disc sprayers. Depending on the type of agricultural practices and the economic development of the area, mist-blowers and power-operated hydraulic or rotary cage sprayers mounted on tractors or aircraft may also be used extensively. In special fields of application, such as hygiene and storage pest control, foggers and fumigating equipment are generally used.

This manual deals in particular with portable apparatus; for larger power-operated hydraulic and air-carrier sprayers, refer to technical literature, such as Pesticide application methods by G.A. Mathews (1974)

and Pest Management (1984) also by the same author. The latter book gives details of sprayers for experiments. Important types of equipment for use by small-scale farmers are the following:

Hand-operated sprayers with hydraulic pumps

* Trombone (or slide-action) sprayers
* Stirrup- or bucket-pump sprayers
* Shoulder-pump sprayers
* Lever-operated knapsack sprayers
  - piston-type pumps
  - diaphragm-type pumps
* Compression or pneumatic sprayers

Air-carrier sprayers

* Motorized knapsack mist-blowers
* Pulsejet fogging machines

Hand-operated rotary dusters and spreaders

Sprayers with centrifugal-energy nozzles
* Hand-carried spinning disc sprayers (rotary atomizers)

Electrostatic spraying equipment
* Electrodyn sprayer

Rope-wick herbicide applicators

Regardless of the size of the spraying machine employed, every operation requires well maintained and calibrated equipment, precautions against contamination of man and the environment, and skilled operators who know the pests and criteria upon which the decision to spray or not to spray should be based. (Photo courtesy of LNV, the Netherlands)

164

Hand-operated sprayers with hydraulic pumps

* Trombone sprayers

These sprayers, also called syringe- or slide-action sprayers, are the simplest and cheapest type of sprayers available. The spray liquid is drawn from a separate reservoir into a pump cylinder by pulling out the plunger; the liquid is then forced through the nozzle on the compression stroke. The pump cylinder and plunger are parts of the spray lance. The pump may be single-action (pumping on the forward stroke only: fig. 10.1) or double-action (pumping on forward and backward strokes: figs. 10.2a + 10.2b).

The only difference between single- and double-action pumps is the hole drilled in the piston half of the pump (see 10 in figs. 10.2a and 10.2b). The part of the cylinder surrounding the piston acts as a temporary storage chamber (or reservoir) into which the spray passes through the hole in the piston wall. On the next backward stroke the liquid trapped in this storage chamber is forced out of the hole and finally through the nozzle. These sprayers do not have a pressure vessel and pumping has to be rapid and continuous to ensure an even spray at the nozzle. The spray stops when pumping is stopped.

Purchase a double-action sprayer to economize on pumping, because pumping requires use of both hands and the trombone sprayer is fatiguing to operate. A long spray lance allowing a long pump stroke may be an advantage, but the longer lance will be more susceptible to bending or breakage.

A knapsack container or bucket may be used to contain the chemical, which should be stirred regularly. These sprayers can produce a fine spray and are often fitted with cone-type nozzles. They are particularly suitable for spraying small vegetable plots. The sprayers should be brass and if possible the ball-valves should be replaceable. With careful maintenance, their useful life can be two years.

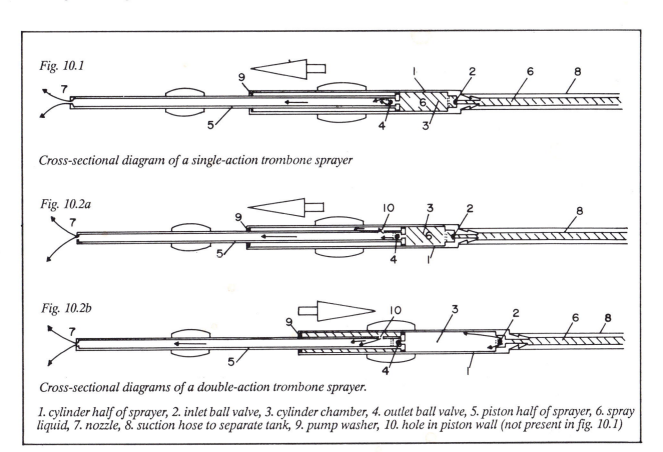

Fig. 10.1

Cross-sectional diagram of a single-action trombone sprayer

Fig. 10.2a

Fig. 10.2b

Cross-sectional diagrams of a double-action trombone sprayer.

1. cylinder half of sprayer, 2. inlet ball valve, 3. cylinder chamber, 4. outlet ball valve, 5. piston half of sprayer, 6. spray liquid, 7. nozzle, 8. suction hose to separate tank, 9. pump washer, 10. hole in piston wall (not present in fig. 10.1)

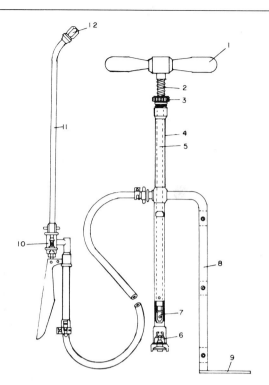

Fig. 10.3: Stirrup or bucket-pump sprayer

1. handle, 2. spring, 3. gland nut, 4. pump barrel, 5. plunger or piston, 6. suction ball valve assembly, 7. plunger bucket with outlet ball valve, 8. bracket, 9. foot rest, 10. trigger cut-off valve, 11. spray lance, 12. nozzle.

* Stirrup or bucket-pump sprayers

The stirrup pump (fig. 10.3) is placed into a bucket holding the chemical solution or is provided with an inlet hose to avoid immersing the pump. The pump is steadied by a foot rest or stirrup on the ground next to the bucket. Two operators are required: one to work the pump while the other directs the lance and nozzle. To facilitate working, the hose is up to 6 m in length; longer hoses are difficult to handle. Spraying is continuous with these double-action pumps as long as pumping is continued. They are strongly constructed and can withstand considerable wear if properly cleaned and serviced after use.

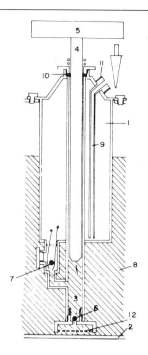

Fig. 10.4: Diagram of typical shoulder-pump sprayer.

1. pressure vessel, 2. tank wall, 3. pump cylinder, 4. plunger or piston, 5. handle, 6. inlet ball valve, 7. outlet ball valve, 8. spray liquid, 9. dip tube, 10. sealing washer, 11. hose attachment, 12. strainer.

* Shoulder-pump sprayers

This type of sprayer (fig. 10.4) has an oval-shaped stainless-steel tank with a capacity of about 10 litres and is carried by a single strap hung over one shoulder. These sprayers use compressed air to generate pressure. A pressure vessel enclosing a pump cylinder is fitted inside the tank, whereas the tank itself is neither pressurized nor air-tight. The sprayer is pumped with one hand, whilst the lance and nozzle are directed with the other. Pumping is continuous at between 20 and 30 strokes per minute and produces a fairly even spray from the nozzle. This sprayer is suitable for spraying of up to two hectares of low crop such as cabbage and may have a useful life of over three years.

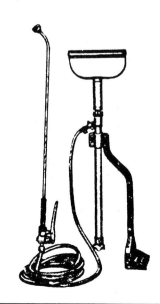

* Lever-operated knapsack sprayers

This type of sprayer consists of a tank of 10-20 litres' capacity, a hand-operated pump, a pressure vessel, a hose and a spray lance with cut-off valve and one or more nozzles. The sprayer is made of stainless steel, brass, hard plastic or an alternative corrosion-proof material. All welded seams and joints and threaded connections should be leak-proof. The sprayer is carried on the operator's back with straps like a knapsack.

These sprayers use compressed air to propel and atomize the spray liquid like the shoulder-pump sprayer, but the construction is different. In lever-operated sprayers, the pump cylinder (piston) and pressure vessel are either separate units or a combined unit, and the pump/pressure vessel may be located either inside or outside the tank. The tank itself is not pressurized and is not air-tight. The filler opening should be wide (e.g. 140 mm) to enable filling from a bucket without spillage. The opening is closed with a tightly fitting filler cap without breather hole in order to prevent leakage onto the operator's back.

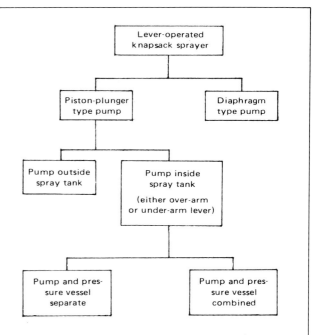

The lever-operated sprayers, which display a wide variety in design, may be divided into categories as shown in the scheme above.

The over-arm lever type is easier to operate than the under-arm type when walking between rows of tall plants, but is more fatiguing than the under-arm lever type. Sprayers may be built for left- or right-handed use or may be adjustable for either. For efficient use of the lever, the sprayer must fit comfortably on the back of the operator with well-tightened straps. The straps are easily attachable to the tank by means of corrosion-proof fastenings or clips. Straps should be at least 5 cm wide, adjustable and made of rot-proof webbing or non-absorbent reinforced plastic.

Fig. 10.5

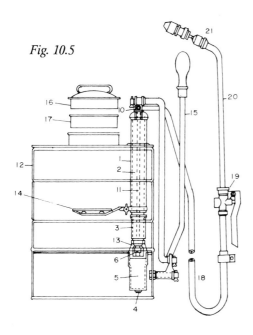

Lever-operated knapsack sprayer with under-arm action. The piston and pressure vessel are combined.

1. combined piston/pressure vessel, 2. pressure vessel, 3. piston, 4. inlet ball valve, 5. cylinder chamber, 6. outlet ball valve, 10. hose attachment, 11. dip tube, 12. spray tank, 13. pump washer, 14. agitator, 15. pump lever, 16. filler-hole cap, 17. strainer, 18. delivery hose, 19. trigger cut-off valve, 20. lance, 21. nozzle.

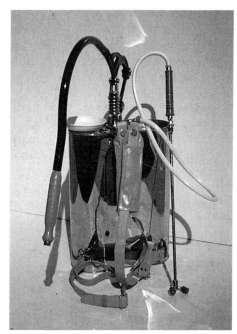

A lever-operated hydraulic sprayer with over-arm type lever and with the pump inside the spray tank. (Photo courtesy of ODNRI)

Lever-operated knapsack sprayers

Piston-type pump inside the spray tank

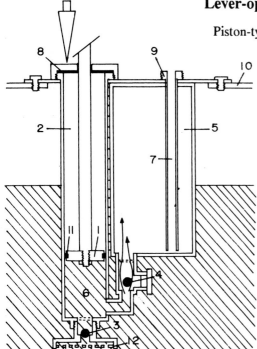

Fig. 10.6: Cross-section of a piston-type pump with separate cylinder and pressure vessel.

1. piston, 2. cylinder chamber, 3. inlet ball valve, 4. outlet ball valve, 5. pressure vessel, 6. spray liquid, 7. dip tube, 8. sealing washer, 9. hose attachment, 10. tank wall, 11. piston O-ring, 12. strainer.

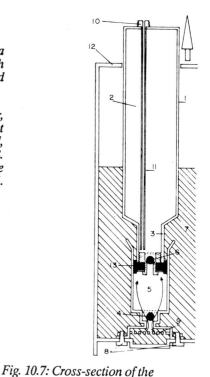

Fig. 10.7: Cross-section of the combined piston/pressure vessel of a lever-operated sprayer as shown in fig. 10.5

1. combined piston/pressure vessel, 2. pressure vessel, 3. piston, 4. inlet ball valve, 5. cylinder chamber, 6. outlet ball valve, 7. spray liquid, 8. retaining bolts, 9. strainer, 10. hose attachment, 11. dip tube, 12. tank wall, 13. pump washer.

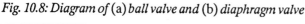

Fig. 10.8: Diagram of (a) ball valve and (b) diaphragm valve

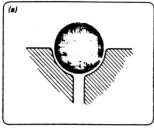 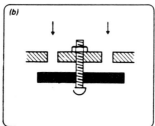

- Piston-type pump inside the spray tank

A piston- or plunger-type pump consists of a wide-diameter piston and cylinder chamber allowing larger volumes of chemical to be drawn in by shorter piston strokes than the narrower piston and cylinder in the shoulder-pump. The piston or plunger carries a handle or lever on one end and a leather cup washer or O-ring on the other. The cup washer is moved up and down in the cylinder with each pumping stroke and must fit the cylinder tightly. On the upward stroke the spray liquid is drawn into the cylinder through an inlet ball-valve at the bottom; on the downward stroke of the plunger, it is forced into the separate pressure vessel through an outlet ball-valve.

Air is trapped in part of the pressure chamber and compressed above the incoming liquid. This compressed air forces liquid from the rpessure chamber through the hose and lance to the nozzle. The size of the pressure chamber varies considerably (160-1,300 ml) but should be at least ten times the pump capacity in order to minimize variations in pressure. The pressure chamber must be strongly constructed to withstand pressures of up to 6 bar.

The ball-valves at either side of the pump must close tightly to prevent flow-back of liquid and air (figs. 10.6 and 10.7). The valves may be either a diaphragm made of synthetic rubber or plastic, or a ball-valve of stainless steel or poly-propylene (figs. 10.8a and 10.8b).

Collected debris in the valve chamber may cause liquid to leak past the valve. Beware not to lose ball-valves when carrying out repairs.

In lever-operated sprayers with a combined piston/pressure vessel, the pressure vessel moves up and down; its lower end acts as the piston. The piston fits tightly with a leather cup-washer in the cylinder chamber which is attached to the bottom of the sprayer tank (figs. 10.5 and 10.7).

On the upward stroke, the liquid is drawn up into the cylinder chamber through an inlet valve at the bottom of the chamber, while the outlet valve at the lower end of the piston remains closed. On the downward stroke, the liquid is forced into the piston/pressure vessel through the outlet valve and leaves through the spray hose attached to the top of the pressure vessel.

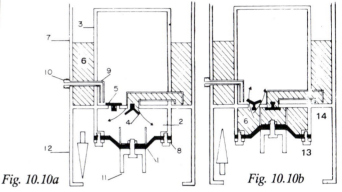

- Diaphragm-type pump

This type of pump consists of a flexible diaphragm, a rigid and sealed diaphragm chamber and ball- or flap-type inlet and outlet valves. The diaphragm is a cone-shaped piece of rubbber or synthetic material, which forms a flexible wall of a closed diaphragm chamber. The diaphragm is moved up and down by strokes of the pump lever. On the downward stroke, the liquid is drawn into the chamber from the spray tank through the inlet valve and is forced out into the pressure chamber through the outlet valve on the upward stroke. The pressure vessel is attached to the bottom of the tank on the inside, and the diaphragm pump is situated directly below, on the outside of the tank (figs. 10.10a and 10.10b).

Fig. 10.10a   *Fig. 10.10b*

Cross-sections of lever-operated sprayer fitted with a diaphragm pump with under-arm action.

1. diaphragm, 2. diaphragm chamber, 3. pressure vessel, 4. inlet flap valve, 5. outlet flap valve, 6. spray liquid, 7. tank wall, 8. diaphragm fixing ring, 9. dip-tube, 10. hose attachment, 11. lever attachment, 12. pump skirt, 13. fixing bolt. 14. tank bottom.

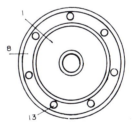

Fig. 10.10c: Overhead view of diaphragm shown in figs. 10.10a and 10.10b.

Fig. 10.9

Cross-section of lever-operated sprayer with under-arm action. The pump assembly with separate piston and pressure vessel is mounted outside the tank.

1. piston, 2. pump cylinder, 3. pressure vessel, 4. pump washer, 5. inlet ball valve, 6. outlet ball valve, 7. spray liquid, 8. strainer, 9. hose attachment, 10. tank wall.

- Piston-type pump outside the spray tank

Generally the piston-type pump gives a higher output than the diaphragm-type, but diaphragm pumps often require less energy and maintenance for operation and withstand wear better.

The pressure vessel, which is available in shoulder-pump and lever-operated sprayers, serves to maintain pressure on the spray liquid between the pump strokes. This is necessary for the delivery of a continuous flow of liquid at the nozzle with sufficient velocity to cause the liquid to break up into fine droplets. Trombone sprayers and stirrup pumps have no pressure vessel and require continuous pumping.

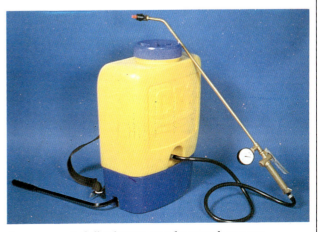

Lever-operated diaphragm-type knapsack sprayer.

* Compression or pneumatic sprayers

This type of sprayer has a cylindrical tank in which air is pressurized with a plunger pump that passes through the top of the sprayer separately or, usually, is screwed into the filler hole as part of the lid (fig. 10.11 and picture right below). The entire tank of the sprayer is air-tight and acts as a pressure vessel (make sure that the threads for the pump are sufficiently deep to give a good seal). The sprayer should be made of corrosion-proof metal or durable plastic and should be leak-proof at welded joints and threaded connections. Less expensive sprayers may have galvanized tanks which are extremely susceptable to corrosion. Corroded tanks are weaker and may burst if pressurized; galvanized tanks may last only a single season, becoming potentially dangerous afterwards.

The sprayer should be filled only with liquid to about 75 per cent of the total tank volume, leaving a space above the liquid to pump in and pressurize the air. The sprayer is pumped before spraying starts; no pumping is done during actual spraying. Ideally, these sprayers should be equipped with a pressure gauge indicating the pressure inside the tank. The maximum working pressure and the full charge in litres of the tank should be permanently marked on the outside. If no pressure gauge is fitted, the operator should be instructed about the number of strokes needed to achieve safe working pressure.

In practice, a single pressurization before spraying is sometimes not sufficient to discharge the whole tank. In such a case, spraying should be interrupted and pressure restored by pumping again to maintain a uniform flow of the liquid to the nozzle. Otherwise, the fitting of a pressure-regulating valve to the tank outlet or lance will make it possible to spray the contents at a pre-set pressure rather than at a falling pressure.

Once all liquid has been sprayed, air pressure may remain in the tank. Unscrewing of the lid should therefore still be done carefully by turning the lid partially until air pressure has been released (hissing sound stops), before opening completely.

Compression sprayers are available in varying sizes: 1-5 litres is the typical volume of hand-carried sprayers; 7.5-15 litres for knapsack or shoulder-slung types. These sprayers are usually fitted with one or two nozzles on the lance but they may have booms added with more nozzles.

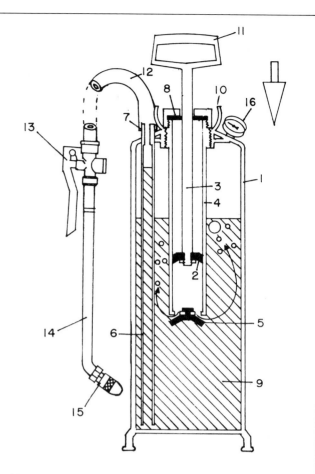

Fig. 10.11: Cross-section of compression sprayer with piston pump screwed into the filler hole.

1. tank, 2. piston washer, 3. pump piston, 4. pump cylinder, 5. flap valve, 6. dip tube, 7. hose attachment, 8. sealing washer, 9. spray solution, 10. filler hole, 11. handle, 12. delivery hose, 13. trigger cut-off valve, 14. lance, 15. nozzle, 16. pressure gauge.

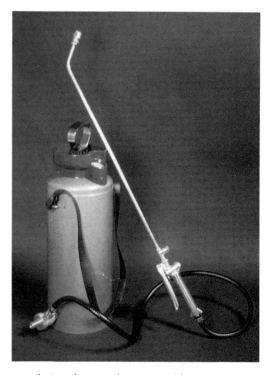

Pneumatic-type knapsack sprayer with pressure gauge and hose attachment positioned at bottom of sprayer.

Air carrier sprayers

* Motorized knapsack mist-blowers

Mist-blowers belong to the group of air-carrier sprayers which, as a common feature, provide an air stream in which droplets are projected towards the target. The air stream is produced by a central fan connected to a light-weight gas-fuel or electric engine or to the power take-off of a tractor. The term mist-blower acutally refers to sprayers which produce droplets in the range of 50-100 μm. Air-carrier sprayers are well suited for spraying trees and have largely replaced the use of hand-held, long spray-lances in tree crops. Therefore, large tractor-mounted air-carrier sprayers have been designed primarily for orchard spraying. Other air-carrier sprayers are exhaust nozzle sprayers on vehicles that were developed for locust control, and aerosol sprayers (see section on fogging equipment) developed for hygiene and storage pest control.

For droplet formation, either the force of the air stream itself may be used or, alternatively, the droplets may be produced by hydraulic- or centrifugal-energy (spinning disc) nozzles positioned in the air stream. In the first case, the liquid is fed through a twin fluid nozzle (see fig. 10.21) into the high velocity air stream. Desintegration into droplets is achieved by impact of one fluid, i.e. the liquid containing the pesticide, in contact with a second "fluid", i.e. the air. When droplets have already been produced by nozzles, an air stream of lower velocity suffices to carry the droplets towards the target.

Air-carrier sprayers are usually employed for low-volume spraying, i.e. the application of the same quantity of pesticide in only one-tenth of the volume normally applied with hydraulic sprayers. They can be adapted for ULV (ultra-low-volume) application by restricting the flow of the liquid to the nozzle.

Motorized knapsack mist-blowers are usually fitted with a 35 cc two-stroke engine, although higher powered blowers with 58-70 cc engines for use on tree crops are also made. The air-cooled engine is connected to a vertically mounted, centrifugal fan by direct drive. Engine and fan are attached by anti-vibration mountings to an L-shaped knapsack frame. The fan produces a high-velocity air-blast which is directed through a flexible hose to the nozzle mounted at its end (see fig. 10.12). Air velocity and volume are regulated by the engine speed which reaches 6,000 rpm or more. The fuel is contained in a small tank.

A spray tank, made of heavy-duty plastic and holding 10 litres or more, is located above the engine/fan unit. It has a large opening for easy filling, with a fine mesh-filter to keep out dirt and lumps that could block the nozzle. The lid must close air-tight with a rubber seal as the tank is slightly pressurized by introducing some of the air-blast into the top of the tank. Pressurization of the tank improves the flow of the spray liquid to the nozzle and enables the operator to raise the nozzle above the liquid level in the tank. Some sprayers are equipped with a small pump driven by the engine shaft to provide extra pressure in the tank for spraying tree crops using extended delivery tubes.

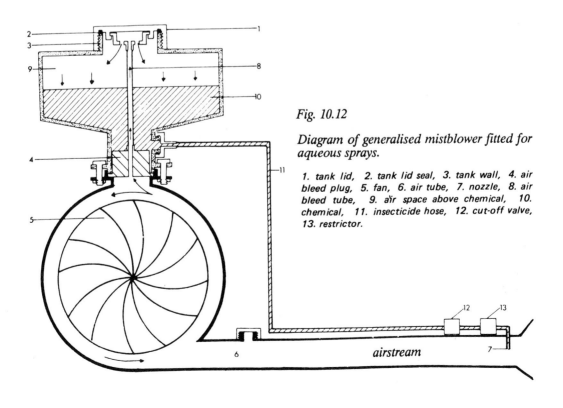

Fig. 10.12

Diagram of generalised mistblower fitted for aqueous sprays.

1. tank lid, 2. tank lid seal, 3. tank wall, 4. air bleed plug, 5. fan, 6. air tube, 7. nozzle, 8. air bleed tube, 9. air space above chemical, 10. chemical, 11. insecticide hose, 12. cut-off valve, 13. restrictor.

airstream

The outside hose is fitted with a cut-off valve and a restrictor regulating the flow rate of the pesticide to the nozzle. There are two types of restrictors for mist-blowers. One is a rotating control with one opening (orifice) or a groove. As the tap is turned, the hole or groove opens up gradually and the quantity of spray passing through the restrictor increases continuously from nothing to maximum. This type of restrictor is suited for high-volume, but not for low- or ultra-low-volume applications, because the restrictor settings and thus the flow rate are very difficult to duplicate. The second type of restrictor has one or more holes bored into a rotating control valve or exchangeable single-hole valves that screw into the restrictor housing on the nozzle. Each restrictor hole of this "fixed orifice type" produces a given and reproduce-able flow rate. The fixed orifice type is suitable for ultra-low-volume application.

Mist-blowers can be converted from liquid to solid application by attaching a tank with sloping floor designed for granules and dusts (figs. 10.13 and 10.14) or modifying the tank for liquid. Remove the spray hose from the tank and insert a wider tube to feed the particles directly down from the hopper through a metering device into the air stream from the blower. Nozzles for dust application may be open tubes of up to two metres in length at the end of the air hose, or long flexible polythene tubes with many holes, up to 30 metres long and carried by two operators. Granules are dispersed through a spreader at the end of the air hose. Because static electricity builds up in the air tube during application of granules and dust in dry air, it

If a mistblower is being used in a closed space such as greenhouse or storage building, the operator should wear a respirator to protect himself against exposure by inhalation. (Photo courtesy of LNV, the Netherlands)

is recommended that the sprayer be earthed by attaching to it a chain that touches the ground.

Motorized mist-blowers need careful and regular maintenance and servicing. The engine runs on mixed fuel according to the manufacturer's recommendation or, if this is not known, on fuel mixed in a ratio of 20 parts of normal grade petrol to 1 part of 30 SAE two-stroke oil.

To start a mist-blower, the on/off switch on the engine is turned to "on", the petrol tap is opened and the carburettor is allowed to fill with fuel. The choke lever is moved to the "closed" position. With the throttle closed, the engine is started by pulling the starter rope sharply but evenly. The rope should be allowed to rewind slowly on the recoil starter and not released to snap back. In new models, the rope starter has been replaced by an electronic ignition which enables easier starting.

When the engine starts, the choke can be moved to the "open" position and the throttle opened up to increase the engine speed. While spraying, the engine should be run at full throttle and not allowed to idle for long periods.

If the sprayer is used for a long period, the operator should wear ear-protectors to guard against hearing loss. The nozzle should be directed downwind so that the droplets are carried away from the operator and do not contaminate him. (Keep the discharge tube at least two metres from the target to allow dispersal of the droplets, as the air-velocity close to the nozzle may reach 80 m/sec.) The operator should walk at an even pace through the crop and close the liquid tap whenever he stops to avoid applying an overdose on part of the crop.

Stop the engine by closing the petrol tap and letting the carburettor run dry. Drain the fuel tank and carburettor completely when the sprayer will not be used for more than three days. Care should be taken to ensure that evaporation of the petrol from the petrol/oil mixture does not leave oil deposits in the engine, which would make restarting troublesome.

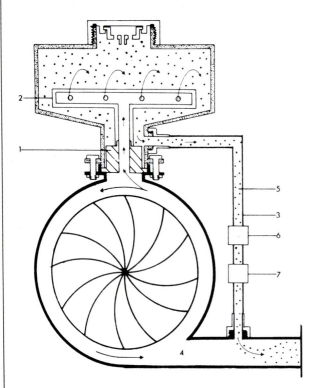

Fig. 10.13

Diagram of generalised mistblower fitted for solid applications.

1. air bleed plug, 2. agitator bar, 3. pesticide discharge tube, 4. air tube, 5. chemical, 6. cut-off valve, 7. restrictor.

The thermal fogging machine shown above pumps its toxic fumes into the interior of a warehouse through a hole in a wooden board fitted into a window frame. The positioning of the fogger outside enables the operator to fumigate a building without risking exposure to the toxic spray mist. (Photo courtesy of LNV, the Netherlands)

Fogging equipment

In thermal fogging machines, an oil-containing insecticide or fungicide formulation is injected into the hot exhaust gas of a combustion engine which vaporizes it. Upon leaving the exhaust pipe, the oil condenses again to form a dense fog consisting of aerosol droplets less than 15 μm in diameter. Also, larger droplets may be formed if the flow-rate from the pesticide tank is too high. The combination of a volatile fumigant fraction and minute aerosol droplets makes fogging very effective for the control of flying insects; eggs and the pupal stages may, however, be less affected. Fogging is carried out mainly in confined spaces such as warehouses, stores and glasshouses. Outdoors, fogging can also be used for the control of vector insects, particularly in the early morning and evening when the air is calm owing to temperature inversion and low wind speed. The fog is released close to the ground and allowed to drift into the target area. Fogging may also show a good efficacy in orchard and tree crops such as cacao and rubber. The foliage must be dry; fogging should not be done when humidity is high. Plants needing water should not be fogged.

Fogging should be done by trained operators. The minute droplets of the fog, which contain the pesticide and a light oil such as kerosine, are dangerous because they are easily inhaled into the lungs. Therefore, a full-face respirator and other protection should be worn. Further, the high concentration of fog droplets in a closed area can be easily ignited by an electrical spark. To prevent explosions, electricity mains should be shut off and naked flames removed before starting the treatment. After fogging, all doors of the treated room or space should remain closed for at least six hours.

* Pulsejet fogging machine

Pulsejet thermal foggers are usually hand- or shoulder-carried machines, but also larger trolley-mounted types are available. Pulsejet foggers consist of a combustion engine with carburettor and spark plug, a fuel and pesticide tank, a hand-operated piston pump and a long exhaust pipe. To start the fogger, the pump is operated to pressurize both tanks and to feed fuel and air to the engine. The engine is started by connecting the spark plug to the battery. Once the engine is running at regular pulses, the spark plug may be disconnected. After warming up of the machine for two minutes, the restrictor valve to the pressurized pesticide tank can be opened to let the pesticide flow into the end of the exhaust pipe. The flow rate may be adjusted by interchanging the restrictor valve. Close the valve to stop the application, but keep the engine running for two minutes to clear the feed line and exhaust pipe of all liquid. Special formulations suitable for fogging are required, which do not decompose during the short exposure to the hot exhaust gas. As fuel, about 1 litre/hour of regular-grade petrol is needed to run smaller machines such as the Swingfog, Pulsefog and Igeba TF 30. Refilling of the pesticide and fuel tanks should be performed in the open without spillage after the fogger has cooled off.

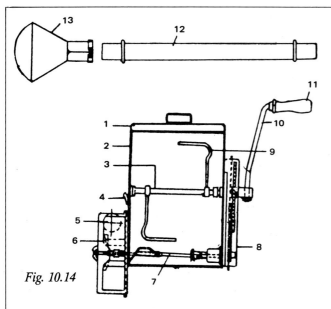

Fig. 10.14

Cross-section of hand-operated rotary duster.

1. lid, 2. hopper, 3. agitating shaft, 4. feed-control lever,
5. dust outlet, 6. blower or fan, 7. blower shaft, 8. gear box,
9. agitator, 10. crank, 11. handle, 12. lance, 13. nozzle.

Hand-operated rotary dusters and granule spreaders

Dusts and granules need not be mixed and diluted with water. Therefore, they offer an advantage in arid areas where water is scarce and expensive to transport. They are also easy to handle and the granules can even be applied by hand. Granular formulations of a good number of pesticides are available, particularly of some very toxic chemicals, because the active ingredient content in the granules is usually below 10 per cent. Granular insecticides are used particularly for the control of soil pests, aphids and stemborers. Also, a number of herbicides are granulated, some of which are widely used in rice. Some special equipment has been designed for precision placement of granules in, for instance, the root zone of plants with the aim of achieving better control with less active ingredient.

A dust and granule applicator consists of a hopper and a blower unit which is a hand- or engine-driven fan. In the hand-operated rotary applicator, the fan is driven by a rotating crank through a gearbox. Agitator bars may be mounted on the elongated part of the crank shaft. They serve to prevent packing of the material in the hopper and to ensure an even delivery of the dust or granules to the metering device. For feeding the dust particles from the hopper into the suction pipe of the blower, some types of dusters have an auger (spiral-shaped threaded revolving cylinder) mounted on the blower shaft, whereas other types are fitted with a feeding brush. The dust flow into the blower can be regulated by a feed control lever (restrictor) or metering device, which alters the opening of the chute. For most applications, this chute between hopper and blower area must be at least half open. Alternatively, the particles may drop through holes, the size and number of which can be regulated. After having passed the metering device, the particles are blown out by the air stream through the discharge pipe as a uniform cloud of dust. Because the metering unit is easily blocked or partially obstructed by caking dust particles, a larger apparatus may be fitted with a displacement rotor for regular delivery. Product characteristics of dust and granules, such as particle size and fluidity, influence the metering system, so each machine must be calibrated for a particular product. Adjustment of the metering system to a lower flow rate requires that the machine be emptied and cleaned, as packing of the granules will block the metering unit.

The hopper is made of plastic or corrosion-proof material and should have smooth sides, preferably sloping down towards the outlet. The hopper has a large opening with a sieve over it to keep out dirt and lumps. The opening is fitted with a lid that protects the content from moisture. The hopper should not be filled to more than three-fourths of its capacity. (Never leave granules or dust in the hopper, otherwise caking and corrosion will occur.)

There are knapsack and chest-mounted dusters and granule applicators, which have the crank handle situated in front of the body. The connection with the blower unit is through a drive chain in a protective metal case. In the knapsack type, the hopper contains about 10 litres of material. In the chest-mounted type, the hopper has to be smaller; the operation is less comfortable and the contamination hazard greater. The discharge tube is usually located on the side opposite the gearbox.

Also, small hand-held dusters are available which have a simple piston or bellow pump instead of a fan. A piston or plunger duster has a bicycle-type pump that blows air into a container. The air agitates the contents and expels a puff of dust through the orifice. The plunger duster is well known in hygiene and household pest control and is suitable for spot treatment in the field. Poor farmers may sometimes fill a cotton bag with dust and shake or beat it above the plants to achieve a similar effect.

Hand-pump bellow-type dusters are commonly used for locust and grasshopper control in the arid countries of northern Africa. (Photo by the author)

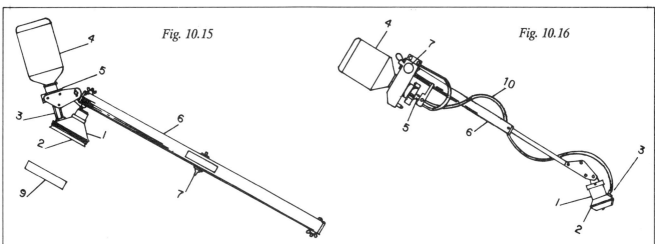

Micron ULVA spinning disc sprayer for ULV-insecticide application (For key to figs. 10.15 and 10.16, see fig. 10.17). (Drawings courtesy of Micronair Ltd., United Kingdom)

Micron 'Herbi' spinning disc sprayer for ULV-herbicide application.

Sprayers with centrifugal-energy nozzles

With sprayers of this type, which are commonly referred to as spinning disc nozzles or rotary atomizers, atomization is achieved by feeding spray liquid to the centre of a fast rotating disc whence it is spread out by centrifugal force towards the edge of the disc. If the rim of the disc is serrated with teeth, which act as zero issuing points, the surface tension forces holding the liquid to the disc edge are minimized. By the same centrifugal force, the liquid is thrown off the pointed teeth as long threads (ligaments) which break into droplets of a fairly narrow range of sizes. Hence, these sprayers are also termed "controlled droplet application" (CDA)-sprayers. The size of the droplets is controlled by the rotational speed (revolutions per minute or rpm) of the disc and the volume of the spray liquid fed to the disc.

* Hand-carried spinning disc sprayers (rotary atomizers)

Some types of hand-held CDA-sprayers are designed especially for application of insecticides and fungicides at ultra-low-volume (ULV) rates of 1-5 litres/ha, with droplet diameters in the size range of 30-130 μm (fig. 10.15). The droplets which are thrown off the periphery of the disc lose their initial velocity within a distance of 20 cm. Further transport of these droplets then becomes totally dependend on the natural forces of gravity and air movement (drift). Other sprayers have been developed for ULV herbicide application of 10-30 litres/ha with droplets of 200-300 μm in diameter. Such droplets have a greater momentum and are less prone to drift than smaller ones. (See figs. 10.16 and 10.17.) Heavier models of centrifugal energy nozzles have been designed for use on aircraft, in air-blast sprayers for orchards and locust control and for fitting onto spray-booms of tractors. Rotary cages of the Micronair type, as used on aircraft, work on the same centrifugal energy principle.

The main components of these light-weight

sprayers are a plastic spray-head with an electrical D.C. motor and rotating disc(s), a screw-on bottle as spray reservoir and a plastic tube-handle holding high-power 1.5 volt (D-size) batteries as used in transistor-radios or torches. As the availability of cheap 1.5 V batteries is often a problem in rural areas, some operators now use rechargeable batteries. The original sprayer of this type, the Turbair-X developed in the 1960s, had two flat discs serrated with 360 teeth. The range of ultra-low-volume applicators (Ulva) for insecticide and herbicide applications were developed in the 1970s. Both the Micron Ulva 8 and Micron Ulva 16, holding one and two sets of eight battteries, respectively, in the tube-handle, as well as the herbicide sprayer "Handy", contained two ridged-shaped toothed discs. The rear disc served to collect and atomize any splashes from the front disc; but the double disc construction caused air resistance and heavy power consumption. To reduce power consumption of the motor, a modified single cup-shaped disc was introduced in later Ulva models.

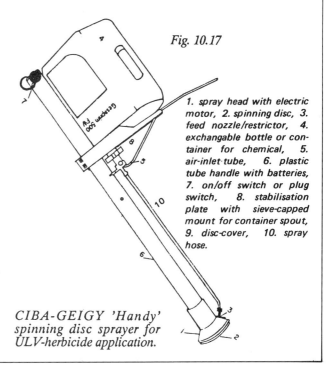

Fig. 10.17

1. spray head with electric motor, 2. spinning disc, 3. feed nozzle/restrictor, 4. exchangable bottle or container for chemical, 5. air-inlet tube, 6. plastic tube handle with batteries, 7. on/off switch or plug switch, 8. stabilisation plate with sieve-capped mount for container spout, 9. disc-cover, 10. spray hose.

CIBA-GEIGY 'Handy' spinning disc sprayer for ULV-herbicide application.

The herbicide sprayer Micron "Herbi" has an 80 mm diameter disc which is flat in the centre, but rimmed with a flange on the inside of which are 360 grooves extending to an equal number of peripheral teeth. The grooves provide a more regular flow towards the teeth and accomodate higher flow rates. A 2-cm deep cup-shaped disc 55 mm in diameter with longer grooves on the inside, has been developed for the production of droplets smaller than 100 μm. This cup-shaped disc is fitted onto the Micron "Mini-Ulva" and "Micron Ulva" (fig.10.15) and produces droplets between 30 and 100 μm in diameter depending on flow rate and power supply. These sprayers use six D-cell batteries for normal use of up to 20 hours of spraying, but are adjustable from two to eight batteries to vary the droplet size. The Micron Ulva and the Mini-Ulva have a metal extension tube which partly slides into the plastic tube-handle and is kept in the required extended position by a locking plate. The tube-handle and extension together are 116-cm long.

Other machines include the "Berthoud C 8" and the "Mini-Technoma T1" which is called "Hi-Spin" in some countries. By incorporating a light-weight electrical fan unit in the spray head behind the disc, the Micron Ulva is turned into an indoor sprayer, the "Ulvafan", for pest control in stores and green-houses were natural air currents are absent. A similar new combination of fan and small spinning cage is currently being developed, but with a thick metal plate perforated with slanted holes placed between the fan and cage for the purpose of converging the cloud of spray drops to achieve better penetration into the crop.

Using fewer batteries or batteries with decreased voltage slows the rotational speed of the disc causing bigger spray drops to form. An increase in the rate of flow at which the spray flows onto the disc has the same effect. The flow rate is controlled by a restrictor nozzle (feed stem) between the bottle cap and the sprayer head. Some models such as the Ulva and C 8 sprayers, have up to five separate restrictor nozzles each with its own colour indicating the diameter of the fixed orifice. Unfortunately however, the colour code system has not been standardized. The Micron Ulva has an orange feed stem giving a flow rate of 30 ml/min or a volume rate of 1 litre/ha. For a continuous flow, it is essential that the metal air-bleed, which is provided in the spray heads, be regularly cleaned of dirt with a thin wire.

The herbicide sprayer "Handy" was developed by adapting the Micron Ulva for lower disc speed and direct fitting onto a standard 5-litre herbicide container. The herbicide sprayer "Herbi" (fig. 10.16) featured a smaller screw-on reservoir and carried an improved cup-disc spray head. The "Herbi" was designed for accurate application of all types of herbicides along a swath of 1.2 metres, applying 10-15 litres/ha. A modified model, the "Herbaflex" herbicide applicator, has a slit opening in the spray head which produces an exact swath width of 10-50 cm. Spray not leaving through the slit, but caught within the spray-head, is sucked back into the reservoir.

The battery-holding handle of Micron rotary sprayers has been redesigned as a common "Microfit" power pack (with four D-cell batteries) into which the spray lances of the Herbi and Herbaflex herbicide and of the Mini-Ulva insecticide applicators can be plugged.

Another development in the range of herbicide applicators is the Ciba-Geigy "Birky" knapsack sprayer (fig. 10.18), which employs the principle of a pneumatically driven spinning disc. The disc is driven by an air

Fig. 10.18: 'Birky' knapsack sprayer

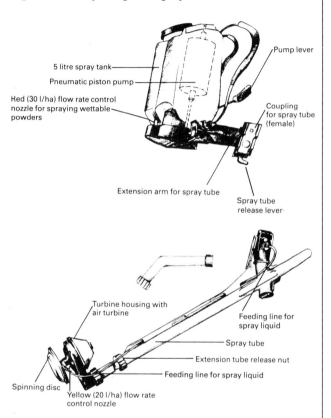

current supplied to a turbine by a pneumatic pump which is located within the spray tank. The turbine, the spinning disc and a flow rate control nozzle are parts of the spray head. The tank of five litres capacity and pump housing are made of plastic moulding and provided with shoulder straps and a back-rest.

The pump is operated by means of an underarm lever; thus, no batteries are required. A plastic spray tube is connected to the frame of the spray tank by means of a special coupling device. The coupling connects the air line and liquid feed line, the latter being controlled by an on/off tap. The spray tube locks in a rearward direction, thus preventing the operator from walking over a treated surface (fig. 10.18). The turbine and spinning disc combination make the disc of the "Birky" spin at the correct speed of 1,200-1,600 rpm, which produces droplets of 250-300 μm VMD. The "Birky" has two flow rate control nozzles: a yellow one for the application of 20 litres/ha and a red one for 30 litres/ha, if a constant swath width of 1.60 metres is maintained. With a walking speed of 1 m/sec, the work output may reach 3-4 ha/day.

* Advantages and drawbacks of CDA sprayers

Normally, hand-held spinning disc sprayers do not require water for spray dilution, which is a major advantage in situations where water is scarce and transport expensive. Also, these sprayers are relatively low in price and are quite easy to operate. However, there are several disadvantages which may impair their efficiency in the field. Major drawbacks concern the following:

* Construction. Owing to pesticide corrosiveness, the components of the spray head are made essentially of plastic instead of metal. This implies that the toothed disc and disc suspension may be easily damaged; moreover, the electrical motor is rather weak. If teeth are broken off, then much bigger drops will splash off the disc.

* Rotational disc speed determining the droplet sizes is dependent on power supply, flow rate and liquid viscosity. Generally, available batteries are expensive, for which reason operators may use fewer or weak batteries. Often, the flow of the spray onto the disc is pulsating and irregular because the reservoir is not pressurized and the air-bleed may be blocked by dirt. CDA sprayers require special ULV formulations which may not be available or the operators may chose to apply other products or blends which might have a much different viscosity.

* Deposit patterns on foliage or other targets depend, among others, on the shape of the spray cloud, distribution by air currents and on evenness in height of release. It is impossible to keep a sprayer at a constant height while walking and thus to achieve an even deposit of herbicide spray. The fine droplet spectrum and total dependence on air movement for distribution easily cause contamination of the operator; the hazard of exposure is aggravated by the commonly high concentration of the pesticide being applied.

Electrostatic spraying equipment

Spray droplets can be charged electrostatically by applying voltage to the nozzle or to the liquid flow system. The effect of the electrostatic charging of spray drops is two-fold. Firstly, all spray drops receive the same charge and will repel each other, thus improving the evenness of distribution. Secondly, the cloud of charged droplets induces an opposite electrical charge on any nearby earthed object to which it is strongly attracted. In agriculture, electrostatic spraying with aircraft and ground equipment has been carried out for many years. Total deposition of charged spray drops on foliage is claimed to be increased substantially with the drops landing on all sides of the target just by spraying from one side. The droplets are also deposited on the underside of leaves and on fruits and stems.

On aircraft, tractor-drawn or engine-operated knapsack sprayers, a small D.C. generator or a 12-volt battery may be mounted to provide the energy for electrostatically charging the spray. Hand-carried sprayers or air-compression sprayers may be equipped with dry cell batteries for the same purpose; charged rotary disc sprayers have also been developed.

* Electrodyn sprayer

A new sprayer is the ICI- Electrodyn nozzle which produces a very narrow droplet spectrum (fig. 10.19). The spray liquid is fed through a narrow gap in the spray-head at a potential of about 25 kilovolts generated by four batteries. An earthed counter-electrode is positioned close to the gap so that the emerging liquid is subjected to a very strong divergent electrical field. The liquid then forms waves, from each crest of which a ligament is produced that breaks up into uniformly sized main droplets and minute satellite ones. The Electrodyn nozzle has no moving parts and uses very little energy.

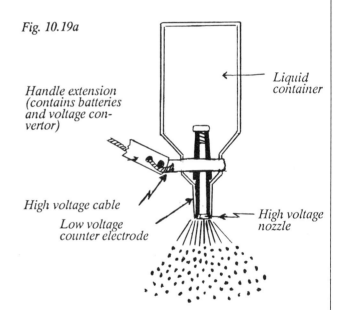

Fig. 10.19a

Handle extension (contains batteries and voltage convertor)

High voltage cable
Low voltage counter electrode

Liquid container

High voltage nozzle

Fig. 10.19b

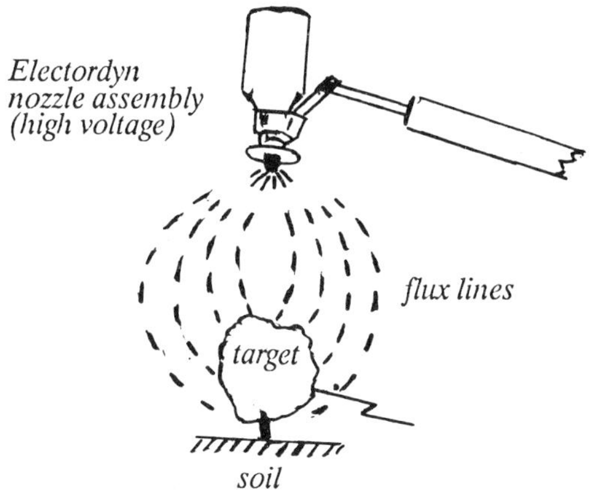

Electrodyn nozzle assembly (high voltage)

flux lines

target

soil

Increasing the voltage causes the production of thinner ligaments and thus of smaller droplets at a given flow rate. Increasing the flow rate has the opposite effect of producing coarser droplets. For the application of insecticides and fungicides, 50-75 μm droplets demonstrate the best efficacy, whereas 100 μm or larger droplets are preferred for herbicide spraying. The sprayer requires specially formulated oil-based formulations of very low volatility at a rate of only 1-0.5 litre/ha. The first special Electrodyn formulation on the market was the pyrethroid insecticide "Cymbush" for use on cotton crops.

A special feature of this sprayer is that pesticide formulations come in a special 750-ml bottle fitted with the particular nozzle pre-set for the chemical to be used. This combination of bottle and nozzle, called a "bozzle", is a closed system that suffices to cover up to 1.5 ha, thus eliminating the need for measuring and mixing the chemical or for calibrating the sprayer.

The bozzle fits into a ring at the end of the handle. When the bottle is emptied, it is recommended that it be disposed off safely. The plastic handle, called a spray stick, contains four U-2 batteries and a high voltage generator which takes the 6 V low voltage input from the batteries and converts this into an output of 25 kV at the HV stud and a zero potential at the earth field electrode. The power requirement of the Electrodyn system is extremely low (less than 0.1 watt); thus the four batteries can give 60 hours of use. This would cover a whole season's need, provided that the batteries are removed from the handle and kept in a dry place between applications.

Operation of the 2.5 kg sprayer is simple. The user merely unties the trailing earth wire, inserts the "bozzle" pack into the holder where the conductive nozzle automatically contacts the HV stud. When ready to spray, the cap is removed, the nozzle is positioned on the crop target, and the switch button is pressed.

Because of the strong electrical field between the nozzle and crop, the charged droplets repelled from the nozzle move at considerable speed towards the foliage and thereby overcome downwind displacement or drift even in a strong wind. Keeping the nozzle at less than 40 cm distance from the crop increases deposition, but holding it too close to the crop interferes with the desired evenness of distribution of the spray on the foliage.

A drawback of electrostatic spraying is that, although the wrap- around pattern of the spray may apply for individual plants, in a closed standing crop the spray cloud is attracted by the heighest and most prominent plant parts and thus the overall deposition pattern may be quite uneven. Also, the penetration into a closed-over canopy is generally poor, unless there is adequate air turbulence. Another disadvantage is that the bozzle is available only with ICI-formulations whereas there is a need for this type of sprayer for the application of other products too. Biologically, however, it is of great advantage that very few charged droplets reach the soil under the crop because, in that way, natural enemies of pests such as spiders, ants and predatory beetles are less affected.

Fig. 10.20: Rope-wick herbicide applicator

Rope-wick herbicide applicators

Another development at the end of 1970s was the rope-wick herbicide applicator. The device consists of a plastic tube, which serves as handle and reservoir for liquid herbicides, and an absorbent wick. The wick runs for its greater part along the outside of the tube, but is fed into and out of the reservoir at regular distances (fig. 10.20). The herbicide is drawn from the reservoir into the nylon wick by capillary action.

The device was developed specifically to apply low volumes of highly concentrated herbicides, such as glyphosate (Round-up) and MSMA, to weeds that grow taller than crops. The herbicide solution is rubbed on any weeds that come into contact with the rope wick.

This inexpensive device without moving parts may be constructed as a small hand-held or larger machine-carried unit. The principle has been used for the construction of a wide range of wipers which use wicks of natural and synthetic fibres, sponge and carpet fabric as absorbent material.

The choice of material for a wick is critical since it should be sufficiently absorbtive, but not dripping nor attracting contaminants such as dust and vegetative matter when rubbing constantly against weed foliage. The small rope-wick "Chemihoe" and the carpet fabric "Sideswipe" and "Roundswipe" applicators are suited for inter-row and spot treatment of weeds in maize, soybeans and other crops. One should realize that an old wick saturated with herbicide residues is dangerous and must be disposed off with great care.

SPRAY DELIVERY ASSEMBLY AND TYPES OF NOZZLE

The delivery parts of hand-operated sprayers with haudraulic pumps, as previously described, generally consist of the following parts.

* Dip tube (or discharge tube)

This tube extends inside the tank to within about 1 cm of the bottom and is connected to the hose at the top. In compressed-air sprayers and lever-operated sprayers with a pressure vessel, the liquid under pressure is forced out of the tank through this tube. In trombone and stirrup-pump sprayers, there are no dip tubes; a suction tube is connected to the pump instead.

* Strainers (or filters)

At two places in the discharge line, between the entry to the dip tube and the cut-off valve, strainers with openings not larger than 0.5 mm in diameter are provided. They prevent passage of large chemical particles or foreign matter which could block the nozzle opening.

* Spray hose

The hose is made of chemical- and oil-resistant plastic or rubber and should preferably be reinforced with one or more plies of synthetic fibre or cotton yarn. Its inside diameter is normally 9.5 mm. One end is connected to the hose attachment on the spray tank and the other to the cut-off valve. These connections should fit tightly or be secured with a strong clip. Check the hose frequently for weak parts, so that it will not burst during spraying operations. Make sure that a sufficient length is supplied with the sprayer for the work you have in mind.

* Cut-off valve

The valve may be a wheel- or tap-type valve, but most commonly is a trigger-type cut-off valve. A trigger valve (fig. 10.11) is preferred because the flow of spray solution is shut off immediately when the handle is released. The handle should fit comfortably in the hand of the operator. Poorly constructed or maintained cut-off valves cause leakage from the nozzle and pose a risk of contamination.

* Spray lance and tail-boom

The lance or extension spray tube, made of brass or other metal, has standard-size threaded connections at both ends. Thus, any number of lance extensions can be screwed together in series and the lance can be connected directly to the cut-off valve housing at one end and to the nozzle at the other. In spraying, an operator usually walks with the spray lance in front of him and moves past plants which have just been sprayed. The wetted foliage touches his legs and other parts of the body, thus contaminating him seriously, when no protective clothing is worn. To prevent this type of risk, spray booms with several nozzles have been developed that are mounted on the back of a knapsack sprayer; they are called "tail-booms". A vertical tail-boom type is used for spraying tall row-crops such as cotton and a horizontal tail-boom for spraying low crops such as rice.

* Nozzle types

There are four main types of nozzle, named after the source of energy used in the spraying system.

1. Hydraulic energy type

The spray liquid is pressurized (hydraulic energy) by pumping air into hand-operated sprayers or by feeding it through a power-operated pump. The pressure forces the spray liquid through the small hole or orifice of one or several nozzles, which causes the liquid to break up into droplets. The hole or orifice is the (hydraulic energy) nozzle.

2. Gaseous energy type

This type of nozzle is used in air-carrier (mist-blower and fogger) sprayers. It is also called "twin-fluid" and "shear" nozzle, because the desintegration of the liquid into droplets is achieved by the impact of a high-speed air stream (first fluid) on the spray liquid (second fluid) either inside or outside the nozzle.

3. Centrifugal energy type

The spray liquid is fed to the centre of a rotating surface, usually a toothed disc, whence it is spread out by centrifugal force towards the edge. The liquid is thrown off the edge in the form of long threads (ligaments), which break down into droplets. Sprayers with spinning discs or cups operate on this principle.

Rotary atomiser

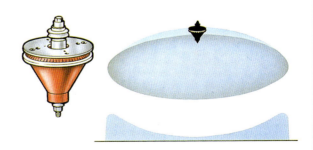

4. Electrostatic energy type

Spray droplets emitted from hydraulic nozzles and rotary nozzles can be charged, but the Electrodyn nozzle itself subjects the spray liquid to an intense divergent electrical field. Under its influence, the charged liquid forms ligaments which break up into mutually repellent (owing to their identical electrostatic charge) droplets.

Hydraulic-energy type nozzles

The success of an application depends to a large extent on the accuracy with which a pesticide is deposited on the target. Since the target may present itself in many different forms and since the conditions under which an application is to be made vary widely, a whole range of hydraulic nozzles has been designed according to the special requirements of the task at hand. The requirements concerning optimum size of droplets, the pattern of spray distribution and the volume of spray are generally different for the application of herbicides, fungicides, insecticides and defoliants. These spray parameters depend on the following:

* Design of the nozzle: Fan-jet, hollow-cone, solid-cone and impact nozzles. The general influence of the nozzle design on the droplet spectrum at the same pressure and same flow rate is as follows:

    - Hollow-cone nozzle: very fine droplets

    - Fan-jet nozzle: fine droplets

    - Solid-cone nozzle: bigger droplets

* The flow rate at the orifice: This depends on the pressure, the diameter of the nozzle opening and restrictor opening, and of the speed of rotation of the spray liquid in the nozzle body.

* Discharge angle of the spray: Applying the same pressure at the same flow rate, a nozzle with a larger spray angle emits finer droplets, thus a 11004 fan-jet nozzle creates finer drops than a 6504 fan-jet nozzle; in this case, a 110-degree versus 65-degree angle.

* Pressure in the sprayer system: Higher pressure results in finer droplets and an increased flow rate.

There are five basic designs of hydraulic energy nozzles; each is described separately:

1. Solid-cone nozzle

This nozzle sprays a circular (conical) pattern of droplets, which are evenly distributed over the whole circle with the centre being filled too. This type is suited for spraying field crops for insect and disease control.

2. Hollow-cone type nozzle

This nozzle produces a circular pattern with almost no spray droplets in the centre. It is suited for insect, disease and weed control spraying.

The solid- and hollow-cone spray pattern depends on two components; see illustrations on this page:

    - Nozzle tip or orifice disc.
    - Core or swirl plate.

Cone nozzle

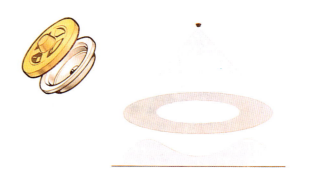

These two separate components may also be integrated in one single unit, a swirl body with slanted groove on its outside; such a unit is known as a conejet spray tip.

Pressurized liquid from the sprayer reaches the core in the cone type. Bored through the core or around the thread of the swirl body, if this is acting as the core, are slanted (or sloping holes). The core does not move but the liquid is forced through the holes into a small chamber known as the swirl chamber (or whirl chamber) which is behind the nozzle tip. The angle of the holes in the core causes the spray to move around the swirl chamber with a circular motion at high speed; as the spray comes out of the nozzle tip, a cone-shaped pattern of spray is formed.

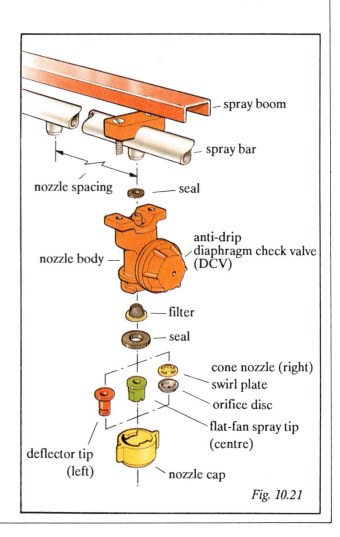

spray boom

spray bar

nozzle spacing — seal

nozzle body —

anti-drip diaphragm check valve (DCV)

filter

seal

cone nozzle (right)
swirl plate
orifice disc
flat-fan spray tip (centre)

deflector tip (left)

nozzle cap

Fig. 10.21

There is no spray in the centre of this cone, thus it is known as the hollow-cone type nozzle.

If the core also has a central hole, the centre of the spray cone will be filled with droplets as well, and is called a solid-cone nozzle.The core and orifice disc of the hollow-cone nozzle can be replaced either by a swirl body and orifice disc or by a cone-jet tip. The latter is designed on the same principle as the swirl body with slanting threads. These components are constructed in standard sizes and can be used with the standard nozzle inlet bodies, strainers and caps.

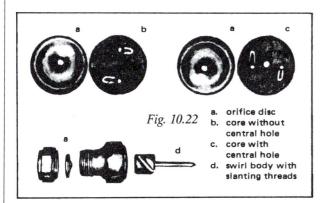

Fig. 10.22
a. orifice disc
b. core without central hole
c. core with central hole
d. swirl body with slanting threads

The amount of spray liquid passing through a nozzle (the flow- or discharge rate) can be increased by using a larger disc tip hole or by fitting a core with more slanted holes in it. The angle of the spray cone depends on the type of core and disc type used. Increasing the pressure increases the flow rate and the angle of the spray cone.

Cone nozzles, either hollow or solid, are best suited for spraying crops. They give maximum uniform coverage in spraying row crops. Therefore, most of the afore-mentioned common sprayer types are supplied with cone nozzles, which are interchangeable if standard threads have been used.

3. Regular flat-fan nozzle

This nozzle with a flat spray-tip sprays a narrow oval pattern with lighter (tapered) edges. The narrow droplet pattern makes this nozzle ideal for spraying flat

Flat fan nozzle

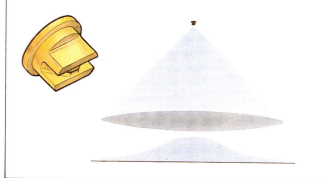

surfaces. It is suited for the application of herbicides and for spraying walls for vector control. Frequently they are used in spraying from tractors where nozzles in series are fitted on a spray-boom with good overlapping of the spray pattern for uniform coverage.

4. Even flat-fan nozzle

This nozzle with an even spray-tip provides uniform distribution across the entire width of the spray band. The even pattern makes it suitable for band spraying in pre- and post-emergence herbicide applications and for the spraying of walls.

The flat-fan nozzle has one single spray-controlling component, the nozzle tip, which is locked onto the nozzle body by a screw cap (see fig. 10.21, at centre below). The shape of the hole or orifice in the nozzle tip determines the shape of the droplet pattern, the angle of the spray and the flow rate. The hole is lens shaped or rectangular.

For herbicide spraying, the pressure should be low and the orifice size large. This produces a spray with large droplets and lowers the risk of damaging adjacent crops with small droplets of herbicides that drift over them. Low pressure is achieved using a pressure control valve. For insecticide spraying, a higher pressure and smaller orifice are needed.

Evenspray fan nozzle

5. Deflector nozzle

The deflector nozzle, also called impact-, anvil- or flooding nozzle, produces a wide-angle flat-fan spray pattern (see next page). More droplets land at the outer edges of the fan and the pattern is rather uneven. Impact nozzles are operated at low pressure for pre- and post-emergence herbicide and liquid fertilizer applications. At high pressures, very small droplets can be produced.

The pressurized spray liquid passes through the relatively large nozzle orifice, impacts on and is deflected by the angled face. The spray leaves the nozzle in a fan-type form. The flow rate is controlled by the orifice size and pressure; the spray angle, by the angle of the sloped surface and pressure. These nozzles are commonly used on knapsack sprayers.

Deflector nozzle

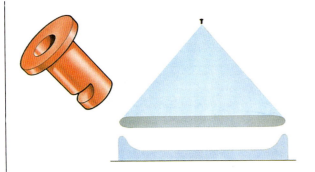

Those needing detailed advice on the selection of nozzles for larger equipment such as tractor-mounted sprayers as well about calibration procedures, nozzle coding etc., should refer to: *Nozzle Selection Handbook,* revised edition 1988, published by the British Crop Protection Council. Write to: BCPC Publications Sales, Bear Farm, Binfield, Bracknell, Birkshire RG12 5QE, United Kingdom.

Materials for nozzle manufacture

Nozzles are subject to much wear and corrosion during their lifetime. For instance, particles in a wettable powder suspension cause considerable abrasion at the edges of the nozzle orifice when the liquid is forced through it under pressure.

The materials most often used for the manufacture of nozzle bodies and caps are:

- Stainless steel. Recommended for use with all spray materials.

- Nylon. Resists corrosion and abrasion; swells when exposed to some solvents.

- Aluminium. Subject to corrosion and has a short life.

- Brass. Not resistant to abrasive materials such as wettable powders.

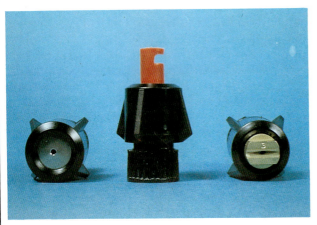

Various types of nylon nozzles; in a type-series, colour coding may be used to indicate the flow-rate.

Materials commonly used for the nozzle tip are:

- Hardened stainless steel. Most wear-resistant of any of the readily available metals.

- Stainless steel. Excellent wear resistance with either corrosive or abrasive materials.

- Nylon. Resists corrosion and abrasion. Swells when exposed to some solvents, but is cheap.

- Brass. Wears quickly when used to apply abrasive materials such as wettable powders, but is cheap.

- Ceramic or porcelain and tungsten carbide. Highly resistant to abrasion and corrosion, but is fragile and expensive.

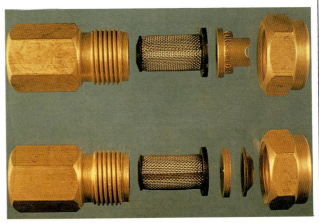

Various components of brass nozzles.

Catalogues of nozzle manufacturers provide detailed instructions and accurate drawings to aid the professional applicator in the selection of correct nozzle components.

MAINTENANCE OF APPLICATION EQUIPMENT

A knapsack-sprayer is an essential piece of equipment for rice farmers. The sprayer may be costly, but with proper care and maintenance, it will give long years of service and it enables farmers to spray pesticides at any time and with good results.

Procedures for maintaining a sprayer are usually given in the operator's manual; they should be understood and followed. When purchasing a new sprayer, obtain from the supplier all available information on its construction, maintenance and operation. The operator's manual for all sprayers should be kept in a binder and stored in one place so that any time repairs or replacement parts are necessary, such important reference material can be found and referred to easily.

All sprayers that are used often should be cleaned regularly, at least once a week. Equipment not regularly used or used with various chemicals should be cleaned thoroughly after use.

* Cleaning of spray equipment

When the applicator starts cleaning his sprayer or removing the empty bottle or package of chemical spray, he must still wear his protective clothing until all the work is finished. Only after the sprayer has been stored away and all waste has been removed should the farmer take off his protective clothing to wash himself and change clothes.

When the spraying operation is finished, the following procedures should be observed:

Steps	Key points
1. Empty the spray tank of all remaining pesticides	a) Spray the content of the tank until it has been emptied on the target crop or on waste land selected for this purpose.
	b) Do not contaminate irrigation canals, streams and other crops.
2. Rinse the tank	a) Use clean water and detergent. Fill the tank one-third full and shake vigorously.
3. Spray with the water-soap mixture	This will clean the dip tube, hose, lance and nozzle. The nozzle may be removed after two minutes to speed up the washing process.
4. Repeat steps 2 and 3 without detergent	Use only clean water this time.
5. Drain the sprayer	Remove as much water as possible from all parts.

* Storing of sprayers

After each day's field work and at the end of the season, complete checks should be made of the sprayer and, where necessary, the engine, before storing the sprayer in a dry place. All sprayers should be kept locked away from children, food and farm animals, and measures taken to prevent rats from chewing hoses and other parts. Many small hydraulic sprayers are preferably stored upside-down with the lid removed to allow complete drainage of the container. If engines are to be stored without use for a prolonged period, the spark-plug should be removed and a small quantity of oil, preferably formulated with anti-rust additives, poured into the cylinder. The engine should be turned over a couple of times to spread the oil. Similarly, it is advisable to rub the metal tank with a mixture of equal parts of kerosene and oil for a prolonged period of storage.

* Routine maintenance procedures

When any part of the sprayer does not function properly, correct the defect immediately. You can take the following steps without the help of a repairman.

Strainers are located at different places in the spray lance and hose. Refer to the operator's manual. Remove the strainers and flush them with water from both sides. When this does not remove all dirt from the screen, it may be necessary to soak the screen in kerosene which will dissolve the pesticide that had been used. If the screen has been damaged or is impossible to clean thoroughly, replace it with a new one.

Cut-off valve should open easily and rapidly and should shut off quickly and tightly. Check the gaskets in the valve and replace the gaskets with new ones if spray leaks from the valve.

The hose should not have any break in the outer covering, particularly at points where the hose is fastened to any couplings. Replace the hose if any break is found. Never carry or store a hose with sharp kinks in it, as this may cause breaks. Screw-driven clips are the best for attaching the hose as they prevent leaks.

Nozzles require special care. Much of the effectiveness of any pesticide application depends on the proper functioning of the nozzle. Nozzle tips are designed to give a specific pattern of spray with specific kind of droplets at a set pressure. Clean the nozzle in the following manner:

Steps	Key points
1. Dismantle the nozzle	a) Unscrew the nozzle cap from the nozzle body.
	b) Remove nozzle tip and strainer
2. Wash nozzle tip	a) Clean the hole in the nozzle tip by back-flushing with clean water.
	b) If the hole cannot be cleaned that way, use a small brush, or copper wire that is softer than the material in the tip itself, so that the tip will not be damaged or enlarged while being cleaned.
3. Clean strainer	As described above.
4. Assemble the nozzle	

Things to remember when cleaning nozzles:

- Never use a nail or sharp object to clean the orifice and screen

- Never put a nozzle or screen to your mouth to blow it clean

- Do not overtighten a nozzle cap on the body

- Replace a nozzle or strainer if it cannot be cleaned easily or if the orifice is damaged by abrasion or otherwise

- Carry a spare nozzle to prevent much loss of time in the field

- Rinse the sprayer and nozzle several times with water after use to keep them free from pesticide deposits and dirt.

Pump assembly. Lubricate plunger or piston leather cups in the cylinder frequently with vegetable oil to soften them and to prevent loss of efficiency in pumping. Lubricate the piston rod also so that it moves freely and easily. Replace leather cup if it becomes hard and worn-out.

* Seals or washers

At the places where sections of the spray lance and cut-off valve join or where the plunger cylinder, dip tube and discharge tube fit into the tank wall, these connections are generally made water-tight by fitting washers (O-rings of leather, rubber, neoprene or other synthetic material) at the joints. Leaking at joints is often caused by drying out or corrosion of the washers. Drying out of leather washers can be prevented by greasing with vegetable oil. Temporary repair can sometimes be made by wrapping fine, greased twine or plumber's tape around the threads before screwing or reassembling the connections. Similarly, seals around the tank lid of pressurized knapsack sprayers must be checked, because their working depends on an air-tight seal of the tank or container. Small air leaks can be easily detected by smearing a soap solution over the washer. Soap bubbles will readily show where air is escaping.

Any worn or damaged washer should be replaced as soon as possible. Shopkeepers should stock a good supply of such inexpensive spare parts.

Problems with motorized sprayers

* Two-stroke engines

Users of motorized knapsack mist-blowers and other equipment with a two-stroke engine frequently complain that the engine is difficult to start. Various causes for the failure to start and other problems are listed together with remedies hereunder. Many of the starting problems could be prevented if the carburettor and engine were drained of fuel after use to avoid gumming up the machine with oil when the petrol has evaporated. This can be done simply by turning off the fuel tap and allowing the engine to continue running until starved of fuel. Preferably, the fuel tank itself should also be drained to avoid increasing the ratio of oil to petrol, especially in hot climates. Starting problems are definitely reduced by ensuring that the correct type of oil is used and that the fuel is properly mixed.

The fuel line from the tank to the carburettor is often made of plastic, which is hardened by the action of the petrol and is sometimes loosened by engine vibration. This plastic tube should be regularly inspected and replaced if necessary to prevent fuel from leaking onto a hot engine and causing a fire.

The spark-plug should be inspected regularly, if necessary, so it should be readily accessible. The spark-plug gap may need adjusting to obtain a good spark before the plug is replaced. The plug should be replaced after 250 hours of use as a routine maintenance procedure. The air filter should also be examined at the end of each day's spraying or cleaned routinely according to the manufacturer's recommendations.

The fuel for two-stroke engines must be a mixture of petrol and oil, usually in the mixing ratio of 24:1. Often the correct mixture is marked on the fuel cap on the fuel tank or in the manual. Use 30 SAE oil or another suitable type, but never multigrade oil because the additives in multigrade oil will cause engine failure. The petrol should be the regular type, unless otherwise stated in the manual.

References:

Figures 10.1 to 10.4 and 10.6 to 10.10, courtesy of J.A. Sutherland.

Matthews, G.A. (1979). *Pesticide application methods*, Longman Group Ltd., London.

Matthews, G.A. (1984). *Pest management*, Longman Group Ltd., London.

Nozzle selection handbook (revised 1988), British Crop Protection Council, United Kingdom.

Sutherland, J.A. (1979). *Non-motorized hydraulic energy sprayers*, Overseas Development Natural Resources Institute, London.

Sutherland, J.A. (1980). *Mist-blowers*, Overseas Development Natural Resources Institute, London.

11

SPRAYING TECHNIQUES

Contents

	Page
Atomization and droplet spectra	187
Behaviour of spray droplets	188
Mechanism of droplet deposition	191
Operating sprayer equipment	191
Preparing equipment for application	191
Calibrating hand-operated knapsack sprayers	193
Calibrating hand-held rotary disc sprayers	195
Calculations for pesticide applications	195
Timing of the application of pesticides	197
References	199

Spraying Techniques

Present development of high-performance spray machines is directed at an improvement of spray deposition and reduction of dosage and volume of spray emitted. This involves improved design and placement of nozzles, stabilization and height-adjustment of spray booms, automatic pressure and volume regulation and cut-off systems, special equipment such as injectors, spreaders, wipers and combinations of tillage/band-application tools for row treatment. (Photo courtesy by John Deere)

ATOMIZATION AND DROPLET SPECTRA

The main function of a spraying machine is atomization of the spray liquid, i.e. the breaking up of the liquid into droplets by means of a nozzle. These droplets are not of equal size; rather the spray contains a whole range of droplet sizes. The size distribution, or droplet spectrum, is typical for the kind of nozzle from which the spray is emitted and is characterized by the following two parameters:

* Volume median diameter (VMD). When a representative sample of spray is divided into two equal halves by volume, then half the volume is contained in droplets larger than the VMD and the other half in smaller droplets.

* Numerical median diameter (NMD). When the droplets in the sample are divided equally by number without considering their volume, then the droplet diameter in one half of them is larger and in the other half smaller than the NMD.

The ratio of these two parameters (VMD:NMD) indicates the spread of droplet sizes; the larger the ratio, the broader the droplet spectrum. If the ratio equals 1, then all droplets are even sized and behave the same way.

Table 11.1 shows data from a comparative test in which the respective droplet spectra of a Fan Jet nozzle and of a rotating disk nozzle are measured with a laser beam analyzer; droplet diameters being expressed in micro-millimetres, abbreviated as μm (= 0.001 mm). The new laser-based measuring technology has greatly improved detection of fine droplets. These instruments direct a laser beam at a spray cone very close to its origin and measure the defraction of light caused by different sizes of droplets. The data are processed by a computer

Table 11.1: Droplet sizes from a hydraulic spray nozzle and from a rotating disk measured with a Malvern Particle Size Analyser

Drop range in μm	Volume (%)	Cummulative volume (%)	Number (%)	Cummulative number (%)
Nozzle: Fan Jet SS 8002, @ 3 Bar, flow rate 700 ml/min				
563 - 262	26.17	61.94	0.07	0.80
262 - 160	35.77		0.75	
160 - 113	17.52		1.36	
113 - 84	8.78	33.84	1.82	8.39
84 - 65	4.81		2.31	
65 - 50	2.73		2.86	
50 - 30	2.66		8.45	
30 - 15	1.12	4.22	21.90	90.81
< 15	0.27		60.46	

Actual droplet size VMD   192 μm
Actual droplet size NMD   12 μm
Ratio VMD/NMD   16
Number of droplets from 1 litre of spray-liquid measuring less than 50 μm: 89,135,700
Number of droplets from 1 litre of spray-liquid measuring less than 10 μm: 0

Rotary disk Mini-Ulva; @ 5,900 rpm, flow rate 30 ml/min.				
563 - 262	0.0		0.0	
262 - 160	0.0	30.78	0.0	9.79
160 - 113	30.78		9.79	
113 - 84	52.59		44.47	
84 - 65	13.38	68.56	26.29	81.82
65 - 50	2.59		11.06	
50 - 30	0.64		6.91	
30 - 15	0.02	0.66	1.37	8.39
< 15	0.0		0.11	

Actual droplet size VMD   104 μm
Actual droplet size NMD   88 μm
Ratio VMD/NMD   1.88
Number of droplets from 1 litre of spray-liquid measuring less than 50 μm: 1,997,800
Number of droplets from 1 litre of spray-liquid measuring less than 10 μm: 0

Source: E. Bals; Micron Sprayers Limited, 1978.

which provides a complete print-out of volume percentage and number percentage of the droplets so measured. The instrument determines the droplet sizes immediately they are issued from the nozzle, before any factor such as evaporation or volatization gets a chance to diminish their volume. Fig. 11.1 shows a typical histogram, from which the per cent volume of spray contained in droplets of any given size can be determined.

The data from the Fan Jet nozzle in table 11.1 indicate that all droplets measuring from 1 to 50 μm in diameter together contain only 4.2 per cent of the total volume of the spray sample, but number 90.8 per cent of the total droplets. About 61.9 per cent of the sample volume is contained in droplets larger than 160 μm, which make up only 0.8 per cent of the total number.

By comparison, the Mini Ulva disk produces a spray cone in which all droplets up to 50 μm in diameter contain a mere 0.66 per cent of the sample volume and only 8.4 per cent of the total number of droplets. However,

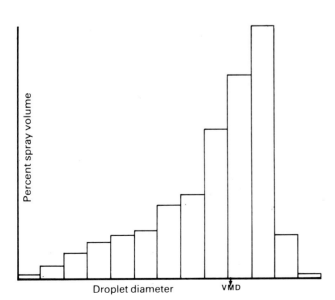

Figure 11.1: Histogram showing the typical distribution of spray in percentage of volume contained in spray drops belonging to any given range of droplet sizes.

over half the volume is contained in droplets of 84-113 μm in diameter. Further, almost 82 per cent of all droplets issued from the Mini Ulva measure between 50 and 113 μm.

The laser particle size analyser reveals for the first time the astonishing mass of very fine droplets from hydraulic nozzles, which could not be detected by older droplet sampling methods using magnesium oxide and kromokote collector sheets or high speed photography. The Fan Jet SS 8002, which had always been regarded as an efficient nozzle, produces a very broad droplet spectrum (ratio 16) and more than 89 million droplets less than 50 μm in diameter. In comparison, the Mini Ulva disk atomizer gives a narrow spectrum (ratio 1.18) and only about 2 million droplets less than 50 μm and none below 20 μm in diameter.

BEHAVIOUR OF SPRAY DROPLETS

Between the moment spray droplets are formed at the nozzle and their actual deposition on any surface, they are subjected to various natural forces and physical processes. Droplet behaviour can be understood somewhat better by analysing each of five phenomena separately as if they occurred in the ideal, theoretical situation of still air, individual drops and obstacles. These phenomena are:

* Movements in the field of gravity

* Movements induced by forces of air

* Movements resulting from the atomization process

* Evaporation of spray liquid

* Deposition on obstacles ·

* A droplet released in still air accelerates downwards by force of gravity until the gravitational attraction is counter-balanced by frictional resistance when the fall will continue at a constant terminal velocity. This terminal velocity is normally reached within a distance of

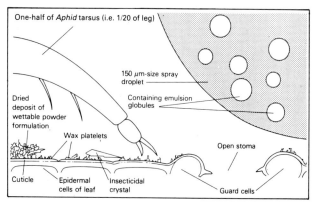

Figure 11.2: The dimensions of a 150 μm spray drop compared with the various parts of an insect and leaf which represent the target of spraying.

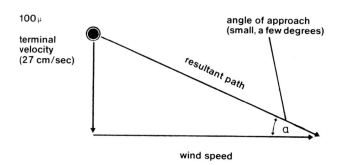

Figure 11.3: Droplet flight path = Resultant of terminal velocity and local wind speed.

less than 25 mm for droplets smaller than 100 μm in diameter and within 70 cm for 500 μm droplets. The terminal velocity is dependent on the size, density and shape of the droplet, and, according to Stokes' law, equals for water drops (density = 1) the following:

Diameter	5 μm	terminal velocity		0.075	cm/sec
"	50 μm	"	"	7.5	cm/sec
"	100 μm	"	"	27.9	cm/sec
"	500 μm	"	"	213.9	cm/sec

Because of their low terminal velocity, droplets of less than 30 μm in diameter will take longer than a minute to reach ground level in still air. However, under field conditions, the slowly sedimenting tiny droplets are all the time exposed to air currents and may be carried by them beyond the target area.

* If only forces such as frictional resistance from an air current affect a falling droplet, then it adopts within a very short time the velocity of that air current, i.e. almost immediately for droplets 100 μm and within a few seconds for bigger droplets of about 1,000 μm. The combination of vertical fall and horizontal air movement results in a side-ways transport of the spray droplets over a distance (S), that can be calculated with the equation:

$S = HU/Vt$ in which H = height of release in metres (m)

$\quad$ U = wind speed in m/sec

$\quad$ Vt = terminal velocity of droplet in m/sec

For droplets released from a height of 1 m at a horizontal wind speed of 1 m/sec, the equation gives the following theoretical results:

Drop diameter (μm):	500	100	50	10	5
Distance (S)　(m) :	0.48	3.5	13.5	340	1,350

Owing to air turbulance and irregular wind velocities, the spray droplets may impact closer or further away, but the above data illustrate how prone small droplets are to being carried away by drift.

Similar calculations can be used to predict the minimum diameter that spray droplets must have in order to impact after release from a certain height within a given target area or swath width if the droplets' own velocity in relation to the surrounding air equals zero. For example, height of release (H) = 1 m, and width of swath = 5 m:

Wind velocity (m/sec):	0.7	1	2	3	5
Minimum diameter (μm):	67	84	128	168	255

* Thus, if all spray droplets are required to be deposited within a point five metres' distant from the point of emission, with wind speeds as indicated, then they need to have a diameter larger than the minimum diameter shown above.

The initial velocity of spray droplets at the point of emission is usually not equal to zero, but depends on nozzle characteristics and type of atomization. For hydraulic nozzles, the initial velocity approaches the flow rate of the spray at the orifice; for rotary disc nozzles, the initial velocity is a function of the centrifugal force. If emitted from such types of nozzle, spray droplets are normally stopped within a very short distance owing to the frictional forces of ambient air currents. For droplets

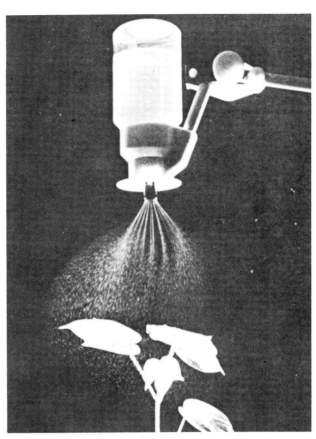

The Electrodyn^{R} Sprayer emits spray drops of remarkably even size which are produced and dispersed under the influence of electrical forces generated by batteries. The spray cloud tends to wrap around individual plants, thereby depositing spray drops on the underside of the leaves also, but on a crop canopy the spray distribution remains uneven. (Photo courtesy of ICI, United Kingdom)

having an initial velocity of 50 m/sec, the following stopping distance can be calculated:

Droplet diameter (μm):	5	10	50	100	200
Stopping distance (cm):	0.33	1.55	38	155	630

Thus, upon leaving the nozzle, droplets impact with the ambient air and thereby rapidly lose their initial velocity, and subsequently are subject to the combined forces of gravity and air currents. In the case of pneumatic nozzles or air-carrier sprayers, the spray drops are carried along by the air-blast over much greater distances depending on the mechanical and aerodynamic characteristics of the sprayer. In the case of electrostatic spraying, the initial velocity of the spray droplets is maintained under the influence of the strong electrical field between the spray-head and crop.

* The problem of drift is aggravated, when the relative humidity is low, by evaporation of the water diluent in the spray liquid, so droplets become smaller; consequently sedimentation takes longer. Table 11.2 shows that, if 1 litre of liquid disintegrates into even-sized droplets of 10 μm diameter, their cumulative surface area would cover 600 m^2. Atomizing 1 liter of liquid into even-sized droplets of 100 μm in diameter gives a total

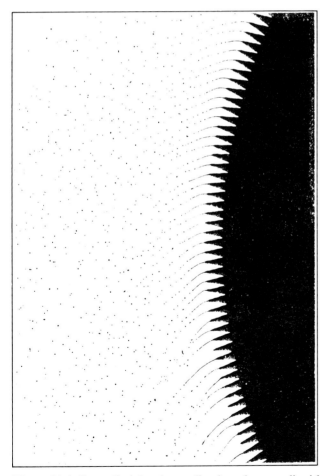

Figure 11.4: Controlled droplet application; the spray liquid is thrown off from the teeth of a fast rotating disc as long threads (ligaments) which break up in a homogenous spectrum of fine particles that together form the spray cloud. (Photo courtesy of Micronair Ltd., United Kingdom)

surface area of 60 m². Smaller droplets thus have a relatively much larger surface on which evaporation takes place all the time. It is a perpetual process of ever-shrinking volume and, consequently, of relative increase of surface area leading to further evaporation. Evaporation takes place very rapidly: e.g., a 50 μm droplet will lose all its water in four seconds. The addition of anti-evaporation agents or thickeners to the liquid to enhance its viscosity would have little effect. While a drop is losing its volatile fraction, its concentration of active ingredient and involatile formulating agents will increase until evaporation ceases. With only 0.8 per cent of involatile material, a 50 μm droplet, from which all volatile fractions have evaporated, will become a 10 μm particle of involatile residue, which will drift away. In general, droplets under 30 microns will remain airborne.

The drifting residue particles consisting of pesticide crystals and involatile agents disappear, their destination unknown. Too often drift is equated solely with deposition of spray droplets outside the target area, such as the chemical drift of 2,4-D causing phytotoxic damage to susceptible crops situated several kilometres downwind of the treated area. However, droplets too small to be deposited on any natural object are lost to the atmosphere and contribute to the general contamination of the environment. This is particularly so if the pesticide has a long persistency of days or months, as is the case with some organochlorine compounds.

On the other side of the droplet spectrum, large drops appear to be an even greater cause of wastage of pesticide. Owing to their high terminal velocity of more than 1 m/sec and large mass, they gain such momentum during their fall that they are not deflected by air currents, but approach the target by a direct route.

At ground level, many large drops may either miss the plant and hit the soil, or even if these drops impact on foliage, they are liable to bounce off, unless the amount of surfactant in the formulation increases retention.

The cube relationship between droplet diameter and droplet volume means that the volume of a 100 μm drop is a million times the volume of a 1 μm droplet (one

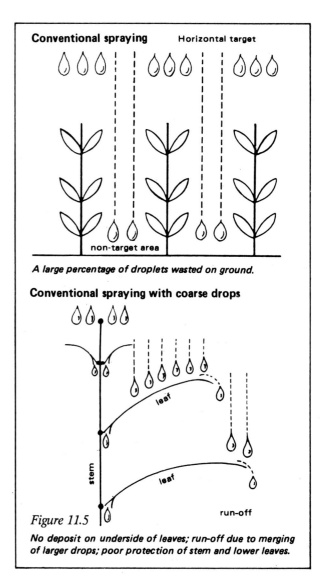

Conventional spraying — Horizontal target

A large percentage of droplets wasted on ground.

Conventional spraying with coarse drops

Figure 11.5

No deposit on underside of leaves; run-off due to merging of larger drops; poor protection of stem and lower leaves.

hundred cubed equals one million). Thus, the variation in pesticide dosage in a 4-μm droplet and a 400-μm droplet (commonly found in a hydraulic nozzle droplet spectrum) is one million. As many of the drops larger than 300 μm are likely to end up on the ground, most of the pesticide dissolved in a watery spray and atomized by a hydraulic nozzle never get the chance to reach the noxious insects or fungi it is supposed to kill. Rather, it is wasted; it may even destroy beneficial organisms living in or on the soil and end up contaminating groundwater.

Table 11.2: Terminal velocity, evaporation rate and surface area as function of droplet size; theoretical parameter for atomization into even-sized droplets of 1 litre of liquid per hectare

Droplet diameter (in μm)	Number of drops/litre ($\times 10^7$)	Terminal velocity (cm/sec)	Dropsurface area per litre (sq. metres)	Evaporation rate (relative)
300	7.1	122	20	1
150	56.6	72	40	3
100	191.0	26	60	8
50	1,527.9	7	120	40
10	190,985.5	0.3	600	-

* Deposition on obstacles. All targets such as leaves pose an obstacle to moving air and, consequently, air tends to pile up in front of the target thus forming a layer of slightly compressed air. This air cushion forces arriving air to move around the target while carrying fine particles along. To leave the deflected air current and penetrate through the layer of compressed air onto the leaf surface, spray droplets need considerable momentum. Only larger drops possess the required combination of mass and speed to achieve impact. Although smaller droplets may still succeed in impacting, most are carried past the obstacle. Owing to eddying currents, some droplets impact on the underside of the leaves, but most move on with the wind.

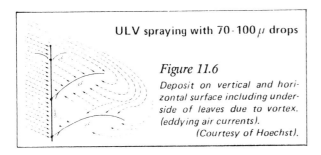

ULV spraying with 70-100 μ drops

Figure 11.6
Deposit on vertical and horizontal surface including underside of leaves due to vortex. (eddying air currents).
(Courtesy of Hoechst).

MECHANISM OF DROPLET DEPOSITION

Spray droplets are deposited upon a target following either straight sedimentation or after initial dispersal by air currents. Since still air is a rare condition in the field, usually the latter mechanism has more significance. If droplets are carried side-ways by wind, they gain speed in a horizontal direction. Their vertical velocity (fall) and horizontal velocity (drift) result in a deflected trajectory as shown in fig. 11.6, along which the spray droplets approach the target.

Evidently, the dimension and shape of the target determines the amount of air resistance and thus the rate of deflection of spray drops. A narrow target such as a hair causes virtually no deflection and is easily wetted by a fine spray mist. The same deliberation applies to moving obstacles such as flying insects. Flying insects reach considerable speed of both flight and wing movements, but because their bodies are small, very small droplets get a chance to impact. Investigation of drop sizes suited for the control of vector insects has shown that flying mosquitoes are best hit by droplets in the size range of 10-20 μm and resting tsetse flies by 30-40-μm droplets. In enclosed grain stores, thermal fogging provides effective control of flying adults of the Indian Mealmoth. Field trials have shown that ULV sprays in the range of 90-130-μm VMD and at an application rate of 1-3 litres/ha provide a biologically sufficient droplet density and crop penetration. Larger droplets over 250 μm are required when sedimentation of drops on horizontal surfaces is required with minimum downwind drift as in the case of pre-emergence herbicide spraying.

Generally accepted optimum droplet size ranges (taking into account constraints in actually achieving deposition) are:

> 10-50 μm against flying insects
>
> 30-80 μm against insects on foliage
>
> 40-150 μm against fungi
>
> 100-300 μm against weeds

OPERATING SPRAYER EQUIPMENT

Spraying is the most common method of pesticide application, although not the easiest one. Quite often farmers do not obtain good results from spraying even when they follow the recommendations on the label or those of the extension agent. Pests are poorly controlled or plants may even be injured by phytotoxic scalding, for example. Sometimes, there may be more than one reason for failure, thus one needs to check systematically all possible causes. Although it may be possible that a pest has acquired resistance against the pesticide selected, it is far more likely that the failure is attributable to miscalculation of the dosage rate, to faulty timing or incorrect method of spraying, or faulty sprayer. To deposit an adequate amount of chemical at a place where and at a time when it can effectively destroy the pest, the farmer should use a good sprayer, obtain training and practise his skill. If a farmer can operate his spray equipment correctly, if he knows the various organisms damaging his crops and understands the timing of control, he will not only do a good job, but he will also save a considerable amount of money in terms of expenses for pesticide, crop losses and work output.

PREPARING EQUIPMENT FOR APPLICATION

Before the actual application can start, the operator has to complete several tasks such as inspecting the pest situation in the field, checking the condition of the sprayer and calibrating it, calculating the dosage rate and getting assistance, if needed. For many years, the spraying of rice, cotton and other important crops has been widely carried out according to a calendar schedule as recommended by plant protection institutions and manufacturers. Such frequent, and usually preventive, applications have furthered the development of secondary insect pests and resistance against one or several pesticides in many kinds of noxious organisms. Because of a better understanding of control mechanisms in nature, a proper restraint in using agro-chemicals and spraying on the basis of need only has recently been

Table 11.3: Collection efficiency (E) for selected droplet sizes on rice and cotton leaves positioned perpendicular to wind direction (wind velocity = 2 m/sec)

Droplet diameter	50 μm	80 μm	150 μm
E for rice leaf	60%	75%	95%
E for cotton leaf	10%	30%	65%

Any decision whether or not to undertake control action against a pest, as well as the choice of an appropriate method, should be based on actual pest occurrence in the field. The farmer should be taught how to relate numbers of pests found to the economic threshold values determined by researchers. On the right, insects are collected from a test plot by means of a D-Vac suction sampler. (Photo left by the author; photo right, courtesy of FAO-IPC Rice Programme)

advocated. This implies regular scouting by plant protection technicians and individual farmers for determining the occurence of damaging insects and diseases in the crops, assessing the need for intervention by comparing the numbers of major pests found in the field to the economic threshold values (see Chapter 1), and wherever possible, treating only the worst infested plots.

Various actions to be taken in preparing sprayers for application are described below; however, for more complete information, the applicator may refer to manufactures' leaflets and seek mechanical help.

Checking application equipment

All types of equipment should be thoroughly checked and, if necessary, repaired before use. With power-driven pumps and sprayers working on the principle of compressed air, weak connections might come loose and damaged hoses could burst, leading to gross contamination. Irregular flow of spray liquid, dust or granules and blocked nozzles affect dosage rate and deposit patterns, while mechanical failure will hold up the work.

Thus, check the sprayer and accessories; connect all delivery parts to the main body by tightening threaded connections and fitting the hose to the tank outlet and lance with strong clips. Check to ensure that all washers and gaskets are sound and replace those in poor condition. Replace old or damaged hoses, straps and strap clips for wear and tear. Check lever-linkages, split pins and other parts of the lever mechanism and grease the lever for smooth operation. Such parts are cheap and easy to renew; do not wait until a break-down occurs. (See Chapter 10 for maintenance advice.)

Motorized mist-blowers and all engine-driven sprayers should be started to check the running of the engine. Refer to technical instruction sheets or consult a mechanical workshop, if there is any malfunction. Spinning disc sprayers and electrostatic sprayers require

a battery check and testing of the electrical motor and the rotational speed of the discs. Generally, select the proper flow restrictor, clean nozzles and air bleed as well as all other parts that influence flow rate and deposit. (See Chapter 10 for details on various types of equipment.)

Pressurizing the sprayer

Hydraulic pump sprayers, as described in chapter 10, work on the principle of pressurized air as the source of energy needed to propel and atomize the spray liquid. To achieve a constant flow and regular spray, the pressure should be kept as constant as possible during spraying. For lever-operated knapsack sprayers, this is achieved by steady pumping with the lever throughout. For compression sprayers, the entire tank of which acts a pressure vessel, a much greater volume of air is pumped to the relatively high pressure of up to 6 Bar (6 kg/cm^2) in one action. As any leakage can be detected by pressurizing the sprayer, a pressure test using water only should always be done following the procedure outlined below:

1 Fill the tank of the compression sprayer 3/4 full with water. Place a strainer over the filler hole or use a funnel with built-in strainer or a clean cloth to remove dirt particles from the water.

2 Tighten the filler cap firmly, because if it is not tight enough, it may loosen further and be blown off by the high pressure, resulting in serious consequences.

3 Pressurize the tank initially to 3.5-4 Bar by pumping the T-handle about 14 strokes per Bar; consult the manufacturer's instructions. Check the pressurized sprayer for leaks on all sides. Minor leaks from threaded joints can be repaired by wrapping twine or plumber's tape around the threads before screwing tight.

4 Make a test spray while holding the nozzle about one metre above the ground. Open the cut-off valve and see if the droplets are fine and uniform; if not, pump more pressure into the tank, but do not exceed the maximum number of strokes prescribed by the

manufacturer (maximum 50-60 strokes) or do not surpass the danger mark on the pressure gauge. With a lever-operated sprayer, the pressure should be kept constant throughout the test by operating the lever at a rate of about 15 strokes per minute. To keep the pressure constant during spraying, the operator needs to practise pumping at an even rhythm. When the spray tank empties, the pumping rate should be increased to 20 strokes per minute to maintain an even discharge rate.

A compression sprayer can be tested while it is standing on the ground; a lever-operated sprayer, while carried mounted on one's back. If the spray pattern remains irregular, one should clean the nozzle properly or replace it if the orifice is damaged or badly worn.

CALIBRATING HAND-OPERATED KNAPSACK SPRAYERS

A sprayer must be calibrated to determine the amount of spray delivered or its discharge rate per minute; this figure is needed for calculating the application rate and the amount of pesticide formulation per sprayer load.

The quantity of spray applied per unit of surface is determined by the following four factors:

* The pressure in the spray tank determines the rate at which spray is emitted. A compression sprayer usually needs repressurization once during a 10-minute period, because the pressure decreases as the tank empties. A pressure regulator is helpful in keeping the discharge rate constant, but many single-lance sprayers are not fitted with such a device.

* The size of nozzle orifice. This opening restricts the volume of spray passing through the nozzle. Damaged or dirty orifices and sieves in the nozzle affect the volume delivered per unit of time.

* The walking speed of the sprayman, which determines the area sprayed per unit of time. Keeping one's walking speed constant, particularly in paddy soil, is another important factor in spraying. With some practice, the operator will be able to maintain a constant speed (1 m/sec in upland areas, 0.5 m/sec in wet paddy fields), while one hand is pumping and the other directing the nozzle. Unnecessarily stopping or slowing one's speed will affect the evenness of the spray pattern.

* Width of spray swath, being the width covered with spray liquid in one pass of the sprayman or in that of larger equipment such as spray aircraft or motorized sprayers. In spraying row-crops or in inter-row applications, the swath is determined by the width of spray dispersion from the nozzle and thus by the characteristic spray angle of the nozzle used (see information under nozzle). Furthermore, the width of the spray pattern depends on the height at which the nozzle is held

A technician calibrates the flow rate of a hand-pumped knapsack sprayer with the help of a graduated "Kali-bottle" (calibration bottle) that can be screwed onto the spray lance.

above the crop or soil. If using a single-lance sprayer, the nozzle should be held at a constant distance above the crop.

In spraying rice fields etc., the operator usually swings the lance to and fro to obtain a swath width of 1.5 metres while walking slower than in spraying a narrow straight swath. A wider swath may leave some sections unsprayed. The spray should have about a 10 per cent overlap. The operator should spray either to his right or to his left side in order to avoid walking through the sprayed crop, or he should use tail-boom delivery gear.

The above factors should be considered when calibrating your sprayer. The following procedure is designed for calibrating exactly a knapsack-sprayer with a single nozzle. The procedure may also be used to calibrate a hand-operated (tail-) boom sprayer equipped with several nozzles and the results may be used to compare the performance of sprayers and types of nozzle.

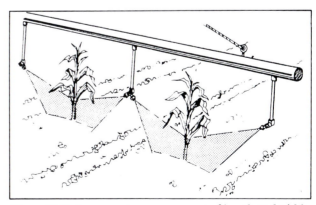

Figure 11.7: An accurate measurement of length and width of the band sprayed in one spray run is needed to enable calculation of the spray volume per hectare. When using typical swivel nozzle arrangements as shown, the width of spray deposition depends on the type of flow rate of the nozzles selected and on the height and stability (swinging) of the spray-boom.

Steps in calibrating knapsack sprayers

1. Determine nozzle discharge.

* Fill the sprayer with water and pump it to working pressure.

* Dip nozzle into bucket or jar and spray water into this container for a precisely measured one-minute period.

* Measure the quantity of water collected in a graduated cylinder in millilitres (ml). This is the nozzle discharge rate expressed in millilitres per minute.

* Repeat this calibration three times to obtain the average nozzle discharge per minute, which should be used in subsequent calculations. A special jar for calibrating sprayers, the "Kalibottle", which can be attached to the end of the spray lance, is manufactured by Hardi Ltd. (Address: St Georges Way, Bermuda Industrial Estate, Nuneaton, Warwickshire, CV10 7QT, United Kingdom). (See picture on previous page.)

2. Determine the walking speed of the sprayman.

* Mark the starting point with a stake in a field planted with the crop to be sprayed.

* The sprayer should be carried on the back and operated by pumping while directing the nozzle at the target. Walk for exactly one minute, while someone else reads the time on a watch. Walk at a normal and constant speed.

* When one minute has expired, mark stopping point with another stake and measure the distance between the first and the second stake in metres.

* Repeat this action three times to obtain average walking speed.

3. Establish the width of the spray swath.

* Spray in the same way as under item 2 above, over a very short distance on a dry path or threshing floor. Measure the width of the swath in metres before the spray dries up.

4. Calculate area sprayed in one minute.

Area sprayed per minute (m^2/min) = width of swath (metres) × walking speed (metres/min)

5. Calculation of application rate for any given area with the following formula:

Volume of spray per area (litres) =

$$\frac{\text{Nozzle discharge (litres/min)} \times \text{Area } (m^2)}{\text{Area sprayed in one minute } (m2/min)}$$

Example: If the nozzle discharge rate is 0.4 litre/min, the area sprayed in one minute is 20 m^2/min, then the volume of spray per 4,000 m^2 (i.e. 0.4 ha or one acre) would be:

$$\frac{0.4 \text{ litre/min} \times 4,000 \ m^2}{20 \ m^2/\text{min}} = 80 \text{ litres of spray liquid}$$

A quicker method for calibrating knapsack sprayers, which does not require measuring the time and would serve the farmer in the field sufficiently well, is the following:

1. Stake out a test area in a field.

2. Place a known amount of water in the spray tank.

3. Establish the width of the spray swath.

4. Enter the test area and make the test run until the tank is emptied, spraying the area at the recommended pressure and speed.

5. After spraying, measure the length of the test area sprayed.

6. Calculate the application rate in liters/hectare.

Area sprayed (ha) =
$$\frac{\text{Width of swath (m)} \times \text{Distance walked (m)}}{10,000}$$

Application rate = Volume sprayed ÷ Area sprayed

Example: Width of spray swath = 4 m
Distance walked = 40 m
Volume sprayed = 5.0 litres

Area sprayed = 4 m × 40 m /10,000 m^2 = 0.016 ha
Application rate = 5.0 litres/0.016 ha = 313 litres/ha

If the farmer has a 10-litre-capacity sprayer, he would need 313 litres/10 litres = 31 sprayer loads to spray one hectare.

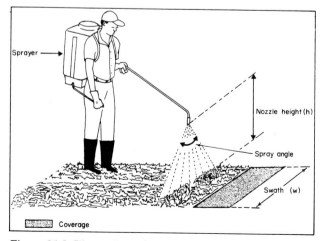

Figure 11.8: Illustration of spray height, angle and coverage

194

CALIBRATING HAND-HELD ROTARY DISC SPRAYERS

The discharge rate of a spinning disc sprayer can be measured in the following way:

1. Fill the spray container with a known amount, e.g. 100 ml, of a simulation fluid having the same viscosity as the product to be applied; failing that, use water as an approximation.

2. With the power switched on and the apparatus held in the spraying position, measure the outflow quantity in ml/min. by indirect or direct method:

- Indirect: Spray into the air and determine the quantity sprayed by subtracting the remaining volume in the container from the original volume.

- Direct: Spray into a plastic bag inverted over the sprayer head and measure the quantity collected.

3. Repeat the calibration three times to obtain an average discharge rate in ml/min.

4. Swaths of 3-4 metres in width are an acceptable assumption when spraying with light cross-winds. Overlapping of the swath as the operator moves up- wind improves spray coverage on the crop.

CALCULATIONS FOR PESTICIDE APPLICATIONS

Recommendations for the proper rate of application (dosage) of an individual insecticide, fungicide, herbicide etc., to control a specific pest are usually indicated on the product label and in technical leaflets. Chemical companies and plant protection research agencies explore, under local conditions, the interactions of specific pests and crops, and in the case of herbicides and soil sterilants, of soil types, on which the dosage recommendations are based. To obtain the full benefit of this research and achieve effective control over the pests in his field, the farmer must know how to calculate the quantity of pesticide required to make a recommended application of a foliar spray, or a proper granular or dust application.

A commercial formulation of a pesticide is a mixture of inert and active ingredients. Only the active ingredients kill animal pests, pathogens and weeds; the inert ingredients are additives in the formulation which improve shelf-life, storability, application characteristics, among other factors. Generally, recommendations for application rates are given in terms of the amount of active ingredient per area and consequently only the active ingredient portion is used in the calculations.

The recommended rate of a pesticide may be expressed in:

* Per cent concentration of the active ingredient (a.i.) in the spray solution, or

* Weight (kg) or volume (litre) of active ingredient per area (hectare) treated.

Before any calculations can be made, the following four factors must be known:

1. The recommended rate in kg or litres of active ingredient(s) per hectare or per cent spray concentration to be applied.

2. Amount of spray liquid per hectare when applying sprays.

3. The per cent active ingredient of the pesticide in the commercial formulation.

4. Area in hectares to be treated.

Calculations for foliar spraying

Foliar spraying is the most common form of insecticide and fungicide application for field crops and orchards. The most common formulation types for foliar sprays are wettable powder (WP) and emulsifiable concentrate (EC).

Example of solving a problem involving an emulsifiable concentrate:

This example applies when a certain spray concentration is recommended for an EC formulation. The four aforementioned factors which must be known beforehand are as follows:

* Recommended concentration is 0.04 per cent.

* 320 litres per ha of spray liquid is desired.

* EC formulation contains 45 per cent active ingredient.

* Area to be treated is 0.5 ha (5,000 m^2)

Problem:

How many litres of the commercial formulation are required to treat the 0.5 ha?

Solution:

* First compute the total spray volume in litres needed to treat the area:

320 litres/ha × 0.5 ha = 160 litres

* Then use this formula:

Litres of commercial formulation =

$$\frac{\text{Amount of spray required} \times \text{per cent spray concentration}}{\text{per cent active ingredient}}$$

$$= \frac{160 \times 0.04}{45} = 0.142 \text{ litre}$$

The amount of formulation needed per sprayer load, if one is going to use an 8-litre sprayer:

Amount of formulation per sprayer load =

$$\frac{\text{Litres of commercial} \times \text{Capacity of sprayer}}{\text{Amount of spray required (litres)}}$$
$$\text{formulation} \qquad \text{(litres)}$$

$$= \frac{0.142 \times 8}{160} = 0.007 \text{ litre (or 7 ml)}$$

Example of solving a problem involving a wettable powder (WP)

We now want to consider the calculations used when the recommendations for a foliar spray are based on the number of kg a.i./ha required. The same formula as used for the per cent concentration problem just solved can also be used for solving this problem, if the four factors are known.

Given:

- The recommended rate is 0.75 kg a.i./ha
- Volume of spray solution is 320 litres/ha (10,000 m^2)
- The WP formulation contains 70 per cent a.i.
- Area to be treated is 0.5 ha (5,000 m^2)

Problem:

How many kg of the commercial formulation are required to treat the 0.5-ha area?

Answer:

Kg of commercial formulation required =

$$\frac{\text{Recommended rate} \times \text{Area to be treated} \times 100}{\text{Per cent of active ingredient(s)}}$$
$$\text{kg a.i./ha} \qquad \text{(ha)}$$
$$\text{in the commercial formulation}$$

$$= \frac{0.75 \times 0.5 \times 100}{70}$$

$$= 0.536 \text{ kg}$$

The amount of WP formulation required per sprayer load can be calculated as above.

Amount of formulation sprayer load =

$$\frac{\text{kg of commercial} \times \text{Capacity of}}{\text{Amount of spray required (litres)}}$$
$$\text{formulation} \qquad \text{sprayer (litres)}$$

If one is going to use a 8-litre sprayer: $= \dfrac{0.536 \times 8}{160}$

$$= 0.027 \text{ kg (or 27 g)}$$

Calculating amount of granules for field application

To determine the amount of a granular formulation, or of a dust formulation required to treat a given area, you must know the following:

* Recommended rate in kg a.i./ha.

* Area to be treated.

* Per cent of active ingredient in the commercial formulation.

Example: You wish to apply pesticide granules at the rate of 0.6 kg a.i./ha to a 2-ha field. The granules contain 3 per cent active ingredient. How many kilograms of commercial formulation are needed to treat this area?

Answer:

Commercial formulation (kg) =

$$\frac{\text{Recommended rate} \times \text{Area to be} \times 100}{\text{per cent a.i. in commercial formulation}}$$
$$\text{in kg a.i./ha} \qquad \text{treated (ha)}$$

$$= \frac{0.6 \times 2 \times 100}{3} = 40 \text{ kg}$$

Calculating amount of granules for seedbed application

This is a different kind of problem, one that involves applications of an insecticide on a small nursery bed that is infested with green leafhoppers. Assume that 3 per cent carbofuran granules are to be applied to protect the nursery bed from tungro infection.

Seed beds of rice (above) or of other crops are usually so small in size that hand-picking of egg-masses and larvae and removal of diseased plants is a feasible way of preventing loss and limiting the need for pesticides applications. (Photo courtesy of FAO-IPC Rice Programme)

Given:

- The recommended rate is 1.0 kg a.i./ha
- Area to be treated is 10 X 5 metres
- The per cent of active ingredient in commercial Furadan (trade name for carbofuran) is 3 per cent
- Two applications are required at 10-day intervals

Problem:

How many kg of Furadan should you buy if you want to treat the nursery bed twice?

Answer:

The number of kg of Furadan required for one application must be determined. Using the same formula as in the previous problem, we have:

$$\text{kg of commercial formulation required} = 1 \times \frac{\dfrac{10 \times 5 \times 100}{10,000}}{3}$$

$$= 0.167 \text{ kg}$$

Since two applications are required, 0.167 kg must be multiplied by 2. Thus, 0.334 kg or 334 g of the commercial formulation are required to treat the nursery twice.

TIMING OF THE APPLICATION OF PESTICIDES

The proper timing of a pesticide application is as essential for achieving good control of the target organisms as the choice of an adequate product and the use of good equipment. One should determine the timing of an application in relation to the following factors:

* The stage of development of the pest(s)
* The growth stage of the crop
* The stage of development of their natural enemies
* The severity of the damage or infestation
* The type of weather and time of the day
* The possible damage to beneficial insects (bees etc.)
* The preharvest interval (safety period)

The development stage of the pest

The susceptibility of organisms to damaging chemical and environmental factors depends very much on their state of development. During periods of rest or reduced biological activity, for instance, as seed and spores, or as eggs, cysts and pupae, they are much more difficult to control than during periods of vigorous growth and voracious feeding. Generally, to kill insect eggs, one needs insecticides with a strong penetrating action (ovicides). The young larval and nymphal stages are normally times of almost continuous feeding and thus of extensive exposure to stomach poisons. The fifth instar is

Five nymphal instars, male and female, of the Rice green leafhopper Nephotettix virescens. *(Photo courtesy of FAO-IPC Rice Programme)*

often difficult to kill, because of reduced feeding behaviour of the last instar before pupation. Moths and butterflies also feed little, but they are well controlled by fumigant action, e.g. the control of Indian Mealmoth in grain stores by fogging. Mobile insects such as caterpillars and flies are vulnerable to contact poisons, but immobile sucking pests such as aphids and San Jose scales are better controlled by systemic chemicals. Insects in their pupal stage are generally well protected and may survive a long fallow period in the stubble of a harvested crop. In such a case, control by burning of the crop residue is much more effective than using any chemical.

The egg clusters of the Cotton Leafworm (*Spodoptera littoralis*) or of various Rice Stemborer species are deposited on the leaves. After hatching from the eggs, the first instar larvae remain outside on the leaves for a short period until they find a cotton boll or the leaf-sheath of a rice stem into which they make their way to feed on the soft inner tissues. Once the larvae are inside, foliar sprays will have little effect; it would be better to apply systemic granular insecticides to the roots.

The growth stage of the crop

The susceptibility of crop plants to damage from insects, nematodes, diseases, weeds etc., usually varies in the course of their lifetime. Seedlings in particular are easy victims for soil pathogens and soil insects; for this reason, seedbeds are often treated with methyl bromide or hot steam. Often, organisms are specific concerning the stage of growth and the parts of the plants they attack or concerning the parts of the plant where the symptoms of infestation become visible. Examples: the Rice Bug (*Leptocorisa acuta*) damages the developing grains at the "milky stage". The symptoms of Smut disease (*Sphacelotheca* spp.) of sorghum appear at the ripening stage of the heads, but the spores are carried over by the seed and germinate in the seedlings. The Rice Bug may be controlled by spraying the ripening ears; however, Smut is controlled by seed treatment with fungicides. Another example: an attack of stemborer larvae in a high-yielding variety of rice in the tillering stage may have

In the FAO-IPC Rice Programme, Indonesian farmers first learn about pests and natural enemies in the field (photo centre). Thereafter, they are tested on their understanding and retention of the training message. An effective aid in testing is the "ballot box" (photo right). One of the pests about which the farmers learn is the dark-headed stem borer, Chilo polychrysis *Mayrick*, of which larvae in a rice stem are shown in the photo at left. (Photos courtesy of FAO-IPC Rice Programme).

the effect of even boosting the yield. In that case, the proliferous growth of new tillers and the effect of thinning on the remaining tillers compensate for the loss caused by the borer larvae; thus, there would be no need for spraying. These examples demonstrate that plant protection extension agents need to advise farmers about growth stages and the behaviour of pests and crops to be treated and their interactions. Control measures should also fit in with the cultural practices such as soil cultivation and intercropping, in order to achieve the optimum effect from the treatment.

Some general guidelines can be applied. Treatment of seeds with fungicides is effective against seed-borne diseases, as is encapsulation of seed with insecticides against soil-living insects. A pre-planting treatment of the soil with a fumigant or steam is effective against soil pathogens, soil insects, nematodes and weed seeds. Fungal diseases may require protective spraying at the time flights of spores are expected. The initial protection of stored produce against storage insects may need to start with spraying of the ripening fruits in the field.

As for weeds, the time of applying herbicide commonly depends both on the stage of growth of the weeds and on the growth of the crop. This is clearly demonstrated by the following terms used for the application of herbicides:

- Pre-emergence for weeds and crop.
- Pre-emergence for the crop and post-emergence for weeds.
- Post-emergence for weeds and crops.

In the case of weeds, the need for spraying is also to a large extent dependent on the "closing" of the crop. Dense foliage or a canopy blocks out sunlight below, lack of which light means that weeds cannot grow. As the competition of weeds and crop is mutual, early herbicide treatments will have a more beneficial effect on yields than late ones. In weed control particularly, soil cultivation, cropping pattern and chemical control should be

programmed together, as effective cultural practices may preclude the need for herbicide application.

The development stage of natural enemies

Similarly, the stage of development and the whereabouts of natural enemies should be taken into consideration. Indiscriminate spraying in the past has decimated locally predatory organisms and parasites to such an extent that pests of hitherto little economic importance suddenly could become a major cause of damage. These so-called secondary pests usually cease to be a problem if proper attention is given to protecting self-regulating forces in nature. The numerous spiders in rice are, for instance, indispensable agents for controlling planthoppers and leafhoppers. Spraying of extensive areas would diminish their chances for rapid re-colonization in treated fields; therefore, such spraying should be avoided.

Besides natural enemies, one should also pay attention to the beneficial insects such as bees and other pollenating insects. Bee-keepers should be warned not to release their swarms on days when spraying is planned.

The severity of damage or infestation

Every pesticide application is costly in terms of expenditure on the product and for hiring labour and in terms of hazards to the applicator and environment. The total cost must be compared with the expected benefits of higher yield or prevention of damage and loss. Experts are establishing economic standards which indicate at what level of infestation a farmer should start applying pesticides. Such standards are quite different for staple crops and high-value crops such as vegetables, flowers and fruits. Since consumers demand fruits and vegetables without any blemishes, fruit and vegetable growers are often enticed to spray frequently and right up to the time of harvest, thereby protecting more the cosmetic appearance of the produce than its nutritional value.

The aforementioned standards are called

"economic thresholds" or "action thresholds", when the economics will be difficult to assess. This means that the level of infestation by pest organisms has been reached when the damage they are causing would surpass the capacity of the crop to tolerate and compensate for it, and the marketing prices would justify the expenditure on crop protection measures. Action thresholds need to be established separately for control of weeds, diseases, insects or other organisms in important crops. Thus, it is a complicated affair in which plant protection services need to advise farmers.

The type of weather and time of day

Adverse weather conditions may spoil the result of a technically good application. For instance, after application, the pesticide should stick to the foliage for good action, but a heavy rain shortly thereafter may wash off the chemical before it can have its effect. For this reason, it is recommended that insecticides not be sprayed when rain is expected within the same day. In the case of herbicide application on the soil, however, rainfall shortly after spraying may improve the penetration of the weed killer into the soil and into the rooting zone of the weeds. Long periods of drought after the application of atrazine herbicide in grain crops or sugar-cane may reduce its efficacy.

Generally, spraying involves the production of numerous fine droplets which depend very much on air currents for their distribution and deposit formation. If a strong wind is blowing, many small droplets and dust particles will be carried far beyond the crop and contaminate the area as a fine "drift spray". If a herbicide such as paraquat is being sprayed on weeds, a drifting spray cloud landing on adjacent crops may cause considerable damage by scorching the foliage. On most days, the wind speed is low in the early morning and evening hours, but much stronger between 9 a.m. and 4 p.m. Also, the direction from which the wind is blowing may change, a factor of particular importance when spraying with rotary disc sprayers.

Spraying during the hottest hours of the day with bright sunshine should be avoided generally, for several reasons: evaporation of the volatile components in the spray liquid is higher; rapid evaporation of the spray droplets on foliage may increase risk of phytotoxic scorching; working under the hot sun leads to heavy perspiration and is fatiguing, both factors increasing the risk of intoxication of the people involved in the application. Therefore, it is recommended that pesticide applications be carried out in the early morning or late afternoon, or on days when the sky is overcast.

One should also take into account the behaviour of the insects to be killed. To control a typical daytime feeder such as the rice hispa (*Dicladispa armigera*), it is better to spray in the morning hours. But to control rice armyworms, or ear-cutting caterpillars, (*Mythimna* spp.), which feed during the evening and night, the late afternoon would be the best time for spraying with an insecticide having a fumigant action also. Spraying or fumigation in stores, glasshouses or plastic tunnel constructions should be planned at the end of the working day or before holidays, when nobody needs to enter the treated area for some period of time.

Good knowledge of the properties of the pesticide used, of the behaviour of the pests to be controlled and of the characteristics of the application equipment are necessary for proper timing of the application in relation to weather and time of day.

Placement and distribution of soil-applied pesticides

The distribution of a soil-applied pesticide varies according to the target. For nematodes and root pathogens, the chemical must be incorporated to a depth of about 15 cm. The vapour phase of nematicides moves through the pores in the soil in an upward direction. Systemic fungicides and insecticides are taken up most effectively by the roots. Some herbicides, however, can enter the shoot of weeds at the soil surface; this means that a degree of movement into the soil is needed for optimum weed control. Application of herbicides near the soil surface enables their translocation with rain or irrigation water, but too much movement may cause damage to a deeper-rooting crop.

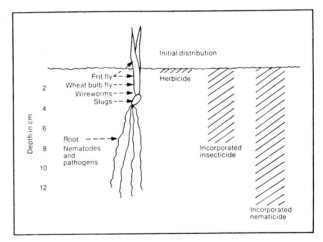

Figure 11.9: The placement and distribution of a soil-applied pesticide varies according to kind and target.

References:

Hance, R.J. *Crop protection chemicals in the soils, some factors affecting their performance*. Journal: Span 28, 1, 1985.

Matthews, G.A. (1979). *Pesticide Application Methods*, Longman Group Ltd, London.

Reissig, W.H. *et al.* (1986). *Illustrated Guide to Integrated Pest Management in Rice in Tropical Asia*, The International Rice Research Institute, The Philippines.

12

CONTROL OF ANIMAL PESTS

Contents   Page

Insects 203

- General morphology of insects 203

- Reproduction and population growth 205

- Taxonomic classification 205

- Problems caused by insect pests in rice 207

- Brief description of common insect pests in rice 209

- Control of insect pests in rice 214

- Biological control of rice insect pests 217

- Chemical control of rice insect pests 219

- Chemical control at various stages of rice growth 223

Mites 225

- Morphology and pest status 225

- Chemical control of mites and spiders 226

Nematodes 227

- Damage by nematodes 228

- Principles of nematode control 229

- Methods of nematode control 229

Molluscs (snails and slugs) 232

- Chemical control 233

Vertebrates 233

- Rodents 234

- Rat control 235

- Birds 240

References 240

Control of Animal Pests

Of the numerous species of animals feeding on living or dead vegetative matter, only a relatively small number compete with man for his sources of food and economic income. Quantitatively, most damage and losses to agricultural production are caused by members of the taxonomic orders of insects, nematodes and rodents, although incidentally other animal organisms may constitute the main local pest.

In this chapter, much attention is given to arthropods, the large group of animals which, in the adult stage, possess articulate or jointed legs and to which insects and spider-like organisms belong. In particular, a brief description of some major insect pests in rice will serve to explain the principles of host-pest and pest-natural-enemy relationships as well as the possibilities of integrated control including the use of insecticides.

One must realise that the field of pest-natural-enemy interactions has as yet been explored only marginally and that the employment of natural enemies alone is rarely sufficient to achieve control.

The application of pesticides will remain indispensable in plant production and plant protection for many decades and probably forever in the control of rodents and stored-product pests. Therefore, in this and following chapters, ample attention is given to the proper selection and use of pesticides in the control of animal pests, pathogens and weeds.

The order of insects include many species which, as vectors, transmit a number of dreaded human- and animal diseases such as malaria, filaria, river-blindness and sleeping-sickness. Research on and control of mosquitoes, ticks, fleas and other vector insects are tasks commonly carried out by specialized organizations in the field of human and veterinary health. Therefore, although vector insects are important targets for biological and chemical control, they are not discussed in this manual.

Examples of animal pests are: Cotton stainer Dysdercus fasciatus *(left).*

Parasitic nematode Acontylus vipriensis *penetrating plant roots and depositing eggs outside (below).*

Egyptian spiny mouse Acomys cahirinus *is a major pest in the Near East (left below).*

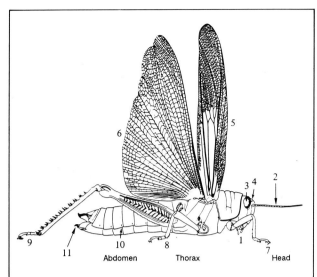

Figure 12.1: *External characteristics of non-specialized,*
adult insect (diagrammatic)
Head with: 1. mouthparts, 2. antennae,
3. compound eyes, 4. ocelli.
Thorax with: 5. forewings, 6. hindwings,
7. forelegs, 8. midlegs, 9. hindlegs.
Abdomen with: 10. spiracle, 11. ovipositor.

INSECTS

Insects are animals which have certain anatomical characteristics in common. The most distinctive trait is that every adult insect has six legs. Estimates of the total number of insect species range from 750,000 to 1.5 million. Out of every five animal species known, at least four species are insects. Insects have inhabited the earth for probably 300 million years. During their very long period of evolution, insects have become immensely diversified and adapted to all kinds of environments. Some insect species live under extreme climatic conditions in deserts, arctic tundras, in water, in underground caves or even on the surface of oceans. Insects are also highly specialized in their functions and behaviour. In size, they range from as small as microscopic bacilli to as large as a pigeon, if one compares their wingspan.

Among all organisms, insects are generally the most serious competitors for foodstuffs, fibres and forest products. Many species of insects, however, are indispensable as pollinators of many plant species or as predators and parasites to control other insects. Such species are regarded as beneficial and they are frequently reared especially as biological control agents or as suppliers of honey, wax, silk thread, shellac etc.

General morphology of insects

The body of a fully developed adult insect is clearly divided into three regions: head, thorax and abdomen (see fig.12.1). The head bears the antennae, the eyes and the mouth-parts. The antennae, always a pair, are borne on the forehead and used for feeling and smelling. The eyes are either compound (facetted) or simple (ocelli). The two compound eyes consist of many wedge-shaped facet eyes, which are each independent and complete. Each facet eye sees a fragment of the insect's emvironment. Together the facet eyes reflect a mosaic type picture. The ocelli are dark, shining epidermal structures on the forehead, with which insects distinguish dark from light. Lower insects and larvae often have only ocelli; adult higher insects usually have both types of eyes.

The thorax is also tri-segmented into the pro-thorax, meso-thorax and meta-thorax. Each of the three thorax-segments bears one pair of jointed (articulate) legs for a total of six legs. Each leg is composed of five parts, of which the extreme one has become adapted for walking, leaping, swimming, digging or gathering. Usually, adult insects have two pairs of wings: one pair each on the meso- and meta-thorax. Some adult insects are wingless or have only rudimentary wings. The abdomen, which is composed of several segments, has no legs. The posterior segment is often extended into an ovipositor with which the female lays eggs.

Insect mouth-parts

Insect mouth-parts are highly diversified but one can distinguish two general types: biting-chewing or piercing-sucking.

Insects with biting-chewing types of mouth-parts use strong, saw-toothed upper jaws (mandibles) to seize and tear off plant tissues in sideways movements. Further below, a pair of lower jaws (maxillae) grind and masticate the food that has been cut by the mandibles. Chewing insects injure plants by eating leaves or boring into roots, stems, buds or fruit. Examples of chewing insects are locusts, beetles, stoneflies, earwigs, termites and silverfish. The larval stages of many insects, such as the larvae of moths and butterflies (Lepidoptera) or those of true flies (Diptera), also belong to the chewing type. Examples include armyworm, earcutting caterpillar,

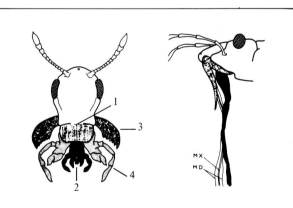

Figure 12.2: *Mouth-parts of insects (diagrammatic)*
Left: *the biting-chewing type*
Right: *the piercing-sucking type*
1. Upper lip (labrum)
2. Lower lip (labium)
3. Upper jaws (MD = mandibles)
4. Lower jaws (MX = maxillae)

looper, stem borer, maggot of the Mediterranean fruitfly and grub of the ladybird beetle.

In insects of the piercing-sucking type, the upper and lower jaws have developed into pairs of hollow needles or stylets. The lower jaws have merged into a joint stylet with two separate channels. When sucking insects such as aphids, leafhoppers and plant bugs pierce plant tissues with their stylet, saliva starts flowing down into the puncture through one channel. Plant sap is sucked from the phloem vessels into the mouth of the insect through the second channel.

Both types of mouth-parts show a considerable variation in shape and mechanism. For instance, butterflies possess a nectar-sucking coiled tube; bees, a licking-sucking, tongue-like tube formed by the lower jaws (maxillae). The beak of licking-sucking flies is a pestle-shaped structure formed by the lower lip (labium).

The insect skin

The bodies, limbs and other protrusions of insects are covered with a very resistant outer skin layer consisting of a hard substance called chitin. The chitin layer, or integument, varies in thickness and serves as a protective outer skeleton. The rigid integument is practically non-elastic and has to be shed periodically during larval growth. Special glands under the skin facilitate the shedding of the old chitin layer and the formation of a new one. Growth, therefore, proceeds stepwise in a series of moults.

The joints between the segments of the insect body are covered by a thick membrane-like skin which gives the insect flexibility.

Growth and development

After hatching from the egg, the growth of an insect is accomplished by a series of moults. At each moulting, the insect sheds the old integument (outer skin layer) and grows a new, larger one. The number of moults varies in most insects from four to eight, with some exceptions of 10 to 12 moults in dragonflies or as many as 20 moults in mayflies. During the interval between successive moultings, an insect is called an instar; the corresponding interval of time is called a stadium.

The change in form of an insect during its development from one instar to another is called post-embryonic development or metamorphosis. Some insects undergo negligble change in form, the young and adults remaining very similar except for size. In other cases, the young and adults differ drastically both in habit and form. There are two types of metamorphosis.

Incomplete metamporphosis

After hatching from the egg, the immature insect gradually passes through several moults and develops into an adult without going through a prolonged resting stage prior to the last moult. During the immature stages these insects are called nymphs if they live on land (e.g. grasshoppers), and naiads if they live in water (e.g. mosquitoes). Both nymphs and adults live in the same habitat and have similar feeding habits and, thus, cause similar damage.

The main changes during growth affect the size, proportions and colouration of the body, wings, ocelli, antennae, mouthparts, abdomen, and internal and reproductive organs. In wingless insects, such as silverfish and springtail, changes are almost imperceptible throughout life. In winged insects with incomplete metamorphosis, wings develop externally and the early instars are usually not able to fly. This gradual type of metamorphosis occurs in termites, aphids, grasshoppers, leafhoppers, bugs, etc.

Complete metamorphosis

There are four distinct growth stages: egg, larva, pupa and adult. The early instar leaving the egg is a worm-like, wingless larva, maggot, grub or caterpillar bearing no resemblance to the adult form. The larva may or may not have legs attached to the thorax-segment; the wings develop internally. Generally, larvae have chewing

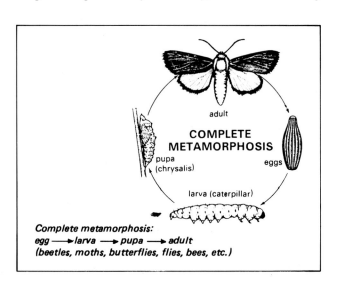

Simple metamorphosis: egg ⟶ nymph ⟶ adult (locusts, grashoppers, aphids, termites, etc.)

Complete metamorphosis: egg ⟶ larva ⟶ pupa ⟶ adult (beetles, moths, butterflies, flies, bees, etc.)

mouth-parts, even in those orders in which adults have sucking mouth-parts (e.g. butterflies, true flies). After moulting several times, the larvae reach full size, cease to feed and then change into immobile pupae. During this resting stage, a profound transformation of the larval organs takes place and the wings develop externally. After the last moult, a mature winged moth, butterfly or beetle emerges from the pupa. This type of metamorphosis occurs among ants, bees, wasps, true flies, beetles, moths and butterflies (order Lepidoptera). In insects such as stemborers, gall midge and cotton bollworm, the larvae feed on plants and cause damage, but the adults neither feed on nor injure plants. In other insects such as hispa, both the larva and adult feed upon the plant.

Intermediate types of development between incomplete and complete metamorphosis occur in thrips, white flies and male scale-insects; for example, in male scale-insects: eggs-crawler-pupa-adult.

According to their outward appearance, insect larvae with complete metamorphosis can be grouped as follows:

Immature stage	Head capsule	Legs	Adult stage
maggot	indistinct	no legs, but creeping pads	flies
larva	well-grown	no legs, but creeping pads	weevils
grub	well-grown	thoracic-, no abdominal legs	beetles
wireworm	well-grown	thoracic-, no abdominal legs	clic beetle
caterpillar	well-grown	thoracic-, abdominal prolegs (non-segmented legs)	moths/ butterflies

Reproduction and population growth

Insect reproductive abilities vary widely; some species may increase their population rapidly under favourable conditions. The rate of population growth depends on various factors, which are either specific for the species or influenced by external conditions, such as:

* Length of the life-cycle to maturity

* Fecundity of the female expressed in number of eggs laid and hatched, and sex ratio

* Availability of sufficient male and female adults for mating

* Supply of food and shelter

* Weather conditions and physical environment

* Risk of being killed by predators, parasites and other natural control factors

Adult green leafhopper Nephotettix virescens. *(Photo courtesy of IRRI)*

The theoretical growth rate within a given temperature range depends on the first two factors, i.e. time to maturity and number of eggs laid and hatched. High fecundity and rapid maturity favour accelerated population growth. A single pair of house flies theoretically may produce more than 3 million offspring within 12 weeks or six generations.

Under actual field conditions, the potential growth rate of insect species is continuously modified by controlling factors such as temperature and humidity. For instance, in the development of *Nephotettix* spp., (green rice leafhoppers), the nymphal stage will last for 21 days at $25^\circ C$, and 15 days at $30^\circ C$. A female green leafhopper of the species *N. virescens* will deposit a total of 140 eggs on average, whereas a female of *N. negropictus* will lay 60 eggs. In the humid tropics - if food is available - leafhoppers remain active year round and produce six or more generations per year. Thus warm and humid weather may, under favourable conditions, induce a green leafhopper outbreak in rice, in particular of *N. virescens*. The combination of all six aforementioned factors generally causes the population density of any insect species to rise and fall several times during the year. The population density may reach explosive proportions only occasionally.

Taxonomic classification

Like other organisms, insect species are commonly classified according to a hierarchy of taxonomic levels. Classification can be made as shown below using man and fly as a point of reference.

Taxonomic level	Man	Tobacco white fly
Kingdom	Animalia	Animalia
Phylum	Chordata	Arthropoda
Subphylum	Vertebrata	Mandibulata
Class	Mammalia	Insecta (Hexapoda)
Subclass	-	Exopterygota
Order	Primates	Homoptera
Family	Hominidae	Aleyrodidae
Genus	Homo	Bemisia
Species	sapiens	tabaci

Table 12.1: Classification of insects (abbreviated)

The class of Hexapoda (= six-legged Arthropoda) or Insecta is divided in three subclasses, each containing numerous orders of insects, of which only a limited number are of agricultural importance either as pests or as natural enemies or as pollinators.

Subclass	:	Apterygota	(wingless insects)
Order	:	Thysanura	silverfish
Subclass	:	Exopterygota	(winged; wings develop externally)
Order	:	Orthoptera	
Family	:	*Gryllidae*	crickets
	:	*Acrididae*	grasshoppers, locusts
	:	*Tettigonidae*	long-horned grasshoppers
Order	:	Isoptera	termites
Family	:	*Termitidae*	termites
Order	:	Homoptera	plant bugs
Family	:	*Cicadellidae*	jassids
	:	*Delphacidae*	leafhoppers, planthoppers
	:	*Psyllidae*	jumping plant lice
	:	*Aleyrodidae*	white-fly, citrus blackfly
	:	*Aphididae*	greenfly, aphids, plant lice
	:	*Pseudococcidae*	mealybugs
	:	*Coccidae*	scales, naked- and waxy scales
Order	:	Heteroptera	bugs
Family	:	*Miridae*	capsids, helopeltis
	:	*Lygaeidae*	cotton-seed bug
	:	*Coreidae*	plant bugs, rice bug
	:	*Pentatomidae*	shield bugs, stink-bugs
Order	:	Thysanoptera	thrips
Family	:	*Thrypidae*	thrips
Subclass	:	Endoptera	(winged; wings develop internally)
Order	:	Lepidoptera	butterflies, moths
Family	:	*Yponomeutidae*	(= Plutellidae), diamondback moth
	:	*Tortricidae*	budworms, codling moths
	:	*Pyralidae*	stem- or podborers, leafrollers
	:	*Noctuidae*	army- or cutworms, bollworms, stalkborers
	:	*Geometridae*	loopers
Order	:	Diptera	true flies (see also table 12.3)
Family	:	*Cecidomyiidae*	gall midges
	:	*Tephritidae*	fruit flies
	:	*Agromyzidae*	dipterous leaf miners
	:	*Muscidae*	shoot flies, house flies
	:	*Anthomyiidae*	root flies, root maggots
Order	:	Hymenoptera	ants, bees, wasps (see also table 12.3)
Family	:	*Formicidae*	ants
	:	*Vespidae*	wasps
Order	:	Coleoptera	beetles, weevils
Family	:	*Scarabacidae*	chafers, white grubs
	:	*Anobiidae*	timber beetles
	:	*Coccinellidae*	ladybird beetles, Mexican bean-beetles
	:	*Silvanidae*	grain beetles
	:	*Meloidae*	blister beetles, oil beetles
	:	*Bruchidae*	bean bruchids
	:	*Chrysomelidae*	leaf- or flea beetles, hispids
	:	*Curculionidae*	weevils

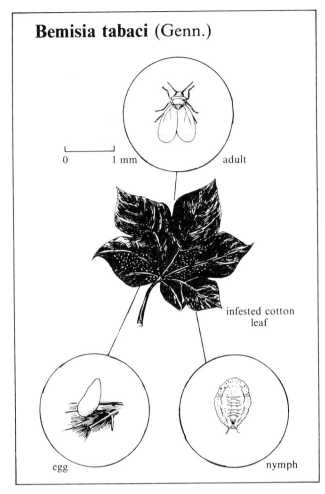

Bemisia tabaci (Genn.)

0   1 mm   adult

infested cotton leaf

egg   nymph

Whiteflies (Bemisia tabaci) *are a common pest of cotton, tobacco and many other crops. They cause damage by sucking plantsap, transmitting viruses and their abundant excretion of honeydew on leaves.*

Scientific names such as *Bemisia tabaci* (Genn) are used to avoid confusion since common names of insects may vary among different countries. This insect is in fact called Tabacco whitefly or Cotton whitefly. The addition (Genn) indicates the taxonomist originally describing the species, while the parentheses show that *B. tabaci* was earlier placed under another genus.

Problems caused by insect pests in rice

A problem that farmers and people engaged in plant protection face almost every day is finding suitable methods for controlling specific pests in field crops and stored products. As rice farmers always have had to cope with ravaging insects, diseases and competing weeds, they have long experience in fighting them. They have learned to keep damage from chronic pests, such as leaf-feeding insects, stemborers, most fungal diseases and weeds, at a tolerable level. However, acute pests such as armyworms, leaf- and planthoppers, blasts, viral and bacterial diseases, and rats may occasionally occur in epidemic proportions that surpass the capacity of farmers to keep the situation in hand. Epidemics, such as from locusts, may cause huge losses to whole regions; the control of such outbreaks is difficult. Severe outbreaks may be associated with extreme weather conditions such as droughts, frost and

floods, which may decimate the natural enemies of rats and insects. Or, hailstorms and typhoons can extensively ravage crops which thereupon become very susceptible to the entry of pathogens.

Since the 1950s, rice production has been greatly intensified, particularly in Asia, although rice culture is expanding rapidly on the other continents too. Production increased enormously owing to fundemental changes in rice technology, which in turn has led to more frequent pest epidemics. Examples of such changes are the following:

* New irrigation systems and the installation of numerous tube-wells have enabled extensive rice cropping in the dry season. Shortening of the rice-free season, during which pest organisms were annually depressed, has favoured the development of more generations per year and, consequently, much larger populations.

* Expansion of farmland planted with rice and a sufficient supply of water to allow multiple cropping secured the availability of food and adequate living conditions for large pest populations throughout the year, whereas previously their numbers were often largely depleted owing to their failure in finding a suitable host. Conversion of large tracts of land into rice farms has enabled isolated pests to migrate from pockets of infestation into new areas or to transfer from wild hosts onto rice.

* New rice varieties developed for their high-yield potential, insensivity to day length, short straw and other desired characteristics have largely replaced traditional varieties which were natural selections on the basis of stable resistance, particularly to diseases. Modern varieties, having a genetically narrower and less stable base of resistance, are generally more susceptible to pests. The higher yield potential also alters the economic threshold level, making pesticide application more profitable compared with most traditional varieties. High yielding varieties of rice are commonly also less capable of withstanding fungal and insect attacks in the post-harvest stage than the traditional varieties.

The planting of new varieties responsive to fertilizer applications and liberal subsidizing by Governments have considerably stimulated the use of agrochemical inputs. Nitrogen induces profusive tillering and lush growth thus providing a rich energy source for rapid proliferation of injurious organisms. Weeds generally grow faster than rice and compete heavily for available nutrients.

The higher incidence of pests in modern rice culture and the greater economic investment for seed, fertilizer and labour to be protected has boosted pesticide usage. Farmers feel compelled to spray more often, but are still making many mistakes regarding choice of product, calculation of the proper concentration and volume of spray, and the timing of the application. For many years, calender-based spray schedules had been

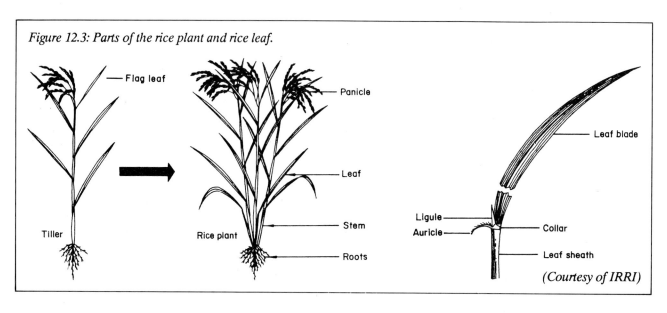

Figure 12.3: Parts of the rice plant and rice leaf.

(Courtesy of IRRI)

recommended by plant protection institutions and the industry without teaching the farmers first to establish the need for spraying on the basis of infestation counts. Improper and ill-timed sprays may fail to kill a pest and may even cause either its number to increase (resurgence) or resistance against pesticides to develop in its population. As pesticides often harm the natural enemies of pests, a formerly minor pest may get the the opportunity to multiply unchecked thus causing a so-called secondary pest outbreak. The outbreak of brown planthoppers as a consequence of large national rice production intensification campaigns is a typical example.

Host-pest relationship

Pests of various kinds often attack specific parts or particular growth stages of plants. Therefore, in describing plant-pest relationships, one should be familiar with the structure of plants and the terminology. For grain crops, one distinguishes three general stages of growth:

* **Vegetative stage**: the period from germination of the seed to panicle initiation, i.e. the time during which the panicle develops inside the sheath of the flag-leaf, creating a bulge at the base of the tiller.

* **Reproductive stage**: the period from panicle initiation to flowering, when the panicle emerges from the leaf sheath (also called heading).

* **Ripening stage**: the period from flowering to mature grain. Ripening grain passes through the stages of milky grain when the endosperm is still liquid, then from soft dough into hard dough grain and finally into the stage of hard, yellow grain.

The seedling stage of a rice plant is the period from emergence of the first leaf (following the sprouting of the coleoptile) until the appearance of the first tiller.

It is remarkable that among most rice cultivars the number of days in the reproductive phase and that in the ripening phase are the same, namely about 35 days and 30 days, respectively. The cultivars differ, however, in the length of the vegetative phase which takes about 55 days in a 120-day cultivar and 85 days in a 150-day cultivar. The term **short-duration cultivar** is used if plants reach maturity in fewer than 100 days. In the tropics, the best yields appear to be collected from cultivars which mature in 140-150 days.

Regarding the occurrence of insects on plants, one sees on the one hand that the majority of the insect pests on food- and cash-crops belong to only a few insect-families and on the other hand that those insects demonstrate a preference for certain plant-families. In the process of evolution, every species has found its specific niche in a hostile environment with its own source of nutrition and its own kind of protection. Thus, borer insects find food and shelter inside stalks or fruits, whereas leafhoppers suck sap from leaves and are somewhat protected by their green colour. Similarly, insects, which as predators or parasites exploit fellow insects as their main nutritional source, may show an equal selectiveness by preying on certain insect families and by being adapted to the same habitat. Frequently, species preference is so exclusive that only one or a few parasitic species are known to relate to a major pest, such as in the case of the cassava mealybug. One should realize that such relationships between host, pest and natural enemies are only relatively well studied for economically important crops, which represent but a fraction of nature.

Important insect pests of rice

About 30 different insect species are considered to be economically important pests of rice in the pre-harvest stage. They are listed in table 12.2 in taxonomic catagories of order and family to demonstrate that only few families of insects are involved, which is an important clue in the search for control methods. Table 12.2 is followed by a section containing brief descriptions of several common rice insect pests. The subsequent paragraphs deal with the general principles of cultural, biological and chemical control of those insect pests in rice. Information on insect pests of rice in the post-harvest stage is given in Chapter 15.

208

The reader may also refer to the *Illustrated guide to integrated pest management in rice in tropical Asia, 1986,* published by the International Rice Research Institute, P.O. Box 933, Manila, Philippines. This guide provides in a systematic, illustrated format for each pest separately, along with biological data and control recommendations.

Insects are recognized as either major or minor pests on the basis of a combination of three criteria. These criteria are:

* Severity of economic loss (high, moderate, low)

* Frequency of occurrence and area affected within the insect's potential habitat:

  - rarely abundant
  - abundant some years in limited areas
  - abundant some years over large areas
  - abundant most years over large areas

* Ease of control (difficult, readily controlled)

Brief description of common insect pests in rice

● **Seedling maggot, rice seedling fly** (*Atherigona* spp.)

Small maggots (legless larvae) of several species of small flies feed within developing rice tillers. Female flies prefer to lay eggs in dry upland nurseries, where they become a pest during the rainy season. Seedling maggots seldom stay in irrigated paddy fields. During a severe infestation, the maggots may injure many young tillers which look like dead hearts owing to stemborer attack. Replanting may be necessary. The larvae pupate in the soil; the life-cycle lasts 2-4 weeks.

● **Rice whorl maggot** (*Hydrellia* spp.)

These tiny, leaf and stem-mining larvae occur in wetland rice. The grey female flies are attracted to young rice instanding water. They lay eggs on the leaves close to the water surface during the first 30 days after the rice has been transplanted. The larvae bore into the leaf tissues and central leaf whorl for feeding and pupate within the leaf. Emerging leaves are tattererd and show white feeding marks.

The adult flies of both aforementioned pests are not attracted by light traps, but the rice seedling fly is strongly attracted by fish-meal bait.

Table 12.2: Insect pests of rice according to taxonomic order

Order : Lepidoptera	(butterflies, moths)
Family : *Pyralidae*	
Yellow stemborer	*Scirpophaga incertulas* (Walker)
White stemborer	*Scirpophaga innotata* (Walker)
Striped stemborer	*Chilo suppressalis* (Walker)
Dark-headed stemborer	*Chilo polychrysus* (Meyrick)
Gold-fringed stemborer	*Chilo auricilius* (Dudgeon)
Rice caseworm	*Nymphula depunctalis* (Guenee)
Rice leaffolder	*Cnaphalocrocis medinalis* (Guenee)
Family : *Noctuidae*	
Pink rice borer	*Sesamia inferens* (Walker)
Ear-cutting caterpillar	*Mythimna separata*
Swarming caterpillar	*Spodoptera mauritia* (Boisduval)
Common cutworm	*Spodotera litura* (Fabricius)
Order : Homoptera	
Family : *Delphacidae*	plant- and leafhoppers
Brown planthopper	*Nilaparvata lugens* (Stal)
White-backed planthopper	*Sogatella furcifera* (Horvarth)
Green leafhoppers	*Nephotettix* species
Zig-zag leafhopper	*Recilia (= Inazuma) dorsalis*
Order : Hemiptera	
Family : *Alydidae*	broad-headed bugs
Rice seed bugs	*Leptocorisa* species
Family : *Pentatomidae*	shield bugs, stink bugs
Rice black bugs	*Scotinophara* species
Order : Diptera	true flies
Family : *Ephydridae*	shore flies
Rice whorl maggot	*Hydrellia philippina* Ferino
Rice seedling maggots	*Atherigona* species
Rice gall midge	*Orseolia oryzae* (Wood-Mason)
Order : Coleoptera	beetles
Family : *Chrysomelidae*	
Rice hispa	*Dicladispa armigera* (Olivier)
Family : *Scarabaeidae*	scarab beetles
White grubs	*Leucopholis* species (Chafer b.)

Larva of Rice seedling maggot (Artherigona oryzae).

Adult of Rice seedling maggot (A. oryzae Malloch). *(Photos courtesy of IRRI)*

The adult tachinid fly, Argyrophylax nigrotibialis (Baranov), lays its eggs inside lepidopterous caterpillars.

The ichneumonid wasp, Amauromorpha accepta metathoracita, *parasitizes plant- and leaf-hoppers.* (Photos courtesy of IRRI)

Biological control takes place through parasitization of eggs by trichogrammatid wasps and of larvae by eulophid and braconid wasps, while spiders and ephidrid flies prey upon adult flies (see table 12.3).

● **Rice gall midge** (*Orseolia oryzae*)

The gall midge is a mosquito-like fly and a major pest of flooded rice throughout South Asia and South-east Asia, with the exception of the Philippines and Malaysia. The yellow-red female flies lay their eggs on the underside of the leaves at night. Hatching larvae move down into the leaf sheaths and attack the growing point of young tillers. This causes a gall to form within the tiller, that develops into a tube-like gall of up to 30 cm long and resembles an onion leaf. Hence its name onion shoot or silver shoot. Early infestation causes excessive tillering, but the new tillers often become infested too and bear few panicles. Infestation takes place from the time the plants are in the seedbed until the end of the tillering stage, when there are no active growing points left to feed upon. The gall midge feeds throughout the year on wild grasses and invades rice

when the monsoon rains arrive. Some gall midge-resistant varieties of rice are available, but they are location specific, as there are several gall midge biotypes.

Biological control of gall midge eggs takes place by predatory mites and parasitizing wasps, while adult flies are preyed upon by spiders.

● **Armyworms and cutworms**

Several species of noctoid caterpillars are important, but localized pests occur in all rice environments throughout Asia. The following species cause the most damage:

* Ear-cutting caterpillar ....*Mythimna (= Pseudaletia) separata*
* Rice swarming-caterpillar ..*Spodoptera mauritia*
* Common cutworm ....*Spodoptera (= Prodenia) litura*

These large leaf-feeding, polyphagous caterpillars come in various types ranging in colour from creamy white

Table 12.3: Natural enemies of insect pests in rice
(Orders and families of insects containing important species of parasites and their favourite prey)

Order/Family	No. on rice[a]	No. of importance[b]	Parasites[c]	Rice pests attacked[d]
Hymenoptera: Wasps				
- *Braconidae*	19	8	3	stemborers, leaffolders
- *Ichneumonidae*	15	10	2	stemborers, leaffolders
- *Mymaridae*	4	1	1	leafhoppers, planthoppers
- *Trichogrammatidae*	9	2	1	stemborers, leafhoppers
- *Eulophidae*	7	3	1	stemborers, skippers
- *Chalcididae*	11	4	2	lepidopterous caterpillars
- *Encyrtidae*	?	1	1,2	plant-lice, scale insects
- *Dryinidae*	6	2	3	leafhoppers, planthoppers
- *Scelionidae*	13	6	1	lepidopterous and noctoid caterpillars, stink-bugs
Diptera: True flies				
- *Tachinidae*	12	?	2	lepidopterous larvae
- *Pipunculidae*	2	2	2	leafhoppers, planthoppers

Notes: [a] *number of species identified on rice pests*
[b] *number of species of economic importance*
[c] *parasites of eggs (1), larvae/pupae (2), all stages (3)*
[d] *main rice pests being attacked*

Nuclear polyhedrosis virus (NPV) in a cutworm (Spodoptera litura) *larva.*

Larvae of a bethylid wasp (Goniozus triangulifer *Kieffer*) *feeding on the body fluid of a Rice leaffolder larva. (Photos courtesy of IRRI)*

to green or dark purple, with or without stripes. The adults are big, greyish-brown moths with marked fore-wings and pale hind-wings. A female moth lays up to 1,000 eggs during its life of about one week. *Mythimna* eggs are hidden in the base of leaf sheaths, whereas *Spodoptera* eggs are laid in clusters on leaves and covered with body hairs. The name "armyworm" stems from the tendency of *Spodoptera* caterpillars to crowd together on the ground, often moving in the same direction; their presence is easily recognized by damaged leaves and cut seedlings. Pupation commonly takes place in the soil of dry fields or on alternative host plants in wetlands. Ear-cutting caterpillars cut young tillers at the base during the vegetative phase, but most damage results from their habit of cutting rice panicles during the reproductive phase. They are particularly a pest of deep-water rice.

Biological control is the most effective means of keeping armyworms and cutworms in check. Many species of wasps and some species of ants and tachinid flies attack the eggs and larvae, while spiders prey on the moths. Also, infection with polyhydrosis-virus kills many caterpillars.

● **Rice caseworm** (*Nymphula depunctalis*)

This small greenish caterpillar lives inside a tube made from cut rice leaves rolled and spun together and filled with water. They spread by floating inside this tube to a neighbouring plant during the daytime. At night, they crawl up rice tillers to feed, still remaining within their cases. When feeding, they skeletonize the leaves, leaving ladder-like patches of removed tissue. Prior to pupation, the caterpillar attaches its case to a tiller above the water and spins its pupal cocoon.

Caseworms are obligate aquatic as they take in oxygen from water; they are a minor pest of nurseries and paddy fields with standing water. They will never attack a dry seedbed. Adults are active during the night with female moths laying batches of eggs on the underside of leaves which float on the water. As caseworms require an aquatic environment, they can be effectively controlled by draining infested paddy fields for several days.

Biological control of caseworm larvae is taken care of by braconid wasps and the larvae of two waterbeetle species, while spiders prey on the moths.

● **Rice leaffolders**

Four species of leaffolders occur in Asia, of which *Cnophalocrocis medinalis* and *Marasmia patnalis* are the most harmful ones. The leaffolder species are very similar in biology and appearance and may occur in mixed populations. The transparent green caterpillars, up to 2.5 cm in length, are a major pest of all rice environments and are more abundant in the rainy season. The larvae fold over the edges of rice leaves and spin them together to make a tube as shelter. They feed by scraping the tissue on the inside of the leaves that causes a scorching type of damage. Larvae pupate after 2-3 weeks inside the leafcase. The small yellowish-brown moths lay rows of 100-300 eggs along the midrib of leaves. Moths can disperse their eggs over distances of several kilometers at night.

Biological control through parasitization of eggs by trichogrammatid wasps and of larvae and pupae by other wasp species has a strong effect. Predation plays a large role too, in particular by crickets preying on eggs, damselflies and beetles on larvae, and spiders on the adult moths.

● **Rice stemborers** (*Lepidoptera*)

Various species of stemborer are occurring as pests of rice in Asia, but they have several alternative hosts such as maize, sorghum, millets, sugarcane, wild rice and grasses. The main species on rice are:

* Dark-headed rice borer ...*Chilo* (= *Chilotraea*) *polochrysus*

* Rice striped borer ....*Chilo suppressalis* (= *C. simplex*)

* Gold-fringed stemborer ...*Chilo auricilius*

* Rice yellow-stemborer ....*Scirpophaga* (= *Tryporyza*) *incertulas*

* Rice white-stemborer .....*Scirpophaga* (= *Tryporyza*) *innotata*

* Pink stemborer ....*Sesamia inferens*

Stemborers are very similar in many aspects of life-cycle, behaviour, type of damage and methods of control. Therefore, they are dealt with as one group, although they differ considerably in anatomy and distribution.

Yellow stemborer adult female,
Scirpophaga incertulas *(Walker).*

Stemborer egg masses of striped (top), yellow (centre) and pink (bottom) stemborer.

Stemborer larvae hatching from egg mass. (Photos courtesy of IRRI)

Stemborers are a pest in all rice environments throughout Asia and southern China, but the white stemborer is confined more to central India, Bangladesh, Myanmar (Burma) and Thailand. Stemborers occur abundantly towards the end of the rainy season.

The first five of the above-mentioned species are pyralid stemborers, of which the adults are yellowish-white moths with distinctive markings on the wings. Female moths of the yellow stemborer, for instance, have two clear dark spots on the fore-wings. The larvae are pale yellow, but those of the striped stemborer show five brownish lines on the back and sides. The moths are active at night and hide during daytime. Female moths, during the four days of their adult-life, lay up to 300 eggs on the upperside of rice leaves. The eggs of pyralid moths are yellowish and disc-shaped and the egg-batches are covered with white hairs.

The sixth species, the pink stemborer, belongs to the *Noctoidae* and its larva has the robust build and strong redish-pink colouring of the armyworm to which it is related. The adults are big, brownish moths, which lay their bead-like eggs in rows hidden in the leaf sheaths of rice plants but do not cover them with hair.

Newly hatched pyralid larvae feed on leaves and thereby show a tendency to disperse. To this purpose the larvae suspend themselves from leaves by a silken thread and are blown to other plants. Other may spin a case from cut leaves on which they float to neighbouring plants. After several days of feeding within the leaf sheath, larvae tunnel into the stem and continue feeding within at the base of the plant. During the larval period of 35-45 days, in which larvae pass through five instars, they may attack several tillers one after another. Finally, they pupate inside the stem near the base. If conditions outside are unfavourable, mature larvae may also hibernate within the stem below soil level. Upon hatching from the pupal case, adult moths leave the stem through an escape window cut by the mature larvae.

By feeding inside the stem, the borer larvae cut the stem at the base causing it to desiccate and to become a 'dead heart' that can easily be pulled out. If the stem is cut after the spikelet has emerged, then the panicle turns white and no grain-filling occurs; empty panicles being called "white heads". Crop losses owing to stemborer attacks are probably on average the highest of all rice pests, being estimated at 3 per cent of the total yield.

Resistance against stemborer attack is one of the goals in breeding improved rice varieties. Some moderate resistance is obtained from chemicals within the rice plants that have a repelling effect on moths and negative effect on young feeding larvae. Another characteristic, a potential for profusive tillering, enables plants to compensate for dead hearts.

Biological control of stemborers is high in an undisturbed environment. Stemborer eggs and larvae are heavily parasitized by several species of wasp. Grasshoppers prey on stemborer eggs; ground beetles, on larvae; and spiders, on the moths. Fungi may attack larvae and pupae in the stem.

● **Rice brown planthopper** (*Nilaparvata lugens*)
 (*= Delphax oryzae*)

This insect is probably the most dangerous pest of irrigated and rainfed wetland rice throughout Asia and southern China. The whitish nymphs and brown adults live and feed at the base of rice tillers. They injure the plants directly by sucking plant sap from the phloem cells, which causes tillers to desiccate and to turn brown. Heavy damage shows up as large circular patches of dried plants and is called "hopperburn". Planthoppers excrete much plant sap as "honeydew" which drops onto the base of the tillers and becomes overgrown with black sooty moulds. Further serious damage may result from infection with viruses, which are transmitted by brown planthoppers. This hopper species is the vector of grassy stunt virus, which causes excessive tillering, and of ragged stunt virus, which causes stunted growth of rice plants.

Adult brown planthoppers of both sexes may have either short wings or long wings; only the long-winged adults are able to fly and disperse. Female adults deposit small batches of up to 16 eggs within midribs or leaf sheaths of rice plants by means of a saw-like ovipositor. Hatching nymphs tend to remain on the same plant which leads to overcrowding.

The use of resistant rice varieties is a very important means of control, but owing to the development of several biotypes of the brown planthopper, such varieties need to be protected by an integrated control programme (see section on cultural control below).

Biological control is usually sufficiently strong to prevent outbreaks, unless the populations of natural enemies are decimated by misuse of insecticides. Trichogrammatid and eulophid wasps parasitize eggs, and drynid wasps and fungi parasitize nymphs and adults. Mirid bugs and mites feed on the eggs while spiders, beetles and dragonflies hunt for nymphs and adults.

● White-backed planthopper (*Sogatella furcifera*)

This planthopper resembles the brown planthopper, but it has a white dorsal band between its wings. It is also a major pest of young rice throughout Asia and may cause hopperburn; however, it does not transmit virus diseases. Biological control is the same as for brown planthoppers.

● Rice green leafhopper (*Nephotettix* spp.)

● Zig-zag leafhopper (*Recillia dorsalis*) (= *Inazuma d.*)

These small leafhopers are a major threat to irrigated and rainfed wetland rice all over Asia because they are the vectors of devastating virus diseases. Green leafhoppers (in particular, *N.virescens* and *N. nigropictus*) transmit tungro-, yellow-dwarf and rice-gall-dwarf virus diseases. Zig-zag leafhoppers are vectors of tungro-, dwarf-, and orange-leaf virus diseases.

Nephotettex adults are pale green with black marks on the head and wings; a *Recillia* adult shows a characteristic zig-zag marking on the wings from which it gets its name. Adults are highly mobile and fly great distances. Females deposit their eggs inside the leaf sheath of young tillers near the base.

Nymphs and adults feed on the upper portions of rice plants in their vegetative stage of growth. The sucking of plant sap retards the growth of rice, but as nymphs tend to disperse they rarely reach sufficiently high population densities to cause hopperburn.

Many rice varieties resistant to green leafhoppers are commercially available. There are few varieties with resistance to virus diseases, but widespread planting of green leafhopper-resistant varieties helps to surpress the incidence of viruses. If tungro disease is prevalent within the vicinity, then the planting of a tungro-resistant variety is recommended.

Biological control is usually strong enough to keep populations of leafhoppers in check. A wide range of wasps, pipunculid flies and nematodes parasitize eggs or nymphs. Various kinds of bugs, empid flies, spiders and dragonflies prey on nymphs and adult leafhoppers.

White-backed planthopper Sogatella furcifera *(Horvath) adult male.*

Rice brown planthopper Nilaparvata lugens *(Stal), long-winged (top) and short-winged (bottom) adults.*

Rice zigzag leafhopper Recilia dorslis *(Motschulsky). (Photos courtesy of IRRI)*

Rice seed bug Leptocorisa oratorius *(F) adult feeding on rice grain. (Photo courtesy of IRRI)*

● **Rice seed bugs** (*Leptocorisa* spp.)

There are several species of true bugs causing damage to rice seeds, which, because of their similarity, are treated as one group of rice seed bugs. They are a minor pest of rice in the reproductive stage, particularly of rainfed and upland rice. Rice seed bugs have a brownish-green, slender body with conspicuous long legs and antennae. When disturbed, the adults fly off while excreting a bad odour. Females lay several hundred eggs along the midrib of rice leaves over a period of 2-3 months. Nymphs and adults have piercing/sucking mouth parts with which they penetrate grain in the milky or soft-dough stage to feed on the liquid endosperm. This feeding results in smaller, discoloured grain that is called "pecky rice". Pecky kernels break easily in the milling process, yielding low-value broken rice.

During the dry season, adults remain in wooded areas where they migrate onto rice at the onset of the monsoon rains. They complete the first generation on wild grasses before moving to rice in the flowering stage. Synchronized planting helps to spread the attack and minimize the damage. As rice seed bugs prefer to rest in grassy areas during periods of bright weather, it is recommended that grassy weeds, in particular *Echinochloa* spp., be removed from the vicinity.

Biological control is by parasitic scelionid wasps on eggs. Grasshoppers prey on the eggs; spiders, on nymphs and adults. Fungi infect both nymphs and adults.

Control of insect pests in rice

Cultural methods are intended to make the environmental conditions in a farming area unfavourable for the development of pest organisms, while maintaining a high production level. Cultural control measures include:

- use of resistant cultivars
- synchronized planting, time of planting
- sanitation and weeding
- fertilizer and water management
- spacing

The principles of cultural control for the rice insects reviewed above are explained below.

* Resistant cultivars

The employment of resistant cultivars of rice alone may not suffice to prevent damage, particularly if one has to deal with high-mutability pests such as the brown planthopper (BPH), white-backed planthopper, green leafhopper, rice tungro virus (RTV) and rice blast disease. Owing to selective pressure within their populations, these pests rapidly generate new biotypes (races) capable of breaking down resistant cultivars. Selection towards new biotypes progresses faster in proportion to the number of genes involved as the resistance is based on one or two genes only. (Narrowly based resistance is called "vertical resistance", as distinct from "horizontal resistance", in which the resistance property is represented in several genes.)

For example, from 1975 to 1979, millions of tons of rice were lost in South-east Asia as a result of the brown planthopper attack. Although resistant cultivars from IRRI were distributed, these cultivars and some others with the same gene for resistance to BPH became susceptible and were hopper burned within two seasons owing to the development of BPH-biotype 2. Replacing these cultivars with the biotype 2-resistant cultivars IR 36 and IR 42 was originally successful, but IR-42 became susceptible on the islands of Sumatra and Sulawesi in Indonesia during the early 1980s. Recently, widely grown national cultivars in Indonesia, such as Cisadane and Krueng Aceh, and IR-42 possessing BRH- biotype 1- and 2-resistant genes, became susceptible in Central Java to a new biotype 3. Thus, pest-resistant cultivars of crops have an important role in control, but they need to be continuously supported by other cultural and chemical measures, which retard or prevent the development of new biotypes of pests.

* Synchronized planting

Synchronization of rice planting, i.e. the transplanting of the whole area within 2-3 weeks, permits a fallow period, during which no rice should be left in the field and all stubble should be removed or ploughed under. As stemborers, for instance, require 3-4 weeks to complete a generation, the fallow interrupts their population build-up, while the removal of the stubble should clear the field of any residual infestation. (See fig. 12.5.)

Traditional rice farming in Indonesia used late maturing (150-180 days) local cultivars for planting in well-irrigated areas one or two times a year. Thereafter, secondary food crops such as soybean, sweet potatoes, groundnuts or vegetables were grown. Although a certain degree of overlapping in rice growing occurred, only relatively mild attacks of endemic rice pests, such as the white stemborer, were recorded.

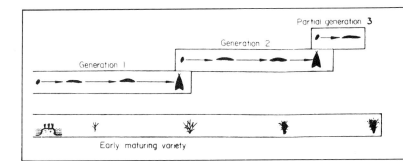

Figure 12.4

In a rice field planted with an early maturing cultivar, stem borers can complete fewer generations than in a late-maturing cultivar. Stem borer populations remain lower on early-maturing cultivars and damage is reduced.

The availability of short-duration rice varieties and improved irrigation led to large-scale intensification of rice production under the 1968/1969 Inmas-programme, which demanded the planting of three crops per year or five crops in two years. Owing to a labour shortage and the extent of the areas under rice, planting was staggered. The almost continuous availability of both younger and older rice plants in the fields enabled pests to proliferate without any interruption in consecutive generations. Thus, outbreaks of brown planthoppers and green leafhoppers occurred for the first time and were followed by epidemics of virus-diseases. Also other pests such as rice gall midge, rice stink-bug and rice stemborer became more harmful than ever before.

Synchronized planting and crop-rotation are now widely practised throughout the main rice production areas of Java and elsewhere, whereby the synchronization of planting time is pursued in irrigation blocks of 300-500 hectares in size. Figures 12.6 and 12.7 depict (I.N. Oka, 1988) a rice cropping system in Central Java, in which a wet-season rice crop is immediately followed by dry-season rice. In fig. 12.6, staggered planting and the use of longer-duration varieties permit a serious build-up of the BPH-population, whereas synchronized planting and the use of a short-duration variety allows one month of fallow in April (fig. 12.7.). The growing of secondary food crops later in the year supresses most rice pests even more (I.N. Oka, 1988).

Owing to strict synchronization of planting and crop rotation, outbreaks of brown planthoppers seem to be under control in Indonesia. Thus, the widely used national cultivar Cisadane, resistant to BPH-biotype 1 and moderately resistant to biotype 2, is still keeping its ground after more than 10 cropping seasons.

* Time of planting

In areas where synchronized planting is difficult to achieve due to water shortage, scarce availability of labour etc., an adjustment in the time of planting to an earlier or later date may help to prevent severe infestation by certain insect pests. For instance, planting of photo-insensitive varieties at the start of the rainy season, allows the crop to pass through the vegetative stage at a time when gall midge populations are still low in the field and are transferring from their wild hosts. However, late planting of rice appears to reduce infestations by white stemborers in Indonesia. The planning of integrated control is often a matter of seeking the best possible compromise between pests occurring in the area, the varieties of rice a farmer wants to plant and other variables.

Usually, farmers have personal experience with the influence of planting dates on pest occurrence in popular rice varieties, and this should be made use of for planning integrated control.

* Sanitation and weeding

Sanitation means the removal of food and sites where pests-organisms can breed and hibernate. This

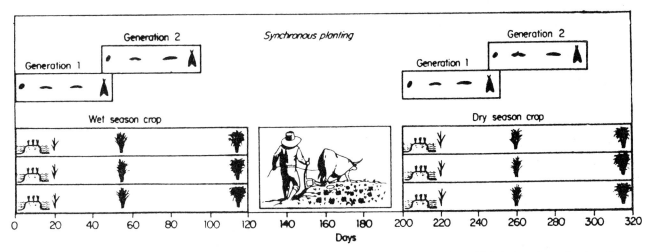

Figure 12.5: By synchronizing the planting of all rice fields in an area within 3-4 weeks, which is less than the time for one stem borer generation, stem borers can complete fewer generations. Stem borers on the first rice crop will not be carried over to the second crop if the stubble is plowed under after the crop is harvested and the ground is left fallow for at least 3-4 weeks between crops. (Photo courtesy of IRRI)

215

Brown planthopper (BPH) population

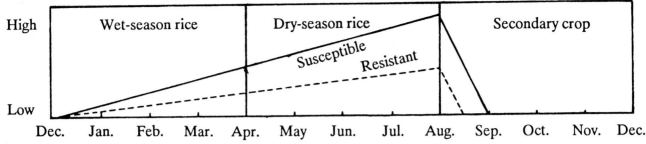

Figure 12.6: Continuous (also called "staggered") planting in well-irrigated areas enables a build-up of Rice brown planthopper (BPH) populations, particularly if susceptible cultivars are used. (Source: I.N. Oka, 1988).

Brown planthopper (BPH) population

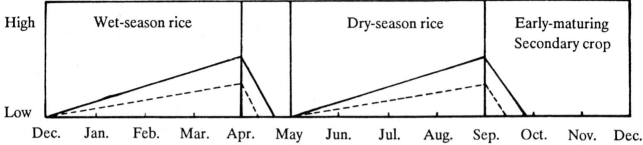

Figure 12.7: A cropping system based on early maturing rice cultivars and synchronous planting in both the wet and dry season proved to be effective in keeping BPH-populations low while still leaving time to plant an early-maturing secondary crop or vegetables. (Source: I.N. Oka, 1988)

requires the study of their behaviour and biology in order to detect the host-range and complete life-cycle. For example, in Japan outbreaks of rice dwarf-virus, transmitted by green leafhopper, were largely controlled within two years by winter-ploughing to control alternative host-weeds such as *Alopecurus* spp.

Well-known sanitation measures are:

- ploughing to bury stubble, ratoon rice and weeds
- cutting of grasses and weeds in adjacent fields and on bunds
- sun-drying of rice straw to kill stemborers and pathogens
- removal of diseased plants or parts thereof.

The burning of plant debris and straw is not recommended, as it destroys many arthropods which assist in the decomposition of organic matter. Burning causes a loss of nitrogen in the plant remains and increases leaking of nutrients from the soil.

Weeding and removal of weeds from bunds and neighbouring fields is effective against rice gall midge, armyworms, cutworms, leafhoppers and stink-bugs. However, the effect of clean-weeding of bunds and fields on the occurrence of spiders, beetles and other beneficial natural enemies has not yet been studied. Weeding of grasses and weeds, which are harmful as competition for nutrients, is a very important cultural method for the protection of productivity. This aspect is discussed in Chapter 14 on weed control.

Removal of diseased plants and plant parts (rogueing) takes away the source of a still limited infestation. For instance, in the case of rice mealybug infestation, the attacked plant(s) should be removed at the first sign of damage.

* Fertilizer and water management

The rate of fertilizer applications has a significant effect on the occurrence of insects, diseases and weeds in rice. Application of nitrogen at high doses appears to stimulate the population growth of rice stemborers, gall midges, rice leafhoppers, brown planthoppers etc. The reason for faster population development may be different for each pest. The increased protein and amino acid synthesis in the rice plant from nitrogen-gifts provides rich nutrition to insect larvae and fertile adults.

Fertilization promotes a lusher growth of improved rice varieties, the wider stems of which allow easier entry for stemborers than do the traditional varieties. Nitrogen-gifts in excess of 90 kg N/ha also appear to increase infestation by diseases such as sheath blight and bacterial blight.

The height of the water and the length of time water remains in the rice paddy also affect the development of the insect population. Rice caseworms, whorl maggots and rice water weevils and brown planthoppers are pests of wetland rice environments, which get worse if the fields remain continuously flooded; green leafhoppers also seem to favour fields with standing water.

Draining all water from the paddy for several days is therefore an effective way of controlling such pests to a large extent. Raising the level of irrigation water periodically drowns the eggs of brown planthoppers and stemborers deposited at the base of tillers and in leaf sheaths. However, the yellow stemborer, being a pest of deep-water rice, is not affected by flooding. The practice of draining and flooding as a means of control, presupposes that water is permanently and plentifully available throughout the cropping season.

* Spacing

The spacing of plants at the time of sowing or transplanting may increase or decrease the density of certain insect pests later on. For example, close planting of rice at 10 x 10 cm distances seems to result in a much larger number of brown planthoppers per plant than wider spacing at 50 x 50 cm (IRRI, 1972). Close spacing provides a moister, cooler and more shaded micro-climate in which brown planthoppers thrive, but most natural enemies do not. Close spacing, causing a thick leaf canopy, also hinders the application (spraying) of insecticides against brown leafhoppers, as the spray should hit the base of the plants where these insects stay and feed.

On the contrary, adult flies of the rice whorl maggot actively search for rice fields with standing water that reflects the sunlight. If the canopy closes too rapidly, the flies would not be able to locate such fields any more. Thus, direct seeding rather than transplanting is recommended for control of rice whorl maggots in heavily infested areas.

Biological control of rice insect pests

Natural enemies, such as parasites, predators and pathogens, play a major role in the regulation of rice insect-pests. The majority of the parasites of rice pests are insects belonging to the orders Hymenoptera and Diptera, but also some nematode species show parasitic behaviour. Eggs are the principal target for parasitization, but also the larval, pupal and adult stages are exploited as

The small red ant, Solenopsis geminata *(F), preying on the eggs of the Malaysian black bug,* Scotinophora coarctata *(F). (Photo courtesy of IRRI)*

Larva of coccinellid beetle, Harmonia octomaculata *(F), preying on a brown planthopper nymph. (Photo courtesy of IRRI)*

sources of nutrition for the parasites' offspring.

Major groups of predators of rice pest insects are spiders, dragonflies, damselflies and certain species of ants, bugs, beetles, grasshoppers and crickets. Many other arthropod species also prey on rice insects, but play a minor role as control agents. Predators usually have a wide range of prey species and may even cannibalize their kin, especially when prey densities are low. As predators frequently attack beneficial insects too, it is very difficult to assess their economic importance. In this respect, it is easier to attach an economic value to parasites, because these behave more selectively and sometimes are species-specific.

The pathogens which infect insects include viruses, fungi, bacteria and protozoa. Some pathogenic bacteria are cultured and available as commercial formulations, e.g. *Bacillus thuringiensis*. Another species of bacteria, *Pseudomonas aeruginosa*, shows high efficacy against caterpillars, planthoppers and rice seed bugs, but probably cannot be used as it causes serious respiratory and wound infections in man and animals.

Fungi infect insects by penetrating the integument and filling their body-cavity with hyphae. In a later stage, hyphae also overgrow the entire insect body and begin to produce conidia or other fruiting-bodies. The coat of fungal mycelium makes the infected insect look like a mummy and the body parts are often locked in a distorted position.

Of all pathogenic infections, one observes infection of rice pests by fungi most often and particularly at the height of the rainy season with heavy fog covering the fields in the morning. The following identifications of fungi are most common:

Erynia delphacis Humber . . . . on green leafhoppers
Hirsutella citriformis Speare . . . . . on rice stemborers
Paecilomyces farinosis Brown . . . on black rice bug

As nature does not make a distinction between injurious and beneficial insects as man does, one frequently also finds spiders (*Clubiona* spp.) covered by mycelium of the fungus *Gibellula leiopus* Mains. This is probably the main fungus species attacking spiders throughout Asia.

Nematodes as parasites of rice insect pests are regularly encountered in rice stemborers. In Peninsular Malaysia, up to a 10-per cent nematodal infestation of *Scirpophaga* stemborers is observed each year. But larvae and adults of other species may be attacked as well. Nematodes are taken in by their host by ingestion and they may multiply within the insects.

Viruses seem to kill many kinds of insects, but few cases are reported on rice insect pests. In Japan, two viruses have been isolated from *Chilo suppressalis* stemborers, namely *Chilo irridescent* virus and *Granulosis* virus.

The families of wasps and true flies listed in table 12.3 comprise many effective parasites. Dryinid wasps are particularly effective control agents, because their reproductive potential is high and a single individual may parasitize many leafhoppers and planthoppers. Some species of the large *Encyrtidae* family reproduce polyembryonically whereby up to 3,000 individuals may develop in a single host. One encyrtid species has been found to parasitize up to 26 per cent of the eggs of rice seed bugs (*Leptocorisa* spp.) Trichogrammatid wasps are common egg-parasites of Diptera, Homoptera and Lepidoptera, to which belong quite a number of pests on rice, maize, sugarcane etc. Trichogrammatid wasps are mass-reared in several countries for use in large-scale integrated control programmes.

The families of parasitic wasps are usually quite large and only a few species have been recorded on rice pests. The order Hymenoptera includes other wasp families not mentioned above, which are either hyper-parasites or predators, or both. For instance, members of the *Ceraphronidae* family are minute wasps living as hyper-parasites (or secondary-parasites) on larvae and pupae of chalcid, dryinid and braconid wasps. Such wasps probably deserve to be regarded as harmful in terms of biological control.

Several non-parasitic wasps are commonly found in sweep-net catches made in rice. Two species, which belong to the *Vespidae* family, build small nests in grass vegetation or ripening crops. These wasps prey on young caterpillars which they feed to their offspring. A third wasp, belonging to the *Sphecidae* family and commonly found around villiges, preys on beneficial spiders and is thus indirectly harmful.

Members of the tachinid family which resemble the housefly comprise many parasites. The adult flies feed on honey or dead material and lay eggs on the skin or inside lepidopterous caterpillars or other hosts. Some species lay their eggs on leaves to be ingested by their hosts; the parasitic larvae develop inside the host. The big-headed, large-eyed pipunculid flies are endoparasites of plant- and leafhoppers (see table 12.3).

Predators are generally not very selective, but may attack several types of insects and other arthropods. They usually feed on all stages, i.e. egg-masses, larvae and adults and some species may also consume dead matter (scavenging) and vegetal matter. Cannibalistic behaviour is also not uncommon. Some families are more effective as biological control agents than others. For example, the carnivorous carabid beetles at night actively run up and down rice leaves in search of leafhoppers and plant-hoppers; they are often quite numerous. But ants, which stay in dry fields and among ripening crops, kill both noxious and beneficial insects indiscriminately.

Spiders

Spiders also are included under the phylum Arthropoda (joint-legged vertebrates), but belong together with the mites and ticks to the class Arachnida. The arachnids are distinguished from the class Insecta by their development of eight legs in the adult stage. Arachnids are further characterized by the absence of wings, antennae and distinct head development; head and thorax being joined to a so-called cephalothorax. The sac-like abdomen is usually unsegmented and, in spiders only, separated from the cephalothorax by a constricted middle-part. Most spiders have eight simple eyes. The jaws are provided with fangs, which are connected to poison glands. The tip of the abdomen is equipped with spinnerets which issue strands from silk-glands.

Spiders prey on insects which they paralize with their poison. The poison also has a digestive action; it softens the tissue. Some spiders suck out the body juices from their prey. Cannibalism occurs frequently among spiders and they may also catch many beneficial insects.

Table 12.4: Important predators of insect pests in rice

Order: Odonata		
- *Libellulidae*	dragonflies	stemborers, leaffeeders
- *Agrionidae*	damselflies	planthoppers
Order: Coleoptera		
- *Coccinellidae*	ladybird beetle	planthoppers, leafhoppers
- *Staphylinidae*	rove beetle	aphids, thrips, egg-masses
- *Carabidae*	ground beetles	
Order: Orthoptera		
- *Tettigonidae*	grasshoppers	polyphagous
Order: Hemiptera		
- *Pentatomidae*	shield bugs	soft-bodied insects
- *Miridae*	plant bugs	planthoppers, leafhoppers
Order: Hymenoptera		
- *Formicidae*	ants	polyphagous (dry fields)
- *Vespidae*	wasps	caterpillars
Order: Dermaptera		
- *Chelisochidae*	earwigs	polyphagous (dry fields)

Wolf spider Lycosa pseudoannulata *(Boes. and Str.).*

Male and female orb weaver spider, Argiope catenulata *(Doleschall), with grasshopper prey.*

Lynx spider, Oxyopes javanus *Thorell. (Photos courtesy of IRRI)*

However, they play a world-wide role in the control of noxious insects in agricultural crops. There are two mechanisms for catching their prey. One group of spiders spins a web to catch their prey, whereas the others hunt their victims down. The web-making spiders cannot be selective, but the hunters can search or wait for their favourite prey.

Table 12.5: Spider families recorded as important predators of rice insect pests

Order : Araneida

- *Lycosidae*	wolf-spiders	leafhoppers, planthoppers
- *Oxyopidae*	lynx-spiders	planthoppers, leafhoppers, borer moths, rice seed bugs
- *Araneidae*	orb-weavers	web-weaving
- *Tetragnathidae*	long-jawed orb-weavers	web-weaving and hunting, leaf-hoppers, plant-hoppers
- *Salticidae*	jumping spiders	adult flies, hopper species

Chemical control of rice insect pests

The introduction, in rice as well as in other crops, of improved varieties and the intensification of production through higher investments in irrigation and drainage, quality seed, fertilizer and mechanical equipment have been accompanied in general by a worsening of the pest situation. For many years, farmers have tried to eliminate pests with the help of widely available, inexpensive synthetic pesticides and they often achieved remarkable temporary success. Gradually, however, problems related to pest resurgence owing to pesticide-resistance in insects, to breakdown of varietal resistance because of the development of new biotypes of pest insects, to the build-up of secondary pest populations and to contamination of the environment, are putting an end to the era of indiscriminate pesticide use.

Economic threshold level

To date, the concept of pest management has been propagated for major food, cash and tree crops. Its goal is to reduce pest populations to levels that are uneconomical to control. As a yardstick on which to base decisions whether or not to intervene with a pesticide application, the concept of "economic threshold" is being introduced (see Chapter 1). Economic threshold is defined as the pest population at which control measures should be taken to prevent pests from inflicting crop losses costing more than the control measures.

Economic thresholds should be established by research institutions for individual pests and locations, and preferably the type of farming system, such as subsistence farming or market-oriented production, should also be taken into account.

The agricultural extension services should teach farmers how to use the established economic threshold values as indicators of the need to commence the application of pesticides. It should also be emphasized that applying pesticides to control injurious insects unavoidably also kills their natural enemies. Therefore, applications of insecticides should be minimized in terms of both size of area treated and number of applications made. Frequently, spot treatment of the parts of the crop actually infested can replace blanket spraying; but

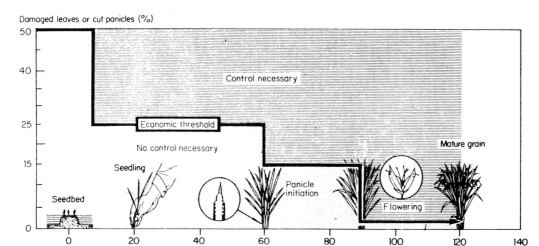

Figure 12.8:

Economic threshold levels at various growth stages of the rice plant for armyworm and rice earcutting caterpillar. Determine the percentage of damaged leaves or cut panicles and apply insecticide when the economic threshold is reached.

adequate dose rates and properly calibrated equipment must always be employed.

However, insecticides need to be applied preventively, to protect rice plants against soil pests such as white grubs, and dipterous pests such as rice seedling maggot and gall midge. If virus diseases transmitted by green leafhoppers are prevalent in an area and a susceptible variety is planted, the crop must be protected in the seedbed and after transplanting. Protective applications can consist of applying systemic granules to the soil, soaking the roots of seedlings in an insecticide solution before transplanting, or seed treatment.

Preventive application is generally not needed if resistant varieties are used. One should, however, be aware of the development of new biotypes of virus-transmitting insects capable of breaking down the resistance in the plant.

Insecticides for use in rice

Over the years, a wide range of insecticide formulations have been recommended for use in rice, many of which had a broad spectrum activity and high toxicity to man or non-target animals. To date, recommendations focus on a few insecticides of acceptable toxicity and persistence, which have a proven effectiveness in controlling insect pests in rice, and discourage the use of other pesticides. Some countries have already taken regulatory steps by banning insecticides for use in rice, primarily because of their hazard to the natural enemies of rice pests and to fish (see Chapter 1).

In a series of lessons on "Control of insect pests in rice", produced in 1979 by the Bureau of Plant Industries of the Philippines and the International Rice Research Institute, about 25 insecticides were mentioned for use on rice. Nowadays, IRRI-scientists are of the opinion that only nine insecticides would be enough to control all major insect pests of rice, provided they are properly applied in an integrated approach to pest control (see table 12.6).

Table 12.6: Recommended insecticides for control of rice pests

	1	2	3	4	5	6	7	8
Sprayable insecticides								
BPMC	*						*	
chlorpyrifos		*				*		
malathion	*							
MICP							*	
monocrotophos		*	*			*	*	*
phosphamidon								*
triazophos		*			*			
Granular insecticides								
carbofuran			*	*				
diazinon						*		

Notes: Insect pests:
1. *Rice caseworm*
2. *Armyworms/cutworms*
3. *Rice seedling maggot, whorl maggot*
4. *Rice gall midge*
5. *Leaffolder*
6. *Rice stemborer-complex*
7. *Rice leafhopper/planthopper-complex*
8. *Rice bugs*

Scouting

Decisions about whether or not to apply insecticides should be based on careful estimates of the rate of infestation in the field concerned. The activity of determining the rate of infestation by counting the number of insects present or degree of damage inflicted is called scouting. It may involve one of the following methods:

* Select both once a week and randomly 20 rice plants (hills) from the field while walking diagonally across it, and record either the number of insects or damaged plant parts per hill.

220

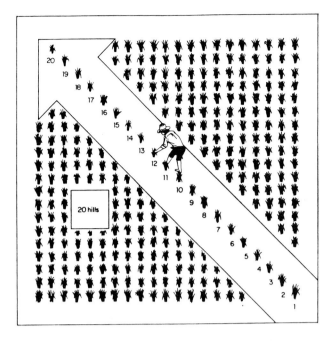

* Sample highly mobile insects such as leafhoppers and rice seed bugs using a sweep-net and record the number caught per 10 sweeps.

* Monitor adult insect activity at night by means of attraction to a light trap and record number caught.

* Monitor adult insect activity by means of a pheromone bait trap.

The counting of insects present may include the recording of egg masses, larvae and adults or adults only, depending on the type of insect one has to examine on the leaves, tillers or roots; the latter may require pulling up of the whole plant from the paddy field. Counting of the number of damaged plant parts usually involves randomly selecting five leaves or panicles per 20 hills, each taken across the field, and recording the number damaged.

The interval for scouting is once a week during the period a pest is likely to infest and damage the crop. This period may be the time on the seedbed and vegetative phase, or the reproductive phase, or some shorter interval thereof, depending on the pest and local conditions.

For instance, in the case of rice stemborers, the rate of population build-up is determined by the survival of eggs from parasites and the speed with which first instar larvae can hide within the tillers from parasites and predators. First instar larvae, however, cannot enter tillers which are either too narrow or have hardened tissues. Actually, only during the intervals of tiller elongation and panicle emergence are the stems wide and soft enough to enable penetration, which leads to dead-hearts and white heads, respectively. Consequently, fields need only to be scouted during these vulnerable growth stages: tiller elongation and panicle emergence.

Chemical control of stemborers should be timed accordingly. Insecticides should either be sprayed or applied as granules to the standing water during the tillering phase or be sprayed only during the interval from panicle initiation to flowering. At other periods of crop growth, no insecticides should be used against stemborers in order to leave full opportunity to natural enemies to destroy eggs and larvae.

Even when the economic threshold for stemborers is reached, one should not spray immediately, but first determine the rate of egg parasitization. To this end, collect egg-masses and keep them in jars until larvae and parasites emerge concurrently. If more larvae than parasites emerge, then spray before the larvae in the field can enter into the tillers. If more parasites than larvae emerge, there is no need to apply insecticides.

If insects are moving away when plants are examined, it may help to hit the plants several times with the hand and to count the number of insects that fall into the water. This method can be used for brown planthoppers and rice leafhoppers.

Scouting for brown planthopper infestation should be carried out weekly from the seedbed to dough grain stage. Only mature nymphs and adults (both brown coloured) are counted, not the whitish immature nymphs also falling into the water. When on average one mature nymph per tiller is counted, the scouting needs to be repeated after three days. If the population of mature nymphs increases more, spray the base of the plants. Scouting for signs of damage to plant parts should take into consideration the time span between infestation or feeding by the insect and the expression of symptoms.

Damage due to rice whorl maggots, i.e. when the tissue of unopened leaves has been eaten, appears too late to serve as an indicator for insecticide application. For this and other insect pests with delayed damage, it is recommended that adjacent fields planted 1-2 weeks earlier be checked. Since whorl maggot adults (= dipterous flies) are attracted to young widely planted fields with standing water, there is no need to scout for this pest in rice fields with densely closed leaf canopy.

The weekly scouting for the various insect pests occurring in the area can be combined in one exercise and the separate data entered on the same recording sheet. The numbers found will tell which pests need more frequent sampling or chemical control.

The damage from leaf-eating caterpillars should, however, be combined, as it is impractical to record early vegative damage from each leaf-feeding pest separately.

Sampling of seedbed and field with a sweep-net is effective when dealing with highly mobile pests such as leafhoppers. The standard sweep-net is 15 inches in diameter; it should be used in a swinging movement brushing about 25 cm deep through the top of the leaf canopy. In rice and other solid plantings, a sweep follows a 180°-path like a pendulum.

In row crops, a sweep is one pass of the net across the plants on a single row. Make 10 sweeps while walking a diagonal line across the field. If scouting for green leafhoppers, take sweep-net samples twice a week from the seedling stage to panicle initiation and count both the nymphs and adults present.

A kerosene lamp with dead moths floating on water surface of pan. (Courtesy of IRRI).

Many adults insects are attracted to a light trap. Some species are caught during a new moon, whereas others are caught during dark nights.

Light traps and pheromone traps

A light trap attracts many adult moths, bugs, beetles and other insects at night and can therefore be employed to monitor the build-up of their populations. The attracted insects are scorched by the heat of the kerosine light and fall into a plate with water placed underneath. The numbers and species caught at night can be recorded the following morning. Some species are more active on nights of the full moon; others on nights of the new moon. Catches of adults correspond to such phases of the moon.

Full moon catches are high for mole cricket, rice planthopper and leafhopper, rice gall midge and rice black bug. New moon catches are particularly high for noctoid moths (army- and cutworms) and pyralid moths (stemborers, caseworm). Not attracted by the light trap are dipterous pests such as the flies of rice seedling maggot and whorl maggot, and thrips.

Monitoring with light traps should be done throughout the interval in which important pests are endangering the rice crop. Flight activity of rice gall midge adults should be recorded from the seedbed period until panicle initiation. Insecticides should be applied after peak light-trap catches.

The fact that rice planthoppers and leafhoppers are attracted to light implies that, in areas with prevalent virus-diseases, the rice seedbeds should be placed away from houses or other illuminated places where virus-carrying rice hoppers may congregate.

Pheromone traps holding synthetic female sex hormones for attracting male adults have been developed for a number of insect-species (see also Chapter 3). In rice, a pheromone trap is recommended for the monitoring of rice leaffolder moths, in which the pheromone held in a slow-release dispenser is suspended over a tray with soapy water. If moths are caught, field sampling for rice leaffolders should be increased to twice a week, examining five leaves of each of 20 plants taken across the paddy-field.

Table 12.7: Economic thresholds for major rice pests

Inspect your field once a week during the stages from seedbed to ripening (hard dough). Walk diagonally across the field to identify the insect pests at the middle as well as the edges of the field. Apply insecticides when threshold is surpassed, unless otherwise recommended, using the formulas below:

Rice insect pest	Threshold value
Root weevil	- 10 per cent infested hills
Seedling maggot	- 10 eggs per 20 hills (from a neighbouring older field)
Whorl maggot	- 2 eggs per hill
Gall midge	no threshold
Caseworm	- 50 per cent damaged leaves (seedbed)
	- 15 per cent damaged leaves (after transplanting)
Cutworms, armyworms	- 25 per cent damaged leaves (up to 60 days after seeding)
	- 15 per cent damaged leaves (from panicle initiation to flowering)
	- 2 per cent cut ears (ripening stage)
Leaffolder	- 15 per cent damaged leaves (from transplanting to appearence of flag leaf)
	- 5 per cent (from appearence of flag leaf to flowering)
Stemborers	- 2 egg masses per 20 hills (from seedbed to panicle initiation)
	- 1 egg mass per 20 hills (panicle initiation to flowering)
Brown and white-backed planthopper	- 1 mature nymph per tiller (seed-bed to ripening stage)
Green leafhoppers	- sweep-net; 2 hoppers per sweep (seedbed to panicle initation)
	- tapping; 5 hoppers per hill
Rice earbug	- 10 bugs per 20 hills
Hispa	- 4 adults and larvae per hill
Black bugs	- 5 adults and nymphs per 20 hills

Chemical control at various stages of rice growth

● On seedbeds

A paddy seedbed rarely exceeds an area of a few hundred square metres and is easy to inspect for insect damage. The most common pests of seedbeds are leaf-feeding insects. Generally, they do not pose a real threat, because paddy seedlings can tolerate much damage either from insects or rough handling. At the time of transplanting, seedlings are pulled up from the seedbeds, bundled and tossed around; many farmers also cut off the tops of the leaves. Yet the seedlings recover well and rapidly. It is recommended that the seedbed be inspected twice a week for insects and that no insecticides be applied unless a definite pest infestation is building up. Such a build-up can often be prevented by searching and destroying eggmasses with the help of children.

Spraying against leaf-feeding worms or caterpillars should be carried out only after 50 per cent of the leaf area has been destroyed by the combined feeding of all kinds of caterpillars.

Spraying against hoppers will be needed only if non-resistant rice varieties are grown and an infestation is building up in the area. The criterion for commencing chemical control of hoppers is the presence (on average) of more than one adult of the brown planthopper (*Nilipanvata lugens*) per tiller or more than two green leafhoppers (*Nephotettix* spp.) per sweep with a sampling net.

The insecticides which may be used to control leaf-feeding caterpillars or hoppers on the seedbed are listed in table 12.6.

● During the vegetative growth stage

After a rice plant has been transplanted and has established itself, it usually passes through a stage of vigorous growth during which many tillers are formed. Any loss of tillers owing to an early stemborer attack is normally compensated by the growth of additional strong tillers to the extent that the crop may ultimately produce an even higher yield.

The insect pests that can cause economic yield loss from the time of transplanting until panicle initiation in flooded fields, are leaf-folders, stemborers, armyworms, cutworms, leaf and plant hoppers, caseworms and whorl maggots. With the exception of the latter two, all these are also pests of dry rice fields.

These insects can be controlled by the appliciation of granules or by foliar sprays. Against the brown planthopper and green leafhopper, spraying of systemic insecticides such as BPMC and MICP is effective; dusts may also be applied against leafhoppers. Granules generally show better efficacy than sprays against white grubs, boring and mining insects such as stemborer larvae,

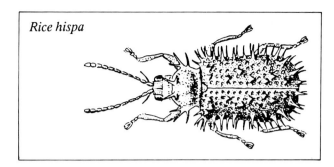

Rice hispa

whorl maggots, gall midge and grubs of the rice hispa. In particular, systemic granules, after being absorbed by the plant roots, are transported with the sap-stream up to the leaf tips. In this way, systemic granular insecticides, such as carbofuran (Furadan and Curaterr), can provide good protection against the whorl maggot and rice gall midge.

The non-systemic granule Diazinon G gives control of whorl maggots and stemborer larvae in the rice-stem, but not of leaf-folders and leafhoppers. Against seedling maggots, granules are inefficient because of the high dose required. Soaking seedlings in systemic insecticide solution is recommended for seedling maggots, whorl maggots and the rice gall midge.

It is recommended that systemic granules be harrowed into the paddy soil prior to transplanting. Carbofuran granules are less effective if broadcast into the paddy water. Non-systemic granules do not have to be incorporated into the soil, but should be broadcast into standing water in the rice field. The reason for this is its mode of action. The active ingredient in these granules floats on the water and moves up the plants in films of water behind the leaf sheaths where the insects reside.

● During the generative growth stage

From the time of panicle initation and booting of the rice crop onwards, there are only a few insects that can cause serious damage. At this stage of growth, the farmer should inspect his field for damage from leaffolders, armyworms, cutworms and rice seed bugs.

Armyworms and cutworms are not common at this stage, except for ear-cutting caterpillars (*Mythimna*) in (deep water) Aman rice throughout South Asia. If panicle cutting is observed, the farmer should spray at once. Volatile formulations such as dichlorvos (e.g. Nogos, Dedevap, Vapona) sprayed in the late afternoon, are very effective against this pest. Spraying against other armyworms and cutworms should start only if 15 per cent of the leaf mass has been removed by feeding.

Rice bugs are a typical late season pest in the sense that they cause harm to the rice grains only for about three weeks from post-flowering to the hard-dough stage of the grains. The yield loss resulting from a rice bug infestation is usually low, because the sound grains in a panicle tend to grow larger if other grains are damaged. Spray with monocrotophos or phosphamidon when the threshold level of 10 bugs per 20 hills is reached.

Integrated control of stemborers in rice

● Insecticide application. *Because stem borer larvae enter tillers, controlling them with insecticide is often difficult.*

1. Tillering stage. If fields are flooded during the tillering stage, sprays and granules are equally effective.

If the paddy water depth is less than 5 cm, sprays can be used.

2. Panicle initiation to flowering. After the crop reaches maximum tillering, granules are not effective and only spray formulations should be used (see fig. 12.10).

● Scouting. *The key to the establishment of stem borers in a crop is the survival of eggs -- mostly the net result of parasite activity -- and the ability of the first-instar larvae to quickly bore into tillers, seeking shelter from predators and insecticide.*

Most first-instar larvae succumb to predators because the tillers of seedling rice are too thin to be entered. But small-mandibled first instars can readily enter wider diametered, older tillers which are actively elongating, particularly under heavy fertilization. After the elongation phase, tissues become more densely packed and hardened, prohibiting entry to most first-instar larvae.

Larval penetration again becomes easier into the soft tissues of the bases of elongating panicles. Successful entry during panicle extension leads to whiteheads (see fig. 12.11).

Fields should only be scouted during the two most vulnerable growth stages: tiller elongation and panicle exsertion (see fig. 12.12).

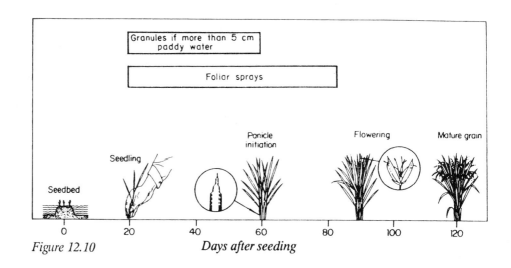

Figure 12.10   Days after seeding

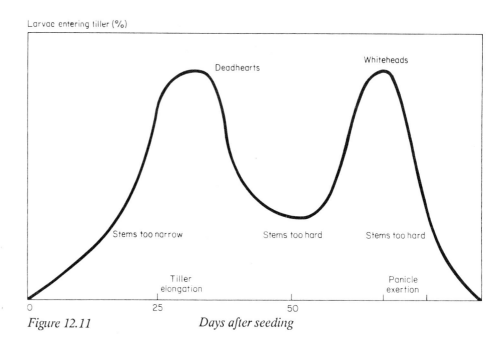

Figure 12.11   Days after seeding

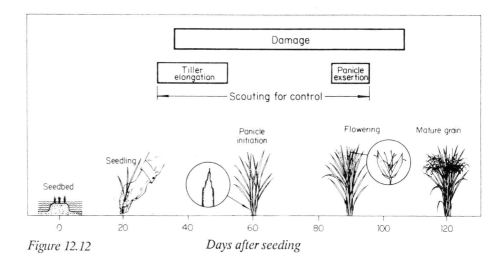

Figure 12.12   Days after seeding

(Courtesy of IRRI)

224

A stylized drawing of a tarsonemid or true spider mite adult. The male and female differ distinctly in body shape but each has four pairs of legs.

MITES

Morphology and pest status

Mites are closely related to insects, but belong together with ticks and spiders to a different class of arthropods, the Arachnida. This class is characterized by the absence of wings and antennae, and by the joined cephalothorax. This frontal part and the much larger unsegmented abdomen together form an oval body. The adults generally possess four pairs of articulate legs. Other characteristics are:

Spiders: See above section about biological control for their morphological description.

Ticks: Usually parasites of mammals and birds, their bite is injurious and may trigger serious reactions in the host. Also, ticks may transmit serious diseases to man.

Mites: Small, soft-skinned organisms with a chitinous skeleton, they occur almost everywhere as scavengers, parasites of plants and animals, or as predators. They rarely exceed a size of 0.5 mm; many are of microscopic size. Mites often spin extensive cobwebs, under which they live. In their development from eggs to mature adults, mites go through one or more active larval stages and several moultings.

A hatching mite larva has three pairs of legs, but an adult has, generally, four pairs of legs. An exception are the gall mites (Fam. *Eriophyidae*); both larvae and adults possess only two pairs of legs. Dry and warm weather favour the development of mite populations, whereas heavy rainfall reduces them rapidly, except for Tarsenomid mites.

Many mites are important parasitic pests of fruits, vegetables and field crops. Some species form colonies on the upper surfaces of leaves and fruits; others may prefer the under surfaces. Mites feed by sucking plant cells, thereby causing discolouration and wilting of leaves, early fall of leaves and fruit, and eventually yield loss or crop failure. Gall mites may induce malformations (galls) on shoots, fruits and leaves. Several predacious mites have been detected and are reared as natural enemies. For example, the species *Phytoseiulus persimilis* controls effectively *Tetranychus urticae* and *T. cinnabarinus*, two major pests of glasshouse crops.

Storage mites feed on stored grains, flour and other such food items. They can spoil its quality completely.

The most important mite pests on plants belong to the following families:

● *Tetranychidae* or "true" spider mites

They can be recognized by their oval shape, arched back, red to green colour and very lively, quick movements. In hot regions, they reproduce throughout the year with rapid population build up. In temperate regions, these mites go through a winter diapause.

Examples:
Tetranychus cinnabarinus .... Carmine (red) spider mite
T. urticae (syn: *T. telarius*) ... Two-spotted spider mite

Cosmopolitan species that infect cotton, peanuts, cucurbitaceous plants and many more species of plants and weeds.

● *Tenuipalpidae* or "false" spider mites

In comparison with the "true" spider mites, the "false" ones have a flat body. They are much slower and more sluggish and they develop more slowly. Colour varies from red to green.

Examples:
Brevipalpus spp.	False spider mite.
(*B. phoenicis*, *B. australis* and *B. obovatus*)	Widely distributed throughout tropical zones; they feed on fruits, sweet potatoes, ornamental shrubs, tea and coffee.

● *Eriophyidae* or gall mites

Very small, worm-shaped, brownish bodies studded with microtubes and possessing only two pairs of front legs. Under tropical conditions, more than 20 generations may develop in a year.

Example:
Eriophyes sheldoni .... Citrus bud mite

They are cosmopolitan in the tropics between 35°N and 35°S. A serious pest of all citrus varieties, they cause injury to buds, flowers, fruits, twigs and leaves.

Spider mites in cotton (*Tetranychidae*, *Acari*)

Spider mites are important pests of cotton. They prefer areas where dry warm weather prevails. Symptoms of spider mite attack are found both on young and old cotton plants. The leaves show a white mottling; later they turn yellow to red and shed prematurely. On the lower side of the leaf, white lustrous webs appear; these spread over the entire plant in cases of severe infestation. Plants may die gradually within a few weeks of the appearance of the first perceptible symptoms.

Cause

Damage is caused by several spider mite species of the genus *Tetranychus*. The females lay eggs into the webs on the plant. After 4-10 days, depending on weather conditions, six-legged larvae, 0.15 to 0.2 mm in length, hatch out of the eggs. During their development from the larval to the adult stages, several mobile and quiescent stages alternate with each other. The total period of development of a generation depends on environmental conditions; for example, under favourable weather conditions, this usually takes 7-10 days. In various cotton-growing regions, 12 to 16 generations develop within a single year.

The following species are pests of cotton worthy of note: *Tetranychus urticae* Koch; *T. cinnabarinus* Boisd.; *T. atlanticus* McG.; and *T. telarius* L. In the tropics, *Tetranychus cinnabarinus* is more important than other species; in temperate zones, *T. urticae*.

Tarsonemidae are closely related to spider mites. They damage cotton in a way similar to *Tetranychidae*; however, they do not form webs, but curl and roll up young leaves. The reactions of the plant to infestation by *Tarsonemidae* often comprise the shedding of leaves, flower buds and young bolls. A prominent polyphagous pest of this group is *Polyphago tarsonemus latus* Banks, which also infects tea plants in addition to cotton.

Chemical control of mites and spiders

Mites differ from insects biochemically as well as physically. Consequently, many insecticides do not affect them. This may cause problems when applications of broad spectrum insecticides kill the insect predators of mites, thus allowing them to become major pests. The dangers of upsetting the ecological balance should be appreciated and avoided. This applies also to the selection of acaricides for integrated mite control programmes. For instance, binapacryl and chlordimeform are highly effective in the control of mite eggs and immature stages, but are also highly toxic to predaceous mites. On the other hand, fenbutatin oxide and propargite are of low toxicity or non-toxic to predaceous mites and thus are useful for integrated control. Similarly, dormant oil, used in orchards in the budding stage, controls European red mite, but favours the survival of predaceous mites.

Compounds developed to control mites are called **miticides** or **acaricides**. Some substances are effective against both mites and insects. Spider mites rasp the tissues of leaves and suck up a relatively large volume of cell sap in the process. Consequently they tend to be vulnerable to systemic organophosphorous insecticides, such as dimethoate, omethoate and thiometon as well as to the specific acaricides. The latter have only a very limited systemic action. Spider mites have a great ability to develop resistant strains rapidly; hence, the useful life of a commercial acaricide tends to be short. Their resistance to insecticides has also contributed to outbreaks and it has been reported that certain organochlorines, carbaryl and some copper compounds can even enhance their reproductive capabilities.

Miticides (acaricides)

* Diphenyl aliphatic compounds

These are very effective non-systemic acaricides, with strong ovicidal action:

bromopropylate	chlorobenzilate
chlorbenside	dicofol
chlorfenethol	metoxychlor
chlorfenson	

* Formamidines

These compounds demonstrate high efficacy against most stages of mites and ticks and are valuable for control of mites and insects resistant to organophosphorous and carbamate insecticides. Also, they provide good ovicidal activity:

amitraz	formetanate
chlordimeform	

* Dinitrophenols

Four outstanding, non-systemic acaricides with good ovicidal activity, also showing fungicidal efficacy against powdery mildews, are:

binapacryl	dinocap
dinobuton	dinoterbon

* Organotin compounds

The organotins are selective acaricides and fungicides with a long residual activity:

azocyclotin	fenbutatin oxide
cyhexatin	

* Organosulphur compounds

These compounds combine a high, non-systemic toxicity to mites with a remarkable low toxicity to insects.

Because of this selective action, these miticides are very useful for integrated pest control:

ovex	tetradifon
propargite	sulphur (elemental form)

* Miscellaneous miticides:

benzoximate	hexythiazox
chlofentezine	quinomethionate

In addition to the aforementioned miticides, there are a number of insecticides which demonstrate good efficacy against mites. They belong in particular to two groups of organophosphorous insecticides, namely:

- Hetercyclic organophosphates: Some of these compounds are very effective against mites and soil insects and have persistant activity. Examples are: azinphos-ethyl, chlorpyrifos-methyl, dioxathion, phosmet quinalphos.

- Aliphalic organophosphates: Such compounds having a systematic action are particularly effective against sucking-type insects, such as mites, aphids, thrips and scale insects, and some have nematicidal activity too. Examples are: demeton (-methyl), dicrotophos, dimethoate, disulfoton, omethoate, phorate, phosphamidon, triazophos and vamidothion.

- Carbamates: Some systemic carbamate compounds show high efficacy against mites too. Examples are: aldicarb, carbofuran, mexacarb, methiocarb, oxamyl and thiofanox.

NEMATODES

Nematodes are small, slender, colourless, round-worms of the Phylum: Nemathelminthes; Class: Nematoda. They are very distinct from insect larvae and earthworms; often the name "eelworms" is used to describe them. About 17,000 species of nematodes are known to man but only a fraction of these have been described as parasites either of plants, insects, or other animals and man. Parasitic plant nematodes are usually shorter than 2 mm in length (average 1 mm). Nematodes are present in water and on land; practically every cubic centimetre of arable soil on Earth contains between 20 and 50 nematodes. Some species can survive in dry conditions for a considerable time, but to be active they need moisture. They either live on plant roots as ecto-parasites or enter plant tissues via the roots and become endo-parasites in roots, leaves or stems. They reproduce inside plant tissue. Ecto-parasites (and some endo-parasites) retain their slender body form and mobility throughout life. Thus, they may migrate from one infested root to another at any time. Only the males and second-stage juvenile stages swell up inside plant tissues, thus they are no longer able to move around.

The life-cycle of most plant parasitic nematodes is rather simple. The female lays eggs from which the juveniles hatch; these pass through four moulting stages to become adults. The average life-cycle takes 20 to 60 days, during which time the females of some species produce more than 500 eggs. The eggs may remain dormant in the soil for long periods. Often the secretion of chemical substances from growing plant roots seems to be a prerequisite to start the hatching process.

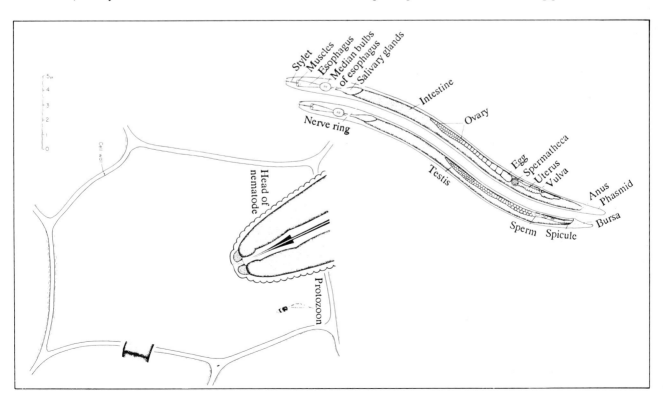

Figure 12.13: Schematic drawing of a nematode penetrating through the wall of a plant cell (below left); male and female adults of a typical plant-parasitic nematode (above right).

The root-knot nematode, Meloidogyne hapla, *causes irregular swellings (galls) on the roots of carrots (right) which disrupt the vascular bundles and nutrient uptake thus leading to poor growth. (Photo courtesy of LNV, the Netherlands)*

Damage by nematodes

All parasitic plant nematodes possess a piercing mouth-piece (stomato stylet) which is used to pierce plant cells and suck their contents. Some species spend their entire lifetime inside plant roots or bulbs. Other species remain in the soil and feed on the epidermis of the roots whereas still others pass from the soil into stems, leaves or flowers. Generally, nematode attacks cause poor crop growth without characteristic symptoms. Nematode damage impairs the proper functioning of the root tissues which may result in yellowing of foliage, rapid wilting in hot weather, decreased drought tolerance and reduced growth and ability to yield. If symptoms of damage become visible, however, it is sometimes possible to identify by group which type of nematode is causing the injury. The most common symptoms are:

Root gall:	malformation of the root; the root-knots may or may not enclose the nematodes
Root lesions:	destroyed layers of cells, which may be invaded by other pathogens
Root rot:	general decay of the root
Excessive root branching:	formation of numerous short lateral roots
Leaf, stem and flower damage:	gall formation; distortion of leaves and stems, abnormal development of flowers or fruits; rot.

Nematode-infested plants become more susceptible to other pests, especially under poor growing conditions. It has been shown that cotton plants and tomatoes, for example, are less resistant to Fusarium wilt disease if the plants have previously been attacked by root-knot nematodes (*Meloidogyne* spp.).

Also some cases of association between nematodes and bacteria have been detected; for example, tundu disease in wheat in India is thought to be caused by *Corynebacterium tritici* and spread by the wheat gall nematode (*Anguina tritici*).

Parasitic plant nematodes of three genera, *Xiphinema*, *Longidorus* and *Trichodorus*, are the vectors of plant viruses. Outbreaks of nematode-borne virus diseases in non-irrigated fields typically occurs in patches which expand slowly because of the very slow movement of nematodes in the soil. Nematodes may spread rapidly with water in irrigated fields; for example, tobacco-rattle virus in tobacco and various other crops, that is transmitted by stubby root nematodes (*Trichodorus* spp.), spreads rapidly in this way.

It is realized nowadays that parasitic plant nematodes cost farmers all over the world several thousands of millions of dollars annually in crop losses and higher production costs. Some important nematode-induced diseases in the Asian and Pacific region are:

Disease	Nematode
Rice	
White tip disease	Rice leaf nematode (*Aphelenchoides besseyi*)
Ufra disease	Rice stem nematodes (*Ditylenchus angustus*)
Unnamed (also found in maize, sugarcane)	Rice root (lesion) nematode (*Hirschmanniella oryzae*)
Wheat	
Ear cockles disease	Wheat gall nematode (*Anguina tritici*)
Coconut palm	
Red ring disease	Coconut palm nematode (*Rhadinaphelenchus cocophilus*)
Citrus	
Slow decline disease	Citrus nematode (*Tylenchulus semipenetrans*)
Banana	
Blackhead toppling disease or banana root rot	Burrowing nematodes (*Radopholus similes*)
Vegetables, potatoes, tea, pepper, ginger, cardamon etc.	
Root gall disease	Root-knot nematodes (*Meloidogyne javanica* and other *M.* spp.)

Classification

A simple, convenient scheme for classifying parasitic plant nematodes is based on the part of the plant affected, i.e. roots, stem, or leaf and buds, and the life-cycle of the nematode:

Table 12.8: Classification of important parasitic plant nematodes

Sedentary root nematodes:	
Cyst-forming nematodes	- *Heteroderinae* spp.
Root-knot nematodes	- *Meloidogyne* spp.
Migratory root nematodes:	
Root-lesion nematodes (endoparasities)	- *Pratylenchus* spp., *Radopholus* spp.
Spiral nematodes (ectoparasities)	- *Helicotylenchus* spp.
Spiral nematodes (ectoparasities)	- *Trichodorus* spp.
Dagger nematodes (ectoparasities)	- *Xiphinema* spp.
Stem nematodes	- *Ditylenchus* spp. *Rhadinaphelenchus* spp.
Bud and leaf nematodes	- *Aphelenchoides* spp. *Anguina* spp.

Principles of nematode control

If a serious infestation by nematodes is suspected based on symptoms of plant damage, one should first carefully identify the nematodes involved. Discovery of nematodes around the roots of an unhealthy plant does not, by itself, prove that the condition is caused by nematodes, since all soils contain large numbers of harmless nematodes. Moreover, different species of parasitic plant nematodes require different control measures.

A careful job of sampling can save much time and money. It may be that only a portion of a field requires treatment and, by proper sampling, the infested area can be isolated for such treatment. A large field may be divided into easily treated blocks and random samples taken from the blocks. Small amounts of soil at a depth of 5-20 cm (2-8 inches) should be taken from at least 25 random locations per hectare or 10 per acre. A soil sample taken from one hectare should total about 1.5 litres of soil (1 pint per acre).

It is sometimes advisable to sample an area of poor plant growth to determine if nematodes are the cause. Nematode damage in fields usually appears as a "bull's eye" target-spot with depressed plant growth at the centre. Generally, populations will be lowest in the centre of such a spot and highest where there is more abundant food. Thus, samples should be taken from the edges of target spots. In established crops, samples should be taken from zones where feeder roots predominate.

The composite sample should be placed in a plastic bag, securely closed and labelled on the outside as to the location of the sampling. Samples may be forwarded to a competent institute of nematology. If none is known, the department of agricultural extension may be asked for advice beforehand.

Interpretation of findings

It is difficult to make a general recommendation as each situation usually requires individual attention. A parasitic plant nematode causes economic damage only when the population density exceeds the tolerance level of the crop. This level, or tolerance threshold, varies according to the host crop and other factors such as moisture, temperature, soil fertility and soil type. These conditions vary from season to season and field to field. It is, therefore, unwise if not impossible to establish an arbitrary level for the control of a certain number of nematodes per volume of soil.

Methods of nematode control

There are several ways in which a nematode problem may be solved. Control often involves the integration of different methods. While chemical control is of primary concern, sound practices which prevent the introduction and spread of parasitic nematodes must be incorporated into a control programme, otherwise chemical treatment may be a costly, but useless exercise.

- **Sanitation and quarantine:** Most species of nematodes spread through the soil at fewer than 30 cms per month, but they are easily relocated by any outside moving force. The most important means of spreading nematodes are transplants, farm equipment and drainage water carrying particles of soil. Transport of farm produce, nursery plants and topsoil may carry nematodes over long distances; thus, sanitation and quarantine must be recommended as a control method.

- **Rotation:** Alternating crops with non-host crops may prevent reproduction of parasitic plant nematodes by denying them access to plants on which they can

The wheat gall nematode, Anguina tritici *(Steinb.), by feeding on the vegetative tips, produces stunted and twisted leaves and stems and in a later phase they infest the grains. (Photo courtesy of PD, the Netherlands)*

reproduce. Rotation with a resistant crop is powerful tool both as a means of controlling a bad infestation and as a means of preventing a rapid build-up of a harmful population.

- **Flooding**: Inundation of the soil over a period of 4-6 weeks strongly reduces the populations of *Pratylenchus*, *Rotylenchus*, *Ditylenchus* and *Trichodorus* species. It is also effective in controlling various soil-fungi such as *Rhizoctonia* and *Sclerotinia* and many weed species.

- **Chemical control**: Nematode populations can be greatly reduced by fumigating the soil with fumigant chemicals or by incorporating nematicides in the soil. These chemicals also frequently kill many types of soil-living fungi and weed seeds. Non-fumigant granular nematicides can be broadcast or applied to row-planted crops by means of mechanical spreaders or injectors or simply by hand. In all cases, great care must be taken to protect the applicator from contamination as nematicides and fumigants are highly hazardous chemicals. The granules should be incorporated into the soil e.g. with a rotary hoe for better activity. The active ingredient is released from the granules by up-take of soil moisture. It subsequently affects, by contact action, the nematodes also living in the soil-moisture. After up-take by the roots of plants, there may be an additional systemic activity. In both cases, nematicides cause a paralyzing effect that does not kill nematodes directly but impedes feeding.

Well-known non-fumigant nematicides, or insecticides with strong nematicidal activity, are:

Non-systemic organophosphates
    dichlofenthion
    ethoprophos - Mocap, Prophos
    triazophos - Hostathion

Systemic organophosphates:
    fenamiphos - Nemacur
    isazofos
    phorate - Thimet

Systemic carbamates:
    aldicarb - Temik

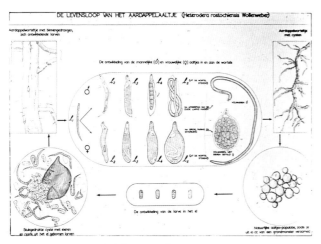

Drawing of Hoplolaimus columbus *nematode, a major endo-parasite on the roots of cotton and soybean.*

Table 12.9: Dose recommendations for non-fumigant nematicides and for fumigants in kg/ha or litres/ha

Nematicide	Sedentary root nematode	Migratory root nematode	Stem-, leaf nematode
aldicarb 10% G	30	30	30-60
oxamyl 10% G	40	50	50
oxamyl 25% WP	20-40	10-20	-
ethoprophos 10% G	100-200	50-100	50-100
ethoprophos 20% G	50-100	25-50	25-50
dichloropropene	150-250 litres/ha for field crops 200-600 litres/ha for horticulture, arbiculture		
dichloropropene/ methyl isothiocyanate	350-850 litres/ha		

aldoxycarb	-	Standak
carbofuran	-	Curaterr, Furadan
oxamyl	-	Vydate

Stem- and leaf-dwelling nematodes may sometimes be controlled by spraying with a suitable organophosphate insecticide, e.g. spraying with phosphamidon controls *Aphelenchoides besseyi*, an ecto-parasite nematode which feeds on young leaves of rice. Also, seed dressing with pesticides helps in preventing nematode damage.

Soil fumigants

Chemicals effective against free-living nematodes, cyst nematodes and root-knot nematodes, but with limited or no efficacy against other living soil organisms, are:

    dichloropropene - Telone
    dichloropropene/methyl isothiocyanate
    ethylene dibromide - EDB

Multi-purpose soil-fumigants effective against nematodes, fungi, bacteria and soil insects as well as seeds of grasses and other weeds are:

    chloropicrin - Tri-chlor
    dazomet - Basamid, Salvo, Mylone
    metham sodium - Monam, Vapam
    methyl bromide - Brom-o-gas, Dowfume
    methyl isothiocyanate - Metifume, Trapex

It is very important to remember that chloropicrin, ethylene dibromide and methyl bromide are highly toxic fumigants. Since all the aforementioned chemicals are gases or highly volatile liquids, great care should always be taken in handling fumigants.

The most widely used methods of control of nematodes involve expensive treatment of the soil with live steam or volatile chemicals and fumigants having a

Heavy injector equipment is used to fumigate soil in large-scale potato-production areas in the Netherlands. The soil is compacted and sealed off with plastic sheets to retain the fumigant action for several days. (Photo courtesy of LNV, the Netherlands)

wide spectrum of toxicity. Since many soil nematodes live in the top-layers of the soil, the results of such treatment are generally good. Plants are particularly vulnerable to nematode attacks in the seedling stage or during the period of renewed growth after transplanting. Thus, elimination of nematodes from the initial rooting zone before sowing or planting may give the best protection.

In general, the best time to apply any chemical control is at the end of the dry or rainy season when soil temperature and moisture are moderate, and when there is plenty of time for the nematicide to kill and thereafter disappear from the soil, without leaving residues.

Soil fumigation

Soil fumigation is used to control many pests, including diseases and organisms such as nematodes, fungi, bacteria, insects and weeds. It is of particular importance for the preparation of seedbeds, nursery beds and plots planted with high-value crops such as pepper, ginger and cardamom.

* **How fumigation works:** Fumigation is the application of a volatile material such as a liquid or solid which converts to a gas in the soil. This gas moves through the air spaces or pores in the soil and is absorbed in the water film surrounding soil particles. The fumigant absorption in the soil water is in equilibrium with the

fumigant gas in the air spaces (pores) in the soil. Eventually the fumigant chemical disappears from the soil. But if the fumigant leaves the soil too soon, there will be insufficient exposure to kill the pests.

If the soil is too wet, there will not be sufficient air space for the fumigant to move through the soil. If the soil is too dry, there will be too little moisture around the soil particles to absorb the fumigant and the fumigant, therefore, will evaporate from the soil too quickly.

In heavy or compacted soils, the fumigant cannot disperse and tillage is usually required to condition the soil prior to fumigation. In light sandy soils, there is often too much air space and the fumigant escapes from the soil before the pest is killed unless the soil is sealed or compacted immediately after fumigation.

Organic matter readily absorbs fumigants, but ties them up. If there is a high level of organic matter or too much crop residue in the soil, larger amounts of fumigant will be needed to do an effective job.

* **Selecting the fumigant:** In selecting from various fumigants, one should consider the purpose and cost of application as well as the equipment available. Some fumigants such as ethylene dibromide, dichloro-propene, and the dichloropropane-dichloropropene mixtures are very specific and control mainly nematodes

231

and insects. Other mixtures such as methyl bromide, chloropicrin, methyl isothiocyanate, methyl isothiocyanate-dichloropropene and metam-sodium have a wide spectrum of activity and control many soil insects, nematodes, weeds and plant diseases organisms. It is important to select the right fumigant or combination of fumigants for the job.

* **Preparing the soil:** Soil preparation is essential for soil fumigation. The soil should be worked into seedbed condition prior to fumigation. This will permit thorough diffusion of the fumigant. The soil moisture should be such that when a handful of soil is squeezed, it will form a compact ball. In drier areas or where soils are sandy, water may have to be applied prior to fumigation to obtain the desired soil moisture content. All crop residues should be completely decomposed or removed. A good practice, where possible, is to plough out the roots of susceptible plants immediately after harvest. This prevents additional nematode build-up, exposes them to the drying sun and wind and breaks down the roots where nematodes may be living.

* **Applying the fumigant:** In small plots, such as nurseries, or for spot fumigation, infection with a hand applicator is possible, but for field application, tractor-drawn equipment is necessary. Metham-sodium (Karbathion, Vapam) can be used as a drench and dazomet (Basamid, Mylone) powder can be incorporated with a rotary tiller; the killing action of both ingredients is effected by their breakdown into methyl-isothiocyanate. Placement of the fumigant generally is at a depth of 20 cm, but may be as deep as 25 cm. It should never be less than 15 cm deep.

Immediately after fumigation, the soil surface should be sealed to delay the escape of the fumigant from the soil. For most fumigants, simply compressing the surface with some device, such as a roller, is sufficient. For more volatile materials such as methyl bromide, a plastic cover is necessary.

Be sure to check the label on your fumigant to determine the best type of seal to use. Depending on the fumigant, the soil should be exposed to the fumigant for about two days to one week, to assure maximum lethal effect on the pests. The cooler the soil, the longer the exposure period should be. Since all fumigants are toxic to plants, the soil should then be allowed to aerate for a recommended period of time so that the subsequent crop will not be injured.

Soil temperature is important for effective fumigation. High temperatures should be avoided by shading, in particular when the treated soil is sealed with plastic. However, it is interesting to note in this context that remarkable results in soil sterilization without chemicals have been obtained by sealing the soil with transparant plastic and allowing the high temperatures generated by solar radiation to complete the required action.

With some fumigants, the soil is free of damaging amounts when the odour of the fumigant can no longer be detected. Two to three weeks is usually enough time; however, if the fumigant persists, the soil may be tilled to speed up aeration.

Since many fumigants are corrosive, equipment should be cleaned thoroughly after use by flushing with fuel oil or kerosene.

MOLLUSCS (SNAILS AND SLUGS)

Snails and slugs are animals without backbone, having a soft, unsegmented body. While snails have a well-developed shell, slugs have only a rudimentary shell often enclosed in a visceral hump. Many molluscs belong to marine species, but the class Gastropoda contains terrestial species, some of which are pests of agricultural importance.

Classification of some families of snails and slugs, with pest status:

Phylum:	Mollusca	
Class:	Gastropoda	
Subclass:	Pulmonata	
Order:	Stylommatophora	- snails and slugs
Family:	*Achatinidae*	- giant African snails
	Arionidae	- slugs
	Helicidae	- garden snails
	Limacidae	- slugs
	Oleacinidae	- carnivorous or cannibal snails
	Subulinidae	- subulina snails and owl snails

Snails and slugs are polyphagous feeders; they attack many types of crops such as cereals, vegetables, ornamentals and fruits in fields or gardens. The harm inflicted may either be in the form of severe damage to or destruction of young plants or a reduction in quality owing to large trails of slime. Injuries inflicted by feeding are also entry points for parasitic diseases.

Molluscs lack good protection against dehydration; hence, they avoid direct sunlight and environments with a low relative humidity.

They hide during daytime in moist places or under debris and feed mainly at night when the temperature drops and humidity rises. Removal of debris and stubble,

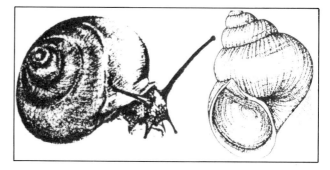

clean weeding and fine tillage of the fields to break soil clods are therefore effective measures in preventing a build-up of mollusc populations.

In molluscs, the head is well developed and has two pairs of retractable tentacles; the posterior tentacles are fitted with eyes at the tips. The slit-like mouth is positioned just below the frontal pair of tentacles and contains a horny, rasping tongue, called radula. The tongue is fitted with many rows of curved teeth which work against a small traverse bar, the jaw, rasping the plant tissues held by the mouth. Snails and slugs have a slime gland, that opens just below the mouth and produces mucus or slime, which offers protection to the animal and eases its movements. They have an elongated foot for moving around (locomotion). The head and foot can be completely retracted inside the shell by powerful muscles.

Snails and slugs are hermaphroditic, each animal having both male and female reproductive organs. The genital opening is at the right side of the body behind the head. Copulation normally results in cross fertilization of both partners. The eggs are laid in moist places such as holes in the soil, whence young molluscs emerge after hatching.

Snails as a pest of rice

Consumption of certain types of snails had been a common practice among Philippine farmers for a long time and therefore the introduction of another edible snail, the Golden Kohol (*Parmacea* spp.), was not researched deeply. Rather it seemed to be an item for potential export and thus could offer another source of income. The snail rapidly spread throughout the islands and became well established. For several years, the snail has been becoming an increasingly severe pest of seedbed and young rice plants up to 20 days after transplanting. Stands are reduced and replanting may be necessary. In 1988, damage affected about 500,000 ha of transplanted rice and the pest is rapidly getting worse.

Control measures being applied are sanitation by collecting snails as feed for domestic animals; collection and destruction of egg masses, which are bright pink in

Bright pink egg masses of the "Golden Kohol" or "cherry" snail on a palm trunk.

colour and deposited on elevated, exposed places; clean-weeding of bunds helps in removing the eggs.

Chemical control has been tried. It involves lowering the water level in the field and spraying snails which congregate in the remaining small pools of water. The molluscicides used for this purpose include Brestan, Bayluscide and Aquatin.

Chemical control

The best known pesticides (molluscicides) for the control of snails and slugs are metaldehyde and methiocarb; these are formulated as granules or soluble powder.

Metaldehyde draws fluid from the body of a snail or slug by contact action and thus disturbs its fluid balance by causing dehydration. The animal tries to protect itself against such contact action by secreting slime. If it can hide in time or is living in a very humid environment, the animal may recover. If the chemical, however, has entered the body by intensive contact with the skin or through the mouth, the animal will die. Therefore, metaldehyde should not be applied during periods of heavy or frequent rain.

In granular formulations, technical metaldehyde has been mixed with bait; however, the bait action is limited. In a dense crop, the spraying of metaldehyde solution may give better control than granules.

The effect of methiocarb is different and results primarily from poisoning. Another advantage of methiocarb is that it retains its effectiveness at lower temperatures better than metaldehyde.

Molluscicides

Common name	Trade name
animocarb	Metacil
fentin hydroxide	Aquatin, Du-Ter
fentin acetate	Brestan, Suzu
metaldehyde	Antimilace, Slugit
methiocarb	Draza, Mesurol
mexacarbate	
niclosamide	Bayluscide

VERTEBRATES

Vertebrates comprise a large group of animals having a spinal column or backbone; included under this classification are mammals, birds, reptiles, amphibians and fish. While only relatively few species of mammals, such as rodents and birds, are agricultural pests of regular economic importance, and others, such as monkeys, bats, deer, rabbits and squirrels may cause incidental damage, many vertebrates are very important predators of pests. For instance, snakes prey on rats and mice; lizards and frogs on insects; many bat and bird species mainly feed on insects and some species of fish feed on water weeds or insect larvae.

Bats are not rodents, but belong to the Order Chiroptera, which contains 950 known species of bats. The majority of bat species are indispensable as pollinators of trees with musty smelling flowers, such as the mango, from which they collect nectar and pollen. Other species feed on insects, consuming several thousands per night. Only very few species cause damage to ripe fleshy fruits. As their beneficial role is far greater than the localized damage they may cause, bats should never be endangered by pest control measures.

Rodents

About 6,000 species of mammals currently have been described; of the total, about 2,500 species are classified as rodents thus constituting the largest group. Rodents belong to the order Rodentia, mammals possessing two pairs of large chisel-like front teeth or incisors specialized for gnawing. These incisor teeth keep growing throughout the animal's life at a rate determined by the rate at which they wear away. The order Rodentia is divided into three suborders and further into families of which the following contain the most injurious species from the agricultural point of view:

Suborder	Family	Pests
Myomorpha	*Muridae*	true rats and mice
	Cricetidae	jirds, gerbils, voles, hamsters
	Spalacidae	mole rat
Sciuromorpha	*Sciuridae*	squirrels, sousliks, marmots
Hystricomorpha	*Hystricidae*	porcupines

The majority of these taxonomic groups of rodents live permanently in fields or in the wild. Some others, however, have become perfectly adapted to living in urban environments. Three of such cosmopolitan, commensal species, which constitute the principal pests of domestic and commercial premises and foodstores, are the Norway rat (*Rattus norvegicus*), the roof rat (*R. rattus*) and the house mouse (*Mus musculus*).

Field rats

Of the more than 500 different species of animals identified as "rats", those of the genus *Rattus* cause the most economic damage and public health problems. Many *Rattus* species occur in Asia, but only a few are of agricultural importance. Owing to their great variation in colour types and patterns, the taxonomy of rats is complicated. However, it may be pointed out that the most important subspecies are: *Rattus rattus*, subspecies *umbriventer* Kellog (synonyms: *Rattus rattus brevicauda* and *Rattus rattus mindanensis* Mearns), the common rice field rat. The rice field rat is predominant in both lowland and upland rice fields; in the Phillipines, they may comprise more than 95 per cent of the rat population in rice.

The rice field rat is a medium-sized rat weighing 130 grams on average and reaching a total length (including tail) of 40 cm; it has a high tolerance for extreme climatic conditions. It prefers as a habitat water-logged rice fields or plots with succulent grass, but it also lives fairly well in extremely dry rice fields.

A very similar species is *Rattus argentiventer*; also called rice field rat, it is predominant in Malaysia. This rat is whitish on the ventral side and brown on other parts of the body. It is a smaller sized species, the adults measuring 16-28 cm in length. It also thrives in irrigated paddy fields.

Rattus exulans, the Polynesian or bush rat, is a species preferring dry upland conditions. It is an excellent climber, weighing 120 grams on average and reaching about 12 cm in length. It is a major pest in upland rice, maize, root crops, vegetables and coconut trees in Southeast Asia and the Pacific islands. In Thailand, *R. exulans* rats are found mostly in houses and markets or close to habitated areas. The roof rat, *Rattus rattus*, is another notorious pest of tree crops such as citrus, mango and oil palms. Dates are attacked throughout the growing season and losses may reach up to 30 per cent. Coconuts are usually damaged when the rat gnaws a hole in the unripe nut near its point of attachment; such damaged nuts fall off. Squirrels may also cause great losses of tree crops, such as coconuts. Such tree-inhabiting species often build their nesting sites high above the ground.

Throughout South Asia, the species *Bandicota bengalensis* is predominant in irrigated paddy and wheat fields and may also readily attack stored products. In the dry season, this rat withdraws to swamp areas. This large brown rat (17-20 cm in length) is referred to as the lesser bandicoot to distinguish it from the even taller but less common *Bandicota indica*.

In arid areas of low and variable rainfall, which supppports crops such as millet and sorghum, the predominant species are desert-adapted rodents, such as jird and gerbils (*Meriones* spp.) and gerbils (*Gerbillus* and *Tatera* spp.). In such areas, agriculture is mainly marginal with farming communities being very dependent upon the yields from the fields and the grass cover on the rangelands for the grazing of their livestock. As a consequence, farmers and tribesman in such areas are very vulnerable to outbreaks of rodent populations.

In the moister, more productive regions of the Near East and central Asia, which have a more temperate climate with cool or cold winters, other species assume greater importance, among which are the social vole (*Microtus socialis*), the short-tailed bandicoot rat (*Nesokia indica*), the mole rat (*Spalax* spp.) and the field mouse (*Mus booduga*).

On the African continent, other species dominate the scene. These are the Nile rat (*Arvicanthis niloticus*) which roams the moister, open areas north of the equator, and the multimammate rat (*Mastomys natalensis*) which is most typical of cultivated land in the Sahelo-Sudan region. These pests attack grain crops and vegetables, and may also penetrate into villages.

The species *Rattus norvegicus*, the common brown-, Norway- or city rat, is common throughout Asia and Africa. Although it normally lives near buildings, it may, however, infest fields near houses. It is a large rat, weighing 200-500 grams, with a heavy tail. Its colour varies from greyish to blackish. (For rodents as major storage pests, see Chapter 15).

Characteristics and behaviour

The majority of pest rodents are essentially nocturnal in activity, but many species may move around at any time of the day when conditions are quiet. Rats rove and feed mainly at night, although they have poor vision. However, their senses of smell, touch and hearing are extremely well developed. Whiskers and guard hairs (elongated hairs in their short fur) guide them in the dark.

Squirrel-like rodents, including ground squirrels such as sousliks, are active during the day, especially in the early morning and evening.

Generally, rodent species do not wander far from their nesting sites. House mice range only a few metres if food is plentiful. Most species stay within a 40-metre distance and only a few go regularly farther away than 400 metres. However, if food gets really scarce and population pressure mounts, rats and other species may migrate over a distance of several kilometres.

If conditions are very favourable, then many species can reproduce abundantly. Sexual maturity may be reached at 2-3 months of age. The duration of pregnancy (from mating until delivery), which is also called the gestation period, is about three weeks. Female rats produce litters of 6-7 young and many mate again on the day of delivery. Consequently, species such as rats and mice can produce several litters without interruption when the food supply is copious. Under very favourable conditions, some species may even increase their reproduction rate by becoming sexually mature a few weeks earlier than normal and by delivering more offspring per litter. Such reproductive potential of certain species may lead to an outbreak causing destruction of agricultural crops and rangelands over vast areas. Species which may cause outbreaks include rice field rats, social voles, jirds, Nile rats and multimammate rats.

Quickly breeding rodents recover rapidly from low population levels and frequently reach high densities when crops grow into the reproductive stage. Multiple cropping, therefore, tends to worsen an existing rat infestation in an area because food is available for a greater part of the year. However, if food supply gets scarce thereafter, such fast-increasing populations may collapse as dramatically within a few weeks leaving only a very low number to survive the fallow period.

In contrast with species which demonstrate such volatile population dynamics, other, usually larger-sized species, live longer, reproduce slower and are generally less dependent on high-energy food such as grain and seeds. Populations of such hardier rodents are more stable throughout the year and fluctuate much less in size in correspondence with food availability and therefore are less capable of recovering quickly from a control operation.

A combination of rapidly and slowly breeding rodent species occupying the same habitat may demonstrate a changing pattern of predominance. For instance, in the rice growing areas of the Sind, the short-tailed bandicoot rat withstands the dry season well, but breeds relatively slowly when irrigation water reaches the field. The lesser bandicoot rat, however, disappears entirely from the fields in the dry season, but reproduces very fast and overtakes the short-tailed bandicoot species by far in number towards harvest time. Similarly, the Nile rat, which utilizes low energy food such as coarse vegetation, is out-numbered by the granivorous multimammate rat during the cropping season only.

Regarding choice of food, rodents are generally omnivorous in the sense that they appear to be able to subsist on various kinds of food. Most species prefer high-energy foods such as grains and seeds, and vary in their ability to survive on less nutritious vegetative matter. The more species are dependent on a diet of grains and seeds, the more their populations fluctuate in a seasonal pattern that corresponds with grain cultivation practices. Many primarily granivorous species dig actively for groudnuts, tubers, succulent roots etc., and several species hoard food in their burrows.

Species that survive well on coarser food such as stems and leaves, include moles, gerbils, Nile- and roof rats. Flowers, immature fruits and, particularly, ripe fruits may be heavily damaged as are vegetables. In colder seasons, rodents frequently inflict heavy damage on trees by stripping the bark in order to feed on the cambrium. Almost all rodents eat insects as part of their diet.

Harbourage and burrows

Rats settle and reproduce rapidly in areas that offer food and natural cover against harsh climatic conditions and protection against predators while allowing space for movement. Hence, cultivated fields with a closed leaf-canopy offer the best opportunity for a rat population to grow. Fields with long grass, wasteland with bamboo or shrubs offer a good alternative. Rats dig elaborate burrows for breeding and shelter in the dikes or bunds around the edges of rice fields and on higher land.

Rat control

The best approach to control is by preventing a build-up of a rat population; a resident population should

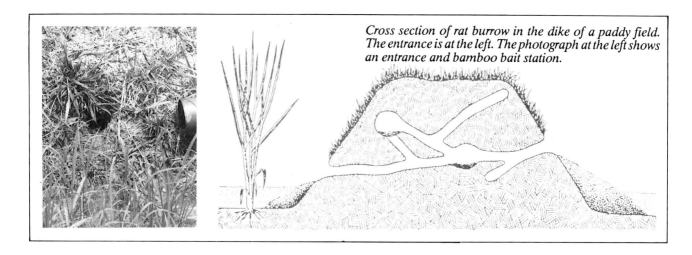

Cross section of rat burrow in the dike of a paddy field. The entrance is at the left. The photograph at the left shows an entrance and bamboo bait station.

be kept at the lowest possible level throughout the year. Control of field rats can never meet with success unless it is undertaken as a community effort. Although rats tend to feed within 200 meters of their burrows, migration over long distances in search of food and shelter occurs frequently. Thus, the work of some farmers who effectively practise rat control may be undone if other farmers within the area allow rats a chance to reproduce freely.

Rat control practices

- Determine if rats are present. Always look for signs of their presence such as footprints, runways, droppings, burrows, gnawings and crop damage. Use snap traps if infestation is low in houses or small stores. Place a larger number of traps than the estimated number of rodents present.

- Reduce food and shelter. Keep nearby waste- and fallow land areas free from long grass, bushes and food residues. Rat proofing of graneries and other food stores, together with removal of wastes such as straw, coconut husk and cornstalks as well as waste material from around the house and other structures are helpful steps.

- Protect trunks of coconut and oil palm trees with aluminium or tin-sheet sleeves to prevent roof rats and squirrels from climbing them to attack the fruits. Remove undergrowth and trailing fronds so that rats cannot bridge the obstacle.

- Keep dikes and dams between irrigated rice fields narrow (less than 40 cm in width) to prevent burrowing.

- Create a nearly rat-free area with an initial large-scale killing operation using either a 'blanket' killing method or acute poisons.

- Maintain a low population throughout the vegetation period by carrying out sustained preventive control with anticoagulant poisons over the whole area.

- Create a functional organization that will keep farmers interested in on-going rat control, help in supplying control material and inspect the effectiveness of the control.

- Killing field rats by the so-called "blanket-system": An infested area of up to 1/4 hectare or 1/2 acre is surrounded by a few dozen people who cut the vegetation working towards the centre in a circular fashion, thereby driving the rats towards the middle of the field. When the circle has been reduced to only a few square metres, the rats milling around the circle can be clubbed to death. Then repeat the same action in an adjacent plot until the whole area has been covered. Swamp areas should be purged of rats in the same way, since such areas are likely breeding places.

Using rat-poisons (rodenticides)

Rodenticides have traditionally been divided into two groups, namely: acute, quick acting poisons, a single dose of which suffices to cause death, and chronic, slow-acting poisons requiring multiple-dose feeding; these compounds have an anticoagulant activity in common.

In recent years new rodenticides have been developed which do not fit into such a strict distinction; however it remains meaningful to describe anticoagulant rodenticides as a separate group.

Anticoagulant rodenticides

These compounds reduce the clotting ability of blood and make the walls of blood vessels porous, resulting in a painless death from internal bleeding. These poisons interfere with the function of vitamine K, that is essential for the synthesis of four blood clotting factors known as "prothrombin". The anticoagulant prevents prothrombin-synthesis and consequently the clotting factors in the blood decrease over a period of days. When, after several intakes, a lethal dose has been consumed, then the healing of the blood vessels is definitely impaired and the animal is beyond recovery.

These poisons are non-specific and toxic for non-target animals also, but as these anticoagulants are commonly applied in the form of low-concentration baits, the risk of accidental poisoning is small. In cases of accidental consumption, vitamin K should be administered under medical supervision until the victim has fully recovered from the haemorrhagic symptoms.

Anticoagulants are more widely used than any other rodenticide because they possess some important advantages. These are:

* Anticoagulants do not cause bait- or poison shyness. Rats and mice do not link their exposure to anticoagulant bait with the ensuing death and thus will continue to feed at the same bait-station. Probably the lapse of time of at least three days between ingestion and the onset of illness is too long for a rodent to associate the bait with the consequences.

* Since anticoagulants do not induce bait shyness, the concentration of active ingredient in the bait can be reduced to a level so low that repeated uptake is required in order for them to have ingested a lethal dose. This also reduces the hazard of accidental poisoning to man, livestock and birds.

A disadvantage of using anticoagulants is that they do not kill immediately but need to be administered by repeated baiting, at intervals of 3-7 days, until the control is successful.

Usually, mortality begins to occur at the end of the first week of treatment, rises to a peak during the second week and takes off during the third to fifth weeks as the rodents are eliminated. Because of this slow effect, inexperienced users may lose confidence in the efficiency of these poisons.

Some species of *Rattus rattus* and *Mus musculus* possess a certain level of natural (genetic) resistance to warfarin-type anticoagulants. Such resistant species occur mainly in Europe and North America.

First-generation anticoagulants

This denomination applies to a group of rodenticides developed before 1965. These include:

Compound	Brand name
warfarin	KG22, Ratoxin, Tikumin
coumachlor	Ratilan, Tomorin
coumafuryl	Fumasol, Ratafin
coumatetralyl	Endox, Racumin

These four compounds are all hydroxy coumarin-derivatives. The first three are formulated in bait at a concentration of 0.025 per cent; coumatetralyl, at the higher concentration of 0.0375 per cent.

chlorophacinone	Caid, Drat, Ratoxin 5
diphacinone	Diphacin, Promar, Ramik
pindone	Chemrat, Pival
valone	Kilgore

These four compounds are indane-dione derivatives. They are all formulated in bait at a concentration of 0.005 per cent.

Currently, warfarin, coumatetralyl and chlorophacinone are probably the most widely used of the first-generation compounds; coumafuryl and valone are superseded rodenticides. Coumatetralyl appears to be toxic to warfarin-resistant *Rattus* species, a characteristic it has in common with second-generation anticoagulants, of which it is the stem compound.

Since a repeated uptake of daily sublethal doses of first-generation anticoagulants is required for obtaining a lethal dose, these rodenticides will give effective control only if applied by a technique known as surplus, unrestricted, continuous, sustained or saturation baiting. Such baiting ensures that the bait is constantly and easily available to all the rodents in an area until they have been eliminated. This technique provides encouraging control after about two weeks and usually achieves complete control after 3-5 weeks of baiting.

Saturation baiting, however, requires a high degree of discipline in application that farmers find difficult to adhere to. It also involves much labour for transporting and distributing relatively large quantities of bait over the bait stations in the fields. Further disadvantages of the first generation anticoagulants are the development of genetic resistance in some populations and the disappointing results against less susceptable species, against which those compounds have not been properly tested in research trials.

Second-generation anticoagulants

Since 1973, five compounds have been developed which control warfarin-resistant *Rattus norvegicus*. These anticoagulants are:

brodifacoum	Klerat, Ratak +, Talon
bromodiolone	Maki
difenacoum	Neosorexa, Matikus, Ratak
flocoumafen	Storm
difethialone	Frap

Except for bromodiolone, these compounds are derivatives of coumatetralyl.

The better activity of these compounds seems to result from a more efficient inhibition of the synthesis of blood-clotting factors. Also, because they are more slowly metabolized and excreted by the rodents, the anticoagulant effects of a single dose are sustained for a longer period of time.

The combined effect of these phenomena is that, with more susceptible species, a single day's feeding on the bait provides a lethal dose and that, with less susceptible ones, fewer and smaller doses of second-generation anticoagulants will still provide good control.

Thus, second-generation anticoagulants have made it easier for applicators to fight rodent infestations effectively.

Rodenticides other than anticoagulants

Although anticoagulant poisons are the most efficacious ones, there remain a number of situations in which the use of quicker-acting rodenticides offer advantages. For instance:

- for treatment of rodents which are resistant to all anticoagulants. So far only the Egyptian spiny mouse (*Acomys cahirinus*) is resistant to all;
- to knock down a heavy infestation, to be followed up by baiting with anticoagulants;
- if only limited time is available for control, such as to eliminate an infestation on ships.

Some widely used or less well-known rodenticides are:

Zinc phosphide

This widely used rodenticide has a quick action and causes death within 24 hours. Upon ingestion, phosphine is released in the stomach, which affects several internal organs and the heart. However, it is potentially hazardous to man and domestic animals and there is no specific antidote. Technical grade zinc phosphide is a heavy grey-black powder which has a characteristic phosphine smell. Zinc phosphide is usually formulated in bait at 0.75-5.0 per cent concentration. Under field conditions, this chemical breaks down within a few days; thus baits need to be well maintained.

The main disadvantages of zinc phosphide are that it causes bait shyness and it sometimes gives unrealiable control, particularly of commensal rodents.

Fumigants

Aluminium and magnesium phosphide are widely used for the gasing of burrows. (For the action of aluminimum phosphide, see Chapter 15). Fumigant formulations can be inserted into the burrows with a long-handled spoon. All burrows that can be detected must be sealed. The recommended rate is 1-2 aluminium phosphide tablets.

These fumigants are very effective and may provide a 85-95 per cent kill rate with a single application. However, fumigation of burrows demands much labour and is therefore costly. Fumigation should be carried out by trained operators.

Alpha chloralose

This is a quick-acting poison that induces unconsciousness in mice within 15 minutes. It is a narcotic that acts by depressing neural activity in the brain, slowing down the heart and respiration and rather radiply causes death by hyperthermia.

A 4 per cent concentration in bait is very effective for control of house mice. However, its activity becomes unreliable at temperatures above $15^{\circ}C$. Therefore, this rodenticide is less frequently recommended for hot tropical countries, although good control of rats in poultry houses has been reported from Nigeria. Birds however, are at risk if exposed to it. In cases of human ingestion, there is no specific antidote, but keeping a victim warm assists in achieving a recovery.

Bromethalin

Bromethalin is a broad-spectrum rodenticide still under development. It affects the nervous system and impairs motoric and sensoric transmissions. Symptoms of poisoning take several hours to appear and death occurs within 1-4 days. Bait at a concentration of 0.01 per cent does not seem to cause bait shyness.

Calciferol

The rodenticide (ergo-) calciferol is a vitamin D-compound. Vitamin D is an essential trace element in the diet, but calciferol uptake causes a lethal overdosing which results in an excessive calcium metabolism and consecutive damage to the blood vessels, heart and kidneys. Its action is relatively slow as it takes 2-10 days to kill rodents.

Application is, therefore, by cummulative baiting as in the case of anticoagulants using a commercial 1-per cent bait formulation based on edible oil to improve the stability of the active ingredient. As this rodenticide causes cessation of feeding after 2-3 days, it is essential to supply plenty of bait from the beginning.

The efficacy against house mice is excellent, but against rats, somewhat variable because of bait shyness. To improve its activity against rats, this compound is often applied in a combination bait formulation containing calciferol 0.1 per cent and difenacoum 0.005 per cent concentration.

Sodium fluoroacetate and fluoroacetamide

The first compound, also known as "1080" (teneighty), is a very efficacious, broad spectrum rodenticide, that is also extremely hazardous to man and animals. The poison inflicts neurotoxic damage that results in convulsions, respiratory and cardiac failure. Symptoms of poisoning first occur after a latent period of about 40 minutes after uptake and death usually occurs within 24 hours. The LD_{50} to mammals and birds is about 0.1-5.0 mg/kg.

Because of its quick action, it is used against rats in ships, as a 0.3 per cent solution, presented in non-spill plastic disks, for a period of 24 hours. For control of rats in sewers, it is formulated in a cereal bait at a 0.25 per cent concentration.

Fluoroacetamide or "Compound 1081" has similar characteristics as a rodenticide and may demonstrate slightly better control than sodium fluoroacetate. It is used equally for rat control in sewers as a 2.0 per cent bait.

Because of their high toxicity, the use of both compounds is in many countries restricted to trained personnel for use in areas where the public and domestic animals have no access.

Baiting technique

Rodenticides are commonly available in various formulations. The main forms are:

- the technical grade active ingredient;

- powder or oil-based concentrates for mixing with a solid bait base;

- solid or liquid concentrates for dissolving in water to form a drinkable bait;

- ready-for-use baits.

All rodenticides contain a strong dye as a warning agent, the most usual colours being blue and red. Other additives include water repellants, anti-oxidants, binders and taste or smell enhancers. The binder is usually a vegetable food oil or technical white oil.

In the case of bait formulation by the user, acute or anti-coagulant poisons are mixed with bait, which generally should comprise the main crop grown in the region. If any doubt exists, different non-poisoned baits should be laid out in weighed portions first to determine the rats' preference. Very often high-quality husked rice or other cereals are preferred.

Farmers should not prepare bait, but leave its preparation to technicians of larger organizations who have been adequately trained. In preparing baits, one must strictly follow the label instructions. The cereal grains may be mixed with the rodenticide and the binding oil in a bucket or drum with a stick or hand tool, but hands should not become contaminated. The mixture may consist of 95 per cent broken rice, 5 per cent oil and the recommended dose of rodenticide.

Baits in ready-to-use formulations may consist either of loose grain or flour with rodenticide contained in a plastic sachet or of paraffin blocks containing the food and the poison. Such blocks and sachets are easier and safer to handle and are more resistant to weathering in the field than loose bait. These ready-to-use formulations are also safer for small birds.

Baiting stations can be made of varying materials and in different shapes, but they should be easy to handle, accessible to rodents and the bait kept dry. Split bamboo tubes of about 50 cm in length, placed horizontally alongside paddy dikes, along runways and near burrows, are inexpensive and very effective. Baiting stations made of plastic covered dishes, with a hole at the side are being

Using a locally made mixing drum improves the blending of the bait ingredients and reduces the danger of contamination for workers. (Photo by the author)

used in some rodent control programmes in Asia. A recent development of the plastic dishes is one that fits like an air cushion around a bamboo stick and floats upwards with the rising or falling water level in rice fields.

Baiting stations can be spaced about 10 metres apart inside a treated area and 10 meters apart along the edges of the area in a grid pattern to prevent infiltration from outside. However, distances within the grid depend on the rat species present, the population density and the kind of habitat. Bait may also be placed near burrow entrances and on runways; this is called "spot" baiting.

Before starting to dispense the poison, one should practise pre-baiting to familiarize the rodent with the bait. For a few nights, place up to 200 grams of fresh untreated cereal (e.g. rice or maize) in each station until nightly grain consumption is regular. It has been found that a field rat consumes on average 10 grams of such bait per night. After 3-5 days of pre-baiting, the untreated cereal may be replaced with poisoned bait using the same cereal base as in pre-baiting, but use at least double the quantity of poisoned bait. Check the baiting station every other day for freshness of the bait; refill as necessary. If all bait has been eaten, there has not been enough in the station so the next portion should be larger. Do not touch the bait with hands, but use a spoon for filling instead.

Keep the baiting station free of ants and vermin and beware of snakes. The best time to conduct (or intensify) rat control with baiting is towards the end of the dry season, when the rat population is low because of the scarcity of food and shelter. In a low population, every kill counts heavily. Hungry rats easily take the bait.

Birds

Birds may cause serious damage to field crops at sowing time by picking seeds and emerging seedlings. They may inflict even worse damage in fields and gardens from the time of flowering to ripening of grains or fruits. In particular, fields with maturing rice, wheat, sorghum or millets may be badly damaged by large flocks of seed-eating birds. The shattering varieties of cereals are more severely affected than non-shattering varieties, as much grain is spilled during attacks. In tropical Asia, weaver birds, sparrows and parakeets are serious pests of ripe grain. Pigeons and crows are a nuisance, especially at sowing time.

For information on the fight against the weaver bird as a major pest of grain crops in the Sahelo-Sudan region of Africa, the reader may refer to "The ecology and control of red-billed weaver bird (*Quealea quelea L*) in Northeast Nigeria", by H.J. Conert (1987), GTZ-Publication, Postbox 36, D 6101 Rossdorf 1, Federal Republic of Germany (English and French editions available).

Bird control

The use of various bird-scaring devices, such as scare-crows, dead birds, blinking or sound-making devices, is widely practised, but with poor results. Some species can be caught in nets at night time. In orchards, gas-operated "canons" are used to produce regular explosions of sound. Chemical control is not practised often in Asia but in some cases, groups of trees where birds roost are sprayed from the air at dawn or dusk.

In Africa, nesting sites of weaver birds may occupy an area of up to 250 hectares and contain some 100,000 nests per hectare. Roosting sites, where birds congregate at sunset, may carry a population density of up to 20 million birds per hectare. In order to control such huge concentrations of birds, special techniques are required. Against large nesting sites, aerial ULV-application, at the end of the afternoon when the adults come flying in, is effective. On roosting sites, semi-permanent telescopic masts equipped with ULV-sprayer devices are erected and spraying is done at night-fall when the birds come in to roost. The birds fly through a cloud of drifting 30-35 μm VMD spray droplets and are killed in high numbers.

Chemicals which can be used for this purpose are parathion (Queletox and other trade names) or diazacosterol hydrochloride (SC-12937, Ornitrol). To protect seeds, the latter avicide is coated onto whole kernel grains, such as maize or wheat.

References:

De Datta, S.K., (1981). *Principles and practices of rice production*, John Wiley and Sons, New York, USA.

Greaves, J.H. (1989). *Rodent pests and their control in the Near East*, FAO plant production and protection paper No. 95, Food and Agriculture Organization of the United Nations, Rome, Italy.

Kranz, J., Schmutterer, H., Koch, W., (1977). *Diseases, pests and weeds in tropical crops.* Paul Parey, Hamburg, Federal Republic of Germany.

Oka, I.N., (1988). *Role of cultural techniques in rice integrated pest management systems*, IARD journal vol. 10, No. 2, 1988.

Oudejans, J.H., Mumford, (1988). *Evaluation of integrated pest management in rice in the Philippines*, Ministry of Agriculture and Fisheries, the Netherlands.

Riesig, W.H., Litsinger, J.H., et al. (1982). *Illustrated guide to integrated pest management in rice in tropical Asia*. (2nd edition). International Rice Research Institute, Los Banos, Philippines.

van Vreden, G., Ahmadzabidi, A.L., (1986). *Pests of rice and their natural enemies in Peninsular Malaysia*. Pudoc, Wageningen, the Netherlands.

13

PLANT DISEASES:
CAUSES AND INTEGRATED CONTROL

Contents Page

Causes of plant diseases 242

- The concept of plant diseases 242

- The host-pathogen relationship 244

- Main categories of plant diseases 245

- Non-infectious diseases (disorders) 245

Symptoms of plant diseases 247

Infectious microorganisms 248

- Fungi 249

- Bacteria 251

- Rickettsias, Mycoplasmas 252

Viruses and viroids 253

Integrated control of plant diseases 256

- Epidemiology 256

- Preventive and curative control 257

- Control methods or control strategies 258

- Examples of preventive control against micro-organisms 261

References 262

Plant Diseases: Causes and Integrated Control

CAUSES OF PLANT DISEASES

All forms of life depend essentially on the capacity of chlorophyll-containing plants to utilize, through a photo- synthetic process, light energy for the production of carbohydrates from CO_2 and water. Green plants use these organic molecules as building material for the synthesis of more complex compounds such as amino acids, proteins (including enzymes), lipids and secondary plant metabolites. These metabolic processes make plants a rich source of nutrition for other organisms.

Over a long period of time, man has selected amongst numerous wild species those individual green plants which attracted his attention as suppliers of food, feed, fibres and medicines. He has domesticated those plants and, in a more or less strong isolation from the complex natural environment, has gradually improved the early selections of cultural varieties. Currently, about 300 plant species are know as the principal producers of food and feed. Of these, a mere 10 species provide about 95 per cent of world food production. These species are wheat, rice, maize, potato, sweet potato, sugar-cane, cassava, haricot beans, coconut and banana.

Such a severe limitation in the number of major food crops and the practice of growing them in monocultures on ever-increasing surfaces, have both strongly upset natural enemy control as well as stimulated parasitism. Every culture is threatened by a particular range of injurious biotic and a similar range of abiotic factors against which it needs to be protected. Of all factors posing a threat to crops or to individual plants, parasitic organisms and viruses that can cause diseases in plants inflict generally the worst damage to agriculture. For thousands of years, the factors responsible for inducing diseases in plants remained unknown. Only in 1728 was it demonstrated that healthy plants can become diseased by fungal infection; around 1870, it was proven that bacteria too were the cause of certain diseases in plants. Owing to this lack of understanding about diseases and the absence of means to control them, epidemics such as late blight (*Phythophthora infestans*) in potatoes in Ireland in 1845 and the Panama disease (*Fusarium oxysporum*) in bananas in Central America in 1906 caused crop failures and famine.

In the twentieth century, considerable knowledge has been acquired about factors which can cause diseases in plants (aetiology). A whole range of disciplines, such as phytopathology, virology, bacteriology and nematology, study the many aspects involved. Over the years, a whole range of methods and means of preventing or controlling diseases has been developed, including also the application of nuclear biological techniques. Among these, the breeding of resistant crop cultivars and the application of fungicides play a major role. However, fungi, bacteria and viruses have an awesome capacity for producing mutations, some of which may overcome the resistance in plants. As it takes much time to breed a new resistant cultivar, fungicides frequently offer the only solution for acute disease problems. Older methods, such as sanitation, crop rotation and cultural control, though very useful, are usually not sufficiently effective to stop diseases from spreading. As in the case of controlling animal pests, the best chances for acceptable results in terms of preventing or controlling diseases as well as of minimizing pesticide hazards seem to be offered by an integrated control strategy.

The concept of plant diseases

A plant is referred to as "diseased" when the entire organism or a part thereof shows a harmful deviation from normal functioning of physiological processes, that is of sufficient duration to cause a disturbance of vital activity.

Usually, a disease causes a progressive and continuous disturbance of cellular activities that eventually become manifest as **symptoms**. A symptom is a visible or otherwise detectable abnormality arising from disease. Frequently, distinct symptoms may either succeed each other in a changing **course of disease**, or they may occur simultaneously thus providing a "**symptom picture**". The totality of effects produced in a plant by one disease, whether all at once or successively, and whether or not easily detectable, is called a **syndrome**. Disease symptoms should be closely observed, as they often give indications about the cause of the disease and thus have a diagnostic value. Environmental factors must always be

An infection by the fungus Ascochyta fabae *on broad bean,* Vicia faba, *shows up on various parts of the plant successively. At the seedling stage, typical black lesions appear on the leaves (left). At the fruiting stage, black lesions show up near the base of the pods (centre). At the ripening stage, the infection appears to have affected the seeds too (right). The disease* A. fabae *is transmitted by seeds; actually, it is one of the most infectious seed-borne diseases. (Photos courtesy of IPO, the Netherlands)*

taken into account in order to recognize possible abiotic causes. Symptoms may either be mild in terms of extension or intensity, or they may be acute or severe, the latter usually leading to the death of plants or plant parts.

The outbreak of a disease is always preceded by an infection, i.e. the entry of a parasitic organism or virus into a host plant and the establishment of a permanent or temporary host-parasite relationship. The site of invasion of the host plant is called the **infection court**. The period of time from natural infection (or artificial inoculation) until the appearance of the first symptoms of disease is called the **incubation period**.

In the case of a fungal infection, the fungus keeps growing until it reaches the reproductive phase and commences to reproduce. The time that passes between the infection and the moment that spores are formed at the infected spot is known as the **latency period**. The latency period is followed by the infectious period, during which the fungus produces a certain number of spores per day. The spores produced are dispersed and they can start a new infection at other parts of the plant, where a new generation cycle of the fungus begins again. An infection may remain **latent** as an inapparent, chronic infection that gives no overt signs of its presence (symptomless infection). In the case of many symptomless virus infections, the viruses may nevertheless be vigorously multiplying within the host. Such hosts, as carriers of the virus, constitute a source of infection to other plants.

Tolerance is the ability of a host to endure infection by a particular pathogen, with little or no reaction, as shown by the more or less complete absence of symptom expression and damage. The term **tolerant** is used when there is an apparently unrestricted and extensive colonization by a parasitic organism or virus without symptom development. Absence of symptoms because infection and colonization have been restricted in extent or intensity by the host's defensive response implies **resistance**, not tolerance.

Parlevliet defines **tolerance** as the property in which relatively little damage is caused by the development of a pathogen. This is tolerance against damage; despite a relatively serious infection, the resulting damage is relativily low. For viruses, the meaning is different. In the case of tolerance, the virus can reproduce itself within the host plant with relatively few disease symptoms. If there are no symptoms at all, symptomless carriers are referred to. This is a tolerance against expression of the symptoms.

The terms **tolerant** and **tolerance**, like the antonyms **sensitive** and **sensitivity**, are applicable also to the response of the host or pathogen to pesticides and other toxic substances. As tolerance does not reduce the cause of the damage but the degree of damage, the term is also used to indicate "tolerance to drought or to heat" (instead of "resistance to drought or to heat").

A parasite may either be carried accidentally by wind, water or insects onto the host plant or it may be applied (**inoculated**) intentionally to the host. One needs to differentiate clearly between the terms "to infect" and "to inoculate". **Infecting** is the activity of a pathogen that enters into a host with the objective of establishing a parasitic relationship. Man and insects can **inoculate**, for instance, by placing fungal spores (the inoculum) onto another host or onto a laboratory medium. An inoculation does not necessarily lead to infection and disease. If a parasitic disease or virus is spread by organisms, such as insects, mites or ticks, nematodes and snails, these transmitting agents are called vectors. Feared human vector diseases are, for instance, malaria, river blindness and sleeping sickness. Virus diseases of plants are often transmitted by aphids, leaf hoppers or other sucking insects.

The spreading of a disease depends on two factors, i.e. time and environment. The time determines the course of the disease as the pathogen population increases with the passage of time. Environmental factors, such as climatic and soil conditions, and cultivation methods, have a great influence on the activity and development of the parasite and on the expression of the disease's symptoms. The study of the complex relationships between organisms and their environment is the subject of a science called **ecology**. And like their host plants, parasites may have active and restive phases that are usually related to seasonal weather patterns.

The host-pathogen relationship

In this relationship, one living organism (the host) harbours another living organism or virus (the parasite), the latter being dependent on the host for its existence. It is obvious that the effects of such intimate co-existence depend not only on the hereditary (the genotypical) properties of each of the partners, but also environmental conditions, which have a great influence. The way an organism manifests itself (the phenotype) is the result of the interaction between its genotype and the environment. The host-pathogen relationship can be differentiated into a number of detailed phenomena each having its own name. These names are explained below:

Aggressiveness (or aggressivity) is the capability of a pathogen or virus to infect a host plant. The term is used particularly in a gradual sense and is measured by the degree of infection. However, the latter is also demined by the susceptibility or resistance of the host.

Susceptibility is the inability of a host to prevent the development of the parasite, thus defending itself against infection by a pathogenic organism or virus. The meeting of an aggressive pathogen with a susceptible host results in the host being affected by the disease. The

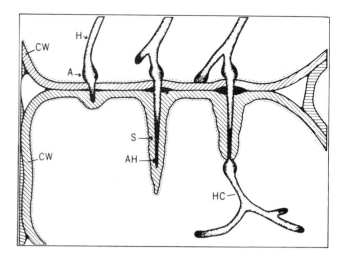

Figure 13.2: Another defense reaction is the formation of a sheath (S) around hyphae (H) penetrating a cell wall (CW). AH = advancing hypha still enclosed in sheath; HC = hypha entering cytoplasm.

degree of disease can be measured either by the quantity of parasitic organisms that develop within or on the host, or by the size of the surface affected. In a symptomless infection, for instance, of a tolerant host, the disease remains invisible and usually can be demonstrated only by a sophisticated detection technique. The opposite of **susceptibility** is **resistance** of the host, i.e. the ability of the host to suppress or retard the activity of the pathogen. A low resistance corresponds with high susceptibility, and high resistance with low susceptibility.

Virulence is the observed capacity of a pathogen to infect certain genotypes of a host plant. In reference to genotypes, which cannot be infected, the term **avirulence** is used. However, virulence also depends on the **sensitivity** (see below) of the host and on the circumstances. The concept of virulence should be distinguished from that of aggressiveness. Virulence and agressiveness together determine the **pathogenicity** of a parasitic organism or virus. The term **pathogenicity** indicates in general the disease-generating capacity of biotical and abiotical factors. A parasite is not regarded to be a pathogen, if it lacks virulence.

Sensitivity is the property of a host organism to react with severe symptoms (yield decrease included), to the attack of a parasite or abiotic factor. The opposing concept of host sensitivity is **tolerance**; thus a higher sensitivity corresponds with lower tolerance. The sensitivity of the host and the virulence of the pathogen determine together the **severity** (or intensity) **of the disease symptoms**. Furthermore, the concept of sensitivity should be distinguished from susceptibility, the latter being the opposite of resistance.

Vulnerability is the inability of the host to defend itself against infection by parasites or infestation by phythophages. The vulnerability of the host plant and the pathogenicity of the parasite are complementary; together they determine the **intensity of the disease** and the extent of the expected damage.

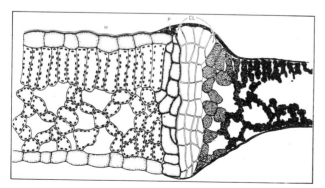

Figure 13.1: Infection of plants by fungi and bacteria frequently induces formation of a cork layer (CL) between infected (I) and healthy leaf (HL) areas.

Main categories of plant diseases

As to the cause of the malfunctioning of processes in plants, a distinction can be made between **infectious** and **non-infectious diseases**.

* **Infectious diseases** are caused by pathogenic organisms or viruses, a **pathogen** being a parasitic organism or virus able to cause a disease in a host plant. Pathogens include parasitic fungi, bacteria and protozoa, as well as mycoplasmas, viruses and viroids.

* **Non-infectious diseases** are malfunctions caused by either hereditary or abiotic factors. Hereditary diseases are, for instance, caused by genetic defects such as chlorophyll or enzyme deficiencies or the presence of genes which can induce necrosis.

Abiotic factors causing disease include adverse soil conditions, deficiency or excess of certain macro- or micro-nutrients, adverse meteorological conditions and environmental pollution by industrial contaminants. For malfunctions caused by such genetic and abiotic factors, the term **disorders** deserves preference.

The main groups of pathogenic organisms and viruses causing infectious diseases in plants will be discussed later, but the various factors causing disorders in plants are discussed in the following section.

Non-infectious diseases (disorders)

Many serious disorders are caused by external physical factors which adversely affect the proper growth of plants. Very often, the disorder is a direct result of a deficient or excessive supply of an essential growth factor, such as light, oxygen, soil moisture and heat. These are indispensable for the various life processes in plants including water and nutrient uptake, respiration and photosynthesis. A strong interaction exists between these growth factors; generally, each plant has the capacity to balance its requirement for each of them. However, abrupt changes in environmental conditions or a severe deficieny or injurious overabundance of one or more of the growth factors may disrupt a plant's life processes and bring about disease symptoms. Since these disorders are not caused by parasites, they are not infectious.

Light: The requirement for light is very variable; some plant species prefer shade, others full sunlight. In many species the total hours of daylight -- long-day versus short-day varieties -- determine the transgression from the vegetative to the generative phase. However, too much light, especially in combination with high temperature and lack of water, produces wilting or scalding in broad-leaved plants. Too little light produces chlorosis and etiolated or "lanky" growth.

Oxygen: An adequate supply of oxygen to above-ground parts of the plant is essential for effective respiration. A shortage of oxygen in the soil owing to poor soil structure, compaction or water-logging hampers the development of a healthy root system.

Water: Regular availability of water is essential for photosynthesis and metabolic processes in the plants. Plants absorb water mainly through the root hairs; this assists in the uptake of nutrients from the soil and their distribution throughout the plant. Evaporation of water from the leaves aids in cooling of leaf surfaces. Water contained within the walls of plant cells lends firmness (turgor) to plants; juicy fruits and succulent leaves may contain more than 90 per cent water.

A low relative humidity (dry air) may accelerate evaporation of water from the leaves and cause wilting when water pressure (turgor) cannot be maintained in the plant by replenishment from the roots. Early stages of wilting may be cured by watering; more advanced wilting may be irreversible and the tissues may be damaged permanently.

Where crops are lodging due to high winds or torrential rains, pathogens may flourish because of the high humidity that prevails in the tangled mass of plants. The infection will be even worse if the foliage and fruits have been damaged by hail storms.

Temperature: The supply of heat, or the lack of it, determines to a large extent the functioning of life processes in plants. Plants have an optimum temperature range for healthy development and a particular total requirement for heat in order to reach their full productivity. Severe summer heat, particularly in combination with a scorching dry wind, can cause extensive injury and death to plants. Plants differ in their resistance to heat or cold; for instance, sorghum may yield well in areas which are too hot and dry for maize, whereas bananas die at temperatures below 10°C. Frost may cause the freezing of water within plant cells, which thereupon may rupture since ice crystals occupy a larger volume of space than water.

Overabundance and deficiency of nutrients may cause a whole range of symptoms from discolouration to rot, die-back or stunted growth. Commonly, the productivety of a plant is lowered and its resistance to parasites weakened. The appearance of characteristic deficiency symptoms on the leaves does not necessarily mean that the nutrient is in short supply in the soil. In fact, it might be sufficiently available, but its uptake may be impaired by an unfavourable acidity rate (pH) or by the shortage of another element. Nutrient deficiencies are very complex problems and leaf symptoms are usually difficult to interpret as they may vary among plant species and varieties. Deficiency symptoms are best described for perennial tree crops where nutrient requirements remain more or less constant over the years and deficiencies have time to show up. But there exists a good **deficiency symptomalogy** that also includes annual crops.

Symptoms of phytotoxicity on wheat caused by drift from herbicide spraying by tractor. A low position of the spray boom and selection of suitable nozzles can greatly reduce spray deposition beyond the target. (Photo courtesy of LNV, the Netherlands)

Large-scale irrigation and intensive fertilization in arid regions has led to a toxic accumulation of soluble salts at the soil surface that causes necrosis of plant tissues or inhibits plant growth altogether. Soil salinity or calcidity are serious problems threatening irrigated agriculture, where no possibilities exist to flush and drain the salt from the top layer of the soil. Continued fertilizing with ammonium sulphate may lead to high acidity of the soil and in time to a toxic accumulation of elements such as iron, manganese and aluminium.

Environmental pollution: The contamination of soil and water from industrial activity and the widespread use of chemicals in agriculture, industry and households is in many places causing damage to crops. Air pollution as a result of burning fossil fuel (coal and oil), and of the destruction of industrial and household waste in furnaces, has become a major cause of damage. Huge quantities of sulphur dioxide, nitrogen oxides, ozone and other toxic gases as well as particles go up into the air later to fall as "acid rain" onto crops, forests, range land and surface water. Acid rain penetrates into the root zone of trees, often killing vital mycorrhizas around the rootlets, which help trees in the uptake of nutrients. The pine and spruce forests of Middle and Northern Europe are seriously affected and numerous trees have been killed.

Contaminants in the air may cause phytotoxic symptoms on all plant parts and thereby reduce their market value, or they may even kill plants. Air pollution may also make crops more susceptible to fungal diseases. For these reasons, in many places, horticulture had to give way to advancing industry.

Pesticides applied in overdose, in a faulty formulation or in a wrong way, may also cause injury to crops. Obviously, traces of herbicides on trees and crops, e.g. stemming either from drift or from the use of a contaminated sprayer, may inflict damage or even cause complete crop loss. Residues in the soil from previous herbicide applications or chemical soil sterilization may cause poor emergence and growth or even complete crop failure. Moreover, they may enhance susceptibility to fungal diseases.

Symptoms of phytotoxicity are generally discolouration of the leaves and buds, followed by necrosis of the tissues along the margins and at the tip of leaves. More severe phytotoxic scalding leads to spreading of the lesions over the total surface and to shedding of the leaves.

Animal parasites

Another group of non-infectious diseases or disorders is the one caused by animals. Many kinds of insects can cause a disease condition in plants by eating large parts of the plant thereby weakening the entire plant, some others by secreting toxic substances, and still others by producing overgrowth such as galls and witch's brooms or other abnormalities in plant parts. Many species of nematode cause diseases in the strict sense of the word. First of all, the plant is weakened by direct withdrawal of food. In addition, tissue functioning is disturbed by mechanical damage; the roots affected cannot absorb as much water and nutrients as healthy ones, and the leaves affected assimilate less well. While they feed, many nematode species release enzymes into the plant tissues which gives rise to physiological and histological changes; this occurs, in particular, among the endoparasitic species which reproduce within the plant. Some nematodes are vectors of viruses or provide an entry point to other pathogens such as fungi and bacteria; this function of transmitting pathogens may be more harmful than their direct feeding activity.

SYMPTOMS OF PLANT DISEASES

The way plant diseases express themselves by way of specific symptoms often holds important clues for diagnosis and treatment. The diagnosis must, however, take into account all environmental conditions, since these affect symptom expression. For instance, soil-borne diseases and nematodes may inhibit root development and thus cause poor growth. Likewise, a hard and compact soil structure near the surface may prevent the development of plant roots with a similar result.

Nutrient-deficiencies produce symptoms which include chlorotic or mottled leaves, redish or purple discolourations of the leaves, wilting or dying of leaves and buds etc.

General kinds of visible disease symptoms

- Necrosis: the rotting or decay of tissues;
- Hypoplasia: underdevelopment of tissues;
- Hyperplasia: enlargement of tissues owing to excessive production of cells;
- Hypertrophy: abnormal overgrowth of tissues owing to enlargement of the cells; and
- Discolouration of leaves, fruits and vasculair tissues.

Necrosis of plant tissues may be general or local, and proceed gradually or rapidly. General decay is usually referred to as rotting. It occurs especially in fleshy organs such as tubers, bulbs and fruits, but it may also affect leaves, stems and roots. There are wet rots and dry rots; soft rots and hard rots; white, brown or black rots, as well as abiotic rots. The bacteria and fungi which cause general rotting usually continue to grow as long as there are healthy tissues left or until the whole fruit or plant collapses, unless such pathogens are checked by outside factors.

Local necrosis normally results in limited lesions of various kinds such as leaf spots and fruit spots, even in susceptible tissues and under favourable conditions. For example, the typical black, sunken lesions caused by bean anthracnose disease on bean pods seldom exceed one centimetre in diameter. Many kinds of local lesion, in particular leafspots, may be so characteristic in shape, pattern or colour as to have diagnostic value; e.g. shot-hole spots and spots showing concentric rings. Some leafspots, especially those near the base of the leaflets, cause defoliation which may decrease yield. Necrosis may also be caused by viruses, such as the Yellow crinkle virus

that produces small "shot hole" lesions in tomato leaves, and Citrus exocortis virus that causes the development of yellow lesions and cracks in the bark of susceptible citrus trees, leading to scaling of the bark over the entire length of the stem.

Many cankers could be regarded as slow rots of the outer parts of stems of herbaceous and woody plants, but they exist on living tissues. Most cankers on woody plants remain small, but some can spread rapidly in the inner bark and may girdle the branches or stem base, resulting in the death of the parts above the canker.

Underdevelopment of tissues may affect the whole plant or only certain parts. Stunting and dwarfing are caused by several virus diseases such as tungro in rice. Several fungi such as cereal rusts cause underdevelopment of grain kernels, although the grains themselves are not infected. Bacterial wilts may stunt the growth if they do not kill a plant completely. Underdevelopment of a plant may also result from damaged root systems, shortage or excess of water or nutrients, and excessive acidity or alkalinity of the soil.

Hyperplasia or the excessive production of cells may, for instance, be induced by bacteria causing crown gall in many kinds of plants that stimilate abnormal cell division. Smut fungi in cereals may cause tumour-like swelling of kernels, which contain masses of spores. Some virus diseases, such as swollen shoot of cacoa and groundnut rosette virus, produce charactertistic deformations.

Abnormal growth may become apparent as tumour-like swellings, deformations, witch's brooms, club roots, swollen shoots or excessive growth of roots or tillers. Certain kinds of insects and nematodes cause galls or root knots.

Discolouration of leaves, flowers and other plant parts may be so specific as to lend its name to the disease. This is particularly the case with viruses, such as those causing chlorotic or distinctly coloured mosaic or striped patterns on leafs and flowers, e.g. Tobacco Mosaic Virus

These roots of the cucumber plant are stunted by the root knot nematode Meloidogyne incognita.

The typical symptom of red ring disease inside a coconut trunk is caused by the nematode Rhadinaphelenchis cocophilus.

(TMV), Yellow Dwarf Virus in barley and *hoja blanca* (Whiteleaf) Virus in rice. Vascular diseases, such as those caused by species of the Fusarium fungus in many crops such as cotton, cabbage and cucurbits, often produce a typical brown discolouration of the xylem vessels and, at a later stage, wilting.

A deficiency of certain essential nutrients may become manifest by a typical change of colour of some parts or of the whole plant, e.g. mottled, chlorotic and redish-purple. Such colour changes together with other symptoms, such as rots and wilts, are difficult to diagnose.

Fructifications of fungi, such as mushrooms, often have characteristic shapes and colours which offer a direct clue to which type of fungus is present, e.g. powdery mildew or smut.

INFECTIOUS MICROORGANISMS

Microorganisms distinguish themselves from higher plants and animals, by the following characteristics:

* Very small size cells; bacteria commonly measure about 1 micrometer. Fungi are usually several times thicker and often much longer than bacteria as they have thread-like cells separated by partition walls.

* Many microorganisms, such as bacteria, yeasts, protozoa and many algae, are one-celled (unicellular) organisms. Although certain bacteria and fungi form irregular agglomerations or chains of cells, each cell remains an individual entity. Owing to this autonomy, each cell in a culture can reproduce itself, thus enabling a very rapid increase in the number of cells.

* Bacteria are procaryotic organisms, i.e. their genetic material, desoxyribonucleic acid (DNA), exists as a single large chromosome in the cytoplasm, without an envelopping membrane. Bacteria thus have no cell nucleus and are always haploid.

In contradistinction to bacteria, fungi and all other microorganisms are eucaryontic organisms; they have got -- just as the higher plants and animals have -- a cell nucleus that is separated from the cytoplasm by an envelopping membrane; the nucleus contains chromosomes.

* Microorganisms demonstrate important differences in physiological properties. For instance, the variation of metabolic processes in microorganisms is considerably larger than in higher organisms. Almost all higher plants are photo-autotrophic, meaning that they are able to synthesize carbohydrates from CO_2 and water using sunlight as their source of energy (photosynthesis). Amongst the microorganism, photo-autotrophic types can be distinguished, such as algae, certain protozoa and some bacteria, which have a metabolism resembling that of green plants. Other microorganisms resemble the heterotrophic higher animals that always require complex organic compounds which they break down after ingestion or absorption in order to obtain their supply of carbon, nitrogen, sulphur and energy. Many bacteria and fungi combine both properties; although being heterotrophic, they are also capable of using only one organic compound (e.g. an amino-acid) as source of nutrition and energy.

The following very divergent groups of micro-organisms can be distinguished (van Egeraat and Deinema, 1983):

- Protozoa (amoeba, flagellates and ciliates)
- Algae and algae-like plants (having chlorophyll)
- Fungi and the related yeasts
- Bacteria, including the actinomycetes and cyanobacteria.

Viruses and bacteriophages (the viruses of bacteria) cannot be regarded as microorganisms because they demonstrate neither metabolic nor division processes.

Protozoa are regarded by some taxonomists as belonging to the lowest class of the animal kingdom because they ingest their food via a mouth-like opening and because their cells, as in higher animals, are surrounded by a membrane and lack a cell wall. However, since a large group of protozoa also possess chlorophyll, these organisms appear to occupy an intermediate position between the animal and plant kingdoms.

The other microorganisms mentioned above absorb their food exclusively -- as do the higher plants -- in a dissolved state via the surrounding cell wall and cell membrane.

A great number of protozoa are parasitic pathogens of man and animals; for example, amoeba cause amoebic dysentry and flagellates, *Trypanosoma* spp., cause sleeping sickness in man and cattle in Africa.

Other protozoa are pathogens of plants and may cause disastrous epidemics; protozoic diseases are particularly rampant in South America in coffee, oil-palm and cocoa-palm.

Fungi

Fungi are heterotrophic microorganisms which have no chlorophyl and thus cannot synthesize carbohydrates, but which obtain their energy by oxidizing organic compounds, such as sugars, synthesized by higher plants or animals. The oxidation process often remains incomplete in which case organic acids are produced. Many yeasts (one-celled fungi) are facultative aerobes; they can also grow under anaerobic conditions whilst getting their energy from fermentation, i.e. by using carbohydrates as H-acceptors.

Fungi live either as saprophytes, which comprise the majority, on dead organic matter, or as parasites on living plants, animals or other fungi. If the parasites exude toxins which kill the host plant, they may also be called perthotrophe organisms. Many parasitic fungi go through saprophytic and parasitic phases; they are facultative parasites. An example is *Armillaria mellea*, the "honey disease" which causes split roots and root-rot in plantation crops such as coffee, rubber and tea and survives in the soil on left-behind stumps and pieces of roots. Only a few taxonomic groups, such as rusts (Uredinales), powdery mildews (Erysiphales), and downy mildews (Perenosporales), are obligate parasites.

Some of the simpler fungi consist of only one cell, but most fungi are composed of many cells and develop a "hyphal" structure. **Hyphae** are thread-like organs that may be divided by partition-walls (septa). But the hyphae are not divided into independent cells, because the partition-walls are perforated, enabling a free passage of cytoplasm and nuclei through the hyphal tube. Hyphae measure up to 5 micrometres in diameter, but they may become very long as they continue to grow in length as long as substrate is available. For that reason cultural media and spoiled food may be covered with an extensive **mycelium**, the fluffy, thread-like network of strongly branched hyphae.

Propagation of fungi takes place by means of **vegetative spores** formed asexually or by **generative spores** which are formed by a sexual fusion process and doubling of chromosomes. The following types of vegetative spores have been described:

* **Zoospores** (although not all zoospores are vegetative): these spores are mobile protoplasts provided with one or more flagellae and formed in a sac-like organ (sporangium). Zoospores can swim around in water and germinate after some period of time;

* **Sporangiospores**: these are immobile spores which also develop in a sac-like structure (sporangium);

* **Conidiospores** or **conidia**: these spores develop externally at the tip of conidia-bearing hyphae (conidiophores);

* **Chlamydospores**: these are cells of the mycelium which grow thicker and thus are able to withstand a period of unfavourable conditions. They may be regarded as resting spores; and

* **Oidea**: these are rod-like spores which are formed by the desintegration of septate hypae.

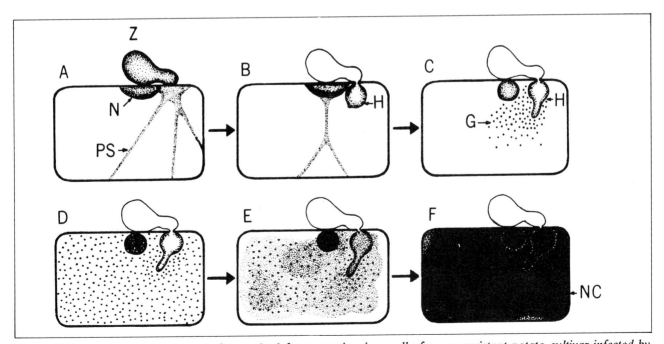

Figure 13.3: Stages of development of necrotic defense reaction in a cell of a very resistant potato cultivar infected by Phytophthora infestans. *N = nucleus, PS = protoplasmatic strands, Z = zoospore, H = hypha, G = granular material, NC = necrotic cell. (Source: P. Agrios, after K. Tomiyama, 1956).*

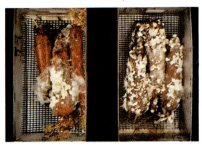

(Left) carrots infected by the fungus Sclerotinia sclerotiorum *are covered by a whitish mycelium from which some hardened, black mycelial tissues (sclerotia) develop. The carrots in the left tray are additionally infected with an antagonistic fungus that slows down the development of* S. sclerotiorum. *(Right) the generative fruiting bodies of* S. sclerotiorum *(apothecia) on the top of which spores are formed and released. (Photos courtesy of IPO, the Netherlands.)*

Fungi often produce spores in abundance, for example, a single wheat grain showing a black mass of the stinking smut fungus, may contain 5-10 million microscopic spores. Spores can be easily dispersed by wind (the most common vehicle) and water (rain, irrigation water etc.), but also by insects, animals and man as vector or on tools and seed and planting material. Fungi may also be spread in the form of mycelium.

Fungi can survive unfavourable periods, such as a winter, in the form of resting spores or sclerotia and in the form of some vegetative or generative organ. They may survive on or in the soil (from where they attack their hosts), or on living or dead organic matter. Some parasitic fungi can bridge unfavourable periods by infecting alternative hosts. This is, for instance, obligatory for a great number of Uredinales.

Generally, the germination of fungal spores and infection primarily depends on free water on the host surface. During the infection process some fungi may get easy access into the plant tissues by way of the stomata or fresh injuries of any kind. Other fungi first penetrate through the healthy cuticle and epidermis. The activily growing hyphae break through the host's defences with the help of enzymes. Host-plants oppose fungal penetration by means of preformed chemical substances, pre-existing barriers, and active host responses such as cork-formation and hyper-sensitivity, i.e. the host cells at the infection site die so rapidly that hyphal growth of the attacking fungus is halted. Thus, the interaction between host and pathogen determines the amount of disease (number and size of lesions) and duration of the incubation period. (See previous page.)

Some fungi have a very restricted range of host plants, but other fungi may infect many different plant species. For example, there exist many species and *formae speciales* of the *Phytophthora* fungus which may cause tuber-rot in potatoes, root-rot in avocado; foot-rot in citrus; stem-rot in soybeans; black-rot in cacao; bud-rot in palms, leaf disease of rubber trees or girdle canker of the stems of cinchona trees.

Taxonomic classification

The kingdom of the fungi contains two divisions, namely: the Myxomycota or slime molds which lack a mycelium, and the Eumycota or lower true fungi. Fungi belonging to the division Eumycota include the four long-recognized classes:

* **Oomycetes**: root-, collar- and stem-rots, damping-off, blights, downy or false mildew;

* **Ascomycetes** (sac fungi): moulds, yeasts, ergots, true or powdery mildew;

* **Basidiomycetes** (club fungi): rusts, smuts, mushrooms; and

* **Deuteromycetes** (fungi which reproduce by vegetative spores only): blasts, blights and rots.

More than 7,500 fungi are either of economic importance or parasites of agricultural crops. The taxonomic grouping of fungi is based on the type of sexual reproductive structures and the classification is based largely on the characters of the spores and the way in which they are produced.

Yeasts are one-celled fungi (usually oval-shaped or round) which reproduce vegetatively by budding or fission, but some also produce generative spores. As in the case of the fungi, the yeasts may belong to the groups of Ascomycetes, Basidiomycetes or to the *fungi imperfecti*.

The reproduction of bacteria seems to be a simpler matter and thus classification is rather straightforward. Bacteria are regarded as belonging to another division, the Schizomyphytae, that contains two classes:

* Schizophycetes (fission algae), and
* Schizomycetes (bacteria).

Reproduction of fungi

Reproduction of plant pathogens serves four important purposes: multiplication, dispersal, survival during unfavourable conditions and genetic diversity. The process of reproduction is a complicated matter, as fungi produce many kinds of reproductive structures. Generative spores are formed after a sexual process in which two or more specialized fungal cells fuse. After these cells have merged, the nuclei -- each with a single set (haploid) of chromosomes -- will also fuse. From the conjugated cells a new mycelium may grow, which remains diploid (having a double set of chromosomes) until a meiotic division takes place that reduces the nuclei to the haploid phase again. Mycelial cells can thus contain a nucleus in the haploid or, mostly transitionally, diploid phase. Further, they may contain one nucleus, or two nuclei (dikaryotic) or several nuclei (polykaryotic).

Most fungi reproduce both asexually and sexually.

In many of them, asexual reproduction results in rapid multiplication and dispersal of the fungus, whereas sexual reproduction results in the production of spores that can survive unfavourable conditions. Genetic diversity can result from mutation, from vegetative fusions in the asexual stage (parasexuality), and from recombination and segregation in the sexual stage.

Some fungi are pleiomorphic; this means that they produce several kinds of spores. Usually, a fungus produces a single kind of sexual spore: for example, ascospores in the Ascomycetes, and basidiospores in the Basidiomycetes. The details of sexual reproduction differ for the different classes of fungi and even for species within the same genus. It has been demonstrated that environmental conditions, such as light intensity and substrate composition, determine -- within the limits imposed by the genetical constitution -- when, where and how the reproduction of fungi will occur. Fungi in which only the asexual forms of reproduction are known are called **Deuteromycotinae**. The reader may refer to specialized textbooks on phytopathology for more information on the reproduction of fungi.

Bacteria

The bacteria or Schizomyphyta form the largest and most varied group of microorganisms. They are one-celled prokaryotic organisms, i.e. lacking a genuine nucleus, but having a distinct cell wall. They reproduce only by means of fission. This means that a partition-wall is formed within a bacterium which splits lengthwise thus causing a division of the cell into two equal daughter cells. These two grow and divide, and the four go through the same process, so that the increase follows a geometric progression. Under favourable conditions, many species of bacteria mature and divide within 30-60 minutes. A single bacterium could potentially produce about 10^{11} new bacteria within 24 hours provided suitable substrate is abundantly available. Optimum growth occurs around 25°C under condition of high relative humidity.

Bacteria appear in one of three basic shapes: the round ball (coccus); the rod form (bacillus); and the bent forms, i.e. vibrio and the spring-type (spirillum), (see figure). The daughter cells of most species separate, but others may remain tied together as multicellular filaments or agglomerations.

A bacterium is surrounded by a strong cell wall which protects the protoplast against outside influences and maintains the turgor of the cell. Many types of bacteria are equipped with one or more flagellae. These are thin proteinaceous threads which are located at one or several sites of the cell wall and serve to facilitate bacterial movement. Finally, some types of bateria may be surrounded by a thick layer of slime, usually consisting of exudated polysaccharids or polypeptides, that provides a certain degree of protection against viruses (bacteriophages), protozoa and adverse conditions.

Only a limited number of bacteria are pathogenic to man, animals or plants. Others, such as saprophytic bacteria which can exude exo-enzymes, are beneficial since they assist in the degradation of complex substances such as proteins, glycogen, cellulose and pectin, thus providing nutrients to micro- organisms that cannot feed on dead organic matter as a substrate. Others are capable of fixing free nitrogen from the air (e.g. cyanobacteria and *Rhizobium* spp.) or of transforming ammonium salts into nitrates or nitrites.

About 200 important plant diseases are known to be caused by bacteria. Some of these are highly destructive such as cotton blight (*Xanthomonas malvacearum*) causing angular leaf spots, blackarm and boll rot, and vascular wilts such as *Pseudomonas solanacearum* on banana, cassava and many members of the family Solanaceae. Bacteria can cause typical symptoms such as wet rot, dry rot, tumors, necroses and vascular wilts (withering of the plant parts owing to the blockage of water transport in the xylem vessels).

Pathogenic plant bacteria are all rod shaped, usually motile by means of flagellae, about 2-4 micrometres long, and more or less aerobic. Plant pathogenic bacteria belong to one of five Gram-negative genera: *Agrobacterium*, *Erwinia*, *Pseudomonas*, *Xanthomonas* and *Clostridium* and to two Gram-positive species of *Corynebacterium* and *Streptomyces*. The most important genera, comprising by far the majority of plant pathogenic species, are *Pseudomonas* and *Xanthomonas* which usually form copiously slimy colonies on media containing sufficient glucose or dextrose.

Note: The terms Gram + and Gram- refer to the staining of bacteria with particular dyes and the subsequent loss of colour in an alcohol solution for the Gram-, but not for the Gram + bacteria. This reaction to Gram-staining depends on typical differences in the chemical composition of the cell walls of the two kinds of bacteria and is, therefore, useful for classifying bacteria.

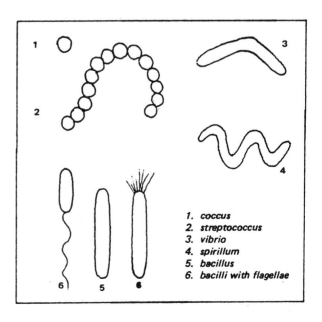

1. *coccus*
2. *streptococcus*
3. *vibrio*
4. *spirillum*
5. *bacillus*
6. *bacilli with flagellae*

Plant pathogenic bacteria fail to penetrate the epidermis of plants directly, but they may enter more passively through injuries, stomata or lenticels. Within the host plant, bacteria are moved along by the sap stream through the vascular tissue and veins. If a bacterial infection occurs outside the veins, it tends to remain localized in inter-cellulair spaces. At first they form discrete colonies, later they may spread individually after dissolving (pectolysis) the parenchym tissue which becomes water-soaked.

If the plant reacts by accumulating a dense material in the inter-cellular spaces and vacuoles between the macerated and healthy tissue, the spread of the bacteria is prevented. This happens in potato tubers infected with *Erwinia carotovora*. The soft rotting *Erwinia*, *Pseudomonas* and *Clostridium* species may eventually cause complete tissue collapse. Many *Pseudomonas* and *Xanthomonas* species, however, show up only as limited spotting or streaking. In the case of *Pseudomonas,* toxins may be translocated through the phloem and result in disease symptoms at a considerable distance from the bacteria, as well as causing haloes of chlorisis round the lesion, e.g. halo-blight of beans. Frequently, bacteria ooze from injuries or hydatodes as a slimy mass, droplet or thread, from where they may be spread over the field by insects, wind and rain.

Bacteria are unable to infect undamaged plants except during cloudy, wet weather when they may enter stomata, leaf scars and fresh injuries. Bacteria do not spread during bright sunny days except by insect transmission. Hence, bacterial diseases are most common in areas with warm, humid (cloudy) conditions.

In the case of a systemic infection, the bacterium may be carried by the sap-stream to places where new seeds are being formed (ovaria), thus infecting the seeds internally. Examples of seed-borne bacteria are: *P. phaseolicola, P. tabaci, X. malvacearum, X. campestris* and *X. phaseoli.* Other bacteria live exclusiviely in or on host plants in a latent form in buds or infected tissues (e.g. *E. amylovora*), except when they are transmitted by means of cuttings and grafts. For instance, *E. carotovora* and *X. manihotis* are spread in infected propagation material. Soil-inhabiting bacteria such as *P. solanacearum* and *C. tritici* are often transmitted by nematodes.

Although only *Clostridium* and *Streptomyces* produce spores that can resist high temperature and unfavourable conditions, the vegetative cells of several *Xanthomonas* species may survive for years in dry leaves and other crop debris. Thus, removal of debris is a sound preventive measure.

Rickettsia (plural: rickettsias)

These are sub-microscopically small, one-celled, prokaryotic organisms which commonly multiply by fission within the cells of the host. Similar to bacteria, the rickettsias are surrounded by a rigid cell wall and thus, they are immotile. Rickettsias are known particularly as pathogens of arthropods, and in some cases also of vertebrates including man; since 1970, they have been found to occur in plants too. Within the plant they are moved along by the sap-stream through xylem and phloem vessels from where they are taken up and transmitted by sucking vector insects, such as plant and leaf hoppers, aphids and jumping plant lice. Recently, several diseases have been attributed to rickettsias and other small micro-organisms, which were previously thought to be caused by viruses. Rickettsias can be grown *in vitro* on special substrates on which they form egg-shaped colonies. As they form a cell wall, rickettsias are, like bacteria, inhibited in their growth by the antibiotic penicillin (which interferes with cell wall-synthesis).

Mycoplasmas

A mycoplasma is the smallest free-living, one-celled, prokaryotic micro-organism that multiplies within the cells of a host. Mycoplasmas are surrounded by a double membrane but, unlike bacteria and rickettsias, lack a rigid cell wall and are therefore variable in shape. Since 1967, mycoplasmas have become known as causal agents of numerous plant diseases which were earlier attributed to viruses. In particular, the group of witch's-broom diseases which are characterized by a stunted growth of the plant, sprouting of normally dormant buds (giving the plant the appearance of a witch's broom), greening of flowers and the replacement of floral parts by leaf-like structures (termed phyllody), and by retarded growth and sometimes yellowing of the leaves.

Owing to their minuteness and changeable forms, mycoplasmas are, like viruses, able to pass through bacterial filtres and to infect plants systemically via the phloem vessels. Like persistent viruses, they are transmitted by insects, such as cicadellids (plant and leaf hoppers) and psillids (jumping plant lice), which feed on the content of phloem vessels. They are also transmitted mechanically through grafting and by parasitic plants such as dodder.

The symptoms of witch's-broom diseases can be, temporarily, decreased by treating the plant leaves and roots with the antibiotic tetracycline or aureomycine. In practice, this treatment is too expensive as it needs to be repeated. Treating affected plants with warm water has sometimes been successful. As they lack a cell wall structure, mycoplasmas are not affected by the antibiotic penicillin.

Researchers have not yet succeeded in growing plant-parasitic mycoplasmas on artificial substrates, as they have been able to do with several related spiral-shaped, motile *Spiroplasma* species. Since the proper identity and taxonomic relationship of the above-mentioned microorganisms have as yet not been clearly established, they are often collectively referred to as **mycoplasma-like organisms (MLOs)**. Rickettsias and MLOs, however, differ fundamentally from viruses by having both DNA and RNA as well as having their own metabolism.

Examples of diseases caused by rickettsias or MLOs are: Citrus greening, Citrus stubborn, Coconut lethal yellowing, Cowpea witch's broom, Sugar-cane grassy shoot and Rice grassy stunt disease.

A cowpea plant infected by Cowpea Aphid-borne Mosaic Virus (CAbMV) shows typical yellowing symptoms.

VIRUSES AND VIROIDS

Viruses are extremely small parasitic particles which cause diseases in man, animals, plants, fungi and bacteria. Many types of viruses lower the yield of infected crops dramatically by causing yellowing and chlorosis, more or less rapid decay and necrosis, or by affecting the quality of the harvested produce. Even "symptomless" viruses reduce a plant's life-span and vitality and raise its susceptibility and sensitivity to other pathogens and stress factors. Viruses are present in all vegetative propagation material from infected plants. An increasing number of viruses appear to be passed on through seeds, although often only in certain host plant species and in a small percentage of the seeds. Many viruses are easily dispersed in the field through vector insects or physical contact, which may lead to a massive outbreak. The rapid spreading of viruses by vectors and through seeds and plant material over often large distances is a matter of international concern. Like viruses which affect the health of humans and animals, the spread of plant viruses throughout the world appears to be unstoppable and virus diseases are likely to become a major problem in third world countries along with rapidly changing agricultural practices (Bos, 1987).

The structure of viruses

Viruses are infectious entities, composed of nucleic acid (either DNA or RNA) and protein, of such small size that they can pass through bacteria-retaining filters. Viruses lack organs and their own metabolism and, therefore, are not recognized as organisms. However, the ability of viruses to replicate and undergo mutation is a reason to consider them to be living matter. Two main types of virus particles can be recognized through an electron microscope:

- a rod-shaped type, measuring 150-750 nm x 10-30 nm; and
- a spherical type, measuring 25-75 nm in diameter. (One nm (nanometer) equals 0.001 micrometer.)

Since virus particles have no metabolic capacity (in the way microorganisms are able to synthesize high molecular cell-components from simple chemical compounds with the help of a whole series of enzymes), they must derive their building material and energy from a living host cell. Thus, a **virion** -- an individual mature particle in a population of virus particles -- lives at the expense of its host cell, which provides the nucleotides, proteins and amino acids needed for its construction. A virion does not form a new virion in the way a cell forms a daughter cell. Owing to their parasitic behaviour, a number of viruses are causing infectious, often very serious diseases. Examples are rabies, influenza, smallpox, infantile paralysis, swine fever in man or animals; or mosaic and leafroll virus diseases in plants. Viruses which are parasites of bacteria are called **bacteriophages**.

A virion consists of a **nucleic acid component** and a **surrounding protein shell**, called the **capsid**. The nucleic acid forms the genome, which consists of a single or double strand of RNA or DNA. This genome contains the genetical information for the formation of the functional and structural proteins which are required respectively during the replication of virus-nucleic acid and as a component of the virus particles. A virion consists of only one type of nucleic acid which is usually DNA for animal viruses and RNA for plant viruses.

The protein shell or capsid is formed from a large number of protein molecules of one or more types which surround the nucleic acid in a characteristic way for each virus (as can be seen with the help of an electron microscope).

The nucleic acid component together with the envelopping capsid comprise the **nucleocapsid** which, for plant viruses, is usually equal to the virion.

A virus, in which the genetical information necessary for multiplication is not located in a single strand of nucleic acid only but divided over a number of different strands, is said to have a **divided genome**. Such divided nucleic acid segments can be contained either within one nucleocapsid (virion) or in separate nucleocapsids (virions). Virus replication can only take place if a susceptible host cell has been infected with the collective genome segments. Therefore, in the case of divided genomes, the infectious agent of one virion alone does not exist, but as a combination of virions, which together contain the complete virus genome.

Virus transmission

A virus is commonly transferred from an infected host to a non-infected one by a vector, which is often an insect, mite or nematode, but may also be a fungus or parasitic flowering plant (e.g. *Cuscuta* spp., dodder). Sucking insects, such as aphids, leaf and plant hoppers, thrips and the whitefly, dominate among the insect vectors, whereas nematodes and fungi play a role in the transmission of soil-borne virus diseases. One particular vector may be able to transmit only one virus or many viruses; one particular virus may have one or more vectors.

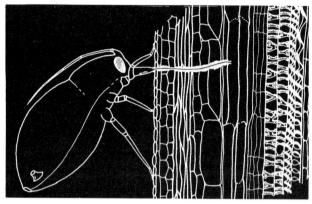

An aphid penetrates the phloem tissue f a plant stem or leaf with its mouth-parts (stylet); by sucking plant sap, it may transmit virus diseases. (Drawing courtesy of IPO/Harrewijn)

The character of the relationship between vector, virus and host enables a differentiation in "persistent" and "non-persistent" viruses. If a virus can be transmitted almost immediately after an insect feeds on a virus-infected plant, and the insect remains infective for only a few hours, then the virus is called **non-persistent**. This means that the insect, after feeding, can infect only a limited number of healthy plants, unless it happens to feed on a diseased plant again. A small group of **semi-persistent** viruses can be retained by their vectors for a maximum of several days only. For example, the "rice tungro" virus and the "rice ragged stunt" viruses are taken up from an infected plant within 30 minutes and retained by their vector, the brown planthopper, for about five days and two weeks, respectively. The length of the retention time depends usually on the length of the **acquisition feeding** time and the efficiency of the transmission on the length of **inoculation feeding** time. The non-persistent viruses seem to be carried externally on the stylets and not to multiply within the vector insect.

If the insect remains infective for a long period (even for its lifetime), then the virus is referred to as **persistent**. In the case of a **persistent virus**, a longer acquisition feeding time of usually several hours duration and a certain **latency period** is required, i.e. the time during which the virus remains inside and sometimes multiplies within the vector insect before it becomes infective. For example, the "grassy stunt" virus can be picked up by brown planthopper nymphs and adults by feeding for 5-10 minutes, but an average of 10 days must elapse after this **acquisition feeding** before the insects can transmit the virus. Persistent viruses pass through the vector's moulting stage, and some may even pass on to the progeny of the vector through their eggs.

Some viruses may also be transmitted mechanically without the intervention of a vector insect. The sap of an infected plant can reach a healthy one either by means of tools used in cutting and grafting, by rubbing one plant against another or, and this is a method often used in experiments, by rubbing the virus-containing sap onto the leaves of a healthy plant. Certain viruses can be transmitted both by insects and through sap transmission; these are usually non-persistent viruses such as the Potato Y Virus. Others can be transmitted with plant sap but not by insects, for example, Tobacco Mosaic Virus. Viruses which belong to the latter group are often very infectious, so that such a virus is already passed on in the field from plant to plant by physical contact between above ground or underground plant parts.

In the first section on viruses it was mentioned that viruses can persist only in living host cells. An exception to this general rule are some very stable viruses, such as the Tobacco Mosaic Virus, which persist in dried tobacco leaves. Many viruses can also pass through a winter period within perennial crops and weeds. Whereas viruses are commonly passed on through vegetative parts, such as cuttings and tubers (as in seed potatoes), they are less often passed on to the progeny through seed. But in the family of leguminous plants (pulses, beans, groundnut etc.) many viruses are seed-borne and able to infect more than one type of crop. Examples are Broad Bean Stain Virus, Bean Roll Mosaic Virus, Pea Seed-borne Mosaic Virus, Peanut Stripe Virus and Soja Stripe Virus. In East Java (Indonesia), for instance, Cowpea Mild Mottle Virus and Peanut Stripe Virus occur very commonly in groundnut crops, the latter virus often infecting the entire crop.

Nomenclature

The names of plant viruses are not yet regulated by an international code. They are based mainly on the common name of the host plant in which the virus was first reported (or in which it occurs most frequently) and on its most characteristic symptom. The name is usually abbreviated in publications after the first mention, for instance, Tobacco Mosaic Virus (TMV). For viruses, a fitting definition of the concept "species" is still lacking and thus there is no possibility of developing a natural system such as for organisms upon which the binary nomenclature is based. However, the International Committee on Taxonomy of Viruses has, since its establishment in 1966, been trying to devise a Latin binominal designation of viruses on the basis of the properties of virus particles and of the group to which they belong.

A collection of viruses or virus strains having all or

nearly all the main characteristic of the type member in common has been termed a "**virus group**". The names of groups of viruses are based either on the best known member of the group, such as the **tabamovirus** group and the **tobravirus** group with, respectively, the Tobacco Mosaic Virus and the Tobacco Rattle Virus as type members, or on the most important properties of the group, such as the **nepo-virus** group for the nematode-transmitted polyhedral particle viruses for which the Tobacco Ring-spot Virus serves as the type member. Sixteen groups have so far been distinguished, three of which comprise closely related strains of a single virus each. Each of the remaining 13 groups is composed of more than one virus; for instance, the potyvirus group with Potato Y Virus as the type member also includes other aphid-transmitted viruses with 700-800-nm-long, rod-shaped particles.

Symptoms of virus-induced diseases

The symptoms of diseases caused by viruses are numerous and higly variable as their expression is influenced by the condition of the plant and environmental factors such as light, temperature and humidity. Visible symptoms may include local necrotic lesions, mosaic or mottled pattern of lighter and darker green leaf-tissues (chlorosis), yellowing or other dis-colourations including ring-spots or other distinct patterns, virescence (greening), vein clearing, reduction of plant growth leading to dwarfing or stunting, distortion due to unequal growth of cells, excessive tillering, flower and fruit variegations, wilting eventually followed by defoliation, and yield depression. A plant may be infected by several viruses at the same time, the symptoms of which may be manifested concurrently, and thereby reinforce or mask one another.

The study of symptoms (termed symptomatology) may be used for diagnosing the virus disease, but one must take into account that similar symptoms may be caused by different viruses, or even by nutrient deficiencies and genetic disorders. The severity of the symptoms depends on the sensivity of the plant and the virulence of the virus. Some plants show a considerable tolerance (have a low degree of sensivity to the attacking virus) and may not even show a visible reaction (**symptomless carriers**). In contrast to symptomless carriers, some other plant species are so very sensitive that they die shortly after becoming infected.

Infected but symptomless carriers of a multiplying virus are a potential source of infection. For instance, the Swollen Shoot Virus of cacao has since long occurred in wild tree species of West Africa, but has remained symptomless. Large-scale planting of cocoa trees, which constituted a new crop for West Africa, brought about an epidemic of Swollen Shoot Virus that killed millions of cacoa trees. In order to avoid such costly failures in the case of introductions and breeding programmes, the possible presence and identity of symptomless viruses must be checked by inoculating a sensitive experimental host. Such **indicator** plants react with characteristic symptoms to certain viruses and thus reveal their presence. The primary leaves of *Phaseolus*-beans and the cotyledons of cucumber are well suited as indicators. The use of detached leaves and leaf-pieces enables testing in standard series under controlled climatic conditions in the laboratory.

Detection and diagnosis of viruses

In addition to using indicator plants, the identity of a virus can also be determined by means of electron microscopy and of **serological** testing with anti-sera. An electron microscope enables the detection of many types of viruses at 40,000 to 100,000 x magnification. An anti-serum as used in serology is the blood serum of an animal such as a rabbit that has been injected with a foreign substance (**antigens**), e.g. a plant virus. The animal's system reacts by forming **antibodies** in the blood against such intruding antigens. Such antibodies are proteins dissolved in the blood, from which they can be collected. The anti-serum (blood serum containing antibodies can be separated from the blood corpuscles by centrifugation. This method can be used for detecting a virus if the tested virus is similar to the one against which the virus was formed. Thus, serology is a means of detection and making a diagnosis of those viruses which are transmitted in plant sap.

Antigens such as viruses (usually consisting of proteins or polysaccharides), react with the antibodies by a typical agglutination, precipitation or, as in the case of the serological method ELISA, by a colouring of the substrate. The Enzyme Linked Immuno Sorbent Assay (ELISA) is employed on a large scale for detection of viruses in crops and in vegetative material such as seed potatoes, grafted fruit trees and ornamental trees and flower bulbs. In principle, propagation material should be free of virus disease, because a virus infection can hardly

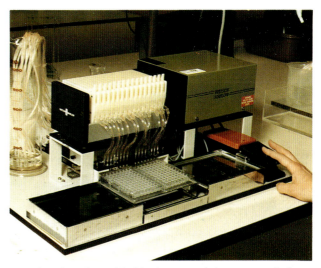

A microtiter plate with 96 tubes containing extracted plant sap and antiserum is part of the standard equipment in the enzyme linked immunosorbent assay (ELISA) serological test for virus detection. (Photo courtesy of LNV, the Netherlands)

ever be controlled in vegetatively propagated plants. In particular, a virus infection in propagation material destined for export must be detected in time; otherwise the whole shipment may be refused by the receiving country.

Further, some species or varieties of plants may be resistant or immune to infection by certain viruses. Their resistance may be based on either of the following two defense mechanisms:

- An immunological defence against viruses, which multiply within the host cells, may be carried out by so-called cytotoxic cells. These cytotoxic cells are able to recognize and destroy the infected cells.

- The second defense mechanism of the host exists of the forming of specific proteins (antibodies) in response to intruding virus particles (antigens).

Serological assays have several advantages such as the following:

* As the reaction is very specific, it is possible to test a suspension containing a mixture of viruses, provided the latter are not related;

* The method can be used for viruses for which no indicator plants are known; and

Table 13.1: Diagnosis of an unknown virus disease

Noticing the diseased plant species

Studying the symptoms

Estimating the place, conditions and degree of incidence

Checking its pathogenicity through:

- inoculation with plant sap
- transmission by insects or mites
- transmission to indicator plants, e.g. by grafting

Inoculation of a series of indicators followed by separate back-inoculations onto parallel series of indicator plants for the purpose of:

- recognizing possible complex infections
- selective isolation
- assessment of a range of host plants and their symptoms

Determining the persistence of infectiousness in crude plant sap

Identification with the help of electron microscopy

Identifation by means of serology

Purification and preparation of specific antibodies

Inoculation of a healthy plant in order to check if the purified sap induces the same virus disease

* Anti-sera can be stored for a considerable time; thus, they can be used to compare identities or concentrations of virus extracts prepared at different times.

Viroids

A viroid is a pathogenic agent, formerly assumed to be a virus, now known to consist solely of a circular, single strand of ribonucleic acid without capsid. The available genetical information of a viroid is insufficient to enable the coding of a protein needed for a capsid shell. The molecular mass of viroids is only a fraction of the nucleic acid mass present in viruses. These minute agents may nevertheless be the cause of serious diseases, such as Potato Tuber Spindle and Citrus Exocortis.

INTEGRATED CONTROL OF PLANT DISEASES

Epidemiology

As in the case of damage inflicted by animal pests and competitive weeds, the purpose of all efforts to control any plant disease is to prevent loss in terms of lower yields and quality, which would affect the farmers' income and the overall food situation in a country. Generally, poorly educated farmers have little understanding about diseases affecting their crops since, unlike damage from insects and rodents, they cannot see what is causing it. The explosive speed at which some diseases spread over large areas has always exerted a strong selective pressure on commonly grown crops towards the development of land races which demonstrate a rather stable disease resistance. Although the yields of land races were generally rather low, diseases were seldom the reason for a complete crop failure. Over the last decades, however, owing to agicultural development and large-scale replacement of old land races by higher yielding cultivars, diseases are becoming increasingly important. Consequently, farmers need guidance from extension agents and pesticide salesmen about available methods for preventing and controlling them.

Behind these advisers in the field stand the research scientists who study the many causes of diseases and possible cures. A particular group among them are the botanical **epidemiologists** who study large-scale outbreaks of plant diseases, i.e. **epidemics**. They try to establish to what degree all those factors, which may further an infection in crops and promote the spreading of a disease, occur and interact. Those factors include the aggressiveness and virulence of the pathogen, the number of infective sources and the amount of inocculum they contain, the number of vectors and their transmission activity, as well as the susceptibility and sensivity of the threatened crop. Epidemiologists try to quantify each of these factors and to unravel their interactions. From the field, they follow a path which may take them into the laboratory to look at the (sub-) microscopic structures and processes occurring at the molecular level. With their findings they go back to the field for verification and practical testing. They try to recognize patterns in the

The dynamics and the spread of plant diseases are closely studied in the laboratory and field. Together with Indian colleagues, the Dutch epidemiologist Prof. dr. J.C. Zadoks investigates symptom expression in a rice field in Kerala, India. (Photo by the author)

occurrence of pests and diseases which might enable them to predict, preferably on the basis of a limited number of observations, whether or not (economic) damage is likely to occur and if control measures should be advised. The monitoring and forecasting aspects of epidemiological research involve the application of mathematics and physics and the use of sophisticated electronic sensors and recorders, telecommunications, computers and remote sensing for the purpose of determining the economic injury and control thresholds.

Originally, epidemiology commenced by studying the population dynamics of the pathogens, which generally pass through a number of reproductive cycles during the cropping season. Then attention was directed at the details of each single cycle. Later, epidemiologists tried to understand how pathogen populations develop over the years and how they relate with host (including weed) populations and vector populations within their particular ecological niche.

In a new method of scientific integration, also called the holistic approach, scientists are now fitting together these many details into an overall picture of cause and effect of a disease outbreak. In this approach, crop development depends on the interactions between host population and pathogen population within a complex environment. Furthermore, these populations appear to be affected by a large number of other species, each of which has its own population dynamics. With regard to the occurrence of infection and damage, this approach also recognizes the often decisive significance of the cropping system adopted by farmers.

The opportunity for the spread of a pathogen

within a crop and for epidemic development depends on the type of planting, i.e. mixed crop, land race, heterogeneous cultivar, homozygous line, vegetatively propagated clone etc. (Bos, 1983).

In this way, analysis provides the parts and synthesis fits them together to provide insight, which, in the end, may help the farmers (Zadoks, 1987). This insight refers particularly to the balance of nature which is maintained by a continuous process of mutual regulation of all kinds of plants, animals, micro-organisms and abiotic factors in an undisturbed environment. Balance creates diversity, from which new species may evolve and, sometimes -- as can be observed in mixed cropping and intercropping -- diversity may lead to greater balance. Modern agriculture, with large areas planted to a single crop and with high inputs, upsets this balance and epidemics may arise as symptoms of the disturbance. Fundamentally, pest and disease outbreaks both stem from a temporary state of unbalance, and thus can be remedied by preventing further disturbance and restoring the efficacy of biological control factors.

A typical example of epidemiological research concerns the effect of nitrogen fertilizer on plant growth and disease development. Liberal gifts of nitrogen are commonly applied to the improved, high-yielding cultivars of rice, maize and wheat, because nitrogen tends to raise yields considerably. However, it has been noticed that in maize the incidence of *Helminthosporium maydis* is reduced but that tropical maize rust (*Puccinia polysora*) is promoted. Similarly, it has been observed in rice that rice brown spot (*H. oryzae*) is reduced, but rice blast (*Pyricularia oryzae*), sheath blight (*Rhizoctonia solani*) and bacterial leaf blight (*Xanthomonas campestris* pv *oryzae*) are promoted by high nitrogen levels in the soil. Likewise, the use of fungicides may cause some unwanted effects such as the resurgence of diseases and secondary diseases. Therefore, as in the case of the management of insect pests, European plant protection experts are successfully investigating how to produce higher yields economically with less nitrogen and far less pesticides (Zadoks, 1989). Comprehensive approaches are necessary in which crop-oriented prevention measures based on a sound knowledge of the prevalent pests and pathogens are supplemented by need-based use of pesticides.

Preventive and curative control

Curative control of pests and diseases aims at destroying the invading parasites in order to save and, when possible, to heal the already affected host. Curative control depends heavily on the application of fungicides which are causing similar negative effects of resistance in microorganisms and contamination of the harvested product and environment as is the case with insecticides. Because the negative effects are nowadays less readily accepted by both scientists and the general public, plant protection research is increasingly being focused on the prevention of pest and disease outbreak. However, it will remain important to treat diseased plants in order to prevent their spreading to healthy crops and to minimize injury and loss. However, owing to the special nature of viruses, virus diseases in crops cannot be cured by pesticides at economically acceptable costs.

Prevention refers to the protection of a crop by hindering parasitic organisms and viruses in establishing themselves in the crop and in infecting their potential host. Such protection can be provided either through an increase of the host's resistance to the infectious agent through removal of the sources of infection, through early control of the pathogens and vectors and also through implementation of protective measures. The latter may include quarantine, production and certification of disease- free propagation material, regulation regarding crop rotation and crop-free (fallow) periods etc. Thus, preventive control can be achieved by a range of methods or techniques which frequently also include the use of protectant pesticides. Examples of the latter are the use of organotin for the prevention of diseases in potato, and of chlortalonil against rusts and leafspot diseases in groundnut.

With rare exeptions, the complete eradication of infectious fungi or other pathogenic microorganisms and viruses cannot be achieved and need not be pursued. Only, if a particular disease has not yet gained a foothold in a country or region, then stringent quarantine measures should be enforced and early eradication of the pathogen tried irrespective of the costs involved. When a disease has already become established within an area, then preventive control is the least expensive and surest way to follow. Any control practice should, however, be evaluated on the basis of costs in terms of money and effort, its expected effectiveness and available alternatives. To this end, economic injury threshold values and warning systems need to be developed for major diseases. Examples of such an approach are the sophisticated European warning system for late blight control in potatoes, and the economic injury threshold system for control of pests and diseases in wheat (the latter being called EPIPRE).

Control methods or control strategies

Methods of controlling plant diseases can be distinguished into categories in two ways according to their objective (direct/curative or indirect/preventive) or according to the character of the methods and techniques employed. They are listed in table 13.2 and discussed further below.

Elimination of sources of infection

Prevention starts with practising sanitation at all times throughout the cropping season. All obvious sources of infection must be removed through weeding, pulling out of diseased and volunteer plants (roguing), destruction of stubble, soil treatments etc. Pathogens may survive for a very long time in plant debris (dry leaves,

Steaming of the soil in a glasshouse for disinfection from soil-borne diseases and nematodes. The soil should remain sealed for a rather long period to obtain good results. (Photo courtesy of LNV, the Netherlands)

stubble, refuse etc.), in reproductive plant parts (stolons, tubers, bulbs), and particularly in alternative hosts. Heaps of rotting potatoes have been recognized as the main source of spores of the late blight fungus (*Phythophthora infestans*). Many fungi and viruses may persist in wild plants, weeds and volunteer plants as their alternative host. Through adequate cropping systems and crop rotations, the population build-up of soil-borne pathogens and virus-transmitting nematodes in the soil can be forestalled. Soil disinfection by steaming and fumigation is suitable for small-scale treatment in nurseries and glass-houses, although the latter method creates residue problems in intensive cultures. Application of nematicides (as described in Chapter 12) is effective in high-value crops, but it is expensive and frequently leads to serious residue problems.

Infection can be avoided or minimized by planting a susceptible crop in an isolated place or outside the regular cropping season. Spatial isolation, for instance, can be achieved by growing test plants in an insect-screened glass-house or by wrapping ripening fruits in plastic bags. Isolation in time can be achieved, for example, by early sowing before the peak of vector development or fungal spore release. Seed potatoes are multiplied in cooler areas (e.g. highlands) and early in the season in order, among other reasons, to avoid virus infection through vectors such as aphids. The possibility of year-round cultivation because of improved irrigation is a major reason for increased infection pressure from pathogens as compared with traditional patterns with a dry season fallow.

The use of vigourous and disease-free seeds and vegetative propagation material is a prerequisite for growing healthy crops. Treating seeds with disinfecting and protectant chemicals has become a standard practice

Table 13.2: Development of control methods or strategies

A. Divided according to objective:

    1. Direct, curative: directed at the pest or pathogen with the purpose of curing the affected plant

    2. Indirect, preventive: directed at the protection of plant and crop

        a. Sources of infestation and infection

            * control of weed and volunteer plants
            * avoidance (growing in isolation)
            * use of healthy propagation material
            * quarantine (avoiding the importation of pests and pathogens, health certification)
            * eradication (removal of sources of infestation)
            * decreasing air pollution (reducing emission of gases and dusts)

        b. Vectors

            * control by chemical, biological or other methods
            * avoidance (growing in isolation)
            * preventing transmission by man (quarantine, quality and health certification)

        c. Increasing crop resistance

            * chosing a genetically resistant variety
            * breeding for resistance (including biotechnology)
            * premunition (viruses - see item 4c below)

B. Divided according to method or technique:

    1. mechanically

        * hoeing (weeds)
        * pruning (diseased tissues)
        * removal of diseased or affected plants (eradication, rogueing, negative selection, meristem culture)
        * catching of pest animals
        * air filtering

    2. chemically; selective or non-selective

        * seed disinfection and seed coating with fungicides
        * chemical disinfection of soil and growth medium
        * fungicide spraying, dusting, granular application

    3. physical

        a. heat treatment

            * propagation material (warm water or warm air)
            * soil (disinfection by means of steam or insolation)

    4. biologically

        a. using parasites and predators (against pests)

        b. using antagonists (against fungi)

        c. premunition (viruses; using a weak strain against a virulent strain)

        d. employing resistant varieties, resistance breeding

    5. supervised control: limited application of pesticides on the basis of monitoring systems and economic injury thresholds, employing cost-benefit analysis

    6. integrated control: employing environmentally sound methods, incuding chemical control if needed.

Source: Modified after L.Bos (1986), Institute for Plant Protection Research, Wageningen, the Netherlands.

for preventing soil-borne diseases in intensive agriculture. It is generally considered to be a Government's responsibility to provide the infrastructure for the production and certification of healthy seeds and propagation material for, at least, the locally important food crops.

The role of man as a vector of pests and pathogens through traffic in plant propagules is important. With the rapid internationalisation of production and the trade in plant material, the risks of spreading pathogens, and in particular seed-borne viruses and viruses in vegetative material, are obvious. Therefore, Governments enact quarantine regulations and enforce their implementation at the national level in order to prevent the import and export of dangerous pathogens and pests. This requires political will to allot an adequate budget for the establishment of a complex infrastructure and for the necessary training and extension. Farmers should be taught how dispersal of pests, pathogen, and weeds through man, plant material and tools can be inhibited at the village level.

Viruses that are endemic but harmless, and thus may occur unnoticed in certain areas, may cause symptoms and damage in crops newly introduced into those areas. The opening up of new areas for large-scale cultivation of new crops may lead to a rapid spread of plant diseases, in particular those caused by viruses. This happened in the East-African Groundnut Scheme after the Second World War. This large-scale undertaking failed owing to an epidemic of Groundnut Rosette Virus. Some years later, the export of groundnuts from Nigeria came to a halt too for the same reason.

Vector

Because of the prominent role of vectors as transmittors of pathogens, preventive measures include their control. The application of systemic insecticides is very effective against sucking insects such as leaf and plant hoppers, whiteflies and aphids which are the main virus vectors. Soil fumigation and the application of nematicides destroys soil-borne pathogens and populations of nematodes and soil-inhabiting vector insects.

Increasing crop resistance

Breeding for resistance and the use of genetically resistant varieties have become the corner-stone in the preventive control of diseases. Often, however, the required resistance is not yet available, or incomplete, or it may be linked to susceptibility for other pathogens or other unwanted properties. Pathogens frequently develop new strains through mutation which may overcome the host's resistance, particularly when this resistance has a narrow genetical basis. Consequently, in intensive agriculture, there is often a need for rapid turn-over of cultivars. When introducing new varieties or cultivars, caution must be exercised because their encounter with

The greenhouse section with isolation booths of the new plant quarantine station of the Plant Protection Service of the Netherlands. (Photo courtesy of LNV, the Netherlands)

endemic infectious microorganisms or viruses may lead to sudden outbreaks of "new" diseases if these introductions are genetically vulnerable. Examples are the outbreak of Rice Yellow Mottle Virus (RYMV) in Africa in the early 1950s when high-yielding, short-stature rice varieties were introduced. Most of the 500 imported lines tested in Nigeria were sensitive, whereas tolerance was common in the old African upland rice cultivars.

Plant breeding is getting new opportunities since biotechnology can help to overcome biological barriers through genetical engineering and the incorporation of resistance genes (see also Chapter 1). But special care needs to be given to minimize the risks involving with transferring biotechnicological experiments to those third world countries where quarantine and inspection procedures are insufficiently developed to enable the neccessary supervision. Another method of increasing a plant's resistance is called **premunity**, a term particularly used in virology. It refers to a degree of non-specific immunity that a plant acquires from infection by a pathogen. Those parts of the plant which contain the virus do not show symptoms if they were previously infected by a closely related pathogen. Intentional infection with viruses in order to get premunity in field crops may, however, give disappointing results.

Mechanical and physical methods

For a long time, the old mechanical and physical methods of disease control were undervalued owing to the ubiquitous availablity of pesticides. Recently, these methods are being re-evaluated and improved since they fit very well into an integrated control strategy aimed at reducing pesticide usage. An example is the use of solarization as a means of killing soil-inhabiting pathogens and pests. Weed control and field sanitation are still carried out largely by hand, but these practices depend very much on the availabilty and costs of farm labour. As few farmers realize that there can be a relationship between host weeds or plant debris and the incidence of diseases, they often neglect weeding and sanitation.

Biological control

Much attention is given to the development of biological control of fungi and bacteria. However, biological control has so far been less successful in phytopathology than in entomology and herbology (weeds).

Chemical control

Control of plant diseases by means of fungicides can be based on direct contact action, systemic action or fumigant action. Further, certain pathogens can be controlled by bactericides and antibiotics. Chapter 2, in the section entitled "classification of fungicides", provides an overview of chemical groups of fungicides and antibiotics and their properties. It also contains an explanation of the differences between protectant, eradicant and systemic fungicides.

Protectant fungicides are applied as a coating over the surface of plants and seeds that should kill the spores upon germination by direct contact (copper and phtalamide compounds) or fumigant action (sulphur compounds etc.). Fungicides are usually applied as a high-volume spray to provide a protective deposit all around, although a complete film coverage is seldom achieved. In cases when there is a water shortage, medium- or low-volume sprays may be applied as long as special attention is given to selecting proper droplet sizes for good coverage. Fungicide dusts are less frequently used in developing countries because of the transport costs involved. The new technique of film-coating of seeds gives, in this respect, an improvement over dust and slurry treatment of seeds. Protectant fungicides are particularly effective in preventing leaf spots, mildews and rusts and post-harvest rots on fruits and vegetables.

Eradicant fungicides can kill fungi in the early stages of infection. Only a few fungicide, such as benomyl, metalaxyl and liquid lime-sulphur mixtures, have such curative action (see Chapter 2).

Systemic fungicides play the most important role in this regard, as they are absorbed by the plant through foliage or roots and can work from within the plant system. The toxic ingredient is distributed through the vascular tissues and kills the invading hyphae by direct contact. Systemic fungicides are powerful protectants of plants and seeds, but may also be applied curatively in the early stages of infection. Antibiotics are used mainly to control bacterial diseases by contact action.

The growth of the fungus Sclerotinia sclerotiorum *on a nutrient medium is inhibited by an antibiotic substance produced by an antagonistic fungus* (Pseuliomyces *spp.*). *(Photo courtesy of IPO, the Netherlands)*

The use of fungicides has created complex problems of resistance, but these can be managed effectively, if the fungicides are applied with deliberation. Resistance develops when the fungicide kills the susceptible strains of the pathogen while resistant ones (e.g. through mutation) continue to survive. If a particular fungicide is used too intensively, contrary to the recommendations, the resistant strain becomes predominant, leaving the grower with a disease problem which he is unable to control using that particular class of product. Fungicide resistance is a relatively new phenomenon. It has become a practical problem with the introduction, since the late 1960s, of systemic products. The older, conventional protectant fungicides, such as inorganic copper, phtalamides compounds and lime-sulphur mixtures, have rarely given rise to serious resistance problems. However, the systemic fungicides offered such advantages to the grower, in terms of safety, curative action, persistence, and lower dosage rates, that they became highly attractive for the farmer to use. Systemic fungicides tend to be rather specific in their action as they inhibit only one or two metabolic processes in the target organism. So, modifications in only one fungal gene can be sufficient to induce change at the site of action, and a resistant strain can develop. The older protectant compounds demonstrate a less specific action, and thus there is less chance for the pathogen to become resistant (Wade, 1988).

Examples of preventive control against micro-organisms

Fusarium wilt in cotton is a soil-borne disease that can survive as a saprophyte for many years; it is spread by water. The fungus can infect cotton plants from the seedling stage. Therefore, seed treatment with protectant fungicides and furrow application with systemic fungicides or organomercurials have a preventive effect. Fusarium wilt often infects cotton roots which have already been injured by nematodes, e.g. those of the species *Meloidogyne*. Crop rotation and fallow are helpful in supressing the nematode population.

Rice blast disease, caused by airborne spores of *Pyricularia oryzae*, is one of the most devastating and wide-spread fungal diseases in rice. The most practical and economic preventive measure is the use of resistant cultivars of rice. At times, even incomplete resistance, supplemented by treatment with fungicides, provides satisfactory control (Chin K.M., 1985).

References:

Agrios, G.N.: *Plant Pathology*, third edition; Academic Press, New York

Ainsworth and Bisby's: *Dictionary of the fungi*, sixth edition; Commonwealth Mycological Institute, Kew, Surrey, 1971.

Bos L. (1983): *Plant virus ecology; the role of man, and the involvement of governments and international organizations*. In Plumb R.T. & Thresh J.M. (eds.). *Plant virus epidemiology*. Blackwell Scientific Publ. Oxford

Bos L. (1986): *Crop protection: what, why and how*: (unpubl. paper) Res. Inst. for Plant Protection, Wageningen.

Chin K.M.(1985): *Response of Pyricularia oryzae populations to integrated disease management*. In B.S. Lee et al. (eds.): *Integrated Pest Management in Malaysia*, Kuala Lumpur, MAPPS, Malaysia, pp. 183-190.

Federation of British Plant Pathologists (1973): *A guide to the use of terms in plant pathology*. Phytopatology Paper No.17. Commonwealth Agricultural Bureaux, United Kingdom.

Kranz J., Schmutterer H. and Koch W. (1977): *Diseases, pests and weeds in tropical crops*. edition Paul Parey, Berlin and Hamburg.

Nederlandse Plantenziektenkundige Vereniging; *Lijst van gewasbeschermingskundige termen*, Gewasbescherming 16 (December 1985), supplement nr.1.

Oudejans J.H. (1982): *Agropesticides: their management and application*. United Nations Economic and Social Commission for Asia and the Pacific, Agriculture and Rural Development Division, Bangkok.

Oudejans J.H., Meerman F. and Takken W. (1990): *Crop protection, vector control and pesticide use in developing countries*. Background paper (in Dutch language) to special report of the Ministry of Agriculture, Nature Management and Fisheries, the Netherlands.

Reissig W.H. *et al.* (1986): *Illustrated guide to integrated pest management in rice in tropical Asia*. International Rice Research Institute, Los Banos, Philippines.

Stakman E.C. and Harrar J.G. (1957): *Principles of plant pathology*. The Ronald Press Company, New York.

Teng P.S. and Heong K.L.(1988): *Pesticide management and integrated pest management in Southeast Asia*. Consortium for International Crop Protection, Maryland 20740, U.S.A.

van Egeraat A.W.S.M. and Deinema M.H. (1984): *Microbiology*, lectures, vol.1. Agricultural University of Wageningen, Dept. of Microbiology, The Netherlands.

Wade M. (1988): *Fungicide resistance: putting the problem in perspective*. Shell Agriculture Journal; nr. 2, 1988.

Zadoks J.C. (1987): *Epidemiology and plant disease management in the tropics*. Labortory of Phytopathology, Agricultural University of Wageningen, the Netherlands.

Zadoks J.C. (ed.), (1989): *Development of farming systems. Report over the five-year period 1980-1984*. Wageningen, Pudoc.

14

INTEGRATED CONTROL OF WEEDS

Contents

Contents   Page

Cultivation systems and inherent weed problems 265

- Shifting cultivation 265

- Labour-intensive continuous cultivation 265

- Draught animal and small tractor cultivation 266

Weed and crop interactions 267

- Economic importance of weeds 267

- Effects of weeds 268

- Factors contributing to damage from weeds 271

Weed biology 272

- Major weed pests 272

- Classification of weeds 272

- Survival of weeds 273

- Dormancy 275

- Germination 275

- Survival by vegetative reproduction 276

- Hardiness of weeds 277

Methods of weed control 277

- Manual and mechanical weeding 277

- Cultural control 278

- Biological and ecological control 279

- Chemical control 281

- Mode of herbicidal action 281

Weed control in crops 285

- Mixed and intercropping systems 285

- Non-irrigated, annual crops 286

- Irrigated annual and perennial crops 286

- Parasitic weeds 288

References 290

Integrated Control of Weeds

About 350,000 species of plants are known to man. Of these, less than 1 per cent or about 3,000 species have economic value and less than 100 species are currently cultivated. But numerous plant species infest these cultures and compete so vigorously for water, nutrients and light that the desired crop may fail completely. Such competitive plants are termed "weeds", but apparently not under all circumstances. For instance, Burmuda grass (*Cynadon dactylon*) is a valuable forage grass in pastures, on canal banks and roadsides, but a very troublesome weed in crop fields.

One 250-year old definition of weeds reads: "A weed is a plant growing where it is not desired." It is a definition of a general nature: it neither identifies a taxonomic category of plants nor specific species as being

Hand-weeding in broadcast-sown crops is a common, but labour-intensive practice. Straight-line seeding enables the use of hoes and animal-drawn tools for weeding between the rows. (Photo by the author)

weeds. It is man himself who decides if and when a plant is to be called a "weed".

A more complete definition would perhaps be stated as follows: "Weeds are plants growing in places where, and at times when, man wants either some other plants to grow unhindered or no plants to grow at all". This implies, that any plant can be called a weed when it interferes with man's effort to grow crops, to manage pastures and forests, to keep irrigation systems and waterways open, or to keep roads, rail embankments and utility yards free of any plants.

From the time that primitive man began to plant and cultivate food in preference to merely collecting edible plants and nuts, he had to struggle with weeds. Every time a plot was cleared to grow food, volunteer vegetation sprang up in fierce competition challenging the crop. The ancient Romans coined a famous phrase that may be loosely translated as "nature does not like empty places." For a very long time man could fight weeds only by pulling them out, cutting them with simple tools, or by using fire to kill them.

CULTIVATION SYSTEMS AND INHERENT WEED PROBLEMS

Over more than 6,000 years, man has developed many different methods for growing crops; these can be grouped into four generalized systems, namely: traditional shifting cultivation, labour-intensive continuous cultivation, draught animal and small tractor cultivation, and modern high-technology cultivation (Unger 1984). A brief explanation of these systems will help to increase understanding of the inherent weed problems connected with each of these systems.

Shifting cultivation

Traditional shifting cultivation, which is still practised on an estimated 7-8 billion hectares of land in Africa, South- and South-east Asia, the Pacific and Latin America, provides most of the food for at least 8 per cent of the world's population. The term covers a wide range of cropping practices on a wide variety of soils with many different types of vegetation and crops grown, varying length of cropping and fallow periods and methods of tillage. They have in common the clearing and cultivation of a plot for a few years, after which it is left to nature to regenerate. Food crops are grown at subsistance level with manual labour and almost no capital imputs. Land clearing and weeding are carried out with the help of fire and some simple implements such as axe, hoe, knife, cutlass or machete.

The intensity of the fire associated with burning to clear the plot determines to what extent weed seeds and tree stumps are killed. Owing to the greater heat developed from burning a clearing in the jungle than one in grasslands, the former will usually remain free from weed and brush regrowth for a longer period. The labour

Table 14.1: Yield losses of certain crops as a result of not weeding in a shifting cultivation system in Africa (from Moody, 1974; in Unger, 1984)

| Crop | Per cent yield loss in the year | | |
	1970	1971	1972
Maize	19	19	28
Cowpea	51	48	59
Yam	-	-	73
Sweet potato	-	-	91
Cassava	-	-	92
Upland rice	100	100	100

requirement for weeding the clearing depends very much on the original vegetation type and climate. Thus, a luxuriant weed growth in humid tropical climates usually demands a high labour input for weed control, i.e. up to 50 per cent of a farmer's working time. Without applying fertilizers, soil fertility is often rapidly depleted. This phenomenon and ever-stronger weed growth after the first season cause the yields to decline and force the farmer to shift to a new plot every three to five years. The effects of weed competition on yields are shown in table 14.1. If the abandoned plot is fallowed for a period of five to ten years, soil fertility may be largely restored.

Labour-intensive continuous cultivation

This group refers to annual cropping on the basis of manual labour and inputs ranging from none at all to low levels of inputs; however, fallow periods of one to two years may be used occasionally. Such continuous cultivation may be carried out at:

* The subsistence level with the purpose of enabling a family to be self-sufficient in growing food. This system replaces traditional shifting cultivation when available land has been completely occupied by an expanding population and land can no longer be fallowed long enough to restore fertility. Unless improved practices are introduced, crop production under subsistence conditions results in serious and irreversible land degradation.

* Low external inputs for sustainable agriculture (LEISA) -- by using suitable practices to avoid soil fertility decreases, erosion and other forms of land degradation.

If the basic goal of growing a stable supply of food for the family has been met and some kind of market for agricultural products is nearby, farmers may produce one or more additional crops for sale. They will try to raise their output either by expanding their plots or by intensified cultivation and multiple cropping.

In continuous land use, greater effort is required to control weeds because weed problems generally increase as the period of cultivation extends. In labour-intensive continuous cropping systems, however, the manual labour that the farmer's family can provide soon becomes a bottleneck for tilling the soil and weeding. This may be easily overcome in areas with a large rural population where labour is plentiful and inexpensive. But, where costs are high or labour is inadequate, there may be no opportunity to raise the level of crop production unless suitable substitutes for labour are available. If farmers have the resources, they can use draught animals or small tractors with suitable implements to perform some essential cultural operations. Commercially oriented farmers usually soon realize that regular weeding and the conservation of soil fertility and water resources are prerequisites for obtaining sustainable production with high yields. They are eager to try new methods which include recycling of organic wastes, mulching to reduce weed competition and herbicides.

Draught animal and small tractor cultivation

Since ancient times, animals have been domesticated and used for transport and traction. Their usefulness was greatly enhanced by the invention of the wheel and the development of carts and implements for soil tillage such as various types of ploughs, mud-boards, harrows, duckfoot sweeps and roller compactors. Although the early primitive implements were gradually improved as better technology and materials became available, no major changes occured until the invention of a steam traction engine that provided mobile power for heavy tillage operations. Nowadays, the availabilty of a wide range of modern, combustion-engine tractors and heavy-duty implements has permitted farmers to largely expand the area under cultivation and to perform intensified tillage operations on a more timely basis.

In many areas of the third world, hand labour and draught animals are still the major means of cultivation, because poverty or land fragmentation (which is a result of repeated partitioning of the family holding among heirs) has left many farmers with small, scattered and odd-shaped plots that are difficult to cultivate efficiently. The use of two-wheel tractors on small holdings can usually only be afforded by more prosperous land owners.

Historically, draught animals enabled farmers to increase the area cultivated and to till the land more rapidly, but not necessarily to increase yields from a unit of land. In Gambia, for example, the use of oxen permitted a 20 to 25 per cent increase in the cultivated area for groundnut, but no yield increases were reported. Gifford (1981) recorded that 60 work-hours of labour were needed to plough a hectare of land using animals and 500 work-hours if the soil was tilled by hand. If water is available throughout the year, the use of draught animals and tractors enables farmers to increase the production on a given area by multiple cropping.

A survey conducted in eight Asian countries in 1977 showed that hand weeding of rice fields is very common throughout Asia, particularly if farms are less than a hectare in size. The use of rotary weeders is very common in Indonesia and Viet Nam and fairly common in Bangladesh. Chemical weed control is practised in the entire rice-growing area of Taiwan province of China, Japan and the Republic of Korea because of the very high cost of weeding. Moreover, in these countries the use of machines for transplanting has aggrevated weed problems in comparison to transplanting by hand; also, it necessitates repeated applications of herbicides soon after transplanting. The use of herbicides is also common in the Philippines and Thailand. In the latter two countries and Viet Nam, farms are generally larger (2-3 ha) but often not so well irrigated. On such farms, weeds are causing considerable problems as there is not enough labour available to cope with them. In those countries, many farmers weed little if at all.

Buffaloes are driven through a rice field to loosen the soil and bury the stubble and weeds. Next, the field should be leveled with a mud board in order to leave it submerged under water for days without allowing dry patches to occur because weeds could then grow. (Photo by the author)

Table 14.2: Weed control practices related to farm size in selected Asian villages, 1971-72

Weed control practice	Practice (%) in farms measuring		
	< 1 ha	1-3 ha	> 3 ha
Herbicides	6	20	29
Rotary weeding	3	20	37
Hand weeding	82	83	87

Source: IRRI 1975; based on survey of 32 villages in South-east Asia (from Barker and Hayami, 1983).

A 1971-72 survey of rice production in relatively progressive Asian villages suggested that even in those villages, where more than 80 per cent of the farmers adopted modern cultivars, fertilizers and insecticides, only a relatively small percentage used rotary weeders or herbicides (see table 14.2). Apparently, rotary weeders and herbicides are not an essential element in modern rice technology, as high yields may also be achieved with the traditional practice of hand weeding. In tropical rice production, herbicides are normally used in combination with hand labour, because chemical control alone is usually not effective against all the types of weeds present. In particular, the increased occurrence of perennial weeds in rice is stimulating the use of herbicides; however, the control of perennial weeds requires integrated methods. (See the section towards the end of this chapter, under weed control in rice.)

Table 14.3: Labour trends in weed contol in four Asian countries or areas (adopted from Barker and Hayami, 1983)

Year	Labour days per hectare				
	Land preparation	Weeding	Other tasks	Harvesting	Total
Shonai Plain, Japan					
1888	50	62	75	113	300
1952	28	46	83	71	228
1976	7	11	41	23	82
Taiwan province of China					
1927	17	20	35	24	96
1961	37	29	42	30	138
1972	10	19	39	16	84
Java, Indonesia					
1878	72	58	40	66	236
1969	35	43	41	40	159
1980	23	42	36	43	144
Laguna, Philippines					
1965	19	11	20	36	86
1978	9	27	22	28	86

Over the longer term, the adoption of tractor power, rotary weeders and herbicides has caused a marked downward trend in labour use for land preparation and weeding in irrigated rice in Indonesia, Japan and the Philippines as well as in Taiwan province of China. Whereas at the end of the nineteenth century land preparation and weeding in Japan and Indonesia required more than 100 labour days/hectare, the labour requirement has steadily declined and is currently less than 20 days/ha in Japan. However, the decline is more obvious with regard to labour input for land preparation as the result of the introduction of power tillers than for weeding. In fact, the introduction of improved cultivars and corresponding fertilizer technology seems to have increased the demand for weeding labour (see table 14.3).

Table 14.4: Average reduction in yield of important crops due to the effect of weeds in India (Recalculated from Mani *et al.*, 1968)

Crop	Yield from weed-free plot (tonnes/ha)	Yield from weedy plot (tonnes/ha)	Reduction in yield due to weeds (per cent)
Rice	3.75	2.19	41.6
Wheat	2.37	1.99	16.0
Maize	3.44	2.07	39.8
Millet	1.83	1.29	29.5
Sugar-cane	94.80	62.32	34.2
Groundnut	13.24	8.76	33.8
Soybean	17.50	13.97	20.1
Potato	25.13	8.04	68.0
Onion	0.80	0.42	47.5
Cotton	2.91	2.02	30.5

WEED AND CROP INTERACTIONS

Economic importance of weeds

In tropical agriculture, vigorously growing weeds exert a direct and negative influence on the productivity of other plants. This surpressing effect of weeds on crops can be clearly demonstrated by comparing the yields of weeded and unweeded plots of some important crops in an African shifting cultivation system (see table 14.1) and in an Indian continuous cropping system (see table 14.4).

The types of weeds and the severity of their competition vary with soil, temperature, geographical location, altitude, tillage system and cultivation practices, water mangement, control methods,etc. For example in rice cultivation, the widespread replacement of traditional tall varieties has increased weed problems throughout the tropics. The traditional rice varieties have droopy leaves; the short-stem varieties erect leaves. Therefore, more light penetrates the crop canopy, and more weeds emerge and survive. The high fertilizer rates applied to modern high-yielding rice varieties worsen weed problems.

A comparison of estimated rice yield losses in several rice producing countries or areas shows the following figures (Smith 1983):

```
India ......... 10 per cent
Philippines .... 11 per cent in dry season
                 13 per cent in wet season
United States .. 15 per cent (of yield and quality)
World ........ 10 per cent
```

In addition to the yield and quality losses, there are losses due to the cost of herbicides, cultural and mechanical control practices, and hand weeding to prevent an even greater loss. Such losses are estimated at about 5 per cent world wide.

Thus, for the world, the total estimated direct losses in rice from weeds and expenditure for their control are 15 per cent annually. The average annual world production of rice at the end of the 1970s amounted to about 372 million tonnes grown on 143 million ha.

On this basis, world-wide losses caused by weeds are about 56 million tonnes of rice annually, valued at $US 12 billion (based on 1977-79 average export prices at Bangkok for 5 per cent broken white rice, Smith 1983).

In addition to the measurably lower productivity of the crop itself, one should also take into account the huge effort expended in the never-ending task of weeding crops. The countless hours spent by a farmer's family on hand weeding are still rarely accounted for in traditional agriculture. It is generally conceded that the recurrent economic damage to agriculture and forestry from weeds far surpasses the more incidental damage inflicted by insect pests, rodents and diseases.

Effects of weeds

Weeds interfere with the production and processing of crops in various ways as is illustrated below:

* Competitive efficiency of plants and weeds

In understanding the enormous competitive power of weeds, one must realize that they have become extremely well-adapted to their environment by a long process of natural selection. The selection pressure on weeds is primarily directed at survival and thus at the development of efficient generative and vegetative means of propagation. On the contrary, cultivated plants were exposed to a long-time selection directed at quality aspects and high productivity and may thereby have lost much of the stress tolerance that wild plants possess.

Black and other researchers (1969) divided crop and weed plants in efficient and non-efficient groups in terms of biochemical characteristics, such as photosynthetic activity (i.e. the process by which carbohydrate compounds are produced by illuminated green plant cells from carbon dioxide (CO_2) and water).

Efficient plants, which are called C_4 plants, have among other characteristics:

* Increased uptake of carbon dioxide as light intensity increases to nearly full sunlight and as temperature increases up to $30°-45°C$; the rate of uptake being expressed as mg of CO_2 per square diameter of leaf surface per hour;

* Photosynthesis uninhibited by oxygen at atmospheric levels (i.e. at 21 per cent concentration); and

* Lower water requirement for the production of dry matter, the requirement being 250-350 grams of water/gram of dry weight.

Non-efficient plants, which are called C_3 plants, have:

* Increased uptake of carbon dioxide as light increases up to about one-tenth to one-third of full sunlight and as temperature increases to $15°-25°C$;

* Photosynthesis inhibited by oxygen at levels well below atmospheric concentration; and

* Higher water requirement of 400-950 grams of water/gram of dry weight.

In general, C_4 plants have higher photosyntetic activity, are fond of higher temperature and more intense solar radiation, and have lower water requirements than C_3 plants. Thus, the efficient plants produce more photosynthetic products, such as phosphorylated sugars and amino acids and eventually the storage compounds starch and proteins, under tropical conditions when similar processes in non-efficient plants become increasingly inhibited. Further, C_4 plants utilize available water more efficiently than C_3 plants. Based on the above criteria, most weeds of tropical dryland crops are classed as efficient plants, whereas most crops, except maize, are non-efficient plants. C_3 plants predominate (more than 80 per cent of crops) in areas with an average temperature lower than $20°C$. (See table 14.5.)

Table 14.5: Photosynthetic efficiency of some major crop plants and major weeds (from Matsunaka, 1983)

Crop plants	C	Weeds	C
Wheat	3	*Cyperus rotundus*	4
Rice	3	*Cynadon dactylon*	4
Maize	4	*Echinochloa crus-galli*	4
Potato	3	*Portulaca oleracea*	4
Barley	3	*Eleusine indica*	4
Sweet potato	3	*Sorghum halapense*	4
Cassava	3	*Eichhornia crassipes*	3
Soybean	3	*Monocharia vaginalis*	3

Because women play a major role in vegetable growing, they should be included in extension training on matters such as timing and labour-saving methods of weed control. (Photo by the author)

* Competition for water

Practically all crops grown in the tropics (C_3 plants) have a relatively high requirement of water as compared with many types of weeds. Moreover, many weeds have an extensive and deep penetrating root system that may be more effective in absorbing water from the soil than that of crop plants.

This may be of particular importance under dryland farming conditions where the crop lives on residual soil moisture during dry periods. In unweeded crop fields or fallow plots the soil moisture reserve will be depleted much sooner than if the fields are kept weed-free.

However, weeds are sensitive to the soil moisture regime in terms of submerged, water-saturated and dryland conditions. In wet-rice experiments it was found that the total dry weight of weeds per unit area was decreased by submergence. The C_4 species accounted for more than 90 per cent of the total dry weight in dryland and water-saturated plots, but for only 10 per cent in submerged plots. C_3 plants are dominant under sub-merged conditions while C_4 plants are dominant under dryland conditions.

* Competition for light

Plants need sufficient light for photosynthesis and growth. When weeds emerge earlier and grow taller than the crop plants, the latter receive sub-optimum quantity of light and become chlorotic and weak as a result. In slow germinating crops such as groundnut, weeds should therefore be controlled after emergence of the weeds and before the crop comes up.

* Competition for mineral nutrients

Strong-growing weeds may accumulate high concentrations of nutrients in their tissues and thus deplete the soil's nutrient reserves at the expense of the crop. Generally, additional application of fertilizer to make up for nutrient shortages caused by weeds, produces less effect than would good weeding of the crop.

As has been demonstrated by numerous experiments, there exists a strong correlation between the type and timing of fertilizer application and the subsequent competition between weeds and the crop. With rice, weed growth and competition are usually stimulated by pre-planting application of phosphorus and nitrogen. Individual weed species may reduce crop yields more or less severely depending on the dosage rate of a certain type of fertilizer. Knowledge of this relationship can help the farmer to manipulate fertilizer gifts to favour the crop and surpress weed growth.

Methods to maintain soil fertility include the use of manure, plant debris, refuse and the growing of soil-improving cover- crops such as *Pueraria phaseoloides* in rotation and other beneficial practices. The use of mulches (organic matter or black plastic sheets) is a very important measure with continuous cropping as they not only reduce runoff, soil losses and evaporation, but also aid in weed control and improve water conservation.

* Parasitism

Some weeds are true parasites that lack green leaves and complete roots and totally depend on their host plant for water and nutrition. Dodder (*Cuscuta* spp.), for instance, develops long, wiry stems with numerous small haustoria, sucking pads, which attach themselves to the host plant. Dodders cannot survive without suitable hosts, such as leguminous plants. Similarly, *Orobanche* (broomrapes) and *Striga* species are obligate parasites (see end of this chapter).

* Increased incidence of pests and diseases

Many weeds are hosts of pathogens and animal pests and can thus constitute a source of infestation for the crop. Weeds may also provide harbourage and breeding places for mites, insects and rodent pests. For instance, *Echinochloa* species and wild rices are alternative hosts for rice green leafhoppers and sources of Tungro virus and dwarf viruses. Bacterial blight and bacterial leaf streak diseases in rice, caused by *Xanthomonas* species, can survive in several grassweeds if rice is not available.

* Reduction of crop and seed quality

Both weeds themselves or weed seeds may spoil the quality of food and feed or other produce. Some examples: nutsedge (*Cyperus rotundus*) in silage or hay makes this feed less acceptable to animals. Some weeds of range-lands or semi-deserts are poisonous and hazardous to grazing livestock, e.g. Johnson grass (*Sorghum halapense*) contains at its tillering stage enough prussic acid to poison cattle; spiny pigweed, *Amaranthus spinosus*, may injure grazing cattle. Seeds of wild rice and grasses, such as *E. crus-galli*, easily mix at the time of threshing with rice grains, a part of which may be put aside as seed for the next season's crop. Many countries have strict certification of crop seed stating that it must be free of the seeds of weeds and of other crops.

![Mikania cordata illustration]
Mikania cordata (Burm.f.) Robinson: 1. habit, 2. head, 3. flower, 4. stamens, 5. achene.

* Interference with farm operations

All movement and mechanical operations become more cumbersome and costly in weedy fields. *Mikania cordata*, and other creepers can severly slow down the picking of tea leaves and harvesting of sugarcane and maize, for example. And dodder, *Cuscuta* spp., impedes combine harvesting of soybeans, for example. Spiny mimosa, *Mimosa pudica*, on fallow fields may injure the hooves of draught animals, whereas a thorny shrub called giant mimosa, *Mimosa pigra*, which grows along waterways, obstruct their access to drinking water.

* Clogging of irrigation and drainage canals

An abundance of aquatic weeds can greatly reduce the flow of water and cause seepage and water-logging near irrigation and drainage canals. Water hyacinth, pickerel weed (*Monocharia* spp.), creeping water primrose (*Jussiaea* spp.) and other water weeds contribute to water losses through transpiration. Proliferation of algae and water weeds in lakes and fish-ponds, owing to enrichment of surface water with nutrients by fertilizer-runoff and effluents, may cause the death of many fish and aquatic organisms. The putrefying dead water weeds (organic matter) rob water of its oxygen content and emit toxic gases such as hydrogen sulphide (H_2S).

* Allelopathic substances

Many plants leach or exude substances from their above-ground parts or living and decaying roots that have a toxic effect on other plants. Examples are the deleterious effect of the ground cover *Salvia occidentalis* on young coffee trees, and the inhibitory effect of *Imperata cylindrica* on maize. Although many tests with plant extracts have shown some growth-inhibiting properties, no practical use of allelopathy as a approach to weed control has as yet been developed. However, allelopathic effects certainly play a role in suppressing the effect of dead mulches on weed growth.

Amaranthus spinosus L.: 1. habit, 2. seed, 3. flower.

Table 14.6: Effects of weed densities on the yield of Lowland IR8 for two planting methods

Weed population	Weed weight* (g/m²) Broadcast	Transplanted	Yield reduction from weeds (per cent) Broadcast	Transplanted
Grasses only	325	285	86	75
Sedges + broadleaf weeds	250	110	24	0
Grasses + sedges + broadleaf weeds	540	330	100	67
Weed-free	0	0	-	-

Note: *Determined at the heading stage of grassy weeds; IRRI, 1968 wet season.
Source: Aadapted from De Datta *et al.*, 1969.

Factors contributing to damage from weeds

Species or group of weeds

The rate of losses inflicted by weeds on crop yields varies considerably with the type and species of competing weeds. Generally, perennial grasses and sedges reduce yields more than the annual species, and grasses tend to do more damage than broadleaf weeds. For instance, experiments have shown that all-season competition of E. crusgalli in rice reduces grain yields more than broadleaf and aquatic weeds together. De Datta (1969) compared the effect of various types of weeds on the yields of rice cultivar-IR8 in broadcast and transplanted stands (table 14.6). Well-prepared wet rice fields are almost completely free of weeds at the time of transplanting. Thus, the transplanted rice seedlings have a considerable head start on emerging weeds, even though their roots and leaves are cut back. Therefore, losses from weeds tend to be higher in broadcast or directly seeded rice.

Weed density

The negative influence of crowding on crop yields is well known from numerous trials on the effect of varying weed densities and plant distances. Commonly, the denser the weed stand and the thinner the crop stand, the greater the yield loss. However, the effect of density varies according to weed species. In trials conducted in the United States, the effect of season-long competition of *Echinochloa crus-galli* on directly seeded rice was studied by Smith (1968). Table 14.7 shows the reduction in rice yield from three densities of weeds in an optimum stand of rice (left side of table 14.7) and from a density of 11 weed plants/per square metre in three different spacings of rice (right side of table).

Duration of the weed competition

The length of the period of weed competition and the growth stage of the crop during competition takes place determine to a great extent the rate of yield losses. Competition from weeds during the early growth stages

Table 14.7: Yield loss in drill-seeded rice due to season-long competition of *E. crusgalli*

Weed density (plants/m²)	Yield loss (per cent)	Rice stand (plants/m²)	Yield loss (per cent)
11	25	32	57
54	49	108	40
269	79	334	25

Source: Adapted from Smith, 1968.

Whilst hoeing cabbage beds at a vegetable research station in Viet Nam, this woman was seen searching for and destroying eggmasses and caterpillars on the young plants. (Photo by the author)

Table 14.8: Yield loss in drill-seeded rice as influenced by duration of competition from *Sesbania exaltata* and *E. crus-galli*

Duration (weeks)	Yield losses (%)	
	S. exaltata	*E. crus-galli*
4	2	8
8	6	35
12	19	43
Season-long	19	70

Source: Adapted from Smith, 1968.

of a crop causes heavier losses than no competition during the early growth stages followed by competition during later growth stages. Therefore, weeding in the early stage of crop establishment is more important than weeding at later stages. Further, the longer the competition lasts the more yields decrease (see table 14.8).

Cultivation method

The way in which crops are sown or planted has a great influence on future weed development and the possibilities of their control. In wet fields, where early weeds can be destroyed by flooding and puddling, the transplanted crop is free of initial competition. On dry land, weeds emerge concurrently with the crop and start to compete immediately. Broadcast seeding enables hand-weeding and herbicide treatment, but no weeding with animal or tractor-drawn implements. In contrast, row-seeding of a crop enables weeding in between the rows with rotary weeders and other mechanical implements, thereby enabling a more timely control of weeds than labour-intensive hand-weeding. Weeding within the rows of crop plants has to be done by hand labour or herbicide application. Also, water management, in terms of the frequency and depth of flooding of a field, appears to have a great influence on the growth of certain weeds.

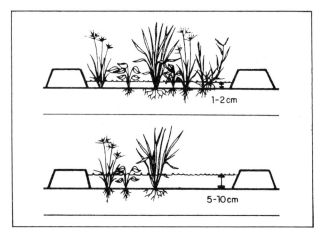

Figure 14.1: Proper management of water depth on seed beds and rice fields can help to control weeds. A layer of 5-10 cm of water eliminates grasses, but a few sedges and some broadleaved weeds may remain. (Drawing courtesy of IRRI)

WEED BIOLOGY

Major weed pests

Although all weeds are troublesome, some have a much greater economic importance than others. Certain weeds are very widespread and occur in several habitats of a country, a region or even world-wide. Others weeds may be restricted to very specific habitats and to one or few crops. Certain weeds, although having a very wide distribution, may be easily controlled, while others of more local importance may be very difficult to eradicate. A deep-rooting perennial grass such as *Imperata cylindrica* or a free-floating waterweed such as water hyacinth are examples of the latter. The 17 weeds listed in table 14.9 have the dubious reputation of being the greatest troublemakers in more countries than any other weeds. The worst of all, *Cyperus rotundus*, occurs in more than 100 countries. All species listed are found in more than 50 countries and in more than 30 crops.

Table 14.9: Ranking of the world's worst weeds

Cyperus rotundus	purple nutsedge
Cynadon dactylon	Bermuda grass, couch grass
Echinochloa colona	jungle rice
Echinochloa crus-galli	barnyard grass
Eleusine indica	goose grass, wild finger millet
Sorghum halepense	Johnson grass
Imperata cylindrica	alang alang, lalang, cogon grass
Eichhornia crassipes	water hyacinth
Panicum maximum	Guinea grass
Chenopodium album	goose foot
Digitaria sanguinalis	hairy- or large crab-grass
Lantana camara	Lantana
Avena fatua	wild oats
Amaranthus spinosus	spiny pigweed
Cyperus esculentus	yellow nutsedge, watergrass
Paspalum conjugatum	sour grass
Rottboellia cochinchinensis	itch grass, Guinea-fowlgrass

Source: Adapted from Holm *et al.*, 1977.

Classification of weeds

Before discussing the biological features of weeds in relation to their control, brief attention will be given to the different ways weeds can be distinguished into groups:

* Narrow-leaved weeds, including primarily members of the monocotyledonous (having one seed-leaf) families such as Gramineae, Cyperaceae and Juncaceae; and

* Broad-leaved weeds, which belong mostly to dicotylodonous (having two seed-leaves) families; but some, such as *Monocharia* spp. are monocotylodonous plants.

Commelina benghalensis *is a creeping herbacious perennial weed that roots at the nodes. Although it reproduces mainly by seeds, its stem fragments, if cut during cultivation, root and grow. (1. habit, 2. flower, 3. seedling, 4. fruits, 5. seeds)*

Another grouping is based on the reproductive charactistics and is often used in combination with the above grouping:

* Annuals. These are weeds which complete their life cycle in one season and reproduce by seed only. Examples are *Digitaria sanguinalis* and *Amaranthus spinosus*;

* Biennials. Such plants require two years from germination to flowering and seed production. Biennials are rare in the tropics, but predominate in the temperate regions; apparently these plants require a low temperature for flower-induction; and

* Perennials. These plants usually produce seeds every season, but sometimes only once in several seasons. They show a renewed growth each year from the same root system. The perennials can be divided into herbaceous and woody perennials. The aerial shoots of herbaceous perennials die each year after flowering, but their underground parts stay alive and produce new flowering shoots in the following season. The aerial parts of the woody perennials continue to grow and they usually produce seeds every year.

Examples: *Lantana camara* is a woody perennial; bamboo species are woody perennial grasses; *Cyperus esculentus*, *Sorghum halepense* and *Mikania cordata* are herbaceous perennials.

A third way of classifying weeds is to distinguish the following groups:

* Seed weeds
* Herbaceous perennials
* Woody perennials
* Aquatic weeds
* Vines, climbers and strangling plants
* Epiphytes and bromeliads
* Parasitic weeds

The occurrence of weeds and limits of distribution

The presence or absence of plants in a certain region can be explained in the first instance by the range of temperatures prevailing in that region. A weed, such as *Eichhornia crassipes* can grow only in the tropics or subtropics because it does not survive periods of frost. Other weeds demonstrate optimum growth at relatively high temperatures; for *Cyperus rotundus*, the optimum range is from 25°C to 35°C, whereas *Agropyron repens* stops growing at temperatures in excess of 25°C and thus is found in the temperate zone. Other factors determining the occurrence of plants in places are water (rainfall, soil moisture, relative humidity of the air), light (intensity and day-length), and soil type (structure, texture, alkalinity and toxic metals and organic-matter content).

Survival of weeds

Plants can become successful weeds if they have, among other characteristics, a high generative or vegetative reproduction capacity as well as competitive

The perennial grass Sorghum halapense *sprouts from underground rhizomes or rootstocks that are difficult to control.*

273

and aggresive habits. In order to survive under temporarily unfavourable conditions, the plants must have certain survival mechanisms.

Reproduction

One of the most striking features of weeds is their abundant means of reproduction. Many weeds show an often prolific production of seeds or spores. In addition to seeds, weeds may possess in considerable quantities specialized organs for vegetative reproduction; these include rhizomes, stolons, runners, offsets, tubers and bulbs. For anyone undertaking weed control, it is essential to know the mode of propagation and characteristics of weeds. If only the visible parts of some weeds are destroyed, while underground reproductive organs remain undisturbed, the weeds will soon reappear in strength and much of the effort will be wasted.

Annual and biennial weeds reproduce mainly by seeds, although some of them can reproduce vegetatively. In general, annual weeds produce an astounding number of seeds, as is illustrated in table 14.10.

Table 14.10: Seed production of some common weeds

Weed species	Seeds per plant (*)
Amaranthus spinosus	196,000
Bidens pilosa	6,000
Commelina benghalensis	25,000
Dactyloctenium aegyptium	66,000
Digitaria sanguinalis	150,000
Eleusine indica	41,200
Portulaca oleracea	10,000
Rottboellia cochinchinensis	3,160

* Average or recorded number of seeds per weed plant

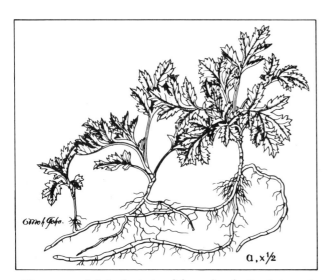

The perennial herb Artemisia vulgaris *L. (a weed occurring on Java, Indonesia, does not set seeds, but propagates vegetatively through its underground stolons. (Drawing courtesy of LPH, Jakarta)*

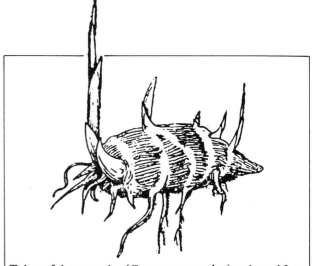

Tuber of the nutsedge (Cyperus rotundus) *enlarged 2 ×.*

Weeds seeds often mature before a crop ripens and thus complete their reproductive cycle before the crop is harvested. For instance, *Cyperus difformis* and *Eleusine indica* complete their reproduction cycle within 4-5 weeks and may each produce up to 50,000 seeds per plant. Shedding of numerous weed seeds before and during crop harvest can cause an enormous accumulation of weed seeds on and in the soil.

Dispersal of weed seeds

Weeds can be dispersed in various ways such as:

* By wind. Very light and small seeds are carried by air currents, while some seeds have special features such as wings or parachute-like structures which facilitate their movement through air currents. Examples: feathery seeds of *Mikania cordata*; several members of the Compositae family.

* By water. Many weed seeds are carried by irrigation and drainage water, by surface runoff after heavy rainfall and by water in canals and streams. Examples: *Monochoria vaginalis*, and members of the Cyperaceae family.

Monocharia vaginalis (Burm.f.) Presl.: 1. habit, 2. flower, 3. vertical section, 4. ovary, 5. stamens, 6. seedlings, 7. capsule, 8. seed.

* By animals. Many seeds and fruits have hooks and barbs with which they stick to the fur of animals. Examples are weeds belonging to the families of Compositae, Gramineae and Boraginaceae. Furthermore, seeds are eaten by animals, and spread through their droppings.

* By man. Weed seeds are transported in crops, crop-seeds, hay, and other agricultural produce as well as by tools, machinery and clothing.

Dormancy

The presence of huge quantities of weed seeds and vegetative plant parts in the topsoil is a key factor in the survival of weeds. In general, every hectare of arable land contains many millions of seeds of the prevailing weed species. The seeds of many species can persist for several years in a state of **dormancy** (suspended development) before germinating and growing into healthy plants. Only a minor portion of seeds germinate within the same year. Some examples of weeds which remain viable (alive) in the soil are given in table 14.11.

Table 14.11: Duration of weed-seed dormancy in the soil

Name of the weed	Seed viability (in years)
Bidens pilosa	>5
Brassica spp.	50
Convolvulus arvensis	>20
Chenopodium album	40
Mimosa pudica	up to 19

Dormancy is a very complex phenomenon that may be induced by diverse factors such as low temperatures, deep placement in the soil as a result of ploughing, genetic disposition or the presence of substances which inhibit germination. Often, seeds cannot germinate immediately after ripening, but may be forced to do so by some special treatment. For instance, treating the seeds of *Convolvulus arvensis* with sulfuric acid (scarification) aids germination that is otherwise inhibited by a hard seed coat. Seeds of some other weeds germinate earlier after having been scalded by brushfire. If a chemical could be developed that would be capable of breaking the dormancy of weed seeds within one season, a concerted effort might clear the topsoil of the bulk of buried seeds. It has, for instance, been detected that treating the soil with ethylene triggers germination in *Striga* species.

Germination

Factors involved in the germination of weed seeds in the soil are:

* Temperature. Weed seeds require specific soil temperatures for optimum germination. Since mechanical disturbance of the soil also stimulates seed germination, cultivation as a control measure should begin when the temperatures are favourable for germination of the main weeds.

* Soil moisture and aeration. Most weeds will germinate when conditions of aeration and soil moisture are adequate. If the moisture requirements for germination of the crop and weeds are known, time of irrigation can be used to favour the crop seedlings. For example, one method to control weeds during the establishment of cotton in a normally dry area is to carry out irrigation and cultivation before sowing and to sow cotton seeds deep in the moist soil. The top-soil layer dries up early and forms a crust in which the weed seedlings are trapped. Once the weeds are killed, one passage to break the crust will enable the crop seedlings to emerge and grow without initial competition. An effective method of killing many weeds is by flooding the fields in order to inhibit normal aeration of the soil and thus rob the weeds of their essential oxygen supply.

* Depth of seed burial. Small-sized seeds usually fail to germinate if they are buried deeper than 5 cm. Deep burying of seeds by ploughing them under can induce dormancy because of limited oxygen supply. However, some weeds may even emerge from a depth below 15 cm. Therefore, soil tillage should not be carried out below the depth from which seedlings of important weeds can emerge in order to avoid bringing a new supply of buried seeds to the surface.

* Light. Many seeds require light for germination, thus weed emergence decreases once the crop canopy starts to close. But other weeds are insensitive to light or darkness and will emerge anyhow.

* Soil nitrate content. High nitrate levels favour seed germination. For this reason, nitrogen fertilizers have been applied to induce germination of *Striga* seeds as a control measure.

Following the processes of germination, a rootlet (radicle) and shoot (plumule) commence to grow from the seed by underground elongation. The seedling obtains the energy needed to grow from the food reserve stored in the seed endosperm. Next, the seedling emerges above the ground either by bringing the seed-leaves (**cotyledons**) above the ground, or by elongation of the **epicotyl** while the cotyledons remain buried. The first is common for dicotylodonous plants, the latter, for monocotylodons. Such characteristics are of importance for chosing the desired herbicidal activity and the method of application.

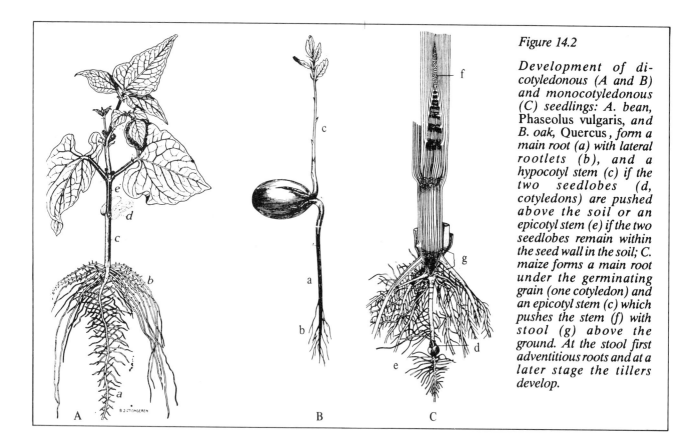

Figure 14.2

Development of dicotyledonous (A and B) and monocotyledonous (C) seedlings: A. bean, Phaseolus vulgaris, *and B. oak,* Quercus, *form a main root (a) with lateral rootlets (b), and a hypocotyl stem (c) if the two seedlobes (d, cotyledons) are pushed above the soil or an epicotyl stem (e) if the two seedlobes remain within the seed wall in the soil; C. maize forms a main root under the germinating grain (one cotyledon) and an epicotyl stem (c) which pushes the stem (f) with stool (g) above the ground. At the stool first adventitious roots and at a later stage the tillers develop.*

The way in which weed roots develop is a further point of consideration in selecting the proper herbicide treatment. Annual grasses have a primary root that dies early and is being replaced by numerous adventitious rootlets which grow near the surface of the soil from the coleoptilar node of the stem, also called crown. The primary root of broad-leaved weeds does not die, but develops in a downward direction as a taproot from which secondary lateral rootlets grow below the location of the seeds. Consequently, surface applications of herbicides give satisfactory control of annual grasses, but for the control of many broad-leaved species and perennials which root in the deeper soil layers, the herbicide has to be brought into contact with their root system by incorporating it into the soil, or a systemic herbicide should be applied.

Survival by vegetative reproduction

The survival of perennial weeds is greatly enhanced by their very effective systems of vegetative reproduction in addition to reproduction through seeds. These survival systems include creeping roots and stems, bulbs and other organs which contain a food reserve. When such plant parts become detached from the parent plant -- owing to discing or harrowing, for example -- they can establish themselves as independent new plants capable of colonizing other fields. For this reason, vegetative parts that come to the surface after tillage and mechanical weeding should be left to dry in the hot sun or be collected and destroyed.

Important structures of vegetative reproduction are as follows:

Stolons and runners

These are slender, modified stems which grow from the axil of the lower leaves and run over or below the soil surface in various directions. The mature parts of these stems usually root and sprout at the nodes. Examples of weeds with creeping stems are: *Polygonum* species and *Commelina benghalensis*, and of weeds forming stolons: *Cynadon dactylon* and *Artemisia vulgaris*. A stem is called a runner or offset if the stem develops anchor roots and new plantlets at the terminal buds and continues to grow in the same direction; for example, the floating water hyacinth (*Eichhornia crassipes*).

Rhizomes and rootstock

These are underground fleshy stems bearing nodes, buds and scaly leaves. A fast-growing, extensive rhizomatous system is a common feature in perennial grasses, such as *Panicum repens*, in sedges (Cyperaceae), and certain broad-leaf weeds such as bindweeds (*Convolvulus* spp.). Rhizomes which grow chiefly downward, such as those of *Sorghum halepense*, are called rootstock.

Tubers

Tubers are the swollen ends of rhizomes. An example that probably is known to every farmer is the nutsedge (*Cyperus rotundus*) which forms numerous tubers deep in the soil. Each tuber is itself a short, complete stem with scaly leaves, internodes and minute buds which give rise to new shoots, roots and rhizomes. One nutsedge tuber left behind in the soil can infest a large field with thousands of tubers in just one season.

Another well-known tuber-forming crop is the potato. In the fully mechanized cultivation of this food crop in western Europe, left-behind potato tubers can disturb a root-knot nematode control programme through crop rotation, if they sprout as volunteer weeds in the next season; volunteer potato plants are normally controlled with herbicides.

Bulbs and bulbils

Bulbs are fleshy, disc-like organs which contain in compressed form a complete plant, including many thick leaves, buds and a flowering apical bud. Examples are wild onion and garlic and *Oxalis* species. Bulbils are buds of the green above-ground portion of a plant which usually separate from the parent plant and drop on the soil where they grow into new plantlets. Example: *Bryophyllum pinnatum*.

Hardiness of weeds

The high persistence or capacity for enduring un-favourable conditions that make weeds so very difficult to control is the result of a combination of several of the aforementioned characteristics, such as:

* Abundant seed production linked with a high rate of seed viability. Provided that flowering is complete, some weeds can produce up to 100 per cent viable seeds even when they are cut before maturity.

* Dormancy, which may last for years.

* Highly effective systems of vegetative reproduction, sometimes more than one and additional to seed production.

* Dispersal by means of specialized features for transport by wind, water or animals.

* Escapism. Characteristics such as bitter taste and thorns repulse grazing animals, underground parts provide protection against fire and other causes of damage etc. In the case of grassy weeds, *Echinochloa crus-galli* and *Leptochloa chinensis*, They look so much like young rice plants that they may be transplanted in numbers with the rice seedlings.

* Numerical impact. Of the more than 8,000 known weed species, there will always be some present as seeds or propaguls to invade a free place, even if the conditions of that place drastically change.

METHODS OF WEED CONTROL

Mankind's long struggle against weeds has yielded many methods of control. However, at least as important as the control of established weeds is the prevention of infestation by newly introduced weeds or of the abundant dispersal of weed seeds and propagules. Some means of preventing infestation are:

* The establishment of and compliance with strict quarantine rules.

* Usage of cleaned and preferably certified seed lots which are free of weed seed and contaminating vegetative matter.

* Not allowing weeds to produce seed or to reproduce vegetatively.

* Burying or burning of residues and pulled-out weeds.

* Keeping levees and irrigation canals free of weeds.

* Cleaning of tractors, machines and tools after use.

* Keeping grazing animals out of crop fields because their droppings may contain many weed seeds.

Important methods of controlling weeds are:

* Manual and mechanical weeding
* Other cultural control
* Biological and ecological control
* Chemical control

Manual and mechanical weeding

The age-old methods of weeding by hand or with the help of simple tools are still practised everywhere, because many farmers in developing countries neither have the money nor the necessary knowledge to use mechanical implements and herbicides. Moreover, manual weeding is so effective that a sensible combination of this traditional method and modern practices often gives a better result than, for example, the use of herbicides alone. In fact, hoes, spades, shovels, sickles and knives are the most common implements for weed control and tillage in labour-intensive cultivation systems.

Hand-pulling of weeds and cutting of the weed roots just below the soil surface are most effective against annual weeds, but less so against perennial weeds which grow from tubers or rhizomes deeper in the soil. Hoeing is more effective against deep-rooting perennials, but all rhizomes should be removed. Often the farmer forces the hoe (*patjol*) too deeply into the soil and pulls up the weed together with surrounding soil clod with its roots remaining more or less intact. Under humid conditions, the weed plant might recommence growing having been in fact merely "transplanted".

During a rainstorm, raindrops impact on bare soil with such force that soil particles are loosened and lifted, and then can be easily carried off with run-off water. Generally, leaving the soil bare can lead to increased surface sealing (crusting), soil erosion by water and wind, and to higher evaporation. Therefore, rather than clean-weeding it is necessary to protect the soil by planting cover crops and applying mulches and, on sloping soils, by constructing ridges and terraces on the contour to control water run-off. Further anti-erosion measures include the planting of certain grasses and leguminous hedge rows on the contour, mixed cropping and strip or ring weeding around tree crops.

Cultural control

* **Tillage**. The working of the soil with hoes, spades or simple mechanical implements serves to break the soil crust, and to mix manure and plant residues into the upper 10 to 15 cm of the soil in order to prepare a seed bed. This action commonly includes weeding, the "earthing up" of tuber and root crops where needed, and is completed with planting the seed, seedlings or cuttings. In tillage, both the timing of the activities and the depth of working the soil are important in terms of weed control.

Delayed tillage means preparing the seed bed and waiting until the weeds emerge before lightly cultivating the soil again and planting the crop seed. The majority of weed seeds in the top soil-layer will germinate soon after seed-bed preparation. The purpose of delayed tillage is, therefore, to destroy the first flush of weeds so that the planted crop can later grow with less pressure from competing weeds.

Blind tillage pertains to the practice of sowing the crop seeds after the normal seed bed preparation. Subsequently, the field is lightly cultivated after the weeds have emerged but before crop emergence. The effect of this method can be improved if the crop seeds are planted deep in the soil with the aim of delaying their germination. In dry-seeded crops, the earlier germination of the weed seeds, in comparison with the crop seeds, is stimulated by light rain or irrigation

Zero-tillage is a relatively new practice that can be adopted when fields have been well-levelled in previous years and have a permeable soil structere. The surface of the soil is hardly worked mechanically and weeds are controlled with herbicides. Zero-tillage is practised on a limited scale on farms in South America, where maize is grown in rotation with beans on non-tilled soils which are sensitive to erosion; mulching is usually included.

* **Mixed cropping**. The growing of a mixture of crops is common practice in self-sufficient crop production. Poor farmers grow a mixture of crops to satisfy their needs, as a safeguard against crop failure from pests and diseases and because of the generally higher total yield. It has been observed that a mixture of crops suppresses weeds better than a single crop does. The usually quite dense canopy of mixed vegetation prevents the germination of light-sensitive weeds. However, decreasing the competition from weeds alone would not suffice to give higher yields from mixed cropping, if other factors are not given proper attention. These factors include chosing suitable types of crops, supplying nutrients and water, performing additional weeding and controlling of insects and diseases.

Growing mixtures of crops in gardens and small fields has lead to the practice of intercropping, usually by planting two annual crops in alternate rows. Intercropping of tall crops, such as maize or sorghum, with low crops, such as beans, sweet potato or watermelon, is common in western Africa. Intercropping also requires favourable growing conditions, suitable combinations of crops having a different growing period and time of ripening, and determining planting distances both in between and within the rows. Intercropping has the same advantages as mixed cropping, i.e. weed suppression, deminishing the risk of total crop failure, and minimizing soil erosion. But since intercropping may cause problems with mechanical harvesting, this method is not commonly practised on large farms.

* **Flooding**. The practice of inundating fields with water to control weeds is standard in rice cultivation and in many irrigation schemes. Flooding controls many weeds by depriving them of oxygen for a fairly long period of time.

* **Mulching**. Weed growth can be suppressed by applying dead mulches, by planting cover-crops as live mulches, and by covering the soil with black plastic sheets. Straw, cut grasses, crop residues and other plant material, manure etc. can all be used as mulches. Mulching smothers weeds, improves biological activity in the soil, preserves soil moisture, prevents crusting and surface

A mulch of cut grass reduces evaporation from onion beds whilst protecting the soil against erosion and suppressing weed growth. (Photo by the author)

run-off of soil particles as a result of rainfall. The planting of cover crops is common practice in estate tree crops.

The planting of leguminous shrubs, such as *Crotolaria* and *Tephrosia* species and *Maghonia macrophylla* in hedgerows, has the double function of fixing nitrogen from the air and supplying material for mulching. A new technique of alley cropping, which has been developed by the International Institute for Tropical Agriculture (IITA) in Nigeria, helps to restore the fertility and structure of the soil.

* Crop rotation. Systematic crop rotation is not usually practised in the (sub-) tropics as a weed control method. But annual rotations help in maintaining soil fertility and preventing a build up of animal pests, diseases and weeds. The growing of different successive crops throughout the year in correspondence with the farmers' need and available residual water has the same effect. The planting of catch and trap crops, particularly in the case of controlling parasitic plants such as Striga, is a form of effective crop rotation for reducing infestations.

The mottled water hyacinth weevil, Neochetina eichhorniae *Warner, from South America, that has been successfully introduced into Africa and Australia, feeds voraciously on leaves of the water hyacinth* (E. crassipes). *(Photo courtesy of CSIRO)*

Biological and ecological control

Biological control of weeds involves the employment of an organism or virus as control agent. Three methods can be distinguished:

- The use of selective organisms or viruses which affect one or a few weed species;

- The use of non-selective organisms or viruses which affect the majority of weed species; and

- The use of competitive plant species, i.e. plants which compete with weeds for growth factors such as light, space and nutrients.

The classical example of successful biological control concerns the control in Australia of the prickly pear cactus (*Opuntia*) that was introduced from Latin America in the nineteenth century. By 1925, more than 25

millon hectares of grassland were infested by this cactus to the extent that more than half of the area became inaccessible to man and animals. This prompted a search for natural enemies in Latin America and more than 150 insect species were identified as potentially affecting *Opuntia*. About 50 species were tried in Australia under quarantine conditions, of which the larval stage of the moth *Cactoblastis cactorum* appeared to be the most promising enemy. From 1926 onward, these moths were released in large numbers and within six years brought the *Opuntia* cactus plague under control so that the grasslands could be recovered for sheep grazing.

Equally successful was the control of the aquatic "alligator" weed that had been introduced into the United States from South America, and by 1963, had infested over 40,000 ha of water in the southern United States. As chemical control failed, the U.S. Department of Agriculture ordered a search for natural enemies in Soutl America. After screening about 40 possible candidates three insect species were brought in and reared ir quarantine. The flea beetle, *Agasicles hygrophyla*, of whicł the female reproduces within 25 days and lays over 1,00C eggs each, was released in 1964 and destroyed in many places the alligator weed population within the same year. Because alligator weeds mostly die in the winter season, the flea beetle population starved from lack of an alternative source of food. However, the moth *Vogtia malloi* also succeeded in building up strong populations which survived well during the winter. In both the United States and in Australia, alligator weeds are currently well controlled by these natural enemies.

In addition to these selective predatory insects, weeds may also be controlled by various non-selectively grazing animals. Well known is the herbivorous grasscarp (*Ctenopharyngodon idella*), a type of fish occurring in Chinese and Siberian mountain rivers which likes to feed on submerged aquatic weeds. Grasscarps can be induced by hormonal injections to reproduce in captivity and the fingerlings can be released in other countries. In a trial in Egypt, grasscarps were cultured successfully and proved potentially effective in controlling waterweeds in irrigation canals.

Grasscarps (Stenopharyngodon idella) *effectively control the lush growth of some major waterweeds in irrigation and drainage canals. (Photo courtesy of CABO, the Netherlands)*

279

Table 14.12: Other examples of weed control by natural enemies, with the year in which their large-scale release commenced

Weed species	Country	Year	Natural enemy
Cordia curasssivica (brush in grassland)	Mauritius	1948	*Schematiza cordiae* (beetle from Trinidad)
Eichhornia crassipes (water hyacinth)	United States Australia, Sudan	1970s	*Neochetina eichhorniae* (weevil from S.America)
Salvinia molesta (aquatic weed)	Papua New Guinea Australia	1980s	*Cyrtobagous* species (weevil from S.Brasil)
Chondrilla juncea (skeleton weed)	Australia	1970s	*Puccinia chondrillina* (rust fungus from Mediterranean region)
Rubus constrictus	Chile	1970s	*Phragmidium violaceum* (rust fungus from western Europe)

Relatively new is the use of plant **pathogens**, in particular of fungi, for weed control. Highly specific fungal pathogens are being used either as classical biocontrol agents or as microbial herbicides. Classical biocontrol involves the importation of a pathogen from the geographical origin of the pest and relies on the ability of the pathogen to self-perpetuate, spread, and reduce weed populations to subeconomic threshold levels (Templeton and Smith, 1982). Two examples of such fungal pathogens which exert classical biocontrol are given in table 14.12.

Microbial herbicides

Microbial herbicides, also called **mycoherbicides**, are usually native (indigenous) plant pathogens with limited capacity to spread and which ordinarily occur at rather low population densities. Such pathogens can be artificially cultured and suspensions of the fungal spores can be sprayed onto weed plants in a way similar to applying chemical herbicides. If the application is caried out at the proper time and under favourable conditions, the spores will infect the weed which will be killed by the fungal disease concerned.

Examples of microbial herbicides (mycoherbicides) are:

* Control of *Aeschynomene virginica* (northern jointvetch) in wet rice and irrigated soybean fields in the United States, with the fungus *Colletotrichum gloeosporioides* f.sp. *aeschynomene*.

* Control of the weed *Morrenia odorata* in citrus plantations in Florida, with commercially available suspensions of the soil-borne fungus *Phytophthora palmivora*.

* Regrowth of *Prunus serotina* shrubs in commercial forests of the Netherlands is controlled by applying mycelium of the endemic fungus *Chondrostereum purpureum* to the cut surface of the stumps.

The nature of the weed problem and the origin of the weed determines if either the classical fungal biocontrol or the mycoherbicide is to be used. The classical tactic is recommended for introduced weeds in rangelands and waterways where other weed control methods are less well suited. But in annual and high-value perennial crops infested with native or introduced weeds, the direct application of mycoherbicides is more appropriate. Furthermore, some test results indicate that different strains of mycoherbicides may be mixed to give a broader spectrum of weed control. For example, the strain of *Coletotrichum gloeosporioides* required to kill *Aeschynomene virginica* can be mixed with another strain of this fungus specific for *Jusiaea decurrens* to control both weeds. The first strain reportedly can also be mixed with the chemical herbicide acifluorfen that controls *Sesbania exaltata* weed.

However, mycoherbicides contain a living organism, the fungal spores, and thus handling, storing and using these formulations require special care and proper conditions. The effect of mycoherbicides may take 4-5 weeks to kill the weeds, and their activity is improved by warm, humid weather conditions. Mycoherbicides are generally destroyed when mixed with liquid fertilizer or herbicides other than acifluorfen.

Competitive plants

The use of competitive plants as a means of weed control is an old practice in tree crops. Cover crops serve a whole range of purposes such as preventing soil erosion, improving soil fertility and structure, and preventing and suppressing weed development. Well-known cover crops are the creeping leguminous species *Indigofera spicata, Pueraria phaseoloides, Centrosoma pubescens* and *Calopogonium mucunoides*. These crops are frequently sown as a mixture after the seeds have been inoculated with nitrogen-fixing *Rhizobium* bacteria. A well-closing canopy of cover crops can control practically all weed growth in perennial and annual crops. It has, for example, been reported from Nigeria that planting of *Centrosoma pubescens* and the wild winged bean (*Phosphocarpus palustris*) in maize provides good control of weeds without adversely affecting the crop yield.

The growing of shade trees too has a suppressing effect on weed development in estate crops such as coffee, tea, cacoa and oil palm. Young cacoa trees, in particular, are very sensitive to weed competition and thus need to be ringweeded and shaded until the trees provide their own shade from a closed canopy. In young cacoa and citrus tree plantations, cover crops should not be employed after the third and fifth year respectively, because their superficial root systems cover almost the entire field and must be kept free of competing plants and weeds. To avoid injury from mechanical weeding, control is usually achieved by applying herbicides. However, in less sensitive tree crops, such as oil palm, the use of cover crops for weed control is very common.

One should, however, be very careful with introducing new species as cover crops as they might run out of control. For instance, *Mimosa pudica* and *Oxalis* species were introduced for this purpose, but currently they are a problem weed in certain places. Beneficial shade trees include *Erythrina*, *Leucaena*, *Albizia* and *Gliricidia* species. Banana, plantain and cassava may also be planted as temporary shade in very young tree crop plantations.

Chemical control

Chemicals are increasingly being used in developing countries because they offer, in many instances, an effective and relatively inexpensive means for managing serious weed problems. Generally, chemicals for weed control are employed in combination with mechanical and cultural methods. Since the 1940s, research has yielded a wide range of herbicides for different purposes and situations and the search for new compounds is continuing with an emphasis on selectivity and minimal residues. Chapter 2 of this manual contains a review of the main groups of herbicides and examples of currently important compounds. It also provides a division into groups of herbicides according to their mode of action.

The use of herbicides has some clear advantages, but also cause some negative environmental effects that need to be avoided or kept as low as possible. Obvious advantages are:

* Herbicides offer a cost-effective way of controlling weeds in situations, such as large-scale farming and rangeland maintenance, brush control in forests, and on industrial sites and utility rights-of-way, where costs of labour would prohibit mechanical control.

* Selective herbicides reduce the need for handweeding and mechanical cultivation in standing crops; these activities often cause damage to the roots.

* Mechanical cultivation between crop rows can be complemented by application of selective herbicides for weed control within the crop rows.

* Pre-emergence herbicides protect a crop against competition from weeds during early stages of growth.

* Chemical herbicides offer the possibility of reducing the need for seedbed preparation in maize and sorghum to minimum (shallow working of the soil surface) or zero-tillage (usually limited to drilling of the seed).

* Herbicides frequently offer the only possibility of effectively controlling annual and perennial "problem weeds".

The following serious disadvantages of herbicides need to be mentioned:

* Herbicides can through long-time use, cause resistance in weeds. This may lead to a change in the composition of the weed flora and a predominance of certain difficult-to-control weed populations which are either resistant to or tolerant of these particular compounds.

* Certain herbicides have a strong negative effect on the biological activity of the soil; their chemical degradation may proceed slowly and leave potentially toxic and persistent residues in the crop, or soil and groudwater. Long-time use of atrazine in maize in western Europe, for instance, has resulted in a build-up of persistent residues in the soil which threaten the quality of ground and drinking water.

* Certain herbicides are very hazardous to the health of workers and consumers, even if formulated in granular form.

* Very few herbicides are recommended for mixed cropping and intercropping systems, because it has been difficult to develop compounds that will control a broad range of weeds without causing damage to the component crops of the mixture. Thus, herbicides hold more promise to those farmers practising monocropping under improved soil and water management conditions.

* Certain herbicides have a carry-over effect by which residues in the soil may cause damage to certain other crops; this phenomenon must be taken into account when composing a crop-rotation schedule.

* Safe application of herbicides requires a relatively high level of training in order to select correctly the proper formulations, dosage rates, timing and method of application to control a specific weed in correspondance with the type of soil and crop. Such training is often not available to poor farmers.

Mode of herbicidal action

It is very difficult to predict the action of herbicides beforehand without carrying out extensive trials with reference to the crop and soil type to test the effect of the chemical to be used. However, from detailed studies of the processes taking place in plants and in the soil the following explanations of the supposed modes of action have been derived.

A chemical demonstrates good herbicidal activity if it is rapidly absorbed by the weed plants and, after entering the plant tissues, it causes a rapid and complete kill. Herbicides have either a short-distance (contact) effect or a long-distance (systemic) effect on plants. Herbicides with contact action kill only the parts of the plants to which the chemical is applied.

Much research has been conducted to discover the processes which control the absorption and translocation to the site of action and the way in which the active ingredient acts. It has been found that about 50 per cent of the herbicidal chemicals affect photosynthesis, about 20 per cent inhibit germination of seeds and the division of cells in seedlings and older plants and about 10 per cent interrupt processes which control respiration and metabolism.

It has been recognized that some herbicides affect more than one of these functions in weeds and thus have a multiple mode of action. However, the mode of action of a relatively large number of herbicides is not yet known.

Systemic herbicides are absorbed or taken up by roots, coleoptile or leaves and translocated within the plant to other tissues. Systemic herbicides, therefore, give much better control of perennial weeds than contact herbicides that do not reach the buried organs in the soil. However, contact herbicides usually have a faster effect than systemic herbicides.

Important modes of action

● Herbicides that disturb the normal hormonal processes in plants (hormone-type herbicides):

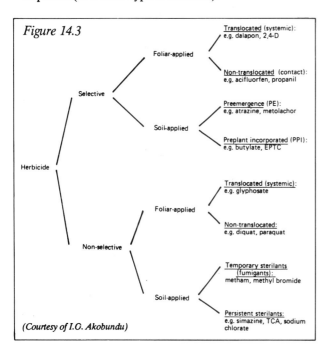

Figure 14.3

Selective
Foliar-applied
Translocated (systemic): e.g. dalapon, 2,4-D
Non-translocated (contact): e.g. acifluorfen, propanil

Soil-applied
Preemergence (PE): e.g. atrazine, metolachor
Preplant incorporated (PPI): e.g. butylate, EPTC

Herbicide

Non-selective
Foliar-applied
Translocated (systemic): e.g. glyphosate
Non-translocated: e.g. diquat, paraquat

Soil-applied
Temporary sterilants (fumigants): metham, methyl bromide
Persistent sterilants: e.g. simazine, TCA, sodium chloride

(Courtesy of I.O. Akobundu)

Phenoxy acetic acid herbicides, such as MCPA and 2,4-D, stagnate top and leaf growth by inhibiting cell division in actively growing meristems, while at the same time increasing cell division in the tissues of the stem base and roots.

In these parts of the plant, such herbicides also enhance protein synthesis and the production of ribose nucleic acid (RNA), thereby causing a further massive proliferation of cells. This excessive cell division consumes most of the available carbohydrates which leads to malfunctioning of the leaves and roots and finally to the death of the plant. The process of cell elongation (the stretching of cell walls) and the uptake of water is also strongly inhibited.

Such effects on actively growing meristems become visible as malformations (sometimes very typical) of the leaves and flowers. Because of these typical results MCPA and 2,4-D are a very good choice for giving demonstrations of herbicidal activity.

● Herbicides affecting the metabolism of nitrogen and the synthesis of protein:

Glufosinate-ammonium inhibits the normal nitrogen assimilation of plants and indirectly also photosynthesis. Inhibition of the synthesis of aromatic amino acids has been reported for glyphosate (Roundup).

● Herbicides that disturb germination or initial seedling establishment:

Carbamates (such as chlorpropham and EPTC), dinitroanilines (such as trifluralin and pendimethalin) and certain amides (such as alachlor and propachlor) inhibit cell division during or very soon after germination.

● Herbicides that inhibit photosynthesis:

Substituted ureas (such as diuron, linuron and fluometuron), triazines (such as simazine, atrazine and prometryn) and uracils (such as bromacil and terbacil) inhibit the process of photosynthesis in green plants. These soil-applied herbicides block a certain step in the complicated process by which carbohydrates are synthesized from CO_2 and water with energy collected by the plants' chlorophyl.

Certain foliage-applied herbicides such as propanil, bentazone, and the dipyridiliums, diquat and paraquat, block the synthesis of carbohydrates. Diquat and paraquat cause a much quicker disintegration of the chloroplast structure than the other photosynthetic inhibitors of photosynthesis owing to their rapid penetration of the leaves.

● Herbicides that hinder respiration:

Nitrophenols (such as DNOC and nitrofen) and benzonitriles (such as ioxynil and bromoxynil) interfere

282

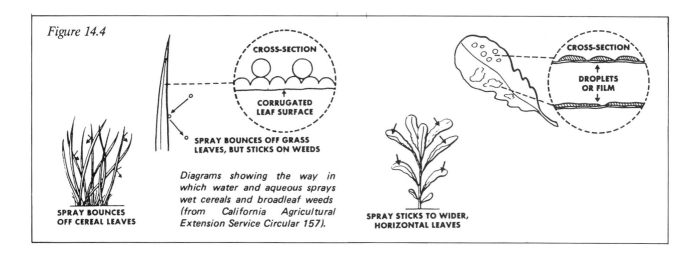

Figure 14.4

CROSS-SECTION
CORRUGATED
LEAF SURFACE

SPRAY BOUNCES OFF GRASS
LEAVES, BUT STICKS ON WEEDS

Diagrams showing the way in
which water and aqueous sprays
wet cereals and broadleaf weeds
(from California Agricultural
Extension Service Circular 157).

SPRAY BOUNCES
OFF CEREAL LEAVES

CROSS-SECTION
DROPLETS
OR FILM

SPRAY STICKS TO WIDER,
HORIZONTAL LEAVES

with respiration processes. During the process of respiration, oxidation of carbohydrates takes place in the cells. At the same time, the synthesis of the very energy-rich compound adenosine triphosphate (ATP) is influenced. This compound is needed for the transfer of energy in all kinds of reactions. When the process of ATP-formation is disrupted, energy is dissipated as heat. As this energy-providing process in plants is identical to that in animals and humans, these herbicides are also very toxic to all forms of life.

● Herbicides that disturb the growth of plants:

Aliphatic acid herbicides (such as dalapon and TCA) cause growth inhibition and eventually death by inhibition of protein and fatty acid synthesis. The latter effect is, however, not the only cause of death. However, further knowledge of the complete mechanism of action of these and many other herbicides is limited or lacking. Herbicides, such as amitrol, can have a pronounced effect on chloroplast development with the result that the photosynthethic process does not function and the plants die. Rapid necrosis can be caused by the scorching action of contact herbicides such as sulphuric acid and ioxynil.

Selectivity of herbicides

The response of plants to different kinds of herbicides differs not only between separate species, but also between individual plants of the same species. Even the reaction of the individuals of one plant species to the same herbicide depends usually on the locality where they grow. It has been found that in many instances the reaction of various plant species to particular herbicides tends to be specific; this may offer an opportunity for selective use. Ideally, herbicides should be developed that would inhibit life processes in certain weeds only, without causing harm to crop plants. The detailed knowledge of metabolic processes and chemical properties of weeds required for such directed chemical research has not yet been acquired. Since the detection of the selective action of sulphuric acid, DNOC, 2,4-D and MCPA, emperical chemistry has yielded several hundred selective herbicides.

Examples of selective herbicides in common use are bentazone, metobromuron, terbacil and cinmethylin.

The selectivity of herbicides can be the result of one or more of the following factors:

- Differences in absorption by foliage

Leaf-applied herbicides, which kill by contact or systemic action, usually injure not only weeds but also the germinating seeds and plants of a desired crop. Contact with the foliage of a crop can be avoided if such herbicides are applied prior to the emergence of the crop or as a directed spray; this would involve spraying between rows of around the base of trees. Contact with crop foliage can be minimized if the herbicide is applied in granular form. With a suitable granular applicator, granules can be safely applied after the emergence of young crops.

The angle and arrangement of the leaves determine to some extent the rate of interception; for instance, the erect leaves of young grain crops collect less spray drops or granules than the more horizontal leaves of broad-leaved weeds (see fig. 14.4). The wax layer on the leaves of cereals causes spray droplets to runoff before they can have any herbicidal effect.

Differences in the way a bud or growing point is protected as in the case of the leaf whorl of grasses or exposed as in the case of dicotyledonous plants can also play a role in selectivity.

- Different availability of soil-applied herbicides

The selectivity of soil-applied herbicides is most often the result of a difference in concentration of these herbicides in the soil at various depths. The movement of herbicides from the top layer of soil in a downward direction as during the process of leaching, depends on factors such as solubility of the chemical in water and the rate of adsorption to soil particles. Low solubility and a high rate of adsorption, in particular to humus and clay particles, go together in most triazines and substituted ureas. In glyphosate and paraquat, however, solubility in

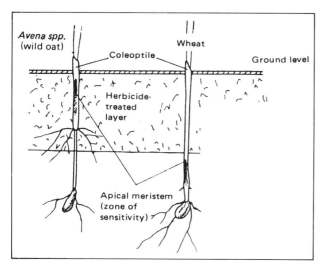

Figure 14.5: In Avena *spp., the mesocotyl extends soon after germination to place the sensitive zone in the band of herbicide-treated soil. Extension of the mesocotyl of wheat is so much less that the sensitive zone remains below the treated layer of soil for a longer period.*

water is high, but their leaching from the soil is countered by their firm adsorption to soil particles. Leaching depends also on the texture and compactness of the soil. Leaching and lateral movement in furrow-irrigated crops is easier in light, loose soils than in heavy soils.

Translocation of herbicides in the soil must be carefully studied in pre-planting and pre-emergence applications in relation to the depth of the soil layer in which the crop seeds are sown and the roots of the crop will grow. Weeds generally germinate in the top layer of the soil, whereas crop seeds are sown at 5-10 cm depth. If a pre-emergence applied herbicide is strongly adsorbed by the top-soil, the germinating weed seeds will be killed while the germinating crop seeds in the layer underneath where the herbicide concentration is low, remain un-harmed. The crop seedlings can then emerge through a weed-free layer of top-soil and gain good advantage in their early development. This type of selectivity, which is called "depth protection," is also used in herbicide treatments in perennial fruit orchards and tree-crops.

Soil-acting herbicides are absorbed not only by the roots, but also by emerging shoots and other parts of the plant. In particular, the meristematic part of cereals is sensitive and thus offers an opportunity for controlling wild oats in wheat. (See fig. 14.5 above.)

- Different rates of penetration

The leaf cuticle is the most effective barrier to the penetration of herbicides deposited on foliage. The permeability of the leaf cuticle to different chemical substances is largely determined by its thickness, the composition of its waxy layer and other features which vary considerably between plant species. Such selectivity can be partly counteracted by the addition of surface active agents (or wetters) to a herbicide spray. These improve the uptake of herbicides through the foliage.

The addition of oils to a spray solution increases the uptake of atrazine and some other herbicides by the foliage. Thus, atrazine sprays (with oil added) can be used for selective post-emergence weed control in maize.

- Different rates of translocation within plants

Differences in movement of the absorbed herbicide to the site of action often plays an important role in determining herbicidal selectivity. For instance, translocation of foliage-applied 2,4-D is much more restricted in some grasses than in broad-leaved weeds, although translocation is relatively high too in grasses which are sensitive to 2,4-D. In wheat and other cereals, the rate of translocation is higher in very young plants than in older ones. Thus, depending on the time of application and concentration of the chemical, it is possible to use 2,4-D for weed control in cereal crops.

Equally, there is a difference in sensitivity between plant species with respect to the rate of translocation after the uptake of soil-applied herbicides. For example, linuron remains stored in the rootlets of a tolerant carrot cultivar, but is translocated throughout the carrot plant in sensitive cultivars.

- Different rate of inactivation of herbicides

Inactivation of a herbicide inside a plant, either by adsorption or conjugation with amino acids or sugars of the toxic molecules during their translocation or by chemical degradation, lowers the concentration of the chemical at the site of action where important life processes are to be inhibited. The major cause of selective activity of herbicides is that such inactivation processes vary distinctly in different plant species or in sensitive and tolerant cultivars. Most frequently the phenomenon of chemical detoxification (decomposition) is the result of a break-down into non-toxic or partially toxic compounds in tolerant plants. Examples include side chain degradation of MCPA by tolerant plants such as *Galium aparine*; conjugation of atrazine in maize, sorghum and sugarcane by an enzymatic process in the leaves; and hydrolysis of propanil by an enzyme in rice.

In this connection, it must be mentioned that besides inactivation of herbicides in tolerant plant species as a factor in herbicidal selectivity, the opposite effect also exists: activation in susceptible plant species by conversion of inactive to active compounds. For example, the wild-oat herbicide benzoyl-ethyl can be converted to benzoyl-prop; this conversion occurs to a greater extent in wild oat than in wheat, for example.

Degradation of herbicides in the soil

The soil medium, which consists of soil particles, organic matter, air and water, reduces the mobility of water and herbicides and affects the availability of herbicides by partial adsorption to soil constituents. In addition to adsorption, other factors, such as herbicide

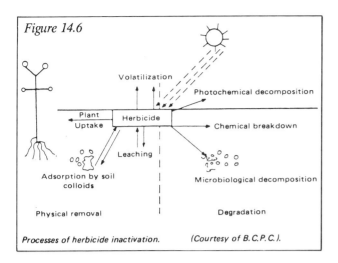

Figure 14.6

Processes of herbicide inactivation. (Courtesy of B.C.P.C.).

solubility, rainfall, soil texture and the degree of compaction, all influence leaching of a herbicide into the rooting zone of weeds and crop plants. It is assumed that only the free solution of herbicide molecules in the soil is directly available for uptake by plant roots or for degradation by micro-organisms. Adsorption is a reversible process. A reduction in concentration of the soil solution following the uptake by plants will bring adsorbed molecules back into the solution (desorption).

Non-biological degradation of herbicides occurs by the chemical processes of oxidation, reduction, hydrolysis and photo-oxidation (in ultraviolet light). Volatilization of herbicides, such as EPTC and trifluralin, also takes place. Adsorption to dry soil, incorporation immediately after spraying, or use of granular formulations can reduce vapour losses. Herbicides are mainly inactivated in the soil by micro-organisms. Algae, bacteria and fungi are capable of degrading certain herbicides very rapidly. Persistent residual activity in the soil is important in connection with the choice of the next crop. (See also fig. 14.5 above)

Timing of application

The timing of a herbicide application depends on many factors such as the type of herbicidal action, the weeds and crops involved, soil structure, temperature and rainfall among many others. The following terms are used to indicate the time at which the treatment should be made (see also fig. 14.7 at right):

- **Pre-plant.** A pre-plant treatment is any treatment made before the crop is planted. Examples of herbicides that can be applied in this way are dalapon, glyphosate and paraquat. A pre-plant incorporated treatment is one in which the herbicide is incorporated into the soil to prevent volatilization or to place the chemical in the zone in which it is needed prior to the planting of the crop. Examples of pre-plant incorporated herbicides are EPTC and trifluralin.

- **Pre-emergence.** Pre-emergence treatment is made after the crop has been planted, but prior to the emergence of the crop or weeds, or both. The herbicide is usually applied to the soil surface. Examples of pre-emergence herbicides are atrazine, metolachlor, butachlor and propachlor.

- **Post-emergence.** A post-emergence herbicide is applied after the crop or weeds, or both, have emerged from the soil. Examples of post-emergence herbicides are 2,4-D, diphenamid and propanil.

WEED CONTROL IN CROPS

From the point of view of weed control it is convenient to distinguish between control in, respectively, mixed cropping and intercropping systems; non-irrigated, annual crops; irrigated, annual and perennial crops; and estate tree crops. Some of the possibilities and limitations for weed control in these cropping systems are discussed in a previous section of this manual under "Methods of weed control". Below, some additional remarks are made, with emphasis on chemical control. The reader should, however, consult an extension officer or knowledgeable pesticide salesman for precise information in individual cases of herbicide use on crops, in order to avoid costly mistakes.

Mixed and intercropping systems

Major cropping systems of the humid (sub-)tropics are based on rice, maize, cassava, yam, banana and pulses, which may occasionally also be grown in combination with young tree crops, such as coffee, cacao, rubber and oil palm. Major cropping systems of the savanna region are based on staple food crops such as maize, sorghum, millet and rice, and the cash crops cotton, oil-seeds and sugar-cane.

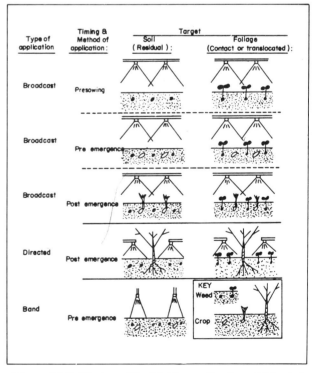

Figure 14.7: Timing of herbicide application. (Adapted from Roberts, 1982)

285

Research on weed control in mixed crops is aimed at selecting early maturing cultivars with a good competitive performance with regard to weeds. Further, on finding suitable rotation cycles of mixed cropping alternating with growing a single crop, different combinations of manual and mechanical cultivation and mulching are tried. In contrast, chemical control is added for the control of hard-to-kill weeds, in particular perennial grasses. Akobundu (1987) recommends the following herbicides to be safe for mixed cropping systems:

- bentazone in maize intercropped with rice, beans, soybean, or groundnut;

- pendimethalin in maize with groudnut and beans;

- metolachlor + metobromuron in maize with beans, groundnut, cassava or yam;

- atrazine + metolachlor in maize with cassava and yam; and

- CDEC in maize with vegetables.

The number of herbicides and herbicide mixtures that can be used in mixed crops without causing damage to one of the component crops is still limited. This implies that the screening of herbicides for crop mixtures must be intensified to gain insight to crop tolerance and herbicide persistence.

Non-irrigated, annual crops

The non-irrigated annual crops grown in the tropics consist mainly of food crops which are, for 70-80 per cent of them grown on small farms under a labour-intensive continious cultivation system. Weeds are controlled mainly by manual and mechanical weeding. On larger farms under intensified, mechanized cultivation systems, as well as in areas where costs of labour are

Soil tillage using draught animals for weed control and ridging of young tomato plants. (Photo by the author)

prohibitive, the use of chemical herbicides is common. Important weeds in non-irrigated crops are the sedges (*Cyperus* spp.), the grasses *Echinochloa* spp. and *Eleusine indica* and the broad-leaves *Portulaca oleracea, Amaranthus* spp., *Bidens pilosa* and *Sida acuta*. The weed spectrum is partly dependent upon the method of cultivation. For example, *Imperata cylindrica* can become a problem for those farmers who depend on weeding with a hoe. *Rottboellia cochinchinensis*, however, is more a problem in mechanized maize production. In the semi-arid tropics, the parasitic weed *Striga* spp. is a major scourge (see below).

Concerning chemical control, quite a number of herbicides are available for application in dry-land crops. In particular, atrazine in mixtures with pendimethalin or metolachlor for pre-emergence application in maize against annual weeds and *R. cochinchinensis*. Atrazines, in both granular or liquid form, are widely used in maize and sorghum because they are well tolerated by these crops. For post-emergence application bentazon and bromoxynil may be used against annual broadleaf weeds. Ametryn, 2,4-D and paraquat can also be applied post-emergence against annual broadleaves and some grasses, but the spray must be directed in order to avoid phytotoxic injury to the crop plants. To prevent a build-up of resistant weeds, herbicides such as alachlor and metolachlor can be used in place of or in combination with triazines. The herbicides EPTC or butylate are soil-incorporated before sowing or immediately after planting to control *Cyperus* spp. and many annual grasses.

For weed control in upland (dryland) rice culture, propanil has been recommended for many years for the control of grassy weeds and sedges. But also butachlor and pendimethalin offer a fair chance for good control of grasses and sedges.

In the case of minimum and zero-tillage, contact herbicides such as paraquat or glyphosate are sprayed onto the weed vegetation shortly after harvest. One light cultivation then suffices to prepare the field for planting a new crop. By using this method, the time-lapse between two crops can be cut back to about one week.

Irrigated annual and perennial crops

Although all crops require an adequate supply of water during the various stages of growth, certain crops demand to be irrigated regularly in order to produce high yields. Examples are cotton, rice, sugar-cane, jute and leafy vegetables.

Because of the general favourable condition of high fertility and ample soil moisture, weeds thrive very well under irrigated conditions and are capable of inflicting sizeable yield losses through fierce competition. There are two main groups of irrigated crops, namely crops such as rice and jute growing under flooded conditions for longer periods of time, and others, such as cotton and sugar-cane, which are irrigated but not flooded. Standing water has a suppressing effect on most grasses and broadleaf weeds, but stimulates the growth of others. Thus, the spectrum of weeds growing under these two systems may differ considerably.

In the case of wetland rice cultivation, which originated in China about 3000 BCE, the gradual development of efficient irrigation, transplanting and weed control practices has enabled substantial gains in production and productivity in comparison to dryland rice cultivation. Cultural practices perfected in China - - and still commonly practised in Asia today -- involve weed

Table 14.13: Some common weeds in tropical lowland and upland rice in South and South-east Asia

Lowland rice

Annual grasses

Echinochloa glabrescens Munro ex Hook f.
E. crusgalli (L) P.Beauv. spp. *hispidula* (Retz.) Honda

Annual broadleaf weeds

Monocharia vaginalis (Burm.f.) Presl.
Sphenoclea zeylanica Geartn.

Annual sedges

Cyperus difformis L.
Cyperus irea L.
Fimbristylis miliacea (L.) Vahl

Perennial grasses

Paspalum distichum L.

Perennial sedges

Scirpus maritimus L.

Upland rice

Annual grasses

Echinochloa colona (L.) Link
Digitaria sanguinalis (L.) Scop.
Eleusine indica (L.) Gaertn.
Dactyloctenium aegyptium (L.) Willd
Rottboellia cochinchinensis (Lour) W.D.Clayton

Annual broadleaf weeds

Ageratum conyzoides L.
Amaranthus spinosus L.
Celosia argentea L.
Commelina benghalensis L.
Eclipta prostrata (L.) L.
Ipomoea triloba L.
Portulaca oleracea L.
Trianthema portulacastrum L.

Annual sedges

Cyperus irea L.

Perennial grasses

Imperata cylindrica (L.) Raeuschel
Paspalum dilatatum Poir.

Perennial sedges

Cyperus rotundus L.

control in two basic steps:

- Puddling and flooding of rice fields; and

- Subsequent control of weeds by manual or mechanical weeding or herbicides.

The more thoroughly the first step, i.e. preparation of the field, is carried out, the fewer the weeds. For effective control, it is necessary to level the soil as well as possible in order to obtain an even layer of water covering all the weeds. The practice of transplanting rice seedlings from a seedbed in weed-free fields give them an advantage of at least two weeks in their early development as compared with competing weeds. Transplanting in straight lines facilitates weeding with a rotary weeder, that can be pushed between the rows of rice plants. The need

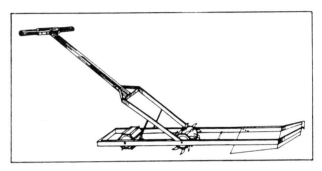

Push-type rotary-weeder for between-row weeding in flooded rice fields.

for weeding later in the season depends greatly on management of the water level in terms of timing and depth of flooding. If fields are kept flooded, few weeds can develop, but alternate drying and flooding of the field stimulate weed growth. Further, flooding the rice field with 1-2 cm of water reduces grasses, but broadleaf weeds and sedges may remain. With 5-10 cm of water, grasses are eliminated, and only a few broadleaves and sedges may survive. Temporary drying of the field destroys aquatic weeds, but farmers usually cannot afford to loose the water.

Chemical control in transplanted rice has received much attention. The herbicides 2,4-D, MCPA and bentazon are commonly used for the control of broadleaf weeds; for the control of grasses, the selective herbicides bifenox, molinate, thiobencarb, fluorodifen, oxadiazon and butachlor can be applied pre-emergence and post-emergence to weeds during the first two weeks after transplanting. Propanil, as a selective, broad-spectrum contact herbicide, can be used in mixtures with fluorodifen or with oxadiazon. Mixtures of a grass-killing and a broadleaf-killing herbicide are often used early post-emergence at 2-3 weeks after rice emergence or transplanting, e.g. molinate + MCPB. Herbicides can be applied as a spray or as granules to transplanted rice; the granular formulation type is easy to apply by hand.

Cotton is usually grown as an irrigated crop planted on ridges in widely spaced rows with the water

Various types of hand-held rope-wick and carpet-fabric wipers for application of contact herbicides such as glyphosate and 2,4-D in sensitive crops such as cotton.

being led in between the rows. Weeds grow abundantly until the crop canopy closes and cuts out light. If ample water is available, the fields may be lightly irrigated to stimulate germination of weeds and to kill the weeds before sowing of the crop. Water can be saved by drilling the cotton seeds below the top layer of the soil from which the majority of weed seedlings emerge. This leaves time for spraying the weeds pre-emergence to cotton.

Many herbicides are available for weed control in cotton. For instance, trifluralin and nitralin applied pre-sowing and soil- incorporated, are effective against grasses, but not against broadleaves. Some urea herbicides, such as monuron and diuron, and triazine herbicides, such as prometryn, applied pre-emergence can give good control if the soil remains moist. However, two weeks of dry weather decreases the effectiveness of these herbicides. Ureas and triazines may also be sprayed post-emergence, provided the spray does not hit the cotton leaves. Cotton is very sensitive to 2,4-D; this compound should, therefore, not be used near cotton fields. More detailed information about cotton herbicides should be obtained from an extension officer or informed distributor.

Parasitic weeds

Some weeds are true parasites of other plants. They attach themselves to the roots, stems or other parts of their host with sucking pads (haustoria) and draw mineral nutrients and water from them. The important parasitic weeds of the tropics and subtropics are described below:

* *Orobanche* species (Broomrapes) belong to a family of annual parasites occurring in dry subtropical and temperate regions. The species *O. ramosa* are parasites of the Solanum family (tomato, tobacco, peppers of the capsicum group and eggplants); *O. aegyptiaca* of lentils, cucurbits, potato and tobacco; *O. crenata* of broad beans, lentils, peas and other leguminous plants; and *O. cernua* of sunflowers. Broomrapes have a tuber-like stem base that is attached to the roots of its host, and a

fleshy stem with rudimentary leaves. They flower within 2-3 months after germination and can produce up to 50,000 seeds per plant.

* *Striga* species belong to the family Scrophulariaceae and are annual parasites occurring in arid tropical regions such as the Indo-Pakistan subcontinent and the Sahel. The main species *S. hermonthica* and *S. asiatica* are obligate parasites of maize, millet, sorghum and other members of the Gramineae family. *S. gesnerioides* attacks most cowpea cultivars. *Striga* plants form very short rootlets which attach to the roots of the host plant by haustoria. They develop below the soil surface during the first 40 days and, during this time, they are totally dependant upon the host and thus inflict the greatest damage. Thereafter, *Striga*-seedlings emerge and form chlorophyl-containing aerial shoots, which can photosynthesize their own carbohydrates. *Striga* species are extremely injurious, probably because they not only parasitize their host but also disturb its hormone functions by secreting toxins. The plant produces bright red flowers and up to 500,000 dust-like seeds.

Both *Orobanche* and *Striga* plants have in common that their seeds can remain dormant in the soil for several years. A particular feature of *Orobanche* and *Striga* seeds is that they are able to germinate only if exposed to a chemical stimulant, which is exuded by the roots of suitable host (and certain non-host) plants. Such germination-stimulants are presumably chemically related to strigol, a compound that was isolated from the

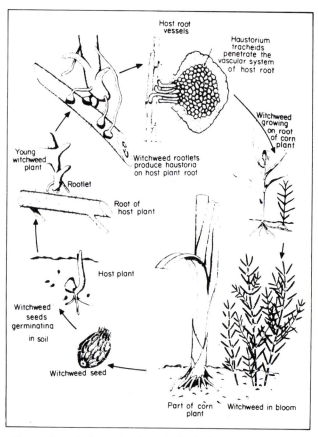

Figure 14.8: Life cycle of parasitic Striga *weed on maize (underground stages left, aerial stages right).*

Colourful Striga hermonthica *(Del) plants parasitize the roots of maize plants which eventually die. (Photo courtesy of A. Pieterse)*

Broomrapes Orobanche crenata *Forssk are often found parasitizing the roots of broad bean* Vicia faba *and other grain legumes. (Photo courtesy of A. Pieterse)*

roots of cotton plants in 1972. Since then, several strigol derivatives have been synthesized which induce germination of *Orobanche* and *Striga* seeds. This phenomenon provides the possibility of sowing suitable, germination-inducing hosts as catch crops, which need to be removed and destroyed before the parasites flower and produce seeds. Other crops, such as cotton and cowpeas in the case of *Striga*, induce germination but are not suitable hosts. In this way they cause the death of the emerging parasitic weeds by starvation; such crops are termed trap-crops. The use of catch and trap crops could be useful in crop rotation systems; elsewhere in monocropping, their practical application seems to be very limited.

The application of ethylene as a chemical stimulant to induce germination of *Striga* is practised only in the United States corn-belt as part of the "witchweed eradication" programme of the State of North Carolina. The method would be too expensive for developing countries.

Chemical control of *Striga* in Gramineae is possible by spraying 2.4-D over the crop plants, but several sorghum cultivars are sensitive. This herbicide cannot be used on dicotyledons and therefore *Striga* can persist on soybeans, for example. In fact, there is no real effective inexpensive chemical control method for *Striga* and *Orobanche*. The *Striga* infestation continues to spread and is inhibiting the expansion of maize cultivation in Nigeria and Sahelian countries of Africa. The method of solarization by covering the soil with plastic sheets, although effective, is also not practical because of the costs.

The selection and breeding of cultivars with resistance and tolerance against *Orobanche* and *Striga* species has yielded some useful cowpea and sorghum cultivars. But no real break-through has been achieved yet.

* *Cuscuta* species (dodders) are stem parasitic weeds which attach themselves to the above-ground parts of the host. They are twining, climbing herbs with very

long, golden-coloured, wiry stems with rudimentary leaves without chlorophyl, and small clusters of flowers.

They are a pest of dicotyledonous plants in tropical and temperate zones such as lucerne, broad beans and tomatoes. *Cuscuta* species reproduce by numerous seeds and small fragments of the stem which become detached.

Control of dodders can be carried out with high-dose foliar sprays of diquat, paraquat or chlorpropham on woody perennials, although the foliage of the host will also suffer. As *Cuscuta* species do not infest cereals, it is recommended to include cereals in the crop rotation scheme in those areas where dodders create a problem.

Parasitic dodder weeds (Cuscuta spp) *strangle their host plant with wiry, climbing stems and suck their nutrients by means of penetrating rootlets.*
(Courtesy of I.S.E.S).

289

References:

Akobundu, I.O. (1987): *Weed science in the tropics: principles and practices*. Publisher John Wiley & Sons, Inter-science, Chicester, England.

Barker R. and Hayami Y.: *Weed control practices as a component of rice production systems*. Proceeding of Conference on Weed Control in Rice, August 1981, p. 37. International Rice Research Institute, Los Banos, Philippines, 1983.

De Datta S.K. (1981): *Principles and practices of rice production*. Publisher John Wiley & Sons, Inc. New York, U.S.A.

Matsunaka S,: *Evolution of rice weed control practices and research: world perspective*. Proceeding Conference on Weed control in Rice, August 1981. International Rice Research Institute, Los Banos, Philippines, 1983.

Pieterse A.H. and Rijn P.J. van (1990): *Tropical Weeds:* Royal Tropical Institute, Amsterdam.

Rijn P.J. van (1990): *Integrated control of grassy weeds in mixed crops*; Lecture at Grassy Weed Symposium, August 1990, Nairobi, Kenya.

Smith R.J.Jr.: *Weeds of major economic importance in rice and yield losses due to weed competition*. Proceeding Conference on Weed Control in Rice, August 1981. International Rice Research Institute, Los Banos, Philippines, 1983.

Templeton G.E. and Smith R.J.Jr: *The role of plant pathogens in weed management in advancing countries*; in: *Improving weed management*, FAO Plant Production and Protection Paper 44, Food and Agriculture Organization of the United Nations, Rome, 1984.

Unger P.W.: *Tillage systems for soil and water conservation*. FAO Soils Bulletin 54, Food and Agriculture Organization of the United Nations, Rome, 1984.

15

CONTROL OF STORAGE PESTS

Contents

Contents	Page
Pests of stored agricultural produce	293
Prevention of storage losses	296
Measures for controlling pests of stored products	299
- Using biological controls	299
- Using environmental controls	300
- Using pesticides in general	302
Insecticides for control of storage pests	303
Using fumigants	309
Storage of seeds	311
References	312

Control of Storage Pests

In warehousing, stacks of bagged produce should be built by experienced workers who know which stacking pattern provides stability. They should leave sufficient space around the stacks to enable moving of stock, cleaning, inspection and treatment of infested stored products. (Photo by the author)

PESTS OF STORED AGRICULTURAL PRODUCE

Once harvested and correctly dried for storage, cereal grains, oil seeds and pulses are in peak condition. However, it is generally accepted that 5 to 15 per cent of the total dry weight of all cereals, oil seeds and pulses are lost after harvest, whereas losses in dry weight for root and tuber crops and vegetables may reach 20 to 40 per cent. There is thus a need continuously to protect the stored produce against deterioration, especially loss of quality and weight during storage.

Quality losses of stored produce can be caused by:

- Respiration of the pests and produce
- Storage at high temperatures and high humidity
- Microflora (mainly moulds)
- Insects and mites
- Rodents and birds
- Contamination with foreign matter
- Chemical degradation as a result of aging, e.g. rancidity

Of these factors, correct management of the physical conditions in the storage environment will reduce the adverse effects of respiration, high temperature and high humidity and, consequently, the microflora.

Deterioration during storage may also be caused by the presence of insects or contamination by rodent droppings and urine. Insects, mites and fungi may cause oxidation and decrease the level of certain nutrients in stored produce; fungi may even form toxic substances such as mycotoxins. Their presence in high-quality produce can result in a considerable financial loss and thus it may be considered economically feasible to commit resources to reducing such loss.

The use of certain types of pesticides is one means of preventing some losses during storage. The choice of pesticide for the control of storage pests, however, is very limited because of the strict requirements imposed for the safe use of pesticides on or near foodstuffs.

Furthermore, the application of pesticides is a waste of effort and money if physical precautions for safe storage have not been taken first. Storage containers, warehouses and silos must be well constructed and clean; the produce must be dried below a critical moisture content and temperatures should preferably be controlled.

Microflora

After harvest, many viable spores of moulds, yeast and bacteria are present on produce, unless it has undergone special treatment: for example, in the case of rice, parboiling. Massive development of such micro-organisms can take place if the moisture content of stored produce is sufficiently high. Bacteria and yeasts thrive in an atmosphere having a relative humidity of over 85 per cent, which is so high that the stored products would spoil whether these organisms are present or not. The growth of moulds, however, begins at lower levels of humidity, the lowest threshold for development being 70 to 75 per cent.

Most species of micro-organisms will grow only within a definite range of temperatures. Some organisms such as aerobic bacteria grow only in the presence of oxygen; others (anaerobes) without oxygen. In the ensilage of corn for fodder, the preference of certain bacteria for a specific acidity range (pH value) is exploited. The intensity of light, the type and condition of the produce are other factors influencing the development of micro-organisms.

Important moulds found on stored cereal grains, root crops and vegetables are:

* *Aspergillus* spp.
* *Penicillium* spp.
* *Mucor* spp.

The damage caused by moulds may involve not only the total spoilage of the produce but also the production of mycotoxins dangerous to those who consume them. There are several types of fungal poisons (mycotoxins):

- Liver poisons which cause liver cirrhosis, hepatoma and necrosis of the liver cells. Most of these poisons, e.g. the aflatoxins, are strongly carcinogenic.

- Kidney poisons which damage the function of the kidneys. Such poisons are produced, for example, by the fungus *Penicillium citrinum*.

- Poisons which damage the brain and the central nervous system. *Penicillium patulum* produces such compounds.

- Poisons which cause dermatitis-like skin lesions in animals and humans who come into contact with crops or food infected with this fungus. Strains of *Sclerotinia sclerotiorum* may produce such compounds.

Insects

Many storage insects have become cosmopolitan in distribution because of the world-wide transport of agricultural produce and foodstuffs. But owing to differences in tolerance to heat or cold, certain species of an order or family of insects may occur more abundantly either in hot tropical climates or in temperate areas. However, (sub-)tropical species may prosper in heated graneries and storage facilities even in cold countries. The number of common storage insect pests in tropical regions is about 40 (see table 15.1).

Insect pests are a greater problem in regions where the relative humidity is high, but they can also reproduce and thrive in dry conditions. Insect development takes

Table 15.1: Common storage pests in tropical climates

Species	Common name	Produce
Order: Coleoptera		
Lasioderma serricorne (F.)	Cigarette beetle	Tobacco products, groundnuts, cacoa
Stegobium paniceum (L.)	Drug-store beetle	Grain products, drugs, dried plants, herbs
Araecerus fasciculatus (De Geer)	Coffee-bean weevil	Coffee and cocoa, beans, spices, maize
Dinoderus minutus (F.)	Powder-post beetle	Bamboo, cassava
Prostephanus truncatus (L)	Larger grain borer	Maize, dried cassava
Rhizopertha dominica (F.)	Lesser grain borer	Cereal grains, pulses, root crops
Acanthoscelides obtectus (Say)	Dried bean beetle	Pulses pre- and post-harvest
Callosobruchus chinensis (L.)	Adzuki bean beetle	Pulses (especially peas and grams)
Callosobruchus maculatus (F.)	Cowpea weevil, pulse weevil	Pulses (especially peas and grams)
Caryedon seratus	Groundnut beetle	Unshelled groundnuts
Zabrotus subfasciatus (Boh.)	Bean seed beetle	Pulses (especially beans)
Necrobia rufipes (De Geer)	Copra beetle	Copra, oilcake, oilseeds, cocoa beans
Cryptolestes ferrugineus (Steph.)	Rust-red grain beetle	Cereal grains
Cryptolestes pusillus (Schoen.)	Flat grain beetle	Cereal grains
Sitophilus granarius (L.)	Granary or grain weevil	Cereal grains
Sitophilus oryzae (L.)	Rice weevil	Cereal grains
Sitophilus zeamais (Motsch)	Maize weevil	Cereal grains
Trogoderma granarium (Everts)	Khapra beetle	Cereal grains, groundnuts
Carpophilus dimidiatus (F.)	Corn sap beetle	Cereal products, dried fruits
Carpophilus hemipterus (L.)	Dried fruit beetle	Dried fruits
Ptinus tectus (Boieldieu)	Spider beetle	Grain products
Ahasverus advena (Waltlus)	Foreign grain beetle	Cereal grains and products, copra, herbs
Oryzaephilus surinamensis (L.)	Saw-toothed grain beetle	Cereal grains and products
Oryzaephilus mercator (Fauvel)	Merchant grain beetle	Cereal grains, dried fruits, oilseeds
Alphitobius diaperinus (Panzer)	Black fungus beetle	Cereal grains and products
Latheticus oryzae (Waterhouse)	Long-headed flour beetle	Cereal grains and products
Palorus subdepressus (Woll)	Depressed flour beetle	Cereal grains and products, pulses
Tribolium castaneum (Herbst)	Red flour beetle	Cereal grains and products, pulses
Tribolium confusum (Duval)	Confused flour beetle	Cereal grains and products, pulses
Tenebroides mauritanicus (L.)	Cadelle beetle	Cereal grains, cacoa, oilseeds
Order: Lepidoptera		
Sitotroga cerealella (Oliver)	Angoumois grain moth	Cereal grains
Corcyra cephalonica (Stainton)	Rice moth	Cereal grains and products, groundnuts, dried fruit, oilseeds
Ephestia cautella (Walker)	Tropical warehouse moth	Cereal grains and products, dried fruits, groundnuts, beans
Ephestia kuehniella (Zeller)	Mediterranean flour moth, mill moth	Milled cereal products, particularly flour
Plodia interpunctella (Hubner)	Indian meal moth	Cereal grains, dried fruits, groundnuts
Phtorimaea operculella	Potato moth	Stored potatoes
Order: Acarina		
Acarus siro (L.)	Flour or grain mite	Cereal grains and other products
Order: Psocoptera		
Liposcelis spp.		Cocoa, maize

place between 17°C and 35°C. The optimum condition for most storage pests is around 30°C and 75 per cent relative humidity. Most species have a rather short life-cycle of 4-6 weeks and a high multiplication rate. They are rather polyphagous, but many prefer produce with high nutritive value, such as grains and beans. Furthermore, some insects select the most valuable portion of the seeds, eating for instance the embryo while others prefer the starchy endosperm. A special group of storage pests, which feed on fungi, is often found in mouldy produce.

A number of storage pests lay their eggs in or near produce. With some storage pests, larval development and pupation take place within the grains or produce; with others, outside.

Several species which are normally associated with stored crops can fly actively. Such species will leave the grain stores, fly into nearby fields and commence egg-laying on the ripening seeds. *Sitophilus oryzae* and *S. zeamais*, and *Sitotroga* infest grain (paddy or maize) in this way. Ripening pulses too can be attacked by bean weevils.

Insect pests of stored products can be distinguished as follows:

* Primary pests which are capable of attacking ripening crops in the field before harvest as well as the stored dry produce. Borer type insects such as *Sitophilus* weevils and *Sitotroga* larvae can penetrate sound grains, and bruchid beetles can bore into beans.

* Secondary pests which normally attack only stored commodities that have already been damaged either mechanically or by vermin; they thrive particularly well on ground matter. Secondary pests include the grain and meal moths such as *Ephestia* spp. and *Corcyra cephalonica*, and the flour beetles such as *Tribolium* spp. However, the moisture content of the produce determines to a large extent the hardness of grains and whether secondary pests can damage whole grains or not. (See drawing above right.)

* Fungi feeding insects. A few beetle species, e.g. *Typhaea* spp., feed exclusively on fungi, but many coleopterous beetles, psocids and mites may supplement their diet by grazing on layers of moulds covering spoiled stored products.

* Scavengers which feed on dead insects, decomposing organic matter and dusts; e.g. psocids and cockroaches.

* Predators and parasites which prey on live insects and mites; e.g. *Tenebroides* spp.

Under tropical conditions, adult storage pests are short-lived generally. However, under cooler conditions or when grain stores are empty, some species, such as *Plodia interpunctella,* are able to survive for many months

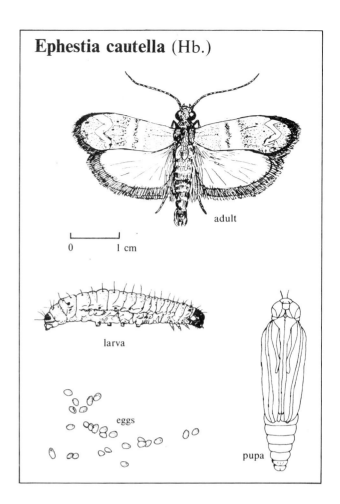

Ephestia cautella (Hb.)

adult

0   1 cm

larva

eggs

pupa

in the fabric of the building, while *Trogoderma* spp. may remain in diapause for several years.

Important parameters, besides climate, in the development and reproduction of stored-product pests are the kind or variety of crop they infest as well as its physical condition in respect of temperature and moisture content. It has been observed that new high-yielding varieties of grains are more sensitive to damage by storage pests than traditional varieties and that similar differences exist between cultivars. Examples of ways in which such host resistance to insect attack demonstrates itself are:

- The presence or absence of good husk cover. Tightly closing husks or sheaths reduce penetration by weevils and grain moths;

- Seeds of some plants may contain chemicals unacceptable to certain insects;

- The hardness of the seed coat, the nutritional value of the endosperm and other such properties influence the degree of infestation and survival of hatching larvae.

The interaction of temperature and relative humidity within the produce and the kind of crop it consists of, determines to a large extent the rate of development of the pest population. For example, at 25°C and 70 per cent relative humidity, *Ephestia cautella* develops in about 30 days on grain, in 43 days on groundnuts and in 53 days on cacao beans.

Mites

Mites have very small, less clearly segmented bodies than insects and have eight legs. They occur world-wide on cereals, dried fruits and other stored produce, particularly after fungal attack or if moisture content is high. They are less common in tropical countries, although they are ever more frequently found in container and pallet-packed produce arriving from the tropics.

Acarus siro has a minimum life-cycle of 17 days under optimum conditions of 20°C and 90 per cent relative humidity. A dormant stage may occur during which it is resistant to desiccation and starvation and is particularly difficult to kill. A massive mite infestation results in losses owing to discolouration and "off" flavour of the produce affected.

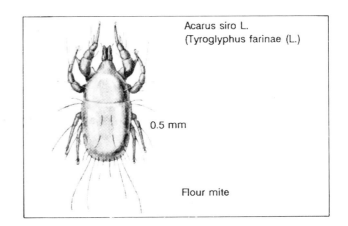

Acarus siro L.
(Tyroglyphus farinae (L.)

0.5 mm

Flour mite

- Burrowing beneath stores causing the collapse of floors;

- Health hazards to people who handle the stored products: certain species of rats and mice are carriers of diseases such as plague, Weil's disease (Leptospiral jaundice), rat-bite fever and salmonella; and

- Loss of feed in intensive poultry- and swine-raising enterprises. (For control of rodents, refer to Chapter 12)

PREVENTION OF STORAGE LOSSES

Precautions at the time of harvesting

The struggle to prevent loss of weight and quality starts in the field so that the crops can be harvested with as little as possible contamination by pests and foreign matter. Certain injurious organisms infest the ripening grain and fruits in the period between the setting of the fruits and maturity, and may develop into large populations during the storage period. Species such as *Sitotroga*, *Sitophilus*, *Rhizopertha* and *Callosobruchus* are active in the field throughout the greater part of the year.

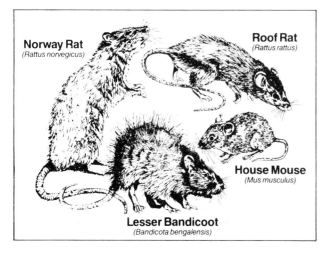

The presence of much foreign matter and unripe grain in the harvested produce causes problems during the drying process and may affect the taste and quality of the produce. In the case of seed production, any contamination with weed seeds or foreign varieties will have a negative effect on its selling price and production potential during the next season. For instance, young grass weeds of *Echinochloa crus-galli* and *E. colona* look like rice seedlings and are often transplanted by mistake. It is important to prevent the seeds of those weeds from reproducing and becoming mixed with rice seed.

During the period of ripening of the grain, its moisture content decreases gradually to 35-40 per cent, the level at which the accumulation of dry matter ceases. After the grain has stopped "growing", it continues to lose water, and changes in moisture content, in response to changing weather, become more marked. Under humid conditions, grain is commonly harvested at 17-22 per cent moisture content (m.c.); during exceptionally dry conditions, it is harvested at 14-15 per cent m.c., although at this time shedding of grains may cause substantial loss

Rodents

Several species of rodents may cause considerable damage to crops in the field as well as to stored produce. Usually, the heaviest losses are inflicted in the field, but in certain areas rodents are the main storage pests, particularly at the farm or village level. The species most harmful to stored produce are:

Rattus rattus L. . . . . Black ship or roof rat
Rattus norvegicus . . . . Berkenhout Common brown, Norway or sewer rat
Rattus exulans . . . . Polynesian or bush rat
Bandicota bengalensis . . . . Lesser bandicoot rat
Mus musculus L. . . . . House mouse
Mus booduga . . . . Field mouse

Storage damage caused by rodents include:

- Loss of volume or weight of produce owing to their feeding;

- Loss in quality caused by droppings, urine, hairs;

- Damage to containers, such as bags, that results in spillage and the cost of repairing and repacking;

among indigenous varieties. In dealing with such varieties having an "early-shedding" characteristic, it may be advisable to cut the ears early when their moisture content is still relatively high. Rain falling on ripe grain may raise grain m.c. by 10 per cent and it may take two fine days to bring it back to its former level. Harvesting in the wet season, which has become an established practice following the introduction of irrigation and multiple-cropping, confronts the farmers with the huge problem of handling larger quantities of grain of high moisture content and uneven maturity. Such a situation makes the necessity of drying the harvested grain even more pressing, but facilities for mechanical drying are scarcely available.

After grain has been harvested, the selection of the appropriate storage method is governed by the amount of additional drying that the crop requires and the volume of produce to be kept. Initially, drying and storage functions may be combined. For example, paddy on the panicle may be heaped in a stack, which is sometimes thatched, or maize on the cob may be placed in a well-ventilated structure (crib). In both cases the objective is to provide some protection whilst permitting drying to continue. Once dried sufficiently, the volume of the commodity to be placed in permanent storage can be reduced by threshing or shelling. The threshed or shelled grain takes up less space and can be stored in bags or in bulk in farm stores, warehouses or silos.

During periods of high relative humidity, the drying in the field may proceed so slowly that the moist paddy in the stacks become heated. Heating triggers metabolic processes in the grain resulting in "yellowing" of the grain. Grain affected in such a manner later tends to crack more readily in the rice mill and to yield a high percentage of "broken" kernels in the polished rice. If harvesting takes place during the rainy season, it is thus better to thresh the panicles shortly after cutting and to transport the moist paddy to drying floors or mechanical dryers. The above remarks concerning the handling of grain crops can be easily amended to apply to any other crop since the principles of drying and storage have general application.

Sun drying on hardened floors or roadsides is the most common way of drying tropical produce. Since the relative humidity of the air increases at night, the produce should be moved indoors or covered at the end of the day. (Photo by the author)

Precautions during storage

Several factors have a bearing on whether or not grain can be stored safely, and they all tend to interact with one another: moisture content, temperature, volume of grain, presence of pests, dust and other contaminants. Presuming that the produce is delivered in a dry and pest-free condition, then it will depend largely on the quality of the store and its management to maintain the sound condition of the stored products.

The storage structure

Stores should be located in dry areas free from flooding. Storage structures should be designed or repaired in such a way that heat and moisture can be kept out, cleaning can be done easily and entry by insects and rodents can be prevented. Air-tight silos with good thermal insulation offer the best protection.

Access to larger stores for trucks should be possible throughout the year and entry into the store from both ends should be possible.

Some points in making a storage building "rat proof" are as follows:

- Floors .... should be made of concrete and joined tightly so that there are no cracks or crevices where they join the walls;

- Walls .... should extend three feet straight into ground, or in a L-shape two feet into the ground with an one-foot lip extending outward;

- Doors .... should fit tightly with a minimal opening between door and threshold. The base of the doors should be covered with tin or metal plates as a protection against gnawing. Doors that are not used should be closed with bricks, or screened permanently;

- Roofs .... should be sound and free from leaks; airways should be screened;

- Windows . if within three feet of the ground, windows should be screened with wire mesh strong enough to withstand gnawing by rodents (1/4" mesh). Protect higher windows and ventilators with fly screening or wire mesh;

- Drains ... waste pipes and culverts should be protected in the same manner as windows;

- Pipes ..... extending metal plates should be fitted as a collar around pipes to repel climbing rats. Indigenous granaries often have flagstones or slabs fitted between upper and lower parts of the supporting posts.

Practising strict hygiene within and around

A good bag store with wide overhanging eaves, window panels in the roof and well-closing doors at both ends. (Courtesy of ODNRI, United Kingdom)

the storage area is the first step in preventing damage and losses from rats.

Any storeroom, particularly for food and foodstuffs, should be kept clean, preferably using an industrial vacuum cleaner to remove dirt and insects from cracks and crevices. It is recommended that the cleaning be performed prior to harvest time, just before new stocks are brought into the store. Left-over stocks from the previous year should be thoroughly inspected for signs of infestation. Stock-keeping procedures must allow for writing off losses and removal of old infested stock from the premises. Infested produce and infested empty bags should be discarded, or if they must be retained, they should be treated carefully with suitable insecticides. Food stores should be swept out at least once a week and the sweepings immediately burned outside the storage building or silo. Spilt grain should be swept up immediately.

When cleaning, particular attention should be paid to machinery and surrounding areas where dust can accumulate. Outside the store, bushes, tall weeds and grass should be cut to discourage the breeding of rodents. Likewise, old containers and waste-disposal units should be removed completely from the storage area.

Stacking of produce

Bagged produce should be stacked on pallets with free space between walls and ceilings. Different products should be stacked separately with sufficient space in between to enable checking and control of infestation and rotation of stocks. Stacking should be supervised by experienced personnel so stacks will be well built, and stable without the danger of collapsing.

Spoilage and hot spots

All agricultural products should be well dried before storage, especially if they are to be kept in silos. During storage, it is essential that produce be kept dry. It is helpful to keep the temperature as cool as possible and to prevent excessive temperature changes by ventilating or insulating the structure.

Modest variations in store temperature during day and night, particularly in metal silos, owing to warming by the sun, set up convection currents in the grain mass. These can cause grain moisture to be redistributed within the silo from a warm zone to a cooler one where it condenses, raising the grain moisture content to a level at which mite and fungal activity can begin. Mould development, bacterial deterioration and chemical changes may then spoil the grain completely. Such spoilage can most commonly be found in the top-layer of bulk grain and in the bottom and top rows of bag-stacks.

Bulk grain with a moisture content of 11-14 per cent, and in apparently good condition except for the presence of insects, often become hot. The heat in such a case is more likely to be caused by the metabolism of the insects than the metabolism of grains and micro-organisms. An insect-caused hot spot begins when insects produce metabolic heat faster than the heat can escape. A small local rise of temperature accelerates the metabolism of each insect and speeds up the rate of population increase. The heating process increases continuously until it reaches a level that is unfavourable to the insects. At this point, the insects move to the periphery of the hot spot, thereby increasing the size of the hot spot and the spread of fungal inoculum and bacteria which they carry with them. Temperature from heating may rise as high as 60°C. Cooling and aeration will slow the development of microflora and insects and diminish heating. But, infected lots or stocks with hot spots should be treated and turned over.

Table 15.2: Recommended moisture content for storing some specific agricultural products

Produce	Moisture (per cent)
Beans, cowpeas, millet	15
Paddy, peas	14
Wheat, sorghum, maize (white)	13, 5
Milled rice, wheat flour, maize flour	11, 5-12
Tapioca, cassava flour and chips	12
Groundnuts (unshelled), sunflower seeds	9-10
Groundnuts (shelled)	7
Copra, palm kernels	5, 8

MEASURES FOR CONTROLLING PESTS OF STORED PRODUCTS

The use of chemicals for controlling pests of stored products, or their use as protectants, should also be compared with the suitability and effectiveness of alternative methods of control if they are available. Non-chemical methods of control are attractive since they neither leave residues in the commodity nor do they cause resistance in insects. However, they require specialized storage structures and equipment involving high capital investment and expert management. By contrast, protectant insecticides and fumigants demonstrate a rapid effectiveness, are relatively simple to apply, can be used in various ways and are immediately available at low cost. There can be little doubt about the fact that control by chemical methods will remain most important in this field during the next several decades.

Control of stored pests should be based on a combination of the following measures:

* Dry adequately the products to be stored;

* Use adequate storage facilities with provisions for aeration and preferably also for cooling; avoid strong temperature fluctuations;

* Practise good store-keeping, including sanitation and inspection;

* Treat warehouse structures with residual insecticides;

* Apply protectant chemicals onto the commodity; or, in the case of storing at the farm level, use effective traditional means, such as the admixing of ashes, vegetable oil or herbs,to grain and beans;

* Kill established infestations by fumigation or with insecticides; and

* Never store fresh produce together with old stock.

Forced drying with heated air is often needed to prevent spoilage of moist produce. Over a period of several days, a Lyster blower forces warm air into a tunnel under stacked bags until the moisture content of the produce has decreased sufficiently to enable prolonged storage.

Using biological controls

The possibility of using natural enemies to control storage insects has been studied for many years. Certain insects of the orders of *Coleoptera* (beetles), *Heteroptera* (bugs) and *Hymenoptera* (wasps) are capable of acting as either predators or parasites of these pests. Also, mites are frequent predators of stored-product insects or other mites. However, thus far, few have been found that might have a potential for reducing pest populations either within an acceptable period of time or to a tolerable level.

The severe losses caused by the larger grain borer (LGB), *Prostephanus truncatus*, to stored maize and dried cassava in East and West Africa, as well as the apparent difficulty in preventing this newly introduced pest from spreading all over that continent, has given a sense of urgency to the search for its natural enemies. An exploration for antagonists of the larger grain borer in various countries of Central America, where it presumably originated, has yielded 23 fungi, 14 bacteria, eight protozoa species and one virus, which were all isolated from dead or live LGB-beetles. Of the fungi, only the species *Beauveria bassiana* and *Metarhizium anisopliae* seemed to be suitable, but require high dosages to provide effective control. The bacteria were not effective as control agents. A previously unknown protozoon, *Mattesia* spp., produced an increased mortality rate in first instar LGB-larvae and general growth retardation.

In Costa Rica, all four predators and parasites of LGB have been discovered. The most promising control agent identified so far is a predatory coleopterous beetle, *Teretriosoma nigrescens* (Lewis). This beetle, which does not occur naturally in Africa, is exceptionally well adapted to the larger grain borer in morphological and physiological aspects. Its size enables it to move through the tunnels bored by the LGB and to feed on the latter's eggs, larvae and pupae. The predator is found only in association with the LGB and appears to be attracted by the aggregation pheromone of this pest. The *T. nigrescens* beetle eats 1.5 LGB-larvae per day on average, whilst its larva eats about five LGB-larvae per day. In laboratory tests on the control capacity of *T. nigrescens* over a period of 110 days, an 87 per cent reduction of a LGB-population was obtained in shelled maize, and a 72 per cent control in maize stored on the cob (unshelled). Similar results have been obtained on unshelled maize stored outdoors in wire cages. Currently, the likelihood of *T. nigrescens* finding alternative prey species under conditions in Africa, its chances of survival and the consequences of its release in that continent are being studied, before importation of this promising predator is approved (Boeye *et al.*, 1988). To date the sterile insect method has not been used for the control of tropical storage pests, although several experiments have been carried out.

The prospect of widespread, successful application of biological control in stored product environments appears to be dim for the following reasons:

(Left) *Because of the low relatively humidity of the air, the loss of millet stored on the cob from insect infestation is generally low in traditional storage structures of Sahelian countries in Africa.* (Centre) *Rice paddy stores of northern Thailand are made of teak wood and placed on concrete poles and therefore last for many years. For rice stored in the hull (paddy), loss from insect damage is low but attention must be given to protection against rodents.* (Right) *Metal silo cells at Tambun, Indonesia, suffer from water condensating at the roof due to heating of the grain mass in day time and cooling at night. Condensed water dripping onto the grain causes it to spoil. (Photos by the author)*

- Predators, parasites, sterile insects and pathogenic fungi would themselves be regarded as contaminants of the produce. Handlers and consumers would object to their presence at population densities that would be sufficiently abundant and persistent for biological agents to establish themselves and become effective.

- The physical conditions required for long-term storage not only impair the development of unwanted insects and micro-organisms, but are equally unfavourable to those beneficial insects, mites and pathogens such as entemophagous fungi and bacteria, which constitute the natural enemy force.

Research on the commercial use of *Bacillus thuringiensis* has shown its formulation to effectivity infect and kill moth populations such as of *Ephestia* spp. and *Plodia interpunctella* at a low dose of 10 mg/kg. Formulations of *B. thuringiensis* are most useful in situations where moths are resistant to insecticides or where insecticide application is not permitted or cannot be carried out properly.

The use of viruses and protozoa has been relatively little studied to date, but would need more attention in the context of an integrated approach. Pheromones of several lepidopterous and coleopterous storage pests have been identified and synthesized, and are currently being marketed. Pheromone traps containing both a moth-attracting pheromone and dichlorvos provide effective control in flour mills and stores where the standard of hygiene is high and moths are the only pest around. Such traps, however, are insufficient where massive infestations of mixed populations occur. Pheromones for *Sitophilus* spp. and other pests are in the development stage.

Using environmental controls

Another way of avoiding chemical residue and resistance problems is to make physical changes in the storage environment in order to render the conditions unfavourable for the development of storage pests. Some measures are traditional and have long been practised by farmers; others involve modern technology. In addition to methods used to conserve agricultural produce in storage, measures designed to limit insect infestation on the ripening crop in the field and to prevent such initial infestations from being carried into the store together with the new crop are just as important as the more direct means of control. Examples of traditional practices are:

- Selecting produce for storage. The likely long-term storage losses on the farm can be drastically reduced by selecting only undamaged, healthy ears with closed husks as these are generally free from initial infestation.

- Maintenance of storage buildings. Traditional-type granaries in villages often show lower losses as a result of pests than modern grain stores in the same area. Shady, dry and well-ventilated farm stores in humid countries, or well-insulated clay containers in arid areas are generally efficient, provided they are in a good state of repair and sealed to prevent the entry of pests.

- Mechanical methods of pest control. A number of activities, which are the result of many generations of farmers' experience, are aimed at destroying or repelling pests or hindering their propagation.]

Simple methods that are widely practised are:

* Picking out pests by hand before the produce is sold is

300

a common practice in markets and sometimes in government grain stores too.

* Sifting out insects with sieves. Since storage insects usually are much smaller than grains, they can be separated by sifting. If grain is regularly sifted, e.g. at weekly intervals, the pest infestation can be considerably reduced. However, care should be taken to destroy the collected weevils, beetles and mites before they can move to other stocks. Special sieves fitted with mesh of the appropriate gauge are an advantage. Sieve units are also being used in modern cereal-cleaning techniques in mills.

* Winnowing. Pouring grain from one basket held up in the air into another placed on the ground, or tossing grain up in the air with a shovel, enables the wind to blow unwanted lighter matter -- straw, husks, damaged or empty grains, dust and insect pests -- away from the sound grains. A winnowing machine performs the same cleaning job.

* Damaging pests by agitation of grains. Adult insects in cereal stocks can be easily injured or even killed if the grain is vigorously agitated, causing the insects to be crushed between the kernels. This method, although labour-intensive, is commonly used by farmers and small merchants. In modern silos, the grain is frequently moved from one bin into another with the help of mechanical augers.

* Solar radiation of grain. Exposing humid grain to the full heat of the tropical sun enables it to be dried rapidly. It also helps in killing adult storage insects -- notably *Sitophilus* species -- that cannot stand the high temperature; however, eggs and larvae inside the grains may survive.

* Admixing of sand or mineral substances. There are many variations on the practice of adding sand, dust or ashes to grains as these substances have in common the ability either to injure the insects or to leave them no room to move by filling out the space left between the grains. Ashes and abrasive or absorbtive dusts adversely affect the protective function of the hard wax layer on the insect cuticle, which is only a few molecules thick, that acts as a barrier to retain body fluids. The insects may thus become victims of rapid dehydration, as they have a relatively large body surface and live in an extremely dry environment.

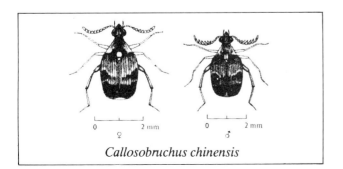

Callosobruchus chinensis

* Addition of vegetable matter. The adding of neem leaves or neem extracts to stored grains and pulses is quite common in Asia as a means of repelling insects. In Central America and India, beans are moistened with vegetable oil by mixing 2-5 ml of oil per kg of beans before storage. This methods is effective against bean weevils, *Callosobruchus* species, as the oil appears to damage the eggs which are laid on the surface of the beans. The adult weevils are not killed, but reproduction is halted.

Examples of modern methods under trial are as follows:

- Regulation of humidity: If the produce has been dried so that its equilibrium relative humidity is less than 70 per cent, then it will be safe from fungal attack and mite infestation. But, because of the tolerance of storage insects to very dry conditions, lowering the humidity level has been found to be ineffective for their control.

- Regulation of temperature: Temperatures under $18^{\circ}C$ suppress the development of micro-flora and insects in stored grain with a moisture content as high as 15 per cent. The aeration of bulk grain in silos is based on this principle. By blowing air through a system of fans and perforated ducts in the floors, a uniform and moderately low temperature is maintained throughout the grain mass. Aeration is complementary to the use of grain protectant insecticides and prolongs the residual activity of the deposit.

- Hermetic storage: When grain is stored in an air-tight container, the oxygen content in the container will decrease slowly owing to the metabolism of the grain, insects and micro-organisms. Eventually, there will not be enough oxygen for any insect development, as insects cannot survive in an atmosphere of less than 2 per cent oxygen. Air-tight storage is an attractive way to protect produce against insects without the use of pesticides, but often the high cost of constructing suitable silos prevents their general use. For air-tight storage on a small scale, clean drums or plastic bags may be used. However, under air-tight conditions, the product's moisture content must be very low, preferably 1 per cent lower than the values in table 15.2.

  Controlled atmosphere storage (CA) is a version of hermetic storage involving the introduction of a gas such as carbon dioxide (CO_2) or nitrogen (N) into the store in order to alter the atmosphere in the grain mass. Even if the oxygen concentration is as high as 15 per cent, storage pests can be killed if a high concentration (36 per cent) of carbon dioxide is also maintained. The relative humidity has a strong influence on the effect of gas concentrations. In general, a low relative humidity increases the mortality caused by low O_2 and high CO_2 concentrations.

- Heat treatment: Exposure of insect-infested grain in a fluidised bed-heating unit to a temperature of $60^{\circ}C$-$70^{\circ}C$ for 2-3 minutes effectively kills weevils and beetles. A pilot plant for disinfecting grain at the rate of 150 tonnes per hour has been installed in Australia.

Using pesticides in general

The application of pesticides for the protection of stored agricultural produce is unavoidable, because the aforementioned biological and environmental methods can achieve effective control of pests only when all environmental conditions can be strictly managed. This is usually not the case in situations where funds and management are lacking. However, by combining the advantage of all available methods in an integrated approach, one will achieve better results with less frequent applications.

The main types of pesticide to be discussed hereunder are:

- Insecticides as protectants and control agents;
- Fumigants for control of existing infestations;
- Rodenticides for killing rats and mice (see Chapter 12); and
- Fungistatic agents.

The choice of pesticides that can be used safely for protection of stored produce is much more limited than the number of those available for application on crops. For storage protection, where pesticides are often applied on or mixed with the produce, toxicity of the formulation, time of degradation, residue levels etc. are important matters that must be considered in making a choice.

It should be obvious that, for the protection of stored produce, only pesticides with residues that easily degrade into harmless compounds and which do not affect the flavour, taste or colour of the product, can be used. These compounds need to have clearance from the FAO/WHO Joint Meeting on Pesticide Residues (JMPR residue tolerance) for use on food grains, so that they are safe for farmers to apply and for consumers to handle and ingest. Insecticides which can accumulate in the human body, e.g. DDT, are, of course, completely unsuitable for use on stored products.

Insecticides. These substances may be used for spraying walls, floors and ceilings of warehouses or storerooms in order to kill a residual infestation. Some insecticides can also be sprayed directly on bagged produce; this may prevent or delay reinfestation of produce made insect-free by fumigation.

Certain insecticides may be mixed with the produce. This can give complete protection for a long period of time and may also kill pests which have already infested the produce. However, where it is possible to achieve gas-tight conditions in warehouses or silos, the best effect is by means of fumigation. Fumigants penetrate whole grain or flour and kill insects within the product. After fumigation there is no residual protection so reinfestation has to be prevented by storing the produce in an insect-proof silo or container or by admixing a protectant insecticide.

Fungistatic agents. The admixture of proprionic acid at a fairly high concentration (between 0.3 and 1.0 per cent by volume of the commodity) affords a method of preserving grain which has a high moisture content. Spoilage by micro-organisms is prevented. However, the method has somewhat limited application: the cost of treatment is high and requires a special applicator in order to spray the corrosive liquid.

Development of resistance

The possibility of the development of resistance to insecticides by insect pests is a constant danger. However, if it is found that a particular treatment is failing to achieve control after a period of successful use, this does not necessarily mean that resistance has, in fact, developed. Failure may be due to insufficient insecticide being available within the grain mass or to rapid reinfestation from outside. The absence of sufficient insecticide may be due to the use of bad formulations, errors in mixing or calculation of the dosage to be applied, or poor application techniques resulting in inadequate coverage. Unfortunately, resistance to organophosphorous insecticides does occur frequently, and resistance to the fumigant phosphine has been confirmed in several Asian countries too.

Insects may survive a fumigation if the intended gas concentration is not reached or maintained for a sufficient length of time. It must be recognized that the survivors of an imperfect insecticide or fumigation treatment are insects possessing an enhanced tolerance which they will pass on to their offspring. The practical implications of this genetically inherited resistance are manifested by an increased difficulty in achieving complete control. The solution is to apply pesticides which have a different chemical structure, such as synthetic pyrethroids. However, it must be stressed that the alternatives are limited. Therefore, it is important to avoid creating resistance in the first place.

Most countries have strict regulations concerning the level of permissible pesticide residues in imported produce. Some countries have a zero tolerance for organochlorine pesticides in produce destined for human consumption. So, before pesticides are used on produce destined for export, information must be gathered about their acceptability in the importing countries.

Winnowing of grain serves to remove debris which would slow down drying and increase infestation.

INSECTICIDES FOR CONTROL OF STORAGE PESTS

For the protection of stored products, only a few insecticides are commonly used. These are:

malathion	fenitrothion
bromophos	lindane
carbaryl	methacrifos
bioresmethrin	permethrin
chlorpyrifos-methyl	pyrethrin
deltamethrin	fenvalerate
dichlorvos, DDVP	methoprene
pirimiphos-methyl	phoxim

A brief review of these insecticides will be given below together with recommendations for dose rates. Methoprene is a juvenile hormone that is still being tested.

Malathion

Malathion is a well-known organophosphorous insecticide that for over 25 years has been widely applied for storage pest control and is still used as a standard for evaluating other insecticides. It is a safe insecticide which can be admixed to or sprayed on shelled (threshed) or unshelled (unthreshed) grains. On stored produce, only premium grade malathion should be used. $LD_{50} = 2,800$ mg/kg (oral, rats); tolerance = 8 ppm for raw cereals (FAO/WHO).

Malathion is weaker against most pests in stored products compared with many of the other organophosphorous insecticides. It is virtually ineffective against moths and requires higher doses to control non-resistant *Rhizopertha dominica*. Strains of some pests have developed "resistance to malathion"; a term used to compare efficacy of malathion with that of other insecticides.

The general recommendation for use of malathion dust is:

- Maize (shelled), paddy . . Mix 100 g of malathion 1% dust per 100 kg of grain (10 ppm).

- Beans (threshed) . . . . Mix 100 g of malathion 1% dust per 100 kg of beans (10 ppm).

- Maize cobs (unshelled) . Apply 100-150 g of malathion 2% dust ground nuts (unshelled) per 100 kg of cobs (20-30 ppm).

Malathion dust has some limitations:

- The stored commodity must be dry (moisture content not higher than 13.5 per cent), otherwise malathion breaks down very rapidly.

- Formulated malathion dust often has a rather short shelf-life (not more than six months).

Soluble formulations of malathion can be used for spraying walls, floors or the outside of stacks. The recommended rate is 1 g of active ingredient (a.i.)/m^2.

When malathion is applied to surfaces with an alkaline character, such as concrete and whitewashed walls, its residual action is shortened considerably.

Bromophos

Bromophos is a non-systemic, halogen-containing organophosphorous insecticide used on crops and animals to control biting, sucking and disease-transmitting insects. It is also used as a stored-product and seed grain protectant. It has a low mammalian toxicity: $LD_{50} = 3,750$ to 5,180 mg/kg (oral, rats). Maximum residue limits: 10 ppm for raw grain and 2 ppm for white flour (FAO/WHO).

In many respects, bromophos is similar but superior to and more persistent than malathion. It is more effective in some pest species, but like malathion, weak against *Rhizopertha dominica* and moths. It is stable on alkaline surfaces (concrete) and therefore useful for treating storage structures.

Bromophos effectively controls some malathion-resistant species and its potency increases at higher temperatures. It penetrates grain rapidly, but residues are destroyed by simple cooking and baking of cereal-based food.

As in the case of malathion, only premium grade bromophos should be used on stored products, because formulations containing impurities may produce an objectionable taint in treated grain. Recommended rates of application are:

- Apply 10-20 g a.i. of bromophos per metric ton of grain (10-20 ppm)

- Apply 8-12 g a.i. of bromophos per metric ton of beans or maize (8-12 ppm)

Bromophos is available as a dust or emulsifiable formulation for treatment of bulk grain, bag-stacks and surfaces.

Carbaryl

Carbaryl is a methyl-carbamate insecticide used extensively on agricultural crops and in animal husbandry as a replacement for low-cost organochlorine insecticides. Though not particularly effective against stored-product pests in general, carbaryl is quite toxic to *Rhizopertha dominica* and, therefore, has been chosen for use in combination with organophosphorous insecticides.

LD_{50} = 500-850 mg/kg (oral, rats). JMPR maximum residue limits: 5 ppm for grain and 2 ppm for wholemeal flour (FAO/WHO).

Suggested rate of application:

- Apply a mixture of 5 g a.i. of pirimiphos-methyl + 5 g a.i. of carbaryl per metric ton of stored produce for effective control of more than six months.

Carabaryl alone gives incomplete control of *Sitophilus*, *Tribolium* and *Trogoderma* species even at the rate of 200 mg/kg and should therefore be used only in combination.

Chlorpyrifos-methyl

This is a broad-spectrum organophosphorous insecticide of moderate persistence and good stability on grain and dried vegetable matter. It is being marketed under the trade name Reldan. LD_{50} = 1,828 mg/kg (in corn oil; oral, rats). Maximum residue limits: raw cereal grain 10 ppm; wheat flour, 2 ppm.

Chlorpyrifos-methyl is more effective than other organophosphorous insecticides against all stored-product pests except against resistant *Rhizopertha dominica*. Therefore, combining it with bioresmethrin (see section on bioresmethrin) would make the mixture a potent grain protectant effective against all species. Chlorpyrifos-methyl has a half-life of 4-5 months at $30^{\circ}C$ and 15 per cent humidity, and gives long-term protection to bulk-stored grain. It is also regarded as a good substitute for malathion for spraying bag-stacks.

Recommended rate of application:

- Mix 2.5 - 4 g a.i. of chlorpyrifos-methyl per metric ton of grain (2.5-4 ppm)

Dichlorvos, DDVP

This is a highly volatile insecticide with short residual action generally of less than one week. LD_{50} = 56 mg/kg (oral, rats); tolerance = 2 ppm for raw cereals (FAO/WHO).

The vapour of dichlorvos is extremely toxic to larval and adult stages of insects within the grain and has an unparalleled knock-down effect. Therefore, it is useful for disinfecting grain brought into the store with live insects. Dichlorvos is very effective for space treatments against moths of the *Ephestia* and *Plodia* species, which are difficult to control with other insecticides.

In protecting stored products, it is important to distribute the active ingredient as evenly as possible throughout the stored material.

Plastic strips impregnated with a slow-release formulation of dichlorvos -- Vapona and Nuvan strips -- may provide effective control of flying insects in small stores. One strip is sufficient for a volume of 30 m³.

For surface treatments, liquid pesticides are used diluted as sprays to wall surfaces, grain bins etc., paying special attention to cracks and corners and other pest concealment places. Knapsack sprayers or power sprayers can be used.

Dichlorvos can also be applied with thermal fogging apparatus -- Puls fog or Swing fog -- producing a fine, clearly visible fog that gives a good effect even at low dosages.

Despite the relatively high volatility of dichlorvos, it is not possible to fumigate cereals with dichlorvos vapour as is the usual technique with methyl-bromide, because the gas is immediately absorbed by the upper layer of grain.

Automatic spraying equipment with a time switch is available for installation in large storage rooms, where a continuous insecticidal concentration is required. Dichlorvos formulations for storage protection include Dedevap, Nuvan and Vapona.

Recommended dosage rate for use as an oil spray or fog is 1-2 mg a.i. of dichlorvos per 1,000 m³ of space. For desinfecting grain, spray or inject dichlorvos emulsion at 5-20 g a.i. per metric ton of grain. Conduct tests to determine the rate corresponding to the severity of infestation.

Pirimiphos-methyl

This is a fast-acting insecticide and acaricide with both contact and fumigant action. Trade names are Actellic, Actellifog, Silosan and Blex. Formulations include emulsifiable concentrate, wettable powder, ULV concentrate, dust, smoke generator and aerosols. LD_{50} = 2,050 mg/kg (oral, rats). Tolerance is 6 ppm for raw cereals and 1-2 ppm for milled rice and wheat flour.

Pirimiphos-methyl is effective against beetles, weevils, moths and mites, including immature stages within kernels, but less effective against some strains of *Rhizopertha dominica*. It is considerably more potent than malathion, bromophos and dichlorvos, and approximately equal in effect to fenitrothion and chlorpyrifos-methyl, but somewhat less effective than methacrifos. Pirimiphos-methyl is extremely stable on grain, having a half-life on wheat of 45 weeks at 30°C and 50 per cent relative humidity. Degradation, however, occurs rapidly.

- **Recommendations for pre-harvest disinfection of stores**. The chemical should be applied to the walls, floors and equipment in stores approximately six weeks before new produce is brought in. If there has been a previous infestation, the application should be repeated at one week before the arrival of the produce. The dosage rate is 50 g of active ingredient per 100 m² for spraying, using a volume of 5 litres of water on concrete, plaster and brick walls and of 15 litres of water on wooden surfaces.

For the treatment of roofs and for control of flying insects, the spraying of walls, floors and machines should be followed by a fog or smoke application. For instance, in thermal fogging one could use 5 ml Actellic 50 EC in 95 ml diluent per 100 m³ of air space. White oil or odourless petroleum distillate is a suitable diluent. For a smoke application, one Actellic No. 20 smoke generator would suffice for a volume of 570 m³.

- **Treatment of stored produce**. Admixture with cereal grains, groundnuts and seeds is recommended for bulk storage requiring long-term protection. For spraying onto grain being fed into bins, atomize 4-10 g a.i in 1-2 litres of water per metric ton (4-10 ppm). Alternatively, mix 200-500 g of a 2 per cent dust formulation per metric ton of grain by means of an automatic seed treater, a rotary drum or by shovel mixing (4-10 ppm).

Bagged grain may be treated by spraying or dusting. Treat the floor and each bag as the stack is built up. Finish by treating the top and the sides of the stack. Spray at a dosage rate of 25 g a.i. per 100 m² surface, using a volume of 5 litres of water, or apply 1.25 kg of a 2 per cent dust formulation per 100 m². These rates will give protection for up to three months; for longer control, the dosages should be doubled.

Formulations of pirimiphos-methyl and fenitrothion (see reference below) have shown excellent efficacy

The worker places a tray with phosphide pellets under large gas-tight plastic sheets which cover a stack of grain bags. The sheet is weighed down by sand bags to prevent the fumigant from escaping. (Courtesy of ODNRI, United Kingdom)

To prevent infestation during prolonged storage, it is necessary to spray the bags of produce with persistent insecticides at the time stacks are being built and every time when a beginning infestation is noticed during routine inspection.

in the control of storage pests in research trials over several years. However, their efficacy against *Rhizopertha dominica* is insufficient (as is the case with other phosphoric esters). If this species occurs in high numbers, the two aforementioned active ingredients should be mixed with bioresmethrin (see section on bioresmethrin). The active ingredient bioresmethrin alone will give good control of *Rhizopertha dominica* only, but not of other storage insects.

Fenitrothion

This is a relatively safe contact insecticide for use on crops, on stored products and for mosquito control. It is formulated as an EC, WP and dust under various trade-marks, such as Sumithion and Folithion. LD $_{50}$ = 500 mg/kg (oral, rats). Maximum residue limits are 10 ppm for raw cereals, 1 ppm for milled rice and 3 ppm for wheat flour.

Fenitrothion can be applied as a disinfectant of stores and for treatment of stored produce in approximately the same way as that described under pirimiphos-methyl. It has a broad spectrum effect against all species, although not fully against *Rhizopertha dominica*. Admixture of bioresmethrin (see section on bioresmethrin) to fenitrothion provides a potent grain protectant that also controls malathion-resistant insects.

Fenithrothion penetrates minimally into the grain so that the deposit is mostly removed either with the husk or the bran. A dust formulation of fenitrothion is thus very suitable for use in tribal grain storage.

Recommended dose rates are:

- The dosage is 50 g a.i. per 100 m² of surface (or a higher dosage when needed) for spraying warehouses.

- Bagged produce may be sprayed or dusted with a dosage of 5 g a.i. per 10 m² of surface.

- Admixture of 4-12 g a.i. per metric ton of grain (12 ppm); maize cobs should be treated with 8-20 g a.i. of fenitrothion per metric ton (8-20 ppm). However, in case of larger grain borer infestation, refer to the section on permethrin.

- Treat with 12 g a.i. of fenitrothion + 1 g a.i. of bioresmethrin per metric ton of grain for protection of up to nine months under favourable conditions.

Lindane

Lindane (or gamma HCH) is a much more persistent insecticide with a higher toxicity than the others previously described. $LD_{50} = 125$ mg/kg; tolerance is 0.5 ppm for raw cereals (FAO/WHO). The use of lindane in storage is currently recommended only for the treatment of floors and walls at a dosage rate of 0.5 g a.i. per 100 m^2 of surface. In Europe, this method of application is no longer permitted. Some European countries have imposed stricter regulations with regard to permissible residues of lindane, particularly on tobacco and on animal feed products. This has influenced the use of lindane in those tropical countries which export produce to Europe.

Attention should be given to avoiding lindane residues in cereals, feed cakes, expellers etc., which will be used for cattle feed, because these residues may be found again in milk or meat.

Methacrifos

An organo-phosphorous insecticide with contact, fumigant and stomach action against insects and mites, it provides good residual protection. It is a promising grain protectant as it controls storage pests such as the *Sitophilus* and *Tribolium* species, which are increasingly resistant to gamma-HCH (lindane) and malathion.

A special property of methacrifos is its ability to penetrate grain, thus killing larval stages inside the kernels. It is extremely potent at lower temperatures and has a pronounced vapour action. Sold under the trade-name Damfin, methacrifos comes as an emulsifiable concentrate, a ready-for-use SO formulation and a 2 per cent dust. $LD_{50} = 680$ mg/kg (oral, rats). Tolerance is 10 ppm for cereal grains and cacao beans, and 2 ppm for wheat flour (FAO/WHO).

Treatment recommendations for South-east Asia are as follows:

- For disinfection of stores, spray 25-50 g a.i. per 100 m^2 of wall and floor surface, or apply 10 g.a.i. per 100 m^3 of warehouse space using ULV atomizers or fine-mist generators, but do not apply with thermal foggers.

- Bags and stacks are sprayed or dusted with 50-200 g a.i. per 100 m^2 of surface area. These rates give 3-8 weeks of protection; the 200-gram dose should be applied only if organophosphorous- resistant insects are to be controlled.

- Bulk stored grain in silos can be protected for up to 12 months by spraying the grain in the conveyor system using a calibrated applicator set at 10-15 g a.i. per 1,000 kg (10-15 ppm). In simple storage bins, methacrifos can

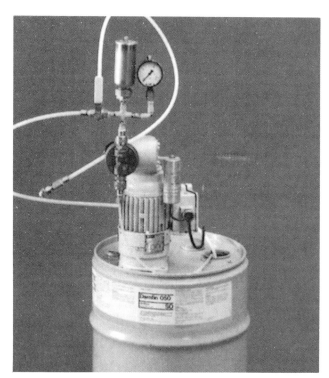

For direct admixture of a liquid insecticide such as methacrifos, a device such as the one pictured sprays the chemical onto bulk grain moving in a conveyor system. A metering pump supplies the exact volume of insecticide from a drum to produce a fine spray through nozzles mounted for distribution over the full width of the conveyor belt. (Courtesy of Ciba-Geigy)

be admixed with bulk grain by spraying or dusting the grain surface after each 10 cm of grain has been added; this is called the "sandwich method" of application, used when organophosphorous-resistant insects are to be controlled.

Bioresmethrin

This synthetic pyrethroid, the trans-isomer of a racemic (i.e. a chemical mixture of two molecules which form one chemical) mixture known as resmethrin, is one of the most potent broad spectrum insecticides currently available.

Bioresmethrin gives a good knock-down performance and, at low concentrations, kills effectively most household and stored-product pests. $LD_{50} = 8,000$ mg/kg (oral, rats); tolerance is 5 ppm for milled cereal products and whole grain (FAO/WHO).

The principal current uses of bioresmethrin are:

- In household aerosols and sprays formulated in combination with pyrethrum and piperonyl butoxide

- As an insecticide for the control of pests in food premises

- In grain desinfection and protection

Bioresmethrin has exceptional efficacy against some insects, particularly organophophorous-resistant strains of *Rhizopertha dominica*. When applied at 1 mg/kg

306

in conjunction with selected organophosphorous insecticides, it enables their dose rates to be considerably reduced without loss of effectiveness. Bioresmethrin is therefore an ideal insecticide to complement organophosphorous grain protectants such as malathion, fenitrothion, pirimiphos- methyl and chlorpyrifos-methyl.

Mixing bioresmethrin with piperonyl butoxide increases its efficacy two- to nine-fold against the more important storage pests, except *Tribolium* spp.

Rates of application cited in the literature are:

- 4 g a.i. of bioresmethrin + 20 g a.i. of piperonyl butoxide per 100 kg of grain provides long-term protection (12 months).

- 12 g a.i. of fenitrothion + 1 g a.i. of bioresmethrin per ton of produce for the control of malathion-resistant strains of *Sitophilus, Rhizopertha, Tribolium* and *Ephestia* species.

Deltamethrin

Being one of the most potent of the synthetic pyrethroids, deltamethrin kills a wide variety of stored-product insects at low dose levels.

LD_{50} = 15,000 mg/kg (oral, rats) for suspensions of wettable powders and flowable formulations. Solutions of deltamethrin in vegetable oils show a higher toxicity, e.g. LD_{50} = 52 mg/kg (sesame oil; oral, rats). Maximum residue limits: cereal grains 2 ppm; white wheat flour 0.5 ppm (FAO/WHO).

Deltamethrin is exceptionally stable on grain; however, as it does not penetrate into the kernels, it can be largely removed on bran at the time of milling. Deltamethrin is very effective against *Sitophillus granarius* at dose levels 20 times lower than those for bioresmethrin.

Recommendations for use are:

- Apply either 0.5 g a.i. of deltamethrin + 5 g a.i. of piperonyl butoxyde, or:

- Apply 1 g a.i. of a deltamethrin WP formulation to 1,000 kg of dehusked maize cobs for long-term storage in cribs (1 ppm). In case of larger grain borer infestation, see section on permethrin below.

- Apply 0.75 g a.i. of deltamethrin formulation to 1,000 kg of grain legumes for long-term protection against *Callosobruchus* spp.

- Apply 1 g a.i. of deltamethrin + 4 g a.i. of piperonyl butoxide per 1,000 kg of grain.

Permethrin

Permethrin is a synthetic pyrethroid with stomach and contact action against a wide range of insects. It has no systemic or fumigant activity, but shows ovicidal effect. It is also used on crops, in horticulture and in animal hygiene. LD_{50} = 2,950 mg/kg in water or 430 mg/kg in corn oil (oral, rats). FAO/WHO - JMPR residue tolerance is 2 ppm on raw cereal grains.

Permethrin kills a wide range of stored-product pests and is particularly effective against *Rhizopertha dominica*, but is weak against *Tribolium* species. It is therefore used in combination with organophosphorous insecticides. It is significantly potentiated by adding piperonyl butoxide as a synergist. Permethrin is stable on grain, 70 per cent remaining after 15 months at 25°C. It is rather insensitive to moisture.

Recommended rates for application in combined with organophosphorous insecticides are:

- 1-2 g a.i. of permethrin + 10 g a.i. of piperonyl butoxide + 4-6 g a.i. of pirimiphos-methyl per metric ton of grain gives protection for at least nine months.

The availability of permethrin is of particular importance for the control of the larger grain borer, *Prostephanus truncatus,* a pest recently introduced from

The larger grain borer, Prostephanus truncatus, *was accidentally introduced in Africa and has become a major pest of stored maize and cassava in eastern and western Africa. Note boring activity into the cob cores.*

Central America into East and West Africa, causing excessive damage to stored maize and dried cassava. *P.truncatus* belongs to the same family of Bostrichidae as *R. dominica*, the lesser grain borer, which is a major cosmopolitan pest of stored cereal grains. The bostrichid beetles are very sensitive to synthetic pyrethroids such as permethrin and to fenvalerate (see section below), but less sensitive to organophosphorous insecticides.

Recommended application rates against the larger grain borer are:

- Application of 2.5-5 g a.i. of permethrin to one metric ton of shelled maize (2.5-5 ppm) kills over 99 per cent of larger grain borers within 14 days of exposure and no living progeny could be found even after six months in maize stocks thus treated.

- Applying a mixture of 1 g a.i. of permethrin + 4 g a.i. of pirimiphos-methyl per metric ton of shelled maize (1 ppm + 4 ppm) gives the same effective results.

Permethrin is available as a 0.5 per cent dust, a formulation most appropriate for small holder farmers. It appears that shelling of the maize before treatment with permethrin gives a much better protection against grain-boring beetles during storage than storing maize on the cobs, even if the sheath leaves are intact and treated with the insecticidal dust.

It was found in East Africa, however, that the practice of storing loose maize kernels tends to favour the development of *Sitophilus* species. For this reason, the use of mixtures of pirimiphos-methyl and permethrin with a synergist added is being recommended.

Permethrin provides effective control of bruchid species in beans and grain legumes too.

Pyrethrum

Pyrethrum is a very safe botanical insecticide with a rapid efficacy against a wide variety of insects. Pyrethrum sprays have the additional effect of stimulating insects to move rapidly and to come out of hiding (flushing effect). LD_{50} = 1,500 mg/kg (oral, rats). Residue tolerance is 3 ppm for cereal grain (FAO/WHO).

In spite of a "knock-down" effect, many insects recover after several hours owing to their ability to detoxify the small amount of pyrethrum which they have received. To increase the effectiveness of pyrethrum, synergists, in particular piperonyl butoxide, may be added to the spray formulation.

The ratio of pyrethrum to piperonyl butoxide employed is usually between 1:5 and 1:10. Within this range, it is possible to achieve a given effect with one quarter or less of the amount of unsynergized pyrethrums, i.e. the factor of synergism is not less than 4.

Piperonyl butoxide also provides synergism for carbamates, organophosphates and synthetic pyrethroids. Since pyrethrum and pyrethroids are costly insecticides, the addition of a synergist to these chemicals enables them to be used at lower doses and thus at a more acceptable price.

Practically all commercial formulations of pyrethrum consist of oil solutions and solution concentrates; wettable powders and dusts have a short shelf-life.

Recommended rates of application are:

- 1.5-2.5 g a.i. of pyrethrum + piperonyl butoxide (1:5) per metric ton of grain (1.5-2.5 ppm).

- Dried fish may be protected for 8-12 weeks against flies and *Dermestes* spp. by immersion in water with 0.02 per cent pyrethrum and 0.4 per cent piperonyl butoxide (on a w/v basis).

Fenvalerate

Fenvalerate is a highly active broad spectrum insecticide, which in many ways is similar to the pyrethroids. LD_{50} = 450 mg/kg (oral, rats). Recommended tolerances are 5 ppm on cereal grain and 0.5 ppm on flour.

Fenvalerate is promising as a grain protectant because of its efficacy at low doses against *Rhizopertha dominica* and, at higher doses, against other stored-product pests.

Rhizopertha dominica is currently the most destructive pest in Australia and has over the last decade mainly been controlled by bioresmethrin synergised with piperonyl butoxide and is used in conjunction with organophosphorous insecticides. Fenvalerate is an alternative for bioresmethrin against this pest; however, it is less suited for controlling *Sitophilus* and *Tribolium* species.

Extensive trials in Australia for protection of bulk stored wheat and sorghum gave good results for six months after applying 1 g a.i. of fenvalerate + 12 g a.i. of fenitrothion + 8 g a.i. of piperonyl butoxide per metric ton of grain.

Methoprene

Methoprene is an insect growth regulator with juvenile hormone activity which provides satisfactory control of mosquitoes and of a wide variety of stored-product pests. It has an extremely low toxicity to mammals: LD_{50} = 5,000 mg/kg (oral, rats). Maximum residue limits are 10 ppm for cereal grains and 5 ppm for white flour (FAO/WHO).

The juvenile hormone does not affect the fecundity of adult insects, but inhibits reproduction when administered shortly before the pupal stage. Direct ovicidal effects have also been observed.

Methoprene causes substantial mortality in the full grown larvae of several storage pest species, such as *Rhizopertha dominica* and *Triboleum* species, but *Sitophilus* spp. appear to be less sensitive to juvenile hormones.

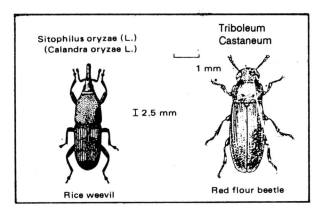

Sitophilus oryzae (L.)
(Calandra oryzae L.)

Triboleum
Castaneum

1 mm

2.5 mm

Rice weevil

Red flour beetle

Rates of application cited in the literature are:

- 5 to 10 g a.i. of methoprene per 1,000 kg of stored produce (5-10 ppm), as a protectant to prevent development of large insect populations.

- Direct spraying of 45-75 mg a.i. of methoprene per m^2 surface of jute bags for control of *Ephestia cautella,* the larvae and eggs of which come into contact with the bags' surface.

Phoxim

A broad spectrum insecticide of low mammalian toxicity, marketed under the trade name Baythion. LD_{50} = 8,500 mg/kg (oral, rats). It is used on crops, as a soil insecticide and seed dressing and against ectoparasites on animals. Its use as a grain protectant has been stopped, but it can be used for disinfection of storeroom surfaces. The chemical has a rapid knock-down effect and high efficacy against coleopterous storage insects.

Rates of application for spraying in storage are:

- 3.2 g of phoxim, 50% EC, per square metre on concrete floors and walls.

- 2.6 g of phoxim, 50% EC, per square metre on jute bags.

USING FUMIGANTS

A fumigant is a chemical which, at the required temperature and pressure, can exist in the gaseous state in sufficient concentration to be lethal to a given pest organism.

In practice, dosages are indicated as the weight of the chemical which is required for a certain volume or space. It has to be realized, however, that the actual concentration that is reached depends on the quantity of gas which is absorbed by the stored products as well as on leakages, temperature and relative humidity.

Any confined space which can be made air-tight may be used for fumigation, e.g. silos, railway cars, trailer-trucks, ship-holds, plastic bags etc. Bagged produce is fumigated mostly under gas-proof sheets. After fumigation, a small amount of unchanged fumigant remains as a residue.

Only two fumigants are widely available and commonly used throughout the world. These are:

Methyl bromide

This gas easily penetrates large stacks of bagged produce but, without a special circulation system, its use

Fumigation of bag stacks with methyl bromide requires careful preparation and good protection with respirators for the operators. This toxic gas is kept under pressure as a liquid in a cylinder from where it is fed through pipes and nozzles into the stack. The bags in the stack have to be rearranged to enable distribution of the gas and the stack has to be well sealed with gas-tight fumigation sheets. After the fumigation has been completed the sheets should be gradually opened leaving sufficient time for airing. (Courtesy of ODNRI, United Kingdom)

Table 15.3: Rates of application of methyl bromide in g/m³ for fumigation of various commodities/ products at temperatures under and above 20°C

Produce	Exposure period (hrs)	Dosage (g/m3) of methyl bromide 10-20°C	Above 20°C
Rice, barley, peas, beans	24	15	10
Wheat, maize, oats	24	35	24
Sorghum, figs	24	55	40
Oilseeds, groundnuts	48	50	40
Oilseed-cakes	48	90	60
Flour	48	50	40

in large silos is limited, because of unsatisfactory distribution of the gas in the bulk grain, except when applied with the addition of CO_2. In flat storage, for instance in barges, this does not apply. Under normal atmospheric conditions, fumigation takes 24 hours; in a vacuum, only a few hours.

Methyl bromide is highly toxic and may also be absorbed through the skin. Therefore, rather sophisticated equipment such as gas cylinders, piping systems, gas masks, gas detectors and special clothing are required when using it. Fumigation has to be carried out by trained personnel. As methyl bromide is odourless, up to 2 per cent chloropicrin, which causes intense irritation of mucous membranes, is sometimes added as a warning agent. Rates of application under atmospheric pressure depend on temperature, commodity and the insect or mite species to be controlled. The simple schedules for fumigation dosage rates adhered to in Europe give CT (concentration X time) products which kill 100 per cent of most common species. For filled containers and under gas-proof sheets, simplification based on volume is possible as shown in table 15.3 above.

Phosphine

This fumigant is a highly insecticidal, strong mammalian poison, which is released from aluminium or magnesium phosphide formulations. Phosphine affects metals (copper, silver, gold) and will damage electronic equipment (e.g. computers) and electric switches. Formulations include:

- Detia Gas-Ex B; in crepe paper bags (sachets) of 34 g to produce 11 g of phosphine;

- Celphos and Phosfumes in 3-g tablets, producing 1 g of phosphine;

- Phostoxin: In tablets of 3 g, producing 1 g of phosphine or in pellets of 0.6 g, producing 0.2 g of phosphine.

These formulations are to be inserted by probe into the produce to be fumigated, or mixed evenly into the flowing grain in the conveying system or distributed otherwise in the room or structure to the fumigated. The tablets can also be placed in or between bags, which then must be covered by air-tight sheets. Phosphine is liberated by absorbing moisture from the air. In contact with water, tablets and pellets will ignite spontaneously. Since the release of phosphine starts an hour or so after application, the use of phosphine is easy from the point of safety, but self-contained gas masks are necessary when aerating large stacks. The duration of fumigation with phosphine will depend to a great extent on temperature, moisture content of the produce and the kind of pest that should be controlled:

At 15-20°C and 10-12 per cent moisture ...4 days
At 15°C and 10 per cent moisture ....5 to 6 days
In cases of heavy infestation with mites ....10 days.

Phosphine has the further advantage of affecting the germination capacity of seeds to a lesser extent than other fumigants such as methyl bromide. A germination test after fumigation, however, remains useful.

Recommended rates of application are as follows:

In silos: 2-5 pellets (1-3 g a.i.) per ton of grain
Bagged goods: 2-12 tablets (0.5-1.5 g a.i.) per m³ of sack pile.

Liquid fumigants such as carbon tetrachloride or mixtures of carbon disulphide, ethylene dibromide or ethylene dichloride and carbon tetrachloride are not widely used currently because of international concern about the toxic residues that remain after fumigation.

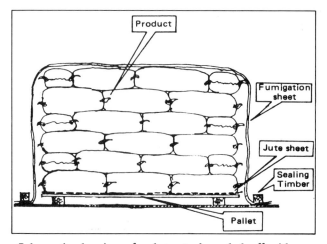

Schematic drawing of a bag stack sealed off with an impermeable sheet that is clamped down at all sides for fumigation. Workers must beware of toxic residues when opening the sheet.

Mechanism of action of Phostoxin

Phostoxin pellets are normally packed in standard aluminium flasks that are a gas-tight and can be sealed again after partial use. As long as Phostoxin is left in its original slightly pressurized packing, the tablets and pellets will remain solid and no hydrogen phosphide will develop. When opening a container, a hissing sound is noticed. The top of a container should be removed only in the open air, or at least near an open window. The pressure is caused by two protective gases, ammonia and carbon dioxide, which develop when only slightly warmed, from an unstable compound, ammonium carbamate. Carbon dioxide and ammonia inhibit inflammation because they dilute the hydrogen phosphide formed during subsequent hydrolysis. In addition, ammonia has an important function as a warning agent, producing a pungent smell.

When this pungent smell becomes noticeable, decomposition of Phostoxin has already commenced, and hydrogen phosphide is also being released. A considerable increase of the carbide-like odour will also be a good indication of this.

Decomposition of Phostoxin tablets. (Courtesy of Degesh)

Commencement of the development of hydrogen phosphide is also indicated by a change in the outer appearance of the tablets or pellets. The shiny grey-green surface first turns matt and finally the tablets or pellets commence "blooming": grey-white dust grows from the surface, the volume of the tablets or pellets increases, and finally there remains a small pile of dust. Depending on temperature and humidity, tablets once exposed to the atmosphere, will completely decompose in appoximately 48-72 hours; decomposition of pellets takes 12-48 hours.

Table 15.4: Some residue levels of fumigants as recommended by FAO/WHO are as follows:

Fumigant	Produce	ppm
Phosphine, PH3	raw cereals	0.1
Hydrogen phosphide	flour, dried vegetables groundnuts, cacao beans, nuts, dried foods	0.01
Methyl bromide (inorganic bromide)	nuts, groundnuts, raw cereals, cacao beans dried fruits bread	100 50 20 0.05

STORAGE OF SEEDS

The viability of seeds may become impaired by a number of often interrelated factors such as temperature, relative humidity, moisture content, physical condition of the seed (breakage, insect and mite damage), gaseous atmosphere around the seeds and pesticide treatment.

Temperature, relative humidity and moisture content

Moisture content is probably the most important factor in seed storage. Research has shown that it is not the moisture content itself, but the fungi which can develop as a result of the high moisture content that cause a decrease in viability of the seeds. Storage fungi which affect seeds grow most rapidly at 30-32°C. Thus, seeds must be stored under dry and cool conditions. Under tropical conditions, however, cooling of stores may become too costly. Two well-known rules for seed storage emphasize the importance of moisture content and temperature:

For seeds with a moisture content of between 5 and 14 per cent, the length of time in storage before seed viability significantly declines is doubled for every 1 per cent reduction in seed moisture content.

For every 5°C reduction in storage temperature, the length of time before seed viability declines is more than doubled.

Physical condition

Machine shelling or threshing may damage seeds, especially when they have a high moisture content or when high cylinder speeds are used. When seeds are damaged, they become more easily attacked by moulds, resulting in a decrease in germinative capacity. Insects and mites may also damage seeds, in particular by boring into the embryonal parts.

Gaseous condition around seeds

Reports on the influence of the gaseous environment of seeds are somewhat conflicting. In general, the viability of seeds with high moisture content is reduced when they are stored under air-tight conditions. When seeds have been dried to the safe moisture content level, air-tight storage does not affect seed viability. A high oxygen content in the atmosphere surrounding the seeds decreases seed viability, but seeds stored in a nitrogen or carbon dioxide atmosphere retain their viability for a long time.

Pesticides for seed protection

Insecticides and fumigants used to control insects in seeds may reduce seed viability. In fact, lindane may be harmful in high dosages. In general, however, insecticides such as lindane, malathion, dichlorvos and pyrethrins have little or no influence when applied in normal dosages on dry seed.

Fumigation with methyl bromide may damage seed viability, especially when a high dose is applied and the seeds have a high moisture content.

Phosphine is considered safe for fumigation of most seeds, but damage to the embryos of leguminous seeds may occur when the moisture content of the stored seeds is too high and when the dose of applied aluminium phosphide is increased. For fumigation of stacked bags containing leguminous seeds, one should not use more than one tablet per cubic metre of produce.

Liquid fumigants may reduce seed viability, especially when the moisture content is high. Repeated fumigations should be avoided no matter what the fumigant.

Seed dressings, which are used to protect the seeds against fungal or insect damage after planting, may also reduce seed viability.

References:

Boeye, J. *et al.* (1980). *The Possibilities for Biologically Integrated Control of the Larger Grain Borer (Prostephanus truncatus Horn) in Africa*. Proceedings of Workshop on the Containment and Control of the Larger Grain Borer, Arusha, May 1988. Ministry of Agriculture and Livestock Development of Tanzania, FAO.

Bond, E.J. (1984). Research Centre Agriculture Canada, *Manual of fumigation for insect control*. FAO Plant Product and Protection Paper No. 54.

Champ, B.R. and E. Highly, (1985). *Pesticides and humid tropical grain storage systems*. Proceedings of International Seminar, Manila, Philippines, May 1985, Australian Centre for International Agricultural Research, Canberra.

Cink, J.H. and Harein, P.K. (1989). *Stored grain pest management*. Department of Entomology, Minnesota Extension Service, University of Minnesota, Saint Paul, Minnesota 55108.

Schulten, G.G.M., F.A. Roorda and V.M. Russel (1980). *Protection of stored products against deterioration*. Royal Tropical Institute, Amsterdam, The Netherlands.

Snelson, J.T. (1987). Australian Centre for International Agricultural Research, A.C.I.A.R., Canberra, *Grain Protectants. ACIAR Monograph No. 3.*

Storage Department, Tropical Development and Research Institute (1985). *Manual of pest control for food security reserve grain stocks*. FAO Plant Production and Protection Paper No. 63, FAO, Rome.

Tyler, P.S. (1989). *Storage Operations and Management*, Overseas Development, Natural Resources Institute, United Kingdom.

Zehrer, W. (1980). *Traditional methods of insect pest control in stored grains*. Post-harvest problems, paper 115; Documentation of a OAU-GTZ Seminar, Lomé.

GLOSSARY

A

abatement:	an attempt to reduce the degree of severity and intensity of pest populations, e.g. mosquito abatement.
abrasion:	the process of wearing away by rubbbing; scraping off or cutting.
abscission:	the separation of fruit, leaves or stems from a plant; the formation of a layer of cells which causes the fruit or leaves to fall off the plant.
absorption:	movement of a substance (toxicant, light, fluid etc.) from a surface into a body; the process by which a chemical is sucked or taken into organisms or other matter.
acaricide:	(= chemical miticide) a substance causing the death of mites.
acetylcholine (ACh):	a chemical conductor of a nerve impulses, formed at the end of nerves, to conduct the impulse over the microscopic gap between nerves or between nerves and muscles or glands. As soon the response has been completd, ACh is destroyed by an enzyme called cholinesterase.
acquisition:	the entry of a virus into, or attachment to, a vector.
acquisition feeding time:	the time that a vector feeds on a virus source in transmission tests.
active ingredient (a.i.):	the chemical component (actual toxicant) of a formulation that is responsible for the pesticidal effect.
activator:	a chemical added to a pesticide to increase its activity.
actual dosage:	amount of active ingredient (not of formulation) applied to a given area.
acute toxicity:	toxicity of a brief, usually single, exposure to an active ingredient, respectively to a formulation thereof; usually expressed as LD_{50}.
---- dermal:	acute toxicity after a single contact with the skin.
---- inhalation:	acute toxicity from a single contact through breathing.
---- oral:	acute toxicity following a single intake through the mouth.
adaptation:	non-inherited change towards increased fitness in an organism.
adhesive:	an adjuvant that helps in making a spray material stick to the sprayed surface, e.g. sticking agent, wetting agent.
adjuvant:	material added to improve some chemical and physical property, hence its effectiveness. Includes emulsifiers, dispersing agents, wetting agents, spreaders, foam suppressants etc.
adsorption:	the process by which chemicals are attracted to and held on the surface of a plant or soil particles.
adulterated:	any pesticide, the strength or purity of which falls below the professed standard or quality stated on the label. Also, a food, feed or product that contains illegal pesticide residues.
aerosol:	(1) a dispersion in a gas of droplets ranging in diameter from 0.1 to $5\,\mu$m, and (2) a certain type of pesticide formulation which is dispersed as a mist by a gas under pressure, e.g. an aerosol spray.
aetiology:	in plant pathology, the science of the cause of disease.
aggressiveness:	capacity of a parasite to infect.
agitation:	stirring or mixing a substance, e.g. pesticide, to prevent it from separating or settling in the spray tank.
agro-ecosystem:	an agricultural area sufficiently large to permit long-term interactions between all living organisms and their non-living environment.
alkaloids:	natural organic bases present in plants.
alternate host:	either of the two hosts on which a heteroecious pathogen or pest must develop to complete its life cycle.
anaerobic:	relating to a microorganism that lives or a process that occurs in the absence of molecular oxygen.

annual:	a plant that completes its life cycle in one year, i.e. germinates, matures, produces seed and dies in the same season.
antagonism:	(1) a relationship between different organisms in which one partly or completely inhibits the growth of the other or kills it, and (2) decreased activity arising from the effect of one chemical or another (opposite of *synergism*).
anthracnose:	common name of a plant disease characterized by black lesions, usually sunken, caused by certain imperfect fungi, e.g. *Colletotrichum* and *Gloeosporium*.
antibiotic:	a substance produced by a microorganism that is able to inhibit the growth of other micro-organisms, or kill them.
antibody:	see antigen.
anticoagulant:	a chemical that prevents normal blood-clotting; the active ingredient in some rodenticides.
antidote:	a treatment given to counteract the effects of a poison.
antigen:	a substance, usually a protein or a polysaccharide which, when injected into the body of a living warm-blooded animal, induces the formation in the blood of proteins (antibodies) with which the antigen reacts specifically, e.g. by precipitation.
antiserum:	blood serum of an animal, that has been injected intentionally with an antigen and consequently contains specific antibodies, with which that particular antigen can be detected or diagnosed.
aphicide:	substance used to control aphids (plant lice).
aquatic:	(plant, animal) growing, living in or near the water.
arboricide:	substance killing trees and shrubs.
aromatics:	solvents containing benzene or compounds derived from benzene.
artificial:	man-made; the converse of natural.
atomize:	to reduce a liquid to fine droplets by passing it under pressure through a suitable nozzle, or by applying drops to a spinning disc, or by passing them through an electrical force field.
atropine (atropine sulfate):	an antidote used to treat organophosphate and carbamate poisoning.
attractant:	a substance that lures pests to a location where they may be destroyed, sterilized or trapped. Usually classed as food-, oviposition- or sex-attractants.
autoecious:	(of a plant parasite or pest) completing the life cycle on one kind of host; *cf* heteroecious.
auxin:	a plant growth substance capable of inducing cell extension in *in vitro* tests on fragments of higher plant tissue.
avicide:	lethal agent used to kill or repel birds.
avirulent:	descriptive of a variant of a pathogenic organism that is incapable of causing severe disease. Synonymous of **non-virulent** which is the preferred term.

B

bacillus:	a rod-shaped bacterium.
bacterium (pl. bacteria):	microscopic unicellular organism (prokaryotic with ribosomes and both DNA and RNA, surrounded by a strong capsule wall).
bactericide:	a substance causing the death of bacteria.
bait:	food or other material used to attract a pest to a trap or toxicant.
bait shyness:	the tendency for rodents, birds or other pests to avoid a poisoned bait.
balance:	dynamic equilibrium. A balanced system is stable; loss of a system's balance leads to instability and, in extreme cases, the system can become self-destructive.
band application:	deposition of a pesticide in a narrow band, usually over or alongside a row of crop plants.
basal application:	application of a pesticide to the lower part of stems.
base:	alkaline chemical material; reacts with acid to form a salt.

314

basic manufacturer:	chemical industry manufacturing the basic, or active ingredient used in a pesticide formulation.
biennial:	plant that springs one year and matures, produces seed and dies during the second year.
bio-assay:	biological test; quantitative and/or qualitative testing of substances regarding their biological activity with the help of living organisms.
biocide:	(1) (noun) substance that kills organisms; (2) (adj.) killing by chemical action.
biodegradation:	the process whereby microorganisms utilize a substance as a source of energy and bring about its destruction; usually applied to biological processes in soil, water and sewage.
biological control:	originally, the use of natural enemies to control the incidence of pests or pathogens. However, it seems rational to include under this term the use of any method that depends solely on biological interactions to control pests or diseases, e.g. host resistance.
biological pesticide:	a pesticide the active ingredient of which consists of a living organism or virus.
biomagnification:	the increase in concentration of a pollutant in animals as related to their position in the food chain, usually referring to persistent organochlorine insecticides and their metabolites.
biorational pesticides:	biological pesticides include pest control agents and chemical analogues of naturally occurring biochemicals (pheromones, insect growth regulators etc.)
biotic insecticide:	usually microorganisms known as insect pathogens that are applied in the same manner as conventional insecticides.
biotrophic:	entirely dependent upon another living organism as a source of nutrient; obligate parasite etc.
biotype:	(1) a group of genetically almost identical individuals but distinguished by criteria other than those of morphology, such as parasitic ability, resistance etc.
blight:	generally, the term refers to any agent causing widespread blackening of leaves and shoots. In plant pathology, the only valid use is in the common names of certain diseases characterized by potentially epidemic spread, e.g. potato blight.
boom:	pipe or tubing with several nozzles to apply chemicals over a wide area at one time.
boom-and-bust cycle:	the cycle of repetitive plant breeding in which a new cultivar is produced to replace one the vertical resistance of which has broken down.
botanical pesticide:	a pesticide obtained from plants, e.g. pyrethrum.
breakdown:	some resistance mechanisms (either vertical or horizontal resistance) fail to operate under extreme environmental conditions.
broadcast application:	application of chemical pesticide or fertilizer over an entire field rather than only along rows and on individual plants.
broadleaf weeds:	plants which possess two seed-lobes (*dicotyledons*) which serve as a food reserve. In many species, upon germination, these lobes become green primary leaves, which fall off when true leaves develop.
broad-spectrum pesticide:	non-selective pesticide effective against a wide range of pests.
brush control:	control of woody plants, shrubs.

C

calibrate:	to determine accurately the amount of pesticide being applied over a given area by means of a sprayer, or other application device.
callus:	parenchymatous tissue consisting of a mass of thick-walled cells that form in response to wounding, infection by pathogens.
cancel:	to discontinue registration of a pesticide.
canker:	a sunken necrotic lesion of root, branch or stem arising from desintegration of tissues outside the xylem cylinder.
capsid:	the protein coat of viruses forming the closed shell or tube that contains the nucleic acid and consisting of capsomeres.
capsomere:	a small protein molecule that is the structural and chemical unit of the protein coat (capsid) of a virus.

carcinogen:	substance that causes cancer in living tissue of man and animals.
carrier:	(1) a material serving as diluent and vehicle for the active ingredient, usually in a dust; (2) an organism harbouring a parasite without itself showing the disease.
cartridge:	the cylinder with absorbing material inserted into a respirator to absorb toxic gases, fumes or vapours.
catalyst:	a substance that accelerates a chemical reaction but is not used up in the reaction.
caution:	signal word used on the label to alert users to slightly toxic pesticides.
certified applicator:	person qualified to apply or supervise application of restricted-use pesticides.
chemical name:	scientific name for the active ingredient in a formulation.
chemosterilant:	chemical substance which controls pests by causing sterilization or by preventing effective reproduction.
chemotherapy:	treatment of a diseased organism with chemicals to destroy or inactivate a pathogen without seriously affecting the host.
chlorosis:	partial or complete absence of normal green colour, affected organ becomes yellow or white. The condition may result from destruction of chlorophyll, or from failure of synthesis of chlorophyll as in iron deficiency.
cholinesterase (ChE):	enzyme necessary for proper nerve function; it is inhibited or damaged by organo-phosphorous or carbamate insecticides.
chronic symptoms:	symptoms that appear over a long period of time.
clone:	a genetically uniform group of individuals, derived originally from a single individual by vegetative propagation and which constitute one kind of cultivar.
colony:	a collection of individual types of bacteria, yeasts etc. growing together, or a growth of hyphae, frequently with spores.
common name:	(1) a name assigned to a pesticide active ingredient by the International Standards Organization (ISO) or adopted by national standards authorities, to be used as a generic or non-proprietary name for that particular active ingredient only; (2) well-known or popular name for weeds and insects.
compatible:	(1) of a relation between a host and pathogen in which a disease can develop. When host and pathogen are incompatable disease does not develop; (2) when two materials can be mixed together with neither affecting the action of the other.
competition:	the converse of **co-operation**. Competition means that two systems co-exist in spite of each other; **co-operation** means that they co-exist because of each other.
complex:	a collective term for all the strains of a virus, or for viruses shown to be related.
concentration:	content of a pesticide in a liquid or dust: e.g. gram/litre.
contact herbicide:	herbicide that causes localized injury to plants where it touches them.
contaminant:	a substance, microorganism or virus accidentally introduced into a culture, substrate or medium; generally, any "unwanted" material in an experimental system.
control:	(1) untreated subjects, or subjects given a standard treatment, for comparison with those given experimental treatment, synonym **check**; (2) to prevent, retard or eliminate a disease or pest attack, or to reduce the ill-effects of a disease or pest to an economically acceptable level.
cooperation:	see **competition**. In a balanced dynamic system, cooperation and competition between all sub-systems are normally equal.
cotyledon:	**seed-leaf**; one in monocotyledons, two in the dicotyledons.
cover, coverage:	the proportion of the surface area of plants or plant parts on which an applied material is retained following dusting, spraying etc.
crop loss assessment:	a new discipline within plant parasitology which concerns the theory and practice of assessing by examination of parasite population and damage the loss of crop due to parasites.
cultivar:	a botanical variety of a cultivated, domesticated crop species. A cultivar is characterized by abnormally high levels in natural survival values contributing to the quantity and quality of the harvestable product. These high levels were gained by artificial selection at the expense of natural survival values contributing to competitive ability in a natural eco-system (Robinson).

316

cultivation:	the creation of an artificial ecosystem in which plants wanted by mankind are protected from much natural competition (Robinson).
cultural control:	method of pest control by means of a skilful combination of agronomic practices such as tillage, planting, irrigation, sanitation and crop rotation so that the environment is less favourable for the proliferation of certain pests.
culture:	a general term for a growth of a microorganism *in vitro* or *in vivo*. Its use in the sense of isolate or "strain" should be avoided.
culture medium:	a prepared substrate (frequently chemically defined food material) for the growth of a microorganism or of other living cells *in vitro*.
cumulative pesticides:	those chemicals that tend to accumulate or build up in the tissues of animals or in the environment (soil, water).
curative pesticide:	a pesticide that can inhibit or eradicate a disease-causing organism after it has become established in the plant or animal.
cuticle:	a membranous layer on the outer wall of epidermal cells consisting primarily of waxy substances.
cyst:	(1) an encysted zoospore (fungi); (2) in nematodes, the carcass of dead adult females of the genus *Globodera* which may contain eggs.
cytokinins:	a group of plant growth-regulating substances that regulate cell division.
cytoplasm:	all the living substance of a cell outside the nucleus.

D

damping-off:	collapse and death of seedling plants resulting from the development of a stem lesion at soil level.
decay:	decomposition of wooden tissue by microorganisms.
deflocculating agent:	material added to a spray preparation to prevent aggregation or sedimentation of the solid particles.
defoliant:	a chemical that initiates abscission.
degradation:	process by which a chemical is broken down or reduced to a less complex form.
deposit:	amount and pattern of spray or dust deposited per unit area of plant surface.
dermal toxicity:	toxicity of a material as tested on the skin; the property of a pesticide to poison man or animals when absorbed through the skin.
desiccant:	chemical that induces rapid desiccation of a leaf or plant part.
detergent:	liquid normally used as a cleansing substance. Some such products may also be used as wetting agents because of their ability to reduce the surface tension of water.
dieback:	necrosis of a shoot, starting at the apex (tip) and progressing towards the main stem and roots.
diluent:	an inert material added to an active ingredient or technical material to reduce its concentration.
disease:	(1) deviation of normal functioning of physiological processes, of sufficient duration to cause disturbance or cessation of vital activity; (2) the adverse effects on a host due to a pathogen (which lacks mouth parts) as opposed to damage which is due to a pest (which possesses mouth parts).
----, incidence of:	the frequency of occurrence of a disease; commonly the proportion of plants affected in a given population.
----, intensity of:	the amount or severity of disease in an individual plant.
----, prevalence of:	the proportion of crops or stands affected in a defined geographical area.
disinfectant:	an agent that (1) frees from infection by destroying parasites established in a host; (2) destroys bacteria etc. on objects and animals not necessarily in situations where infection is present.
disinfestant:	an agent that kills or inactivates organisms present on the surface of plants, or in the immediate environment, e.g. in soil.

disorder: a harmful deviation from normal functioning of physiological processes, arising from causes other than pathogenic organisms or viruses, e.g. mineral deficiency or toxicity, genetic anomaly etc.

dispersing agent: a material or substance that reduces the attraction between like particles.

dissemination: (1) transfer of inoculum from its source to healthy plants; (2) to scatter, as in sowing seeds.

dormant: being in a state of reduced physiological activity.

dormant spray: chemical applied in winter or very early spring before treated plants have started active growth.

dose, dosage: (1) the quantity of a pesticide applied per individual or per unit area, volume or weight; (2) the quantity of a pathogen used to inoculate living hosts, particularly in quantitative response studies.

drift: movements by wind of fine particles of spray or dust to an area not intended to be treated.

drop spectrum: the distribution (by number or volume of drops) of spray into intervals of drop size.

dust: finely divided particles (of plant protectant) with or without diluent, for application in the dry condition.

E

EC$_{50}$: see median effective concentration.

ecdysone: a hormone secreted by insects essential to the process of molting from one stage to the next.

ecological niche: the environment favourable to an ecotype.

ecology: a branch of biology concerned with organisms and their relation to the environment.

economic injury level: the lowest population density of a pest that will cause economic damage (loss).

economic threshold: the density of a pest at which control measures should be initiated to prevent an increasing pest population from reaching the economic injury level.

ecosystem: the interacting system of all the living organisms of an area and their non-living environment.

ecotype (ecological type): a form arising by selection in a particular habitat.

ectoparasite: a parasite feeding on a host from the exterior.

ED$_{50}$: see median effective dose.

emetic: something wich causes vomiting and is used as a first aid for some types of poisoning.

emulsifier: surface active substance used to stabilize suspensions of one liquid in another; for example oil in water.

emulsion: suspension of miniscule droplets of one liquid in another.

encapsulated formulation: pesticide enclosed in capsules (or beads) of thin polyvinyl or other material, to control the rate of release of the chemical and extend the period of diffusion.

endemic disease: a disease permanently established in a moderate or severe form in a defined area, commonly a country or part of a country.

endoparasite: a paratsite which enters a host and feeds from within.

environment: (1) all the organic and inorganic features that surround and affect a particular organism or group of organisms; (2) all factors which are external to a system but which nevertheless influence that system.

enzyme: a protein produced by living cells that can catalyze a specific organic reaction.

EPA: Environmental Protection Agency of the United States of America that is responsible for formation and enforcement of rules and regulations concerning pesticides.

epidemic: a widespread temporary increase in the incidence of an infectious disease.

epidemic cycle:	the positive growth of a parasite population from minimum to maximum, followed by its negative growth to minimum. Most epidemic cycles are seasonal.
epidemiology:	the study of the factors affecting the outbreak and spread of parasites including pathogens and insect infestations.
epidermis:	the superficial layer of cells occurring on all plant parts.
epinasty:	twisting or curling of leaves and stems caused by uneven growth of cells. This characteristic reaction results from treatment with 2, 4-D and other plant growth regulators.
eradication:	(1) the elimination of a pathogen or pest from the host or from the host's environment, or both; (2) the complete removal of a host for the control of a pest or disease.
eradicant:	see fungicide, eradicant.
exterminate:	often used to imply the complete extinction of a species over a large continuous area such as an island.
extinction:	this term normally involves two processes. The first is a loss of behaviour, which is called death; the second is a loss of structure which is called decay or decomposition.
exudate:	material that has passed from within a plant structure to the outer surface or into the surrounding medium, e.g. by difusion and not usually through an aperture, as in root exudate.

F

face shield:	transparent piece of protective equipment used to protect the face from exposure to pesticides.
fermentation:	oxidation of certain organic substances in the absence of molecular oxygen.
field scout:	a person who samples fields to detect and record pest infestation.
flowables:	sometimes referred to as colloidal formulations, these are suspensions in a liquid or solid material in a finely divided state. Because of smallness of particular size, specific gravity of the suspending medium and the presence of suspending and dispersing agents, the formulation remains almost indefinitly in a suspended stage.
focus:	a site of local concentration of diseased plants or disease lesions.
fogger:	aerosol generator.
foliar application:	application of a pesticide to the leaves or foliage.
food chain:	sequence of species within a community, each member of which serves as food for the species next higher in the chain.
forma:	(Latin) (1) a subdivision of a species below the rank of variety; (2) subdivision of a species of a parasitic or symbiotic microorganism distinguished primarily by adaptation to a particular host.
forma specialist (f.sp):	pl. **formae speciales:** a taxon characterized from a physiological (especially host adaption) standpoint but scarcely or not at all from a morphological standpoint.
formulation:	way in which a technical pesticide material is prepared for practical use. Includes various solid, liquid and gaseous preparations.
fructification:	production of spores by fungi; also, a fruiting body.
fumigant:	a volatile chemical (insecticide, fungicide, nematicide etc.) that forms toxic vapours and that is used to disinfest certain areas.
fungicide:	a substance that kills fungal spores or mycelium.
----, eradicant:	(= **curative fungicide**) a fungicide used to control disease after infection has occurred.
----, protective:	a fungicide used to protect an organism against infection by a fungus.
----, systemic:	a fungicide which is absorbed through a plant surface and is translocated away from the side of application.
fungistatic:	preventing the growth of a fungus without killing it.

G

gall: a localised proliferation of plant tissue producing a swelling or outgrowth commonly having a characteristic shape. Galls are usually formed in response to the action of a pathogen or pest.

gene: a material substance in a chromosome which determines or conditions one or more hereditary characters.

gene-for-gene concept: the concept that corresponding genes for resistance and virulence exist in host and pathogen respectively.

genome: the minimum complete set of genetical material in a gamete, zygote.

genotype: the hereditary potential of an organism forms a genotype; it is a theoretical unit because the physical unit that exists (the phenotype) is the result of the interaction of the genotype and the environment.

general-use pesticide: a pesticide that can be purchased and used by the general public without undue hazard to the applicator and environment as long as the instructions on the label are carefully followed.

germination inhibitor: a natural or synthetic substance capable of inhibiting the germination of a seed, spore or other plant propagule.

granule: coarse particle of inert material impregnated or mixed with an active ingredient (insecticide, herbicide or fertilizer) for application to the soil.

growth inhibitor: substance inhibiting the growth of an organism; the action may be toxic i.e. killing the plant or cells or at low concentration merely acting as growth retardant.

growth regulator: organic substance effective in minute amounts for controlling or modifying (plant or insect) growth processes.

growth retardant: a substance that decreases the rate of growth (of a fungus or higher plant).

H

habitat: the place where an organism grows or lives naturally.

half-life: used of a plant virus inoculum for the time taken to lose half its infectivity on ageing, or in heat inactivation studies. Also used to express the progressive decline in transmission of some viruses by their vectors as the period between acquisition and inoculation feeding increases.

harvest interval: period between last application of a pesticide to a crop and the harvest as permitted by law.

herbaceous plant: a higher plant that does not develop woody tissues.

heteroecius: completing the life cycle on two unlike host species, such as crop plants and weeds; *cf* autoecius.

heterotrophic: utilising organic materials (formed by other organisms) as nutrients.

holistic: this word means "entire, whole". The holistic approach emphasises the entire system rather than its components.

hormone: a product of living cells that circulates in the animal or plant fluids and that produces a specific effect on cell activity remote from its point of origin.

host: a living organism which provides nutrients for a parasite but which is not normally destroyed by that parasite in the way a prey is destroyed by a predator.

host range: the range of plants known to serve as hosts for a given parasite.

hydrolysis: chemical process of (in this case) pesticide breakdown or decomposition involving the splitting of the molecule and addition of a water molecule.

hydrophilic: strong affinity for water.

hydrophobic: aversion to water.

hygroscopic: readily absorbing and retaining moisture.

hyperparasite: an organism parasitic on a parasite.

hypersensitivity: the violent reaction of an organism to attack by a pathogenic organism or a virus resulting in prompt death of invaded tissue, thus preventing further spread of the infection.

hypertrophy: excessive growth due to an enlargement of individual cells, as in gall tissues.

I

immune: exempt from infection; anything less than immunity is resistance.

immunity: freedom from disease due to lack of qualities permitting or to possession or acquirement of qualities preventing the operation of the pathogenic organism or virus.

imperfect fungus: a fungus that is not known to produce sexual spores.

incompatible: see compatible.

indicator plant: one which reacts to a certain pathogen or environmental factor with obvious symptoms and is used to assist in the detection of the pathogen or the environmental factor. Synonymous with indicator host when applied to pathogens.

inert ingredient: any substance in pesticide product having no pesticidal action. This does not necessarily mean that it is an inactive ingredient.

infect: to enter and establish a permanent relationship with an organism.

infected: containing an organism or virus which has established a parasitic relationship. A plant which is infected is usually said to be diseased only when symptoms become evident.

infection: the entry of an organism or virus into a host and the establishment of a permanent or temporary parasitic relationship.

infection court: the site of invasion of a host by a parasite.

infective: (1) of an organism or virus able to attack a host and cause infection; (2) of a vector carrying or containing a pathogen and able to transfer it to a host organism causing infection.

infest: to over-run the surface of a plant, or to be dispersed through soil or other substrate. Used especially of insect and other animal pests.

inoculate: to apply inoculum to a host or into a culture medium.

inoculation feeding time: the length of time a vector feeds on the test host in transmission experiments.

inoculative: of a vector carrying or containing a pathogen and able to introduce it into a host plant.

inoculum: material containing microorganisms or virus particles to be introduced into or transferred to a host or medium.

insect-growth regulator: chemical substance that disrupts the action of insect hormones controlling molting, maturity from pupal stage to adult, and other physical processes.

integrated control: the complementary use of biological and chemical methods to control pests and pathogens.

integrated pest management: a management system that uses all suitable techniques and methods in a manner as compatible as possible to maintain pest populations at levels below those causing economic injury.

interaction: the efects of a parasite on a host and vice versa; the results of such interaction are called disease or damage.

intoxication: a state of giddiness caused by the intake of or exposure to a toxic substance.

invasion: the penetration and colonization of a host by an organism.

invert emulsion: one in which the water is dispersed in oil rather than oil in water. Usually a thick mixture results.

in vitro: literally "in glass"; frequently in the sense "not under natural conditions", e.g. in the laboratory; in experimental culture.

in vivo: literally "in that which is alive". Properly applicable to physiological and biochemical activities in living cells and tissues and to the growths of parasites, pathogens etc. on host organisms, but commonly used more losely in the sense of growth in natural conditions, in contrast to growth in culture, i.e. *in vitro*.

J

juvenile hormone: a hormone produced by a specific gland in an insect that regulates normal metamorphosis during immature stages and subsequent development to the adult stage of the insect.

juvigen: a synthetic substance that induces an effect in an insect that is similar to that of an insect juvenile hormone, e.g. methoprene.

L

latent period: (1) elapsed time between infection and the appearance of disease symptoms; (2) the period after acquisition of virus by a vector before it becomes infective.

larva: the immature life stage of insects, mites and nematodes.

larvicide: chemical substance used for controlling (mosquito) larvae.

LC_{50}: see median lethal concentration.

LD_{50}: see median lethal dose.

lesion: a localised area of diseased or disordered tissue.

lysis: the enzymatic dissolution of a cell or tissue.

M

macroscopic: visible without the aid of a magnifying lens or microscope.

masked virus: one carried by a plant which does not show symptoms of its presence.

mass median diameter (MMD): the figure dividing a total volume of spray into two equal parts: one half of the mass of the spray is contained in droplets of smaller diameter than the MMD, and the other half is contained in droplets of larger diameter.

median effective concentration (EC_{50}): the concentration (ppm or ppb) of the toxicant in the environment (usually water) that produces the designated effect in 50 per cent of the test organisms exposed.

median effective dose (ED_{50}): the dose of a material required to bring about a prescribed effect, e.g. retardation of growth in 50 per cent of the individuals in a population of a given species.

median lethal concentration (LC_{50}): the concentration of a toxicant required to kill 50 per cent of a large group of individuals of one species.

median lethal dose (LD_{50}): the dose of a toxicant killing 50 per cent of a large group of individuals of one species.

metabolism: the process by which cells or organisms utilize nutritive material to build living matter and structural components or to break down cellular material into simple substances to perform special functions.

metabolite: a compound inside an organism resulting from the breakdown of a complex chemical, e.g. a pesticide, into simpler substances. Metabolites are sometimes more toxic than the parent compound.

microbial insecticide: a microorganism applied in the same way as conventional insecticides to control an existing pest population.

minimum tillage: practices which utilize minimum cultivation for seedbed preparation and may reduce labour and damage to soil structure (erosion).

mold, mould: a mycelial or profusely visible fungus growth on damp or decaying matter or on surfaces of plant tissue, e.g. green or blue mould caused by *Pennicellium* species.

molt: the shedding or casting off of the cuticle by insects etc.

molluscicide: a chemical used to kill or control snails and slugs.

monoculture: the contiuous cultivation of either one crop species or one cultivar in both time and space.

mosaic: a leaf symptom in which numerous small areas of discolouration stand out against a background of a different tint, e.g. yellow-on-green, tending to have a clearly defined boundary delineated by veins.

mutagen: substance causing genes in an organism to mutate or change.

mutant: an individual possessing a new, heritable characteristic as a result of a mutation.

mutation: an abrupt appearance of a new characteristic in an individual as the result of an accidental change in genes or chromosomes.

mycelium: the hypha or mass of hyphae that make up the body of a fungus.

mycoplasma: a genus of free-living pathogens intermediate in size between viruses and bacteria possessing many virus-like properties and not visible with a light microscope. They lack a rigid cell wall and are therefore highly pleomorphic; plural: mycoplasmas.

mycorrhiza: a symbiotic association of a fungus with the roots of a plant.

mycotoxins: toxic substances produced by several fungi in infected seeds, feeds, or foods and capable of causing illness of varying severity and death in animals and humans that consume such substances.

N

natural control: the collective action of environmental factors, physical and biotic, to maintain over a period of time the numbers of a population within certain upper and lower limits.

natural enemies: an animal or plant which kills or injures another animal or plant, e.g. predators and parasites (insects or diseases) that attack pests.

necrotic: dead, usually the central area of dead tissues and the surrounding zone of damaged tissues in leaf spots.

negligible residue: a tolerance which is set on a food or feed crop permitting an ultra-small amount of pesticide at harvest as a result of indirect contact with the chemical.

nematicide: chemical used to kill nematodes.

nematode: generally microscopic, worm-like animals that live saprophytically in water or soil, or as parasites of plants and animals.

neurotoxicity: toxic effect causing defects in nerve tissue.

non-virulent: a variant of a pathogenic organism that is incapable of causing severe disease.

nozzles: devices which control droplet size, rate, uniformity, thoroughness and safety of a pesticide application. The nozzle type determines the ground pattern of coverage, e.g. tapered flat spray, even flat spray, flooding off centre, atomizing, broadcast, solid stream, full cone, and hollow cone nozzles.

nuclear polyhydrosis virus (NPV): a disease-causing virus of insects, cultured commercially and sold as a biological insecticide.

nucleus: the dense protoplasmic body found in all cellular organisms and which is essential in all synthetic and developmental activities of a cell.

nymph: immature stage of insects which undergo incomplete metamorphosis.

O

oral toxicity: toxicity of a compound when given by mouth; expressed as milligrams of chemical per kilogram of body weight of animal when given orally in a single dose that kills 50 per cent of the test animals.

orifice: the opening or hole in a nozzle through which a spray liquid is forced out and broken up into fine droplets. The size of the orifice controls the volume, whereas the shape of the orifice controls the spray pattern.

ovicide: substance that destroys eggs.

parasite: an organism or virus existing in intimate association with a living organism from which it derives an essential part of the material needed for its nutrition. Parasites are usually classified either as pathogens (which lack mouth parts) or pests (which possess mouth parts).

----, facultative: one able to live as a saprophyte and to be cultured on laboratory media.

----, obligate: one capable of living only as a parasite.

parasitism: partial or complete nutritional dependence of one organism or virus on the tissues of another living individual.

pathogen: an organism or virus able to cause disease in a particular host or range of hosts. The term can be applied to genera, species, sub-species and races of fungi, bacteria etc., and to strains of viruses.

pathogenicity: the quality or characteristic of being able to cause disease.

pelleting: coating of seed with inert material, often incorporating pesticides, to ensure uniform size and shape.

penetrant: an additive or adjuvant which enables the pesticide to move through the outer surface (cuticle) of plant tissues.

perennial: plants that continue to live from year to year. The plants may be herbaceous or woody.

perfect stage: the sexual stage (e.g. fruiting bodies) in the life-cycle of a fungus.

percolation: filtering of water through the soil's porous system.

permitted tolerance: the maximum content of toxicant allowed in foodstuffs for human consumption.

pest: (1) an animal (commonly insect, mite, nematode) which causes damage to, or destruction of, plant parts, plant products etc.; (2) a plant parasite eating the host by means of mouth parts.

pest resurgence: the rapid reappearance of a pest population in injurious numbers, usually brought about after the application of a broad-spectrum pesticide has killed the natural enemies which normally keep a pest in check.

pesticide: an "economic poison" used for preventing, destroying, repelling, or mitigating any pest, pathogen or weed.

phage: a virus parasitic on bacteria.

phase: some defined period, condition, activity etc., which is part of a process or a sequence of events.

phenotype: the observable characteristics of an organism constitute its phenotype, and are the product of the interaction between the hereditary potential (*genotype*) and the internal and external surroundings.

pheromones: highly potent insect sex attractants produced by insects.

photosynthesis: the process by which carbon dioxide and water are combined in the presence of light and chlorophyll to form carbohydrate.

phototropic: obtaining energy for growth etc. from inorganic substances, utilizing light; *cf., chemotrophic.*

physiologic race: a taxon of parasites (particularly fungi) characterized by specialization to different cultivars of one host species.

physiotype: a population of a pathogen in which all individuals have a particular character of physiology (but not of pathogenicity) in common.

phytophagous: feeding on plants, e.g. phytophagous mites.

phytosanitary certificate: a certificate of health of plants or plant commodities to be exported.

phytosanitation: measures requiring removal or destruction of infected or infested plant material likely to be a source of pathogens and pests.

phytotoxic: toxic to plants or plant growth.

piscicide: chemical used to kill fish.

plant pathology: the branch of science for plant disease.

ploemorph: having more than one independent form or spore stage in the life-cycle.

population dynamics: the term covers the normal changes which occur in both host and parasite populations in the course of an epidemic and, particularly, those associated with growth, reproduction and dissemination.

post-emergence: applied after the emergence of a specified weed or crop.

ppb:	parts per billion (parts in 10^9 parts).
ppm:	parts per million (parts in 10^6 parts).
pre-emergence:	applied prior to emergence of specified weed or crop.
precipitate:	a solid substance that forms in a liquid and settles out.
pre-plant application:	treatment applied on the soil surface before seeding or transplanting.
propagule:	that form or part of an organism by which it may be dispersed or reproduced.
propellant:	liquified gas in a pressurized pesticide container that disperses the pesticide formulation.
protectant:	substance applied to plant surface to prevent penetration and subsequent infection by an organism.
protein:	a high-molecular compound consisting of amino acids. It may be a structural protein or enzyme.
protoplast:	the organized living unit of a living cell; the cytoplasmic membrane and everything in it.
pupa:	(plural: **pupae**) the stage between larva and adult in insects undergoing complete metamorphosis, a non-feeding and usually inactive stage.

Q

quarantine:	all operations associated with prevention of dissemination of unwanted organisms within and between territories.

R

race:	(1) a genetically, and as a rule geographically, distinct mating group within a species; (2) a physiologic race.
random sampling:	the most commonly used method of scouting for insect numbers and damage whereby samples are taken without plan (haphazardly) with good field coverage.
ratoon:	a crop which is produced by allowing a harvested crop to grow again.
re-entry (intervals):	waiting interval required by law between application and entrance of workers into the treated crop without protective clothing.
registration:	acknowledgement of approval of a pesticide by a legislative body concerning the uses stated on the label.
repellent:	substance used to repel insects, rodents etc.
residual herbicide:	a herbicide that persists in the soil and injures or kills germinating weed seedlings over a relatively short period of time.
residue:	trace of a pesticide and its metabolites remaining on and in crop tissue, soil, water or other parts of the environment after a given time.
----, maximum limit:	the maximum concentration (in mg/kg or ppm) of a residue that is legally permitted in or on food at harvest or at a specified post-harvest processing stage up to the point of consumption.
resistance:	(1) the ability of an organism to withstand or oppose the operation of or to lessen or overcome the effects of an injurious or pathogenic factor; (2) the ability of a host to suppress or retard the activity of a pathogenic organism or virus; (3) natural or genetic ability of an organism to tolerate the poisonous effect of a toxicant.
----, acquired:	non-inherited resistance conferred by some predisposing treatment.
----, horizontal:	resistance which is evenly spread against all races of the pathogen (equivalent to race-non-specific resistance).
----, race specific:	resistance to some races of the pathogen, but not to others.
----, vertical:	resistance to some races of a pathogen but not to others (equivalent to race-specific resistance).
respiration:	a series of chemical oxidations within a cell controlled and the oxidations catalyzed by enzymes in which carbohydrates and fats are broken down, releasing energy to be used by the cell or organism in its various functions.

restricted-use pesticide: a pesticide that should be applied only by certified applicators, because of their inherent toxicity or potential hazard to the environment.

retention period: the time following acquisition feeding during which a vector is capable of transmitting a virus.

RPM: revolutions per minute.

rodenticide: pesticide applied as bait, dust or fumigant to destroy or repel rodents (mice, rats etc.).

roguing: the critical examination of a crop and removal of unhealthy or otherwise unwanted plants, e.g. those that are not true to type.

rot: desintegration of tissue as a result of the action of invading fungi or bacteria; a disease so characterized.

run-off: (1) the process of shedding spray from plant surfaces during and immediately after application; (2) the quantity of spray which runs off a unit area of plant surface.

rust: (1) a disease that gives a "rusty" appearance and that is caused by a fungus of the order Uredinales; (2) a fungus of the order Uredinales.

S

safener: a chemical that reduces the toxicity of another chemical.

sanitation: the removal and burning of infected plant parts, decontamination of tools, hands etc.

saprophyte: a non-parasitic organism using dead material as food.

scorch: any symptom, such as a lesion or a system of lesions that suggests the action of a flame or fire on the affected part, especially as affecting extended organs such as green leaves or petals.

secondary pest: a pest which usually does little if any damage but which can become a serious pest under certain conditions, e.g. when insecticide application destroys the pest's predators and parasites.

seed dressing: the process of cleaning seed; also used to refer to the coating of seed with fungicide.

seed treatment: the process of coating or impregnating seeds with a fungicide etc.

selective pesticide: one that, while killing the individual pests, spares much of the other fauna or flora, including beneficial species, either through differential toxic action or through the manner in which the pesticide is used (formulation, dosage, placement, timing etc.).

senescence: process or state of growing old.

sensitive: reacting with severe symptoms to the attack of a given pathogen.

serology: the science of serum reactions: the study of reactions between microorganisms or viruses and antibodies in the serum of immunized animals (including man).

serum: the watery portion of the blood remaining after coagulation.

signal word: a required word that appears on every pesticide label to denote the relative toxicity of the product. The signal words are *Danger-Poison* for highly toxic compounds, *Warning* for moderately toxic compounds, or *Caution* for slightly toxic ones.

silvicide: herbicide used to control undesirable brush and trees.

slow-release formulation: formulation from which the active ingredient dissipates slowly and thus remains effective over a long period of time.

slurry: a semi-fluid mixture of solid particles in a liquid, usually water; fungicides are applied to seeds as slurries to produce a thick protective coating.

smut: a disease caused by a fungus of the order Ustilaginales; it is characterized by masses of dark, powdery spores.

soil incorporation: mechanical mixing of herbicide with the soil.

spore: (1) the populations of air-borne (or water-borne) particles of plant or animal origin; (2) the reproductive unit of a fungus consisting of one or more cells; it is analogous to seeds of green plants.

spray: (1) to apply (water, pesticides etc.) in the form of droplets suspended in the air; (2) a solution or suspension (usually in water) dispersed under air- or water-pressure as droplets.

sprayer:	a machine for producing and applying a spray.
----, air-carrier:	a machine employing air to propel the spray.
----, hydraulic:	a machine depending on liquid energy both to form and to propel the spray.
----, logarithmic:	equipment for applying a spray in concentrations decreasing exponentially with time or distance.
spreader:	ingredient added to a spray mixture to assist in its even distribution over the target or plant surface.
sterilization:	the elimination of pathogens from surfaces or soil by means of heat or chemicals.
sticker:	ingredient added to spray or dust to improve its adherence to plant surfaces.
stomach poison:	a pesticide that must be eaten by an insect or other animal in order to kill or control it.
strain:	(1) the descendants of a single isolation in pure culture; (2) a culture of bacteria having some special economic significance; (3) also of viruses.
streak:	a disease characterized by elongated lesions or areas of discolouration, usually of limited length, on leaves with parallel venation or on stems.
stripe:	a disease characterized by elongated areas of discolouration, of indefinite length, on stems or on leaves with parallel venetion.
stylet:	a long, slender, hollow feeding-structure of nematodes and some insects such as lice.
stunting:	the reduction in height of a vertical axis, resulting from a progressive reduction in length of successive internodes or a decrease in their number.
substrate:	(1) a chemical substance acted upon by an enzyme; (2) the material forming the growth medium for a microorganism.
suction hose:	hose used to pull water from a stream or pond or other source into the sprayer tank, or to draw liquid from the tank into the sprayer pump.
surface tension:	the phenomenon by which a drop of liquid tends to contract rather than to spread out as a film.
surfactant:	abbreviated form of "surface active agent"; ingredient that enhances the surface-modifying properties of a pesticide formulation, e.g. by affecting surface tension, (wetting agent, emulsifier etc.).
susceptible:	non-immune, i.e. subject to infection.
susceptibility:	inability of an organism (host) to oppose the operation or to overcome the effects of an injurious or pathogenic factor.
suspension:	finely divided solid particles dispersed in a liquid.
symbiosis:	a mutually beneficial association of two different kinds of organism.
symptom:	a visible or otherwise detectable abnormality arising from a disease.
syndrome:	the totality of effects produced in an organism by one disease, whether all at one time or successively, and including effects not directly detectable to the unaided eye.
synergism:	(1) the association of two or more organisms acting at one time and effecting a change which one only is not able to make and of which the total effect is greater than the sum of the parts; (2) increased activity of certain mixtures resulting from the effect of one chemical on another (of pesticides and non-toxic materials).
synthetize:	production of a chemical compound by joining various elements or simpler compounds.
systemic:	(1) chemical compound that is absorbed and translocated throughout the plant or animal; e.g. systemic fungicide; (2) (of a plant pathogen) occurring throughout the plant.

T

tank mix:	mixture of two or more pesticides in a spray tank at the time of application.
taxon:	any taxonomic group.
tenacity:	the property of a deposit to resist removal by weathering.
teratogenic:	(1) substance that causes physical birth defects in the offspring; (2) causing gross structural abnormalities, e.g. witches' brooms.

tilth:	physical condition of the soil and its potential for cultivation of plants.
tissue:	a group of cells of similar structure which performs a special function.
tolerance:	(1) the ability of an affected organism to endure invasion by a pathogenic organism or virus with little or no reaction, as shown by the more or less complete absence of symptom expression and damage; (2) amount of a pesticide residue permitted to remain on or in a crop at harvest or in processed produce. Expressed as parts per million (ppm).
tolerant:	able to endure infection by a particular pathogen, without showing severe disease.
toxic:	of a poison: caused by, or acting as a poison.
toxicant:	a poisonous substance such as the active ingredient in pesticide formulations that can injure or kill organisms.
toxicity:	the ability to act as a toxin or toxicant.
toxin:	a naturally occurring poison produced by plants, animals and microorganisms, e.g. snake venom, botulism toxin and mycotoxins.
trade name (trademark):	name given by its manufacturer or formulator, distinguishing it as being produced or sold exclusively by that company.
translocation:	transfer of nutrients or other substances such as a herbicide from one plant part to another.
transportation:	the loss of water vapour from the surface of leaves.
trivial name:	name in general usage; e.g. nicotine.

U

ultra-low-volume (ulv):	very low application rates: also a type of pesticide formulation as supplied by the manufacturer.
ultraviolet (uv):	that portion of the light spectrum beyond the violet band that is invisible to man but both visible and highly attractive to nocturnal insects.

V

vapour (vapor):	steam, gaseous form of a normally liquid or solid substance.
variability:	the property or ability of an organism to change its characteristics from one generation to another.
variety:	(1) a subdivision of a species below the rank of subspecies and above the rank of form; (2) cultivar.
vascular:	term applied to a plant tissue or region consisting of conductive tissue; also, to a pathogen that grows primarily in the conductive tissues of a plant, e.g. vascular disease.
vector:	an organism (e.g. insect, nematode etc.) that transmits pathogens to man, animals or plants.
vermin:	noxious animals or pests; usually rodents or insects.
vessel:	a xylem element or series of such elements, the function of which is to conduct water and mineral nutrients throughout the plant.
viricide:	a substance that inactivates a virus completely and permanently.
virion:	a complete virus particle.
viroid:	a pathogenic agent, formerly assumed to be a virus, now known to be a ribonucleic acid of low molecular weight and not a ribonucleic protein.
virosis:	infection by a virus, or a disease so caused.
virulence:	(1) relative capacity to cause disease; (2) degree or measure of pathogenicity of a parasitic organism or virus.
virulent:	strongly pathogenic.
viruliferous:	of a vector which carries or contains a virus.
virus:	a nucleoprotein entity able to pass through bacterium-retaining filters and to replicate within the living cells of a host.

viscosity: the property of stickiness of fluids and gases that hinders flowing; power of resisting a change in the arrangement of the molecules.

volatilize: to vaporize.

W

wettable powder: pesticide formulation of toxicant mixed with inert dust and a wetting agent that mixes readily with water and forms a short-term suspension (requires tank agitation).

wetting agent: compound that causes spray solutions to contact plant surfaces more thoroughly by decreasing the contact angle of the liquid.

wilt: (1) to lose turgidity; (2) a disease characterized by loss of turgidity and collapse of leaves, typically caused by vascular infection by a fungus.

witholding period: the recommended minimum period that should elapse between the last application of a pesticide to any crop, pasture or animal and the harvesting, grazing, cutting or slaughtering thereof, or the collection of milk and eggs for human consumption.

wood preservation: preserving the usefulness and structural strength of wood and wood products by chemical treatment to repel harmful parasitic organisms.

Z

zoospore: a spore bearing flagella and capable of moving in water.

References:

Federation of British Plant Pathologists (1973): *A guide to the use of terms in plant pathology*. Phytopathology Papers No. 17. Commonwealth Agricultural Bureaux. ISBN 0 85198 290 5.

Robinson R.A. (1976): *Plant pathosystems*. Springer Verlag, Berlin, Heidelberg, New York.

Agrios G.N.; *Plant pathology*; Academic Press, New York, second edition.

Commission for Terminology of the Netherlands Society of Plant Pathology. *List of crop protection terms*. Journal "Gewasbescherming" 16 (December 1985) Supplement No.1.